THE CREATION OF INEQUALITY

THE CREATION OF INEQUALITY

HOW OUR PREHISTORIC ANCESTORS SET THE STAGE FOR MONARCHY, SLAVERY, AND EMPIRE

Kent Flannery

Joyce Marcus

HARVARD UNIVERSITY PRESS

Cambridge, Massachusetts

London, England

2012

Printed in the United States of America

Library of Congress Cataloging-in-Publication Data

Flannery, Kent V.

The creation of inequality : how our prehistoric ancestors set the stage for monarchy, slavery, and empire / Kent Flannery and Joyce Marcus.

p. cm.

Includes bibliographical references and index.

ISBN 978-0-674-06469-0

1. Prehistoric peoples. 2. Anthropology, Prehistoric. 3. Human evolution. 4. Social evolution. 5. Equality. I. Marcus, Joyce. II. Title.

GN740.F54 2012
569.9—dc23 2011039902

Man is born free, and yet we see him everywhere in chains.

J.-J. Rousseau, *The Social Contract* (1762)

Contents

Preface

In the autumn of 1753 the celebrated Academy of Dijon proposed an essay competition. The prize would go to the author who best answered the question "What is the origin of inequality among men, and is it authorized by Natural Law?"

An iconoclast from Geneva named Jean-Jacques Rousseau took up the challenge. His entry, "A Discourse on the Origin of Inequality among Men," did not win, but 250 years later it is the only one still remembered. So influential was Rousseau's essay that many historians believe it provided the moral justification for the French Revolution. Still others consider Rousseau the founder of modern social science.

In less than 100 pages Rousseau presented a framework for the development of human society that preceded the writings of Charles Darwin and Herbert Spencer by more than a century. Rousseau's effort was all the more remarkable because he could not draw upon anthropology or sociology, two sciences that did not yet exist. Nor was he able to draw upon archaeology, since it would be another 120 years before Heinrich Schliemann created it.

To understand the origin of inequality, Rousseau argued, one had to go back to earliest times—to a "state of nature" in which the only differences among human beings lay in their strength, agility, and intelligence, and individuals worked only to satisfy their immediate needs. Rousseau believed that all the unpleasant characteristics of the human condition derived not from nature but from society itself as it developed. Self-respect, vital for self-preservation, was the rule at first. Unfortunately, as society grew, this attitude gave way to self-love, the desire to be superior to others and admired by them. Love of property replaced generosity. Eventually, a growing body of wealthy families imposed a social contract on the poor, a contract that institutionalized inequality by providing it with moral justification.

What makes the influence of Rousseau's work all the more impressive is to consider how few reliable facts he possessed when he wrote it. His entire description of "natural man" was based on the anecdotal accounts of travelers. Rousseau had heard of "savages of the West Indies" who were superb archers and "savages of North America" who were celebrated for their strength and dexterity. He had heard of the natives of Guinea, the east coast of Africa, the Malabars, Mexico, Peru, Chile, and "the Magellan lands." He knew of the Khoikhoi people of the Cape of Good Hope but referred to them by the politically incorrect term "Hottentots."

It would be easy to list all the details Rousseau got wrong, but that would be like criticizing Gregor Mendel for not knowing about DNA. More useful is to build upon Rousseau's essay by using two more recent sources of information. One source is the vast archive of archaeological information on ancient peoples. The other source is the archive of anthropological information on recent human groups. In a nutshell, here is what those two bodies of information tell us.

Anatomically and intellectually, modern humans were already present during the Ice Age. By 15,000 B.C., they had driven their closest competitors to extinction and spread to every major landmass on earth. Our Ice Age ancestors typically lived in small foraging societies whose members are believed to have valued generosity, sharing, and altruism. As anthropologist Christopher Boehm points out, hunting-and-gathering people usually work actively to prevent inequality from emerging.

Not all of our ancestors, however, continued to live that way. Slowly but surely, some of them began to create larger societies with greater levels of social inequality. By 2500 B.C., virtually every form of inequality known to mankind had been created somewhere in the world, and truly egalitarian societies were gradually being relegated to places no one else wanted.

Evolutionary biologist Edward O. Wilson has compared the appearance of complex human societies to hypertrophy, the exaggerated overgrowth of structures, such as the tail of the peacock or the tusk of the elephant. The growth of complex human societies, however, did not require genetic change. It involved changes in a unique social logic that characterizes every human group. We learn the details of this logic through social anthropology, and we discover the long-term results of its changes through archaeology.

In the pages that follow we document our ancestors' creation of inequality by drawing on both archaeology and social anthropology. Several widespread regularities become apparent. First, out of the hundreds of possible varieties of human societies, five or six worked so well that they emerged over and over again in different parts of the world. Second, out of the hundreds of logical premises that could be used to justify inequality, a handful worked so well that dozens of unrelated societies came up with them.

For whom did we write this book? Not for our fellow archaeologists and social anthropologists, although they contributed much of the information we use. Instead, we wrote this book for the general reader who is curious about his or her prehistoric ancestors but has neither the time nor inclination to wade through the social science literature.

Because the book is designed for the general reader, we give the dates of ancient events in two familiar and accessible forms. In the case of remote periods, for which dates can never be more than approximations, we give them in "years ago." For more recent events, dated by Mesopotamian, Egyptian, Maya, or European calendars, we present our dates in the familiar "B.C." or "A.D." system, with which all readers of newspapers and news magazines are familiar.

In this book we refer frequently to both archaeology and social anthropology. One could liken their relationship to that of zoology and paleontology. By examining living amphibians, reptiles, and mammals, zoologists give us detailed knowledge of their anatomy and behavior. By examining the fossil record, paleontologists demonstrate to us that amphibians preceded reptiles and likely gave rise to them; that reptiles preceded mammals and likely gave rise to them; and so on.

Paleontologists are at a disadvantage because they usually have only the skeletons of ancient species to work with. Often, however, the bones reveal attachments for ligaments, tendons, or muscles that zoologists can link to specific behaviors. For their part, zoologists are at a disadvantage because they are limited to those creatures that still live among us. Often, however, paleontologists can fill in the blanks with the skeletons of creatures that lived long ago. Both fields are therefore empowered when there is feedback between them.

Archaeology and social anthropology also work best when they work together, but their relationship over the years has been uneasy at best. Archaeologists turn frequently to social anthropologists for help in interpreting prehistoric evidence. Many social anthropologists, however, cannot imagine that there is anything to learn from archaeology. They consider it a form of manual labor.

One social anthropologist who understands the contribution of archaeology is Robin Fox. "Old-fashioned as it may seem," Fox once wrote, "archaeology is really interested in the *truth* about the past, however elusive this may be." That is because, he adds, archaeology "must always come back to face the brute facts of physical remains, its subject matter. This is a strength, not a weakness."

Social anthropologists are rarely forced to face the brute facts of physical remains. For many, this means that the possibilities of what might be true are limitless. Social anthropologists are free, if they wish, to believe that the past is merely a "text" that we can interpret any way we want. They can even believe, should they choose, that there were no repetitive patterns in the way that human societies developed over time, and that any attempt to detect order in the infinite variety of societies is misguided.

Archaeologists are denied this luxury. They must, for example, face the brute fact that there were no monarchies 15,000 years ago, and that when monarchies finally arose on different continents, they left behind some remarkable similarities in their physical remains.

Today's archaeologists are just as interested in social and cultural behavior as social anthropologists are. They are at a disadvantage because they have only the skeletal outlines of past societies. By reading the work of social anthropologists, however, archaeologists learn what to look for in order to reconstruct the perishable structures of society. At the same time, archaeologists must have the common sense to realize that not every theory developed by social anthropologists can be successfully applied to archaeological remains.

The anthropological and archaeological evidence to which we refer in this book is only a fraction of what we could have used. Out of the hundreds of social anthropological studies available, we chose those that could be most readily used to interpret the archaeological evidence. We also looked for studies that either captured an important moment of social change or made explicit the logic of inequality.

Given a choice, we turned to studies by the first social anthropologists to contact a specific society, that is, those who were able to describe it before it was hopelessly altered by colonialism or globalization. Many of those classic studies are currently underutilized, because they do not showcase the anthropological theories considered trendy today. What these early studies offer the archaeologist, however, are descriptions of indigenous behavior that can no longer be observed. One day in the future many types of societies, despite having once been widespread, will exist only as archaeological remains. Perhaps when that happens, many of the writings by pioneer anthropologists will be rediscovered.

As for the archaeological studies we use, we have been just as selective. Out of the hundreds of possibilities, we chose those from which we could infer actual social behavior. Every archaeological site yields artifacts. Not every site, however, provides evidence of residences, public buildings, ritual features, or burials that show some aspect of inequality.

In the course of writing this book we were frequently made aware that it was not always the most recent studies that were the most useful. Good archaeological evidence can come from any decade. The same is true of the theories and explanatory models we encountered. This should not have surprised us. The theory of natural selection was published in 1859 and is still used today. The same could be said of the theory of relativity, first published in 1905. There is, in other words, no "shelf life" for a truly useful theory.

Theory is indispensable in science, because it makes sense out of disparate facts. At the same time, there are limits to how much theory ought to appear in a book for the general reader. There is probably no bigger "buzzkill" than a long, ponderous chapter on competing hypotheses.

We have been guided, in this regard, by a wise old archaeologist named Richard S. MacNeish. "Theory," MacNeish once told us, "is like perfume. Put on the right amount and the suitors will swarm around you. Put on too much and they'll think that you're covering up the smell of bad data." We trust that the theory in this book is just a dab behind the ear.

I

Starting Out Equal

ONE

Genesis and Exodus

We were all born equal, and our birthplace was Africa. Whoever we are, wherever we live, whatever language we speak, whatever our customs and beliefs, whatever the color of our skin, at some point in the last two million years our ancestors lived in Africa.

It took several emigrations to get us to the four corners of the earth. One exodus, beginning 1.8 million years ago, brought some of our distant ancestors out of Africa but no farther than the warmer parts of continental Eurasia. Joined by African game like the rhino, the hippo, and the elephant, they made it to the northern and eastern coasts of the Mediterranean. From there, some of them reached the Caucasus, while others continued on to India and China. From the Mediterranean, they spread west and north into Europe, reaching the British Isles between one million and 700,000 years ago.

Our distant ancestors did not rush into colder latitudes and had no watercraft capable of reaching places like Australia and New Guinea. But 400,000 years ago they already had wooden spears and throwing sticks for hunting and stone tools for digging, cutting, chopping, and scraping. Innovation does not seem to have been their strong suit. The change in their tools was unimaginably slow, and there is little evidence that they wore clothing or ornaments, imagined a spirit world, or engaged in art or music. More often than not, the raw materials from which they made their tools came from within 30 to 35 miles of their camping places. This would have been about a two-day trip for twentieth-century foragers.

Some 200,000 years ago, the people just described were in decline. The newcomers who replaced them were more "modern-looking" than their predecessors, though far from anatomically uniform. Biological anthropologists see them as consisting of at least two distinct groups of people: those they call Neanderthals and those they call, by that wonderful oxymoron, "archaic modern humans." Geneticist Svante Pääbo and his colleagues have analyzed more than 60 percent of the Neanderthal genome and compared it to that of modern humans. They conclude that from time to time there may have been exchanges of genes between Neanderthals and archaic moderns.

Of these two types of humans, the classic Neanderthals of Europe were the most powerfully built. Their skeletons indicate that they broke their bones more often than archaic moderns. Classic Neanderthals tended to develop skeletal pathologies in their 20s and 30s, seldom living beyond 40 years of age. Their tooth wear suggests that they sometimes used the mouth as a vise. Some scholars believe that the Neanderthals used strength to perform tasks that archaic moderns performed with improved technology. They argue that as the pace of Stone Age technology sped up, archaic moderns used earth ovens to cook their food longer; developed the spear-thrower, the boomerang, and eventually the bow and arrow; learned how to convert plant fiber into string for snares and fishing nets; began to ornament themselves; and accelerated the improvement of tool kits based on flint, wood, bone, antler, and ivory. All these technological improvements are thought to have reduced the need to maintain larger teeth and more powerful muscles.

Clues to the relations between Neanderthals and their more modern-looking relatives can be found in a group of skeletons buried in Israeli caves. Some skeletons, laid to rest 110,000 to 90,000 years ago, look like archaic versions of ourselves. However, later skeletons from the same Mt. Carmel region, buried only 70,000 years ago, are more Neanderthal-looking. We can draw two conclusions from this evidence. First, at this remote period it was not yet clear which of these two types of humans was going to emerge as more successful. Second, if there were exchanges of genetic material between Neanderthals and archaic modern humans, the Near East is one of the places it might have happened.

As interesting as the skeletons themselves are the details of their burial. One of the archaic modern skeletons, from Qafzeh Cave in Galilee, was buried with possible seashell ornaments and sprinkled with red ocher pigment. Another of the archaic moderns, this one from Skhul Cave in the cliffs of Mt. Carmel, had a wound in his pelvis made by what appears to have been a wooden spear. All this evidence hints that the more modern-looking occupants of the Mt. Carmel caves might have worn shell ornaments, incurred spear wounds from enemies, and received preparation for an afterlife by having their corpses painted red. However, archaic moderns would continue competing with their Neanderthal neighbors for tens of thousands of years.

Beginning at least 100,000 years ago, this competition among early humans took place during a period of global cooling called the Ice Age. Authorities on climate point to evidence that the world's temperature was falling dramatically 75,000 years ago. The evidence comes from studying deep-sea sediments, air bubbles trapped in glaciers, and grains of pollen from the clays in lakes. The coldest temperatures occurred between 30,000 and 18,000 years ago. Finally, 10,000 years before the present, world temperatures had rebounded, and the Ice Age was essentially over.

To many anthropologists the Ice Age seems like the kind of stressful environment in which a more resourceful type of human—clever, more resilient, and more able to adapt to difficult conditions—might come to the fore. Others believe that such a scenario relies too heavily on the environment. They prefer to believe that our archaic modern ancestors succeeded by using social skills to create larger networks of kinship, alliance, and mutual aid.

THE NEANDERTHALS CHECK OUT

The Neanderthals dispersed widely over the landscape of Eurasia but generally avoided places as cold as Scandinavia. At the peak of the Ice Age, when their environment included reindeer, woolly mammoths, and woolly rhinos, the European Neanderthals often camped in caves, heating their living space with campfires. Their raw materials came generally from within 60 miles of their encampments, double the distance typical of their predecessors. Roughly 30,000 years ago, however, the Neanderthals vanished, possibly driven to extinction by their more modern-looking neighbors.

Edward O. Wilson has pointed out that once our ancestors were left with no closely related competitors, they had achieved "ecological release." Now they were free to exhibit greater behavioral diversity, uninhibited by rivals to whom they would have to adjust.

THE DISPERSAL OF ARCHAIC MODERN GROUPS

Even before the disappearance of the Neanderthals, our more modern-looking ancestors had been on the move. Now their exodus would carry them to every part of the Old World, and their descendants would eventually colonize the New World and the islands of the Pacific. It is to this emigration of our more recent Ice Age ancestors that we now turn.

To begin with, archaic modern humans seem to have been less heavily built than the classic Neanderthals. They broke their bones less frequently and enjoyed a greater life expectancy. The classic Neanderthal body required plenty of calories for maintenance. Because of their more graceful build and improved technology, more of our archaic modern ancestors could be supported on the same number of calories. Anthropologist Kristen Hawkes has also argued that their greater life span was an adaptation for more intense food gathering. For example, older women could provide child care, freeing women of childbearing age to spend more time harvesting.

The possible results of longer and more efficient harvesting can be seen in its effect on slow-moving prey. Two examples given by archaeologist Richard Klein are the angulate tortoise and a marine mollusk called the Cape turban shell, both native to South Africa. The angulate tortoise grows slowly throughout life. As early as 40,000 years ago, tortoises from archaeological refuse in South Africa had begun to show a steady decrease in size, perhaps because they were now being harvested in such numbers that most did not live to old age. The Cape turban shell showed a similar size decrease, possibly the result of overpicking.

Around 80,000 to 60,000 years ago, two important archaeological sites in South Africa document increasingly sophisticated tool technology. At these two sites, Blombos Cave and Klasies River Mouth, ancient hunters had learned how to convert flint nodules into many more inches of long, sharp blades by using a hammer of softer and more controllable material. They turned some of those blades into scrapers for freeing animal hides from fat. They used chisel-like flints to make slots in bone or wood so that tools could be set more efficiently in a handle. They sharpened bone splinters into awls for perforating hides, allowing for tailoring with string or sinew. They also produced tiny flints that served as barbs for composite weapons.

As early as 80,000 to 70,000 years ago, from Morocco in the north to Blombos Cave in the south, ancient humans had begun drilling seashells to string on necklaces and were decorating themselves with red ocher and white pipe clay, two naturally occurring pigments.

While the humans of 1.8 million years ago had concentrated much of their effort on procuring large-game animals, our more recent ancestors had broadened their idea of food to include fish, perhaps because they now had string for making nets. And one of their most significant improvements in food procurement was recorded at Klasies River Mouth, in archaeological deposits dated from 75,000 to 55,000 years ago.

Klasies River Mouth lies in a zone of vegetation that today's South Africans call *fynbos* (literally, "fine bush"). Included among the plants of the fynbos is a flower called watsonia, a member of the lily family. Like its relative the gladiolus, it has a sizable corm, or bulb, which in the case of watsonia is edible. When fynbos vegetation is deliberately burned off, watsonia grows back with its density per acre increased five to ten times. It seems that the occupants of the region had discovered that fact, because some archaeological layers at Klasies River Mouth have dense accumulations of burned watsonia and other fynbos plants.

What is exciting about this discovery is that it reveals the people of that era to have had what economists would call a delayed-return strategy. Rather than restricting themselves to plants or game whose harvest yielded immediate food, the occupants of Klasies River Mouth were willing to invest labor in activities that would yield no food until the next growing season. At that future time, however, their effort would be rewarded by a harvest five to ten times larger than before. To state it differently, some early humans had learned not merely how to take food out of the environment but to engineer the environment itself. Almost certainly they were able to do this because they had become astute observers of nature and, like the nineteenth- and twentieth-century hunter-gatherers studied by anthropologists, could name hundreds of plants and animals and rattle off the details of their habitat preference and behavior.

From that point on, the evidence for human interference in the environment, sometimes called "ecological niche construction," is repeated at other archaeological sites in widely scattered regions.

Consider, for example, the following case. In the Egypt of 20,000 years ago the level of the Nile River was 50 feet higher than today's—and rising. At flood stage its surge of water, carried north toward the Mediterranean from Lakes Victoria and Albert, was sufficient to drown the arid canyons that entered it from the Egyptian desert. Just north of present-day Aswan, a dry canyon known as the Wadi Kubbaniya enters the Nile from the west. From June to September the flooding Nile backed up into the canyon's lower course, submerging all but its tallest sand dunes. This flooding created a rich, localized environment where catfish and tilapia gathered and where water-loving plants like sedges and rushes grew abundantly.

Such is the clamor of spawning catfish that their mouth-and-tail slapping can be heard hundreds of yards away. That sound is probably what attracted groups of Stone Age foragers to Wadi Kubbaniya, where they camped on exposed dunes. As the floodwater receded, tilapia could be caught by hand in their spawning holes, and catfish could be brought to the surface by driving oxygen from the water with kicks and splashes.

The number of fish caught was too many to eat at one time. An expedition led by archaeologist Fred Wendorf found more than 130,000 catfish bones at one camp, along with evidence that the heads of the fish had been removed and the bodies smoked or dried to be eaten later.

By October, as the water fell, dense mats of sedge called purple nut grass were exposed, as were thousands of club-rushes. The sedge mats were ten feet wide, and every ten-square-foot section could produce an incredible 21,200 tubers. The foragers harvested thousands of nut grass tubers with digging sticks and complemented them with the tubers and roasted nutlets of club-rush.

The plant collectors at Wadi Kubbaniya had learned that during October and November the nut grass tubers were small and tender; by February and March they had grown larger and harder and were filled with bitter alkaloids. Even these mature tubers, however, could be rendered edible by grinding them on stones and roasting them. Like smoked fish, the sedge and rush tubers could be stored, and they were high in carbohydrates.

From 19,000 to 17,000 years ago, Wadi Kubbaniya became a rich target for foragers who had developed techniques of drying, smoking, grinding, roasting, and storing, allowing them to stretch the Nile's temporary abundance into months of food. And like their predecessors at Klasies River Mouth, these people had discovered the advantages of engineering the environment: the greater the number of mature nut grass tubers removed, the more densely the new ones would grow back the next year.

Judging by the skeleton of a young man buried at Wadi Kubbaniya, these fish-and-tuber collectors were anatomically modern, resembling today's residents of Nubia and the Sudan. The youth had an asymmetrically developed right arm, suggesting that he had been a strong, right-handed spear-thrower. A chip of flint from a past wound was embedded in his shoulder, and a healed fracture of his forearm revealed that he had once used that arm to ward off a blow. His death came as the result of a spear-sized projectile that had left two flint barbs between his ribs and lumbar vertebrae.

Archaeology thus gives us two insights into our ancestors of that era. Both Klasies River Mouth and Wadi Kubbaniya show that they were keen observers of nature, with a rapidly improving technology and the foresight to modify their environment. The spear wounds in the Skhul Cave burial and the Wadi Kubbaniya youth show us something else: even as our ancestors improved their social skills, there were times when neighborly contact resulted in homicide. In other words, our ancestors were behaving more and more like us.

During the second half of the Ice Age, our modern-looking ancestors spread all over the world. This second major exodus was aided by the fact that much of the earth's water was by then locked up in ice. Having so much water frozen into glaciers significantly lowered sea levels, temporarily turning large areas of shallow ocean floor into bridges between formerly separate landmasses. Now our ancestors could colonize places that their predecessors could not have reached.

With remarkable speed, some of them hiked east through the warmer parts of the Old World, including India and Southeast Asia. Once in the Far East, they took advantage of the fact that lowered sea levels had created the Sunda Shelf, an area of exposed former ocean floor that linked Cambodia and Vietnam to Sumatra, Borneo, the Philippines, Java, and the Celebes. All these regions could now be colonized.

Farther to the south, lowered sea levels had temporarily created an 11 million-square-mile continent called the Sahul Shelf, which incorporated Australia, New Guinea, and Tasmania into a single landmass. Some 40 miles of open sea still separated the Sahul Shelf from the Sunda Shelf, but our Ice Age ancestors now had watercraft with which they could island-hop to the Sahul. They reached Australia more than 45,000 years ago and proceeded on to Tasmania. Their island-hopping took them to the Bismarck Archipelago 40,000 years ago and on to the Solomon Islands within another 12,000 years.

The later isolation of Australia and Tasmania created a cornucopia of information for anthropologists. Groups of foragers spread over both landmasses and then—as temperatures warmed, glaciers melted, and sea levels rose—the low-lying parts of the Sahul Shelf disappeared. From that point on the Australians and Tasmanians were cut off from the rest of the world for thousands of years, and their foraging way of life remained unaffected by most changes taking place on mainland Southeast Asia. To be sure, the natives of Australia created their own unique way of life and kept modifying it over time. The point is that any changes they made were indigenous and not the result of influence from mainland Asia, where innovations such as the bow and arrow, agriculture, and hereditary inequality were eventually to appear.

As exciting as the colonization of South Asia, the Sunda Shelf, and the Sahul Shelf may have been, it was no more exciting than the colonization of glacial northern Europe. Here our ancestors penetrated farther than any previous humans, in part by protecting themselves from the cold with fur clothing. They entered a Europe so cold that reindeer herds were roaming what is now southern France, and they camped along the migration routes used by those animals, hunting and eating them as they went along. But they also increased their hunting of fur-bearing animals such as the wolf, fox, bear, mink, and marten, and, using bone awls and needles, they tailored their fur into multilayered garments. Thanks to archaeology we know a great deal about these colonists of the frozen north, some of whose technology resembled that of recent Arctic hunters. And we have reason to believe that the color of their skin, and that of their Neanderthal neighbors, underwent change.

In 2007 it was reported that DNA had been extracted from the skeletons of two Neanderthals, one from Spain and one from Italy. Both DNA samples included a pigmentation gene called *MC1R*. This gene causes red hair and pale skin in children who inherit it from both parents.

A different group of scholars had previously reported the discovery of a gene called *SLC24A5* on human chromosome 15. This gene reduces the melanin, or brown pigment, in the human epidermis, leading to fairer skin. Today it accounts for 25 to 38 percent of the difference in skin color between European and African populations.

Anthropologists generally agree that when our ancestors lived in sunny Africa, their skin would almost out of necessity have been brown, because melanin gave them protection from cancer-causing ultraviolet radiation. Once our ancestors entered the cold, foggy, overcast environment of Ice Age Europe, however, dark skin offered them little advantage. In such an environment, pale skin facilitates vitamin D absorption, arguably outweighing the protection given by melanin. What this suggests is that while our African ancestors would have been dark-skinned, the overcast conditions of Ice Age Europe would have favored fair skin.

It is worth considering that the contrast between white and brown skin—a miniscule genetic difference that some recent societies have used to justify extreme social inequality—may simply have been nature's way of protecting some people from skin cancer and others from vitamin D deficiency.

THE BRIDGE TO A NEW WORLD

Our ancestors who colonized the cold northern regions of Asia also underwent genetic change. One change was a lightening of the skin similar to that seen in Europe but caused independently, DNA experts believe, by a different set of genes.

The land surface of northern Asia, like that of Southeast Asia, became more extensive during the lowered sea levels of the Ice Age. For example, one land bridge connected Siberia to the islands of Sakhalin and Hokkaido, allowing for the colonization of Japan. Still farther to the north, lowered sea levels created another land bridge across the Bering Strait, linking Siberia to Alaska. The Aleutian Islands chain was also connected to Siberia by dense ice packs. At least 20,000 years ago Siberian hunters followed the game across the Bering land bridge into Alaska and found a whole new continent waiting for them.

Archaeologists now believe that the peopling of the New World involved several waves of immigrants. Some moved south through ice-free corridors into what is now Canada and the United States. Others may have moved even more rapidly down the Pacific coast with watercraft, reaching Patagonia before the Ice Age had ended.

Some linguistic evidence for the Siberian origins of Native American people seems to have survived. In 2008 Edward Vajda concluded that Ket, an indigenous language of Siberia, could be linked to a Native American language family called Na-Dené. Speakers of Na-Dené languages include the Athapaskans of Canada's Yukon and Northwest Territories, the Apache and Navajo of the U.S. Southwest, and the Tlingit of Alaska.

Fifteen thousand years ago the New World was populated from Alaska to Patagonia, though nowhere densely. And when the glaciers melted back at the end of the Ice Age, sea levels rose and the Bering land bridge disappeared. Later arrivals with watercraft, almost certainly including the ancestors of the Eskimo, may have island-hopped along the now ice-free Aleutian Islands. Like Australia, the New World remained relatively isolated from the Old World until the visits of the Vikings and Christopher Columbus. Over more than 15 millennia, the Americas became a wonderful laboratory for social change, witnessing multiple independent cases of the emergence of inequality.

The direction of settlement in Aleutians seems to have been westward.

LIFE ON THE ICE AGE TUNDRA

The longer the Ice Age lasted, the more our ancestors appear to have behaved like the living hunting-and-gathering people studied by anthropologists. And the more like living groups these Ice Age foragers were, the greater the chance that archaeologists will be able to reconstruct their social behavior.

Some 28,000 to 24,000 years ago the plains of central and eastern Europe had been converted to a tundra, or cold steppe, by falling world temperatures. The good news was that this steppe had more resources than today's Arctic tundra because, owing to its more southern latitude, it enjoyed more hours of sunlight. In addition, the premier game animal of central Europe was the woolly mammoth, a creature providing up to eight tons of meat. The bad news was that one had to hunt such mammoths on foot, armed only with a wooden spear.

Into this tundra strode the Gravettians, cold-adapted people whose tool kits resembled those of the recent Inuit, or Eskimo. Like the Eskimo, archaeologist John Hoffecker tells us, the Gravettians inhabited a land without trees and were forced to burn substitute fuels like mammoth bone. They used mattocks made of mammoth tusks to dig ice cellars, in which meat was preserved by freezing. The Gravettians created lamps that burned animal fat and used knives like the Eskimo woman's *ulu*.

At warmer latitudes our ancestors lived in ephemeral windbreaks of branches and grass, but the tundra was far too cold for that. At places like Gagarino in Ukraine, the Gravettians dug into the earth to create warmer semi-subterranean houses. Based on the number of hearths, archaeologists think that some Gravettian camps may have been occupied by 50 or more people for most of a season. Families came together to hunt mammoths and reindeer, then dispersed for a time, maintaining social networks through visiting, cooperating in ritual, and exchanging raw materials over hundreds of miles. Like the Eskimo, the Gravettians carved figurines in ivory, favoring images of women with huge breasts and hips.

The plains of Europe were at their most bitterly cold between 24,000 and 21,000 years ago, and the Gravettians did not hang around to see how much worse it would get. We pick up the story next at the site of Kostienki, on the great Russian plain east of Ukraine.

At Kostienki a sizable group of mammoth hunters lived in an enormous communal longhouse, 119 feet long and 50 feet wide. A row of ten hearths ran down the central axis of the structure, and along its periphery were pits filled with the bones of butchered animals. What this pattern suggests is that perhaps as many as ten related nuclear families, each with its own hearth, cooperated in the construction of a warm communal shelter and shared in the hunting, storing, and eating of mammoths. Archaeologist Ludmilla Iakovleva reports that Sungir, another camp of the same period, had human burials accompanied by more than 3,000 bone beads, not to mention pendants, stone bracelets, and ivory figurines.

Some 18,000 to 14,000 years ago the bitterest cold of the Ice Age had ameliorated, and people were drifting back to some of the areas abandoned by the Gravettians. The landscape was changing from tundra to an environment called taiga, essentially a brushy steppe with evergreens, willows, and birches.

* * *

On a promontory overlooking Ukraine's Rosava River, just southeast of Kiev, lies a campsite of this period called Mezhirich. Mammoth and reindeer hunting would have been optimal here from October to May. During that season a number of families converged on Mezhirich, building smaller and more widely spaced houses rather than living in one large communal shelter. The houses were roughly 20 feet in diameter, and the total population of the camp may have been 50 people.

The houses at Mezhirich were unlike anything seen previously. The structures were probably framed with birch or willow poles and roofed with mammoth hide, none of which has been preserved. What survived were the walls built of mammoth bones stacked to the roof. Each family had its own architectural design. Dwelling 1 was made from 95 lower jaws of mammoths, laid chin-down using a herringbone pattern. The builders of Dwelling 4 alternated layers of chin-up and chin-down mandibles.

According to archaeologist Olga Soffer, the hunters of Mezhirich used mammoth tusk ivory to make figurines, pendants, bracelets, and scrimshawed plaques. Most remarkable, however, was the monumental work of art found in Dwelling 1. Here someone had propped up a mammoth skull and painted its forehead with dots, parallel lines, and branching designs in red ocher. Given how much floor space this work of art needed, we wonder if, after its use as a residence was over, Dwelling 1 might have been turned into a building where hunting magic was practiced.

At least 20 houses of this era, sometimes clustered in groups of three or four, have so far been found in Ukraine and Russia. They remind some scholars of the whalebone houses built by the twentieth-century Eskimo. In addition to butchering mammoths and trapping fur-bearing mammals, these ancient occupants of the taiga had turned orphaned wolves into companion animals. They may deserve credit for creating the world's first dog, or, if one prefers, Ice Age man's best friend.

Raw materials used by the occupants of Mezhirich came from a distance. Some of their chipped stone was quartz crystal, brought 60 miles from the southeast; some beads were of amber, brought 60 miles from the northwest; still other beads were of fossil marine shell, brought 200 miles from the south. The distances involved make it likely that these materials were being traded from group to group.

With art (and perhaps hunting magic) already documented, let us now move westward into Spain, France, Belgium, and Germany. Some 15,000 years ago, during the last stages of the Ice Age, that part of Europe was a cold steppe with dwarf birches and willows.

Archaeologists call the people of this land and era the Magdalenians, after a site in France. Their dependence on reindeer has been compared to the dependence of the Barren Grounds Eskimo on caribou, which is essentially the same animal.

The Magdalenians had the cutting-edge hunting technology of 15,000 years ago: bow and arrow, spear-thrower, and harpoon. Reindeer provided them with meat, fat for lamp fuel, skins for clothing and tents, sinews for thongs, and antler and bone for tools. Apart from the reindeer, the Magdalenians hunted wild horse and European bison, trapped Arctic hare, ptarmigan, and grouse, and fished for salmon, trout, and pike. Like the earlier fishermen of Wadi Kubbaniya, they may have smoked fish to lengthen the season of availability. Magdalenians moved with the game, occupying caves in the winter and riverside camps in the summer.

Some 15,000 years ago, the archaeological evidence reveals a full-blown complement of art, music, and ornaments. The Magdalenians played flutes carved from animal bone, made figurines depicting humans and animals, and decorated themselves with beads and pendants of bone, ivory, and animal teeth. Their most celebrated forays into the humanities, however, can be found on the cave walls of Lascaux in France and Altamira in Spain. There these late Ice Age people painted realistic scenes of deer, bison, mammoth, humans carrying bows, and humans and animals penetrated with arrows. Even the most cautious archaeologists concede that the Magdalenians must be considered fully equivalent to the hunting-and-gathering groups of the recent past. And that opens the door to a huge archive of detailed information on living foragers, collected by anthropologists over the last century.

Our search for the origins of inequality can, therefore, take 15,000 B.C. as its starting point.

WHY DOES EVIDENCE FOR A "MODERN MIND" NOT APPEAR EARLIER?

Most observers agree that the behavior of the Magdalenians reflects a mind as fully "modern" as the one possessed by the archaeologists who dig them up. An increasing number of scholars, however, pose the following question: If anatomically modern humans have been around for at least 100,000 years, making ornaments for 80,000 years, and carving figurines for 25,000 years, why was it not until 15,000 years ago that we finally see overwhelming evidence for a "modern" mind?

There is no widely accepted answer to this question, but a few suggestions have been offered. One popular view holds that growing population density was the reason. Proponents of this view argue that the ability to generate art, music, and symbolic behavior was probably there throughout the Ice Age but remained latent as long as people were expanding into unoccupied wilderness. Once the world had become more extensively occupied by groups of hunters and gatherers, or so the argument goes, there would have been increasing pressure to use symbolism in the creation of ethnic identities and cultural boundaries. After all, one of the activities that regulate interaction among neighboring ethnic groups is ritual, and ritual often involves art, music, and dance.

We concede that population growth took place throughout the Ice Age. We suspect, however, that there was another process taking place, one that explains why the archaeological evidence for symbolic behavior appears discontinuous—strong in some localities and weak in others. It has to do with an important difference between two types of hunting-gathering groups, recently emphasized by anthropologist Raymond Kelly. The difference hinges on whether a group of foragers has, or does not have, permanent social groups larger than the extended family.

In Kelly's words many foragers—including the Netsilik and Caribou Eskimo of the Canadian Arctic, the Hadza of Tanzania, and the Basarwa of Botswana/Namibia—once manifested "only those social groups that are cultural universals, present in every society, and nothing more." These societies had both nuclear families and extended families, but the extended families rarely persisted beyond the death of the parental pair. Most significantly, families were not grouped into larger units of the type anthropologists refer to as clans or ancestor-based descent groups.

Other foraging societies, however, did feature larger units, each of which contained many families. The Aborigines of Australia had many levels of units beyond the family. Foragers with lineages, subclans, and clans often do have higher population densities than clanless foragers and have moved beyond the informal ways in which extended families can be organized. Essentially they created large groups of people who claimed to be related, whether this was true or not. For this purpose they used language to extend their terms for different kinds of relatives to a much larger group of people.

The division of a society into such units can take many forms. Sometimes each unit reckons descent through one gender only, either the father's line or the mother's. Early anthropologists, needing a term for such multigenerational units, borrowed the word "clan" from the ancient Scottish Highlanders. In other cases, one social unit may reckon descent from a real or mythical ancestor, without weighing one gender more heavily than the other. Both clans and ancestor-based descent groups can be made up of smaller units called lineages.

Kelly has reconstructed the way that society might have been modified to create clans. In the case of descent through the male line, for example, the original founding families were most likely headed by the sons and sons' sons of a set of brothers. In effect, clansmen built upon the bonds that already existed between brothers in clanless societies. Expanding an earlier social premise, that "brothers should hunt together and cooperate with one another," they established that any brother in an antecedent generation would be considered equivalent to any other, serving as an enduring link between living men and the lineage's alleged founder(s). Each clan, in turn, was made up of related lineages or subclans.

Why would the creation of multigenerational lineages and clans during the late Ice Age have escalated the use of art, music, dance, and bodily ornamentation? The answer is, although one is born into a family, one must be *initiated* into a clan. That initiation requires rituals during which clan secrets are revealed to initiates, and they undergo an ordeal of some kind. To be sure, even clanless societies have rituals, but societies with clans have multiple levels of ritual, requiring even more elaborate symbolism, art, music, dance, and the exchange of gifts.

Still other rituals are used to establish each clan's unique identity and to define its relationship with other clans in the same society. Ideas about incest are often extended to the clan level; in such cases, members must marry outside their own clan. When such marriages take place, both the couple and their respective clans often exchange gifts, and the groom may even have to pay a "bride-price." All these rituals provide contexts in which music, dance, art, the exchange of valuables, and the decoration of human bodies are carried out on a scale beyond that of clanless societies.

We suggest, therefore, that even without the pressures of growing Ice Age populations, the creation of larger social units would have escalated symbolic behavior—in effect, launching the humanities. This scenario could explain why the archaeological evidence for symbolic behavior appears at different moments in different regions. Simply put, not all Ice Age societies made the transition to units larger than the extended family.

Among the Ice Age societies described so far, we suspect that the mammoth hunters of eastern Europe may have had clans or ancestor-based descent groups. It seems even more certain that the Magdalenians had them, and that the painted scenes deeply hidden in the caves of France and Spain were visual aids by means of which initiates were taught the origin myths and accepted behaviors of their social unit. Archaeologist Leslie Freeman points out that some scenes painted in Spanish caves could be seen only after initiates had crept on their bellies through constricted passageways, making the experience a more memorable ordeal.

Societies with clans enjoy advantages over those without them. They have created large groups of people, claimed as relatives, on whom they can rely for defense from enemies, for amassing the foodstuffs and valuables needed for major rituals, or to assemble the resources needed to pay off a bride's kinsmen.

The advantages of clan-based society may even tell us something about the disappearance of the Neanderthals. Neanderthals displayed low population densities and show no archaeological evidence for social units larger than the extended family. In face-to-face competition for territory, they probably stood little chance against archaic modern humans organized into clans. We find this likely because by the twentieth century, most hunting-gathering societies without clans had been relegated to the world's most inhospitable environments. They were pushed there by groups with more complex social organization.

The popular press likes to suggest that Neanderthals simply were not smart enough to compete with our more modern-looking ancestors, but that view sounds racist to us. The Neanderthals may simply have gone the way of most foragers who had no social units larger than the extended family.

Before we begin congratulating our Ice Age ancestors for creating clans, however, bear in mind the fact that they had taken a step with unintended consequences. Clans have an "us versus them" mentality that changes the logic of human society. Societies with clans are much more likely to engage in group violence than clanless societies. This fact has implications for the origins of war. Societies with clans also tend to have greater levels of social inequality. Later in this book we will meet societies in which clans are ranked in descending order of prestige and compete vigorously with each other. The germ of such inequality may have been present already in the late Ice Age.

TWO

Rousseau's "State of Nature"

Rousseau felt that to understand the origins of inequality, one had to go back to a long-ago time when nature provided all human needs, and the only differences among individuals lay in their strength, agility, and intelligence. People had both "anarchic freedom" (no government or law) and "personal freedom" (no sovereign master or immediate superior). Individuals of that time, which Rousseau called the "State of Nature," displayed self-respect but eschewed self-love.

Most anthropologists do not like the phrase "State of Nature." They do not believe in a time when archaic modern humans had so little culture that their behavior was directed largely by nature. While conceding that the *capacity* for culture is the result of natural selection, anthropologists argue that humans themselves determine the *content* of their culture. Many anthropologists, therefore, bristle when evolutionary psychologists presume to tell them which parts of human social behavior are "hardwired into the cerebral cortex."

Suppose, however, that we pose a less controversial question to anthropologists: What form of human society, because of its highly egalitarian nature, best serves as a starting point for the study of inequality? In that case, many anthropologists would answer, "those hunting-and-gathering societies that possess no groupings larger than the extended family."

In this chapter we examine four such societies: the traditional Caribou and Netsilik Eskimo, who lived in a setting as cold as Ice Age Europe, and the traditional Basarwa and Hadza, who lived in a world of African game like many of our earliest ancestors. We do not look at the twenty-first-century descendants of those ethnic groups; we look, instead, at the way they lived when anthropologists first contacted them. The less altered by contact with Western civilization any foraging group was when first described, the more useful that group's description is to our reconstruction of ancient life.

Some of the first Westerners to visit clanless foragers considered them Stone Age people frozen in time. This idea was so naïve and demeaning that it triggered a backlash. Soon revisionists were claiming that recent foragers can tell us nothing about the past, because they are merely the victims of expanding civilization. That revisionism went too far, and now the pendulum is swinging back to a more balanced position.

Some of the most eloquent spokespersons for the balanced position are anthropologists who have spent years among foragers. The late Ernest S. ("Tiger") Burch Jr., who devoted a lifetime to Arctic hunters, conceded that the industrialized nations' tendency to swallow up ethnic minorities has left few foraging societies unaltered. This situation does not mean, however, that we cannot make use of recent foragers to understand their prehistoric counterparts. What we need to do, according to Burch, is to select a distinct form of society—clanless foragers would be one example—and create a model of that society that can be compared to both ancient and modern groups. If we do our work well, some aspects of our model should apply to all clanless foragers, regardless of when they lived. In other words, if one finds that the foragers of 10,000 years ago were doing something that their counterparts were still doing in the year 1900, that behavior can hardly have resulted from the impact of Western civilization.

One of the most important behaviors we look at in this chapter is the creation of widespread networks of cooperating neighbors. We also examine the archaeological record for comparable networks in the distant past.

SURVIVING THE ICE

Archaeologists have often compared the Gravettians and Magdalenians of Ice Age Europe to the recent Eskimo (or Inuit, as they call themselves). The Ice Age preservation of meat by freezing, the shelters built of animal bone, the knives resembling the Eskimo woman's *ulu*, the animal-oil lamps, the ivory carvings, and the heavy dependence on reindeer all invite comparisons to living Arctic peoples. To be sure, the Ice Age in Europe ended 10,000 years ago. But as recently as 1920 there were still indigenous foragers at the top of the world, largely unaffected by the industrialized West, who earned their living under conditions reminiscent of the Ice Age.

The Eskimo were not the first people to enter Arctic America. The archaeological record shows that some of the earliest occupants of that region were boreal forest hunters whose behavior resembled that of the later Athapaskan people of Canada. Some 4,000 years ago, however, an archaeological complex called the "Arctic Small Tool Tradition" foreshadowed later Eskimo culture. The people using these small tools kept warm in semi-subterranean houses with tunnel entrances, essentially a sod-covered version of the later igloo.

About 2,500 years ago, from the Mackenzie River on the west to Hudson Bay on the east, a new and more convincingly proto-Eskimo culture spread over the Canadian Arctic. Called the "Dorset Culture," it was created by hunters who lit oil-burning lamps on Arctic nights, used snow-cutting knives to build igloos, made bone shoes for sled runners, used antler or walrus-ivory spikes to walk on ice, and left behind models of what were probably kayaks.

Thousands of years would pass, however, before the full richness of Eskimo culture was revealed to the West. The great pioneer of Eskimo anthropology was the intrepid Knud Rasmussen. Raised in a Danish settlement in Greenland, Rasmussen learned to converse with the local Eskimo as a child. By the 1920s he was using his language skills to study the Eskimo of the Canadian Arctic. Every Eskimo expert of the last 80 years owes him an intellectual debt.

The Eskimo of the 1920s, of course, knew nothing of the Dorset Culture. Like every society of which we know, they had their own cosmological explanation of how their world had come into being. And that cosmology provided the moral justification for their social logic.

The Netsilik Eskimo of the central Canadian Arctic, for example, believed that Earth had always existed. During mythological time, however, humans lived in perpetual darkness like that of the Arctic winter—no sunlight, no animals, and no pleasure or suffering existed. Then Nuliajuk, an orphan girl pushed from a kayak, sank to the ocean bottom and became Mistress of the Sea. She then created all animals, all hard work, and all pleasure.

During this primordial era, both humans and animals could speak, and there was little difference between them. Finally the Arctic hare cried out for "day," and there was light; but day was forced to alternate with night, because the nocturnal Arctic fox cried out for "darkness."

At first there were no Netsilik, only the Tunrit, or "Old Ones," superhuman beings who rendered the land inhabitable. The Tunrit left when the Netsilik arrived, but not before they had invented the leister for spearing fish, the long stone walls for driving caribou on foot, the weir for fishing, and the craft of hunting caribou from kayaks.

In the cosmology of the Netsilik, a gradual process of differentiation created deities, humans, and animals out of primordial chaos. Pleasure and suffering were established to give life meaning. With the crystallization of the visible world came the establishment of moral order, a series of prohibitions to control wickedness and promote restraint.

Because humans and animals had breath, they could interact with the Spirit of the Air, giving them more power than plants or rocks. The difference between humans and animals was that humans had names; one's name was so magical that it made one superior to a caribou or seal. The Spirit of the Air, whose actions could be felt in wind and weather, was one of three spirits more mysterious and powerful than any others. The second of these was the Mistress of the Sea, already mentioned, who controlled the souls of all sea creatures. The third was the Spirit of the Moon, who controlled the souls of all animals on land.

Like many foraging societies, the Eskimo believed that humans would be reincarnated. They were stoic in the face of death because they saw it as a recycling of the soul from one body to another. An occasional elderly Eskimo, tired of the struggle against hunger and cold, took his or her own life, confident of returning as a newborn. So strong was this belief that babies might be given the name of a deceased elder, or even referred to as "grandmother" or "grandfather."

Eskimos loved children, but in a land where starvation was endemic, their belief in reincarnation made infanticide a pragmatic decision in times of stress. When a family knew that it could not afford another mouth to feed, it might simply leave a newborn outside to freeze—in effect, saying to its spirit, "Please come back later; this is not a good time." This act was performed before the infant had been named, so that the recycled spirit within it would not be insulted; indeed, without the magic of a name, the tiny creature was not yet human. If another family heard the infant's cries and felt that they had the means to feed it, it was theirs to adopt.

Most Eskimo groups of central and eastern Canada lived in societies without clans, dividing their year between hunting caribou, harpooning seal, and catching fish. While men killed the seal and caribou, the onerous task of processing the carcasses into meat, hides, bones, and useful organs fell to the women. Marriage was first and foremost an economic partnership, maintained in spite of low population densities and incest prohibitions that sometimes made it hard to find a spouse.

Among the Caribou Eskimo of Hudson Bay, years of female infanticide might leave marriageable women in short supply. Some families therefore betrothed their children in infancy, hoping to ensure that each hunter would one day have a wife. Eskimo marriage had to be a flexible relationship, adaptable to a variety of economic situations. Its four major forms, according to Tiger Burch, were as follows. In one type of marriage, familiar to Westerners, a man and a woman became attracted to each other and married. A second type of marriage, however, involved one man and two wives. Such polygamous unions were favored in cases where a gifted hunter was killing enough game to support two wives and needed both just to process the carcasses. His second wife was added only after the first relationship was stable, and the older wife usually remained the dominant one.

In regions where years of female infanticide made eligible brides scarce, there was a third type of marriage: one wife with two husbands. Processing animals for two hunters was a great burden for the woman, but no man was expected to process them for himself. The two husbands served as co-fathers to the children.

Finally, some Eskimos practiced a kind of co-marriage in which two couples shared sexual partners. This usually happened when two men hunted together and became close friends. Sometimes the wives were coerced into co-marriage, and sometimes they joined willingly. One advantage of co-marriage was that it made siblings out of all the children born to both couples, setting up long-term obligations of food sharing, protection, and mutual support.

THE CARIBOU ESKIMO AND THE EGALITARIAN ETHIC

When anthropologist Kaj Birket-Smith visited the Caribou Eskimo during the period 1921–1924, some 437 of them occupied 60,000 square miles to the west of Hudson Bay. Their land was an Arctic heath of lichens and low bushes, and their staple foods consisted of caribou, seal, walrus, Arctic hare, ptarmigan, salmon, trout and pike—not unlike the diet of the Ice Age Magdalenians.

Men built igloos in winter, hunted, fished, and drove sled dogs; women built tents in summer, tended fires, and tailored clothing from skins. As with so many foragers, no one amassed a surplus. No one claimed exclusive rights to the land. Traps and weirs were communal property. During famines, all food was shared with neighbors. After a successful hunt, the actual slayer of each caribou was identified by the markings on his arrow. The meat was then divided by rule, with the slayer receiving the frontal portion and his hunting companions the rest.

So crucial was food sharing that the Eskimo used ridicule to prevent hoarding and greed. Anyone who has seen Eskimos singing satirical songs about greedy individuals or dancing in masks to ridicule stingy neighbors realizes the crucial role that humor plays in human society. Troublemakers were given the silent treatment and might even be left behind when a camp moved. It was expected that a truly dangerous, aggressive individual would be killed by his own family. If, however, a neighbor did it, he might have to flee to avoid the family's revenge.

Life in the Arctic was stressful, but the behaviors just described are not unusual for a clanless society. It was a truly egalitarian society in which the slightest attempt to hoard or put oneself above others was discouraged. A skilled hunter and good provider might be universally respected, but even he was expected to be generous and unassuming.

The Eskimo lived in a world where magic existed alongside practical knowledge and the spirits of humans and animals never really died. Some rock outcrops were considered places where people or animals had been turned to stone, and because they now lacked breath, they could no longer interact with the Spirit of the Air.

THE NETSILIK ESKIMO AND THE CREATION OF LARGER NETWORKS

The caribou that winter near the tree line in Manitoba migrate north to the Arctic coast in spring. By September they are headed south again. In days of old they had to pass through the land of the Netsilik Eskimo, who used long fences of boulders to divert the caribou toward bowmen.

Anthropologist Asen Balikci describes Netsilik territory as a tundra with lakes and rivers and, near the Arctic coast, saltwater bays. Ringed seals were common in these bays and were easier to hunt in winter, after the caribou had moved south. During that season, with the bays frozen over, seals could be killed at their breathing holes by men waiting with harpoons. The Netsilik also fished for the migratory salmon-trout, or Arctic char, with spears and leisters; they captured seabirds by hurling bolas of stones and thongs.

Like many other Eskimo groups of central and eastern Canada, the Netsilik lived in a society without clans. Female infanticide was frequent but could be forestalled when women were in short supply. At such times the parents of infant girls might be asked to betroth them early. This is a good example of a contradiction in social logic, which can be expressed in the following principles:

1. Male infants are valued because they will become hunters.
2. Female infants are expendable because the Arctic has few plants for women to gather.
3. Hunters need wives to process their caribou and seals.
4. At the moment, there are not enough girls to provide wives for all the young men in the region.
5. Premise 4 trumps premise 2, so female infants are no longer expendable and might even be worth bride service.

SEAL-SHARING PARTNERSHIPS

We come now to a very important Netsilik social strategy called *niqaiturasuaktut.* That awesome word is the name of a Netsilik meat-sharing partnership, first described in 1956 by a priest from the Pelly Bay Mission, and it has implications far beyond the Netsilik.

Early in the life of a Netsilik boy, his mother chose for him a group of male partners, ideally 12. Close relatives and members of the group who camped with the boy's family were not eligible; his mother's goal was to choose individuals who, under ordinary circumstances, would have no close relationship with her son.

Eventually the time came when the boy in question had become a hunter. Waiting silently by a breathing hole in the ice, he saw his chance and harpooned a seal. Ritual demanded that the animal be placed on a layer of fresh snow before being carefully skinned (Figure 1). Though dead, it was given water so that its soul, when reincarnated, would be grateful and allow the seal to be killed again.

Next, the harpooner's wife cut the seal open lengthwise and divided the meat and blubber into 14 predetermined parts. Twelve of these parts would go to the partners chosen for him. The last two parts, the least desirable, would go to the harpooner himself. The first partner—addressed by the term *okpatiga,*

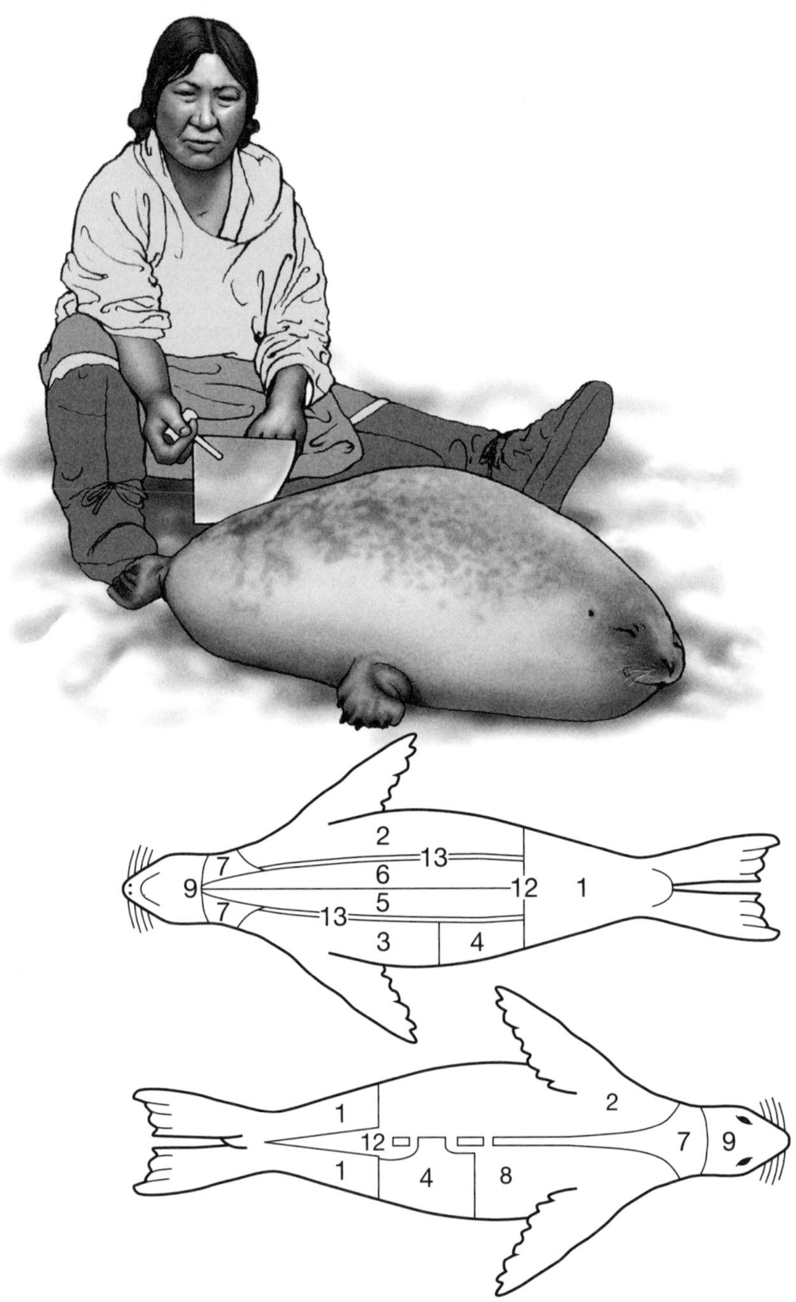

FIGURE 1. Netsilik Eskimo families created social networks through the sharing of seal meat. After laying the dead seal on a layer of clean snow and offering it a drink of water, an Eskimo woman would use her *ulu* knife to skin it and cut it into 14 portions. (Only 11 of the portions are shown on the diagrams, as the other three are internal.) Twelve of the 14 portions were then given to meat-sharing partners.

"my hindquarters"—would receive the *okpat* or hindquarters of the seal. The second partner—addressed by the term *taunungaituga,* "my high part"—would receive the *taunungaitok* or forequarters. Subsequent partners received the lower belly, the side, the neck, the head, the intestines, and so on.

Netsilik meat-sharing partnerships could become hereditary. When two adult hunters became habitual partners, they often arranged for their sons to be future partners. If one partner were to die, he could be replaced by someone with the same personal name. This act followed logically from the premise that any two people who shared the same name were magically linked.

Let us now consider the implications of seal-sharing partnerships. The Netsilik did not have clans or, for that matter, any social grouping larger than the extended family. Clearly, however, they felt the need for a widespread network of allies on whom they could rely to share resources when they were scarce. They created such a network using only their language and the magical power of the name, choosing respected acquaintances to be their sons' "hindquarters," "kidneys," and so forth. And once that network was operating, they allowed parts of it to become hereditary.

Twelve meat-sharing partners is admittedly a small group compared to a clan. But when we consider how many partnerships there were, and the likelihood that a set of brothers might belong to several, we can picture a mutual aid network covering thousands of square miles.

PROVIDING ARCHAEOLOGICAL TIME DEPTH

The Netsilik data suggest that foragers without clans sometimes created extensive networks of cooperating nonrelatives. Can archaeology detect similar networks among ancient societies?

To answer that question we turn to the prehistoric Folsom culture that occupied Colorado 11,000 years ago, near the end of the Ice Age. Colorado had no caribou, but it did have *Bison antiquus,* a creature some 20 percent larger than today's buffalo. Assuming that its habits were similar to the modern bison, this beast would have migrated north to the High Plains in the summer and back to the southern plains in the winter.

Archaeologists Mark Stiger and David Meltzer have excavated a possible winter camp made by Folsom hunters, 8,600 feet above sea level near Gunnison, Colorado. There they have found the remains of circular huts, roughly the size of the mammoth hunters' houses at Mezhirich. The best-preserved one had a basin-shaped floor sunk more than a foot below ground surface; a ring of postholes suggests that the hut had a conical roof, made of branches and daubed with clay to protect the occupants from wind and snow. The fact that the huts found so far were laid out in an arc makes one wonder if they were arranged around an open area, set aside for communal or ritual activity. Such an arrangement was common among foragers.

Some 2,000 feet lower, in the foothills of the mountains near Colorado's border with Wyoming, lies the legendary Folsom site of Lindenmeier. Eleven thousand years ago it was the scene of multiple hunting camps on a low ridge, overlooking a wet meadow that attracted migrating bison. Lindenmeier, excavated by Frank H. H. Roberts in the 1930s, was painstakingly reanalyzed by Edwin N. Wilmsen. One of Wilmsen's most exciting conclusions was that at least two different groups of bison hunters, each with its own style of tool manufacture and its own widespread network of partners, had converged on Lindenmeier to collaborate in the hunt.

Folsom hunters possessed the spear-thrower known by its ancient Mexican name the *atlatl.* The flint points of their spears are considered technological masterpieces. Folsom hunters made the edges of the point sharp and symmetrical by delicately removing tiny parallel flakes. As a final touch, they skillfully struck off a long channel down each face of the point, making it easier to insert in a wooden shaft.

Both groups at Lindenmeier made points of this type. The hunters in one encampment, called Area I, gave theirs gently rounded shoulders and trimmed the edges with small overlapping flakes, taken off at a 90-degree angle to the long axis of the point. Hunters who were camped 330 feet away in Area II, however, gave their points more abruptly angled shoulders and trimmed the edges with nonoverlapping flakes, taken off at a 45-degree angle to the long axis. Each group, in other words, appears to have come to Lindenmeier from a region with its own tradition of point-making.

Nor did the differences end there. While each of the encamped groups made most of its stone artifacts from flint, each had brought to Lindenmeier a small quantity of obsidian, a naturally occurring volcanic glass. Obsidian, while less readily available than flint, would have been preferred when hunters wanted extremely sharp cutting edges.

One can track obsidian to its source by analyzing its chemical trace elements; this was done to the Lindenmeier obsidian, with surprising results. The obsidian in Area I was mostly from a volcanic flow near Jemez, New Mexico, 330 miles to the south of Lindenmeier. The obsidian in Area II was mostly from a flow at Yellowstone Park, 360 miles to the northwest. Not only did the groups camped at Lindenmeier have different point-making styles, they evidently belonged to different networks of trading partners that gave them access to distant resources.

Archaeologists have found no evidence that Folsom society possessed clans. In the case of the Lindenmeier site, Wilmsen reconstructs the individual camps of Area I as consisting of 14 to 18 people; those of Area II he estimates at 13 to 17 people. Clearly these extended families had created networks of trading partners and allowed other families to share their best hunting locales. We do not know if their meat-sharing resembled that of the Netsilik. We have seen enough, however, to conclude that widespread networks of nonrelatives are not a recent development but a long-standing behavior of small-scale foraging societies.

WARM-WEATHER FORAGERS

Not all foragers, to be sure, had to cope with ice. Even during the bitterest cold of 30,000 to 18,000 years ago, equatorial Asia and Africa would have been temperate or even frost-free. At warm latitudes there were thousands of edible plants, and the economic role of women was different from that in the Arctic: often it was the women who harvested the bulk of society's food.

Until about 1,700 years ago, much of Africa south of the Sahara Desert was occupied by foragers. Anthropologists believe that many of them spoke languages whose consonants included clicking sounds that we are forced to write with punctuation marks. Today the speakers of these "click languages" are largely confined to regions no one else wants. They were unable to defend their territory against larger-scale societies, beginning with the Iron Age farmers, herders, and metalsmiths of the so-called Bantu migration. In the rest of this chapter we look at two foraging groups whose lands have been reduced by tidal waves of more complex societies.

THE BASARWA AND THE MAGIC OF THE NAME

Once there may have been 200,000 speakers of click languages in southern Africa. By the 1970s they were reduced to 40,000, many occupying the Kalahari Desert on the border between Namibia and Botswana. Anthropologists love them but cannot decide on a politically correct name for them. Everyone agrees that they should not be called "Bushmen," as they were for centuries. So scholars began referring to them by their local group names, such as !Kung San, Dobe !Kung, and so forth. Eventually someone decided that they could all be referred to by the supposedly neutral term "Basarwa." By the time you finish reading this page, of course, Basarwa will probably have become politically incorrect.

By any name, the Basarwa are among the most thoroughly studied foragers on earth. Anthropologists such as Lorna Marshall, George Silberbauer, and Richard Lee were among the pioneers of the 1950s and 1960s. Unfortunately, by the 1980s, many Basarwa had been converted to an underclass in Botswana society. In our description of Basarwa society, therefore, we lean most heavily on the earliest studies.

The Kalahari provided the Basarwa with endless vistas of brush and savanna, sandy plains, dunes, and low hills. The bedrock was pitted with sinkholes that became watering places, crucial landmarks in a world with only ten inches of rain. Broadleaf trees grew on the land that South Africans call *bushveld,* acacia trees on the drier *thornveld.* Sand dunes supported ricinodendron trees whose nuts supplied the !Kung people with half their plant food. Areas of hardpan supported commiphora plants, the preferred food of the beetles whose pupae the !Kung converted into poison for their arrows.

The Nyae Nyae !Kung believed that the Earth of long, long ago supported a supernatural Trickster who rose to become an omnipotent being in the eastern sky. Using a magic called *n/um,* the great Trickster created the sun, moon, stars, rain, wind, lightning, waterholes, plants, animals, and humans. These first humans, "the Old Old People," had their own variety of n/um. The !Kung eventually lost this magic, although certain of their ritual songs still had n/um and could move people deeply. Other things that had n/um were ritual fires, the sun, rain, elands and giraffes, ostrich eggs, bees and their honey, blood, milk, and certain medicinal plants.

When the Old Old People died they became *//gauwasi,* the spirits of the dead. These spirits lived in the Upper Sky, doing the bidding of the Trickster. People prayed to the //gauwasi to evoke their sympathy, exhorted them in anger when things went wrong, and feared their wrath. Thus just as in the creation myths of so many foragers, the !Kung were taught morals and proper conduct by a previous race whose authority came from a celestial being.

Like many hunter-gatherers, the !Kung arranged their conical huts in a circle, leaving space in the center for a fire around which they danced on certain nights (Figure 2). Each hut was mainly for storing belongings; only during rains did the !Kung sleep inside. Some huts held nuclear families, others widows or widowers, and still others adolescents of the same sex. When a successful hunter had two wives, each built her own shelter of branches and thatch.

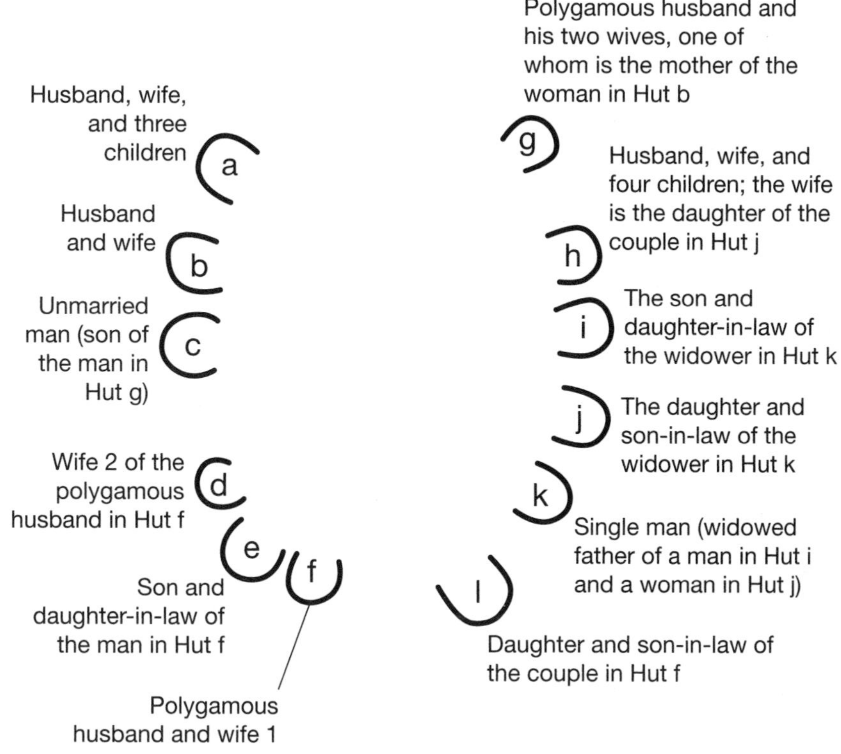

FIGURE 2. Basarwa foragers of the Namibia/Botswana borderlands had extended families but no clans. During the dry season of 1968, 35 !Kung foragers lived briefly in the camp shown in this diagram. The 12 men, 10 women, and 13 children in this camp belonged to three different extended families. They built 12 huts, arranged in a circle around an open area where dances and other ritual activities could be carried out. The men in Huts f, g, and j were the heads of the three extended families, which were linked by blood or marriage. Although the senior males in this camp were respected, none had any real power or authority; their society was as egalitarian as any ever studied.

A groom was expected to live near his wife's parents for several years, until he had supplied them with enough game to work off his bride service. That service was cut short if he took a second wife. Many first marriages were arranged by the parents, sometimes with the early betrothal of children.

In the 1950s the !Kung had no groups larger than the extended family, although there are hints that they had once inherited membership in a larger unit called a *!ku-si*. We do not know exactly what a !ku-si was, but it might have been a lineage or clan. We return to that possibility later in this chapter.

A good-size !Kung camp consisted of four or five extended families, linked to each other in the eldest generation by sibling relationships or marriage ties. The head of the camp was usually a senior man, referred to as *kxau*. The kxau was not considered the owner but, rather, the custodian of a 100–250 square-mile territory called a *n!ore*. His job was to make sure that only the people of his group were using the plant foods of the n!ore, although hungry neighbors could petition to share it. Often camps or water holes were named for the kxau. These headmen had no coercive authority and accepted no privileges for fear of arousing jealousy.

!Kung social logic included a premise that we encounter over and over in this book: "We were here first." The people who had lived at a certain water hole longer than anyone enjoyed the privilege of deciding who else could live there.

Within each n!ore, the !Kung pursued the following schedule. Women collected plants, an activity with a low risk of failure, and shared the harvest only with their immediate family. Men hunted big game, an activity with a high risk of failure, and when they killed a prized animal, they were expected to share the meat with everyone in the camp.

Men—hunting alone or in groups of cooperating brothers, cousins, or in-laws—tried to get within 30 feet of a kudu, wildebeest, or eland, shoot it with a poison arrow, and then track it until it dropped. Each hunter fashioned his own distinctive arrows, and the man whose arrow killed the beast was allowed to decide how the meat would be allocated. In a ritual reminding us of the Netsilik, the carcasses were divided into 11 packages: the breast, left haunch, right haunch, upper back, and so on.

!Kung hunters, like the men in Rousseau's State of Nature, were unequal in strength, agility, and marksmanship. Thus there was always the potential that one skilled bowman's arrows would repeatedly be found to have killed the kudu. It is notable how hard the !Kung worked to prevent a meritocracy of good hunters from arising. First, using a system of reciprocal gift-giving called *hxaro,* they exchanged arrows with each other. Richard Lee once examined the quivers of four men who were hunting together. All but one had arrows made by four to six different men, and two men had literally no arrows that they themselves had made. Thus each hunter would eventually have one of his distinctive arrows credited with a kill, whether he himself had fired it or not.

A second behavior, already seen among the Eskimo, was the use of humor to prevent feelings of superiority. A hunter asking for help in hauling a magnificent kill to his camp would be told, "You think this skinny bag of bones is worth carrying?" No hunter was allowed to boast, and refusing to share would not have been tolerated.

Again we see one of the most basic premises of egalitarian society: If one wants to be well thought of, he will be generous. If he strays from this ideal, he will be reminded of it with humor. If he persists in not sharing, he will be actively disliked.

Unlike the foragers of Klasies River Mouth or Wadi Kubbaniya, the !Kung had an immediate return economy in which storage was actually discouraged, since it could lead to hoarding. When it came to the hxaro gift-giving system, however, delayed returns were expected.

Each !Kung had a series of partners with whom he exchanged gifts. According to anthropologist Pauline Wiessner, about 70 percent of these partners were relatives or in-laws, while 30 percent were unrelated persons, treated (and addressed) as if they were honorary kinsmen. One gave his partner a gift with the expectation that within two years he would reciprocate with a gift of equal value. Not to reciprocate would have been as reprehensible as not to share meat.

It was not the intrinsic value of the gift that mattered. The real value of hxaro was that it created a network of partners, often living up to 50 miles away, whose camps one could visit in times of scarcity. Whole families could visit other camps for up to half a year, sharing food and water, after which the visitors were allowed to build their own shelter and forage in their hosts' n!ore.

The willingness of clanless foragers to host nonrelatives during lean years was presumably crucial to survival. Some scholars, especially those with a background in animal behavior, refer to such generosity as *altruism;* a few have even suggested that there might be a gene (or genes) for such behavior. We wonder, however, if there might not be an alternative way to interpret such hosting. We wonder if clanless hunters, forbidden by society from accumulating surplus food, may instead have been accumulating social obligations. Their alleged altruism, in other words, could be seen as a self-serving investment, a way of obligating their guests to host them in the future when their situations were reversed.

As for hxaro exchange, it is significant that the !Kung were forbidden to reciprocate with a gift more valuable than the one they had received. A hunter with many hxaro partners was greatly respected, but he was not allowed to shame his partners with gifts so large that they could not be matched. We mention this because in later chapters we will begin to encounter societies with clans that used lavish gifts to create the impression of inequality.

NAMESAKES

Hxaro was not the only system used by the !Kung to create networks of partners. There were also networks of *!gu!na,* "namesakes," built on the premise that names were magic.

Among the Nyae Nyae !Kung, Lorna Marshall discovered 46 names for men and 41 for women. First-born children were often named for a grandparent of the same sex. So magic were names that if an unrelated person received the same name as someone's sibling, that person would be treated as if he or she *were* that sibling. If one deliberately named one's child after a distant relative or in-law, one was allowed to address all that namesake's relatives by the same terms that would be used for the child's relatives.

The network of name-partners among the Nyae Nyae extended 80 miles to the south, 60–80 miles to the east, and 115 miles to the north. Arriving at a distant camp, a visitor needed only to give his shared name to be welcomed by the family of his !gu!na.

Once again we see clanless foragers creating a network of mutual aid, cooperation, and food sharing far larger than their home territory. They did this not with expensive gifts but with language and the magic of the name.

The !Kung therefore show us one of the adaptive contradictions of clanless foraging society, expressed in these three points:

1. Having no unit larger than an extended family allowed for great flexibility. As resources waxed and waned, camps could move and families could aggregate or disperse within their own n!ore as needed.
2. On the other hand, there were times when survival depended on being able to leave one's n!ore and seek the hospitality of unrelated neighbors. Under these conditions, having no kin group larger than the extended family put one at a disadvantage.
3. The !Kung dealt with this contradiction by creating two extensive networks of honorary kinsmen: hxaro partners and !gu!na namesakes. For the !Kung, both networks were egalitarian. Later in this book, we will see agricultural societies turn both gifts and magic names into sources of inequality.

THE EASTERN HADZA: SETTING THE STAGE FOR LARGER SOCIAL UNITS

Lake Eyasi is a body of brackish water in Tanzania's Great Rift Valley. To the east of the lake is a dry, rocky savanna with scattered baobab and acacia trees. Once, in an era of unspoiled environments, this savanna would have supported herds of elephants and rhinos, zebras and giraffes, and smaller game such as the eland, impala, and gazelle.

The Eastern Hadza were once 400 strong, occupying 1,000 square miles of the Lake Eyasi savanna. According to James Woodburn, who began studying them in 1958, the Hadza saw their world as divided into four quadrants, occupied by "the people of Sipunga Mountain," "the people of Mangola River," "the people of the West," and "the people of the Rocks." Each of these regions was home to 50–100 Hadza, but all social groups were extremely fluid. During part of the year, families were dispersed and lived in smaller camps. When the fruits of the cordia and salvadora bushes became ripe, families converged on these groves and lived in larger camps. Dry season camps tended to be larger because this was the season of big game, and everyone wanted to live near the most skilled hunters.

Because of this endless coming and going, some anthropologists consider the camp (rather than a permanent band) the only meaningful social unit for clanless foragers. An average Hadza camp had 18 adults, but camps could be as small as one hunter or as large as 100 people. The Hadza moved their camps for a multitude of reasons: to be near large, recently killed game; to abandon a camp where someone had died; to obtain wood for arrow shafts or poison for arrows; or to trade with other Hadza.

The nature of Hadza camps varied with the season. During the summer rains they might occupy rock shelters. In the dry season they lived outdoors in ephemeral huts or windbreaks. Their most substantial camps lay usually within a mile of water, often among groves of trees or protective rock outcrops. At these camps Hadza women made conical shelters thatched with grass, rarely spending more than two hours at the task. A young woman usually built her shelter close to the one occupied by her mother—close enough so that her husband could perform bride service, supplying meat to his mother-in-law, yet not so close that he would violate the custom of "mother-in-law avoidance."

In addition to supplying his mother-in-law with meat, a young man was expected to give her long strings of bride-price beads. Good hunters, with two wives, needed twice the beads. If similar social obligations existed during the Ice Age, it would explain why so many seashells were turned into beads and traded over great distances.

Like the Eskimo and !Kung, the Hadza respected generosity and hospitality. None of the quadrants of their world were considered exclusive territories. Anyone could hunt, collect wild plants, or draw water anywhere. An individual living at a camp where he normally did not belong was called *huyeti*, "visitor," but no one really objected to him. In fact, some visitors ingratiated themselves with their hosts by bringing them what they most desired: honey.

Although sharing was important to the Hadza, anthropologists Kristen Hawkes, James O'Connell, and Nicholas Blurton Jones point out that men were under more pressure to share than women. As with the !Kung, women engaged in low-risk plant collecting and kept most of the harvest for their family. Men engaged in high-risk big game hunting and were expected to share with everyone. Any big game killed would be presented to the whole camp; a refusal to do so would bring supernatural punishment and social retribution.

Game was eaten immediately because the Hadza, unlike some of the Ice Age foragers we saw earlier, did not smoke or store meat. Like the !Kung, they were immediate return strategists whose meat-sharing built social bonds. Meat and honey made up only 20 percent of the Hadza diet by weight, but no other foods did as much to strengthen the fabric of society.

Like the other foragers in this chapter, the Hadza had an egalitarian, consensus-based society in which leadership was noncoercive, really amounting to no more than the advice of a few respected senior men. The composition of larger Hadza camps, however, provides us with a possible scenario for the origins of lineages and clans.

Hadza women, as we have seen, built their huts close to their mothers'. Monogamous Hadza couples, in fact, were five times as likely to share a camp with the wife's mother as with the husband's. This means that a Hadza camp often consisted of a senior woman, her married daughters and their husbands, and her married granddaughters and their husbands. Woodburn points out that this social grouping has the same genealogical composition as a *matrilineage,* that is, a lineage whose descent is reckoned in the female line. In other words the occupants of some Hadza camps may have constituted the raw material out of which, under the right conditions, a lineage could have crystallized.

When a society feels the need for a large, multigenerational group of allies, lineages and clans can be an even better solution than the seal-sharing partners of the Netsilik or the name-sharing partners of the !Kung. We have mentioned the possibility that the !Kung may once have had larger social segments called !ku-si, membership in which was inherited. At certain times in the past, therefore, the genealogical groupings seen in large foraging camps may occasionally have turned into multigenerational social units.

In the case of the !Kung, such larger units might later have dissolved back into their constituent extended families when conditions changed for the worse—when, for example, the !Kung were driven into the Kalahari Desert by more powerful neighbors. Archaeologists should therefore be alert to the possibility that some Ice Age foragers developed lineages or clans for a time, only to lose them when conditions deteriorated.

THE PRESERVATION OF EQUALITY IN SMALL-SCALE FORAGING SOCIETY

In this chapter we have looked exclusively at societies whose largest unit was the extended family. All used social pressure, disapproval, and ridicule to prevent anyone from developing a sense of superiority.

Such attempts to keep everyone equal provide a notable contrast to the behavior of our nearest living relatives, the great apes. Humans share 98 percent of their DNA with chimpanzees, and chimpanzees are anything but egalitarian. They have alpha males who physically abuse their rivals and beta males who bully everyone but the alphas.

One of the goals of anthropology is to understand the differences in behavior between apes and men, in the hope of learning what it means to be human. In a now-classic study, anthropologist Marshall Sahlins proposed that sex, food, and defense of the group were the three basic needs for both apes and early humans. What differed, Sahlins argued, was the priority placed on each need.

For apes the highest priority is sex, followed by food and defense. Chimps are extremely promiscuous, and males compete constantly for mates. Males are willing to share portions of monkeys that they have killed, but they do not like to share females. Strong hormones drive males to fight viciously over the troop's alpha position and the females who go with it.

For human foragers, on the other hand, food is the first priority, followed by defense and sex. Marriage—a food-getting partnership rather than a hormone-driven sexual liaison—has replaced the promiscuity of ape society. For example, we have seen that two Eskimo husbands could share the same wife without displaying the jealousy and violence of male chimpanzees.

Precisely because human marriage is an economic partnership, it showed great flexibility from the beginning. No traditional forager would accept the argument that marriage must be restricted to one man/one woman in order to "preserve the family." Traditional Eskimos, for example, knew that a family could be one man/one woman, or one man/two women, or one woman/two men, or even two men/two women. Far from threatening the institution of the family, this flexibility strengthened it by allowing it to adjust to any economic situation.

That is not to say that male foragers never competed in any aspect of sexual activity. There was one arena in which they competed, but it was out of focus to them because of their cosmology.

Anthropologist William S. Laughlin once lived with a group of 78 traditional Eskimo hunters in Alaska's Anaktuvuk Pass. Part of his research included a study of their blood antigens. The group was led by a senior male, described by Laughlin as a superior caribou hunter. Antigen studies showed that this man had fathered seven of his group's children, who in turn produced ten grandchildren, for a total of 17 living descendants. In other words, this one respected senior male had contributed his genes to some 20 percent of his group's population.

Evolutionary biologists would say that this caribou hunter had competed with others to pass on his genes, and that he succeeded admirably. It is doubtful, however, that he was even aware of competing. To him, each baby born to his group would have been the reincarnated spirit of an ancestor.

This Anaktuvuk anecdote returns us to this question: Just how much of human behavior is controlled by our genes? Fortunately, we now know the answer: a great deal less than hard-core evolutionary biologists claim but probably more than most anthropologists want to admit.

THREE

Ancestors and Enemies

Among clanless foragers like the Basarwa and Hadza, homicide was an individual matter. The assassin might be killed by his own relatives or a member of the victim's family. In other cases, the perpetrator might go into hiding, while his relatives placated the victim's relatives with food and valuables.

An important change in social logic, however, took place with the formation of clans: a kind of "us versus them" worldview seems to have been created. If someone from Clan A murdered someone from Clan B, it was considered a crime against the victim's entire clan. This required a group response. As the result of a principle Raymond Kelly calls "social substitutability," Clan B could avenge its member's death by killing anyone from Clan A, even women or children who were innocent of the original murder. Sometimes, in fact, merely doing something that Clan B interpreted as an insult—trespassing on their territory, for example—could get members of Clan A killed.

How far back can "social substitutability" be detected in the archaeological record? The answer is as far back as the late Ice Age, a time when other evidence for clans or ancestor-based descent groups was accumulating.

JEBEL SAHABA

Our oldest archaeological evidence for group violence comes from the Nile Valley. Jebel Sahaba is a sandstone mesa to the east of the river, two miles from Wadi Halfa. At the base of the mesa, archaeologist Fred Wendorf found a late Ice Age cemetery. This was an area where, as at Wadi Kubbaniya, foragers relied heavily on seasonally flooded embayments, places where catfish and tilapia gathered and dense mats of purple nut grass grew.

Some 15,000 years ago, several different groups of foragers occupied that stretch of the Nile. Each of these groups could be distinguished by its style of stone tools, like the two groups identified by Wilmsen at Lindenmeier. But while the two Lindenmeier groups seem to have coexisted peacefully, it is clear that the groups competing for Nile embayments did not.

The cemetery at Jebel Sahaba contained 58 skeletons of men, women, and children. Twenty-four of the skeletons showed signs of violence. Some 116 flint artifacts, the majority of them parts of spears or arrows, had entered the bodies of these people; in some cases, they remained embedded in the skeleton. Some victims had literally been riddled with arrows, while others had old, healed wounds that would suggest a history of violent injury. Included were defensive fractures of the forearm, broken collarbones, and an arrow point in one man's hip.

Examples of overkill were frequent. A middle-aged man, Burial 21, had been hit with 19 flints; projectiles were found in the top of his pelvis, his forearm, his lower leg, his rib cage, and the base of his neck. One young woman, Burial 44, had been hit with 21 flints; three of these were probably the barbs of a spear that penetrated her face and reached the base of her skull. Nor were children spared. Two youngsters interred together, Burials 13 and 14, had flints in their neck vertebrae, each probably representing a *coup de grâce.*

According to Kelly, armed conflict among groups with clans falls into one of two categories: confrontation and ambush. Confrontation takes place between all-male war parties, and there is no element of surprise. Ambush, on the other hand, often kills unsuspecting men, women, and children. Kelly considers Jebel Sahaba clear evidence for ambush. He describes several victims as having been "pincushioned" with arrows fired by multiple enemies, a common phenomenon in ambushes.

Wendorf agrees with Kelly and suspects that two factors led to a history of ambushes at Jebel Sahaba. First, the Nile embayments of that era would have been rich but circumscribed environments, surrounded by desert. Second, there were several groups of people competing for the fish and nut grass of each embayment, apparently ready to defend them with violence if necessary.

Ambushes of this kind are too small-scale to be considered war, but they can be nasty, and once begun they can turn into blood feuds that can last for years. While blood feuds tend to arise with clans, there are exceptions to this rule of thumb. In areas of rich resources like those of the tropical Andaman Islands, described later, even clanless foragers could "afford to make enemies of their neighbors," as Kelly puts it. In the case of Jebel Sahaba, however, we believe that we are dealing with societies with clans. One reason is that the site was a cemetery.

To be sure, even Neanderthals sometimes buried their dead. Only at later sites like Jebel Sahaba, however, do we see actual cemeteries created so that multiple generations of men, women, and children from the same social unit could be laid to rest together. In an era without written deeds, creating a cemetery allowed foragers to lay claim to a rich embayment by arguing that their ancestors had been buried there since time began. It was their way of saying, "We were here first."

THE ANDAMAN ISLANDERS

The Andaman Islands are located in the tropical Bay of Bengal. They were colonized by hunters and gatherers thousands of years ago, perhaps by watercraft from Burma (modern Myanmar).

When first contacted by Western explorers, the foragers of the Andamans did not have permanent social groups larger than the extended family. In spite of this, they displayed some of the behaviors characteristic of larger-scale societies. They built bachelors' huts for unmarried youths and reserved special types of burials for highly respected elders. They also engaged in ambushes like those of societies with clans. Some anthropologists think that this group violence was permitted by the islands' rich food resources, which made it less crucial to stay on good terms with one's neighbors.

Anthropologist A. R. Radcliffe-Brown visited the Andaman Islanders from 1906 to 1908. He learned that there had once been 5,000 of these foragers living on the largest island, which covered 1,660 square miles. By the 1920s, only 800 of them were left.

The nuclear family was the basic unit of Andaman society, but with this unusual twist: many children were adopted, usually around age ten, by friends. A family would adopt as many children as they could support, allowing the biological parents to visit their children often. At puberty, a boy left home to share a bachelors' hut with other youths, but he remained obedient to both his biological and adoptive parents. Clearly, on the Andamans it *did* "take a village to raise a child."

The group sharing a camp might consist of 20 to 50 individuals, but with an interesting architectural dichotomy. Inland groups created a beehive-shaped communal house, called a *bud,* by packing every family's shelter into a circle so tight that all of it could be roofed over with mats. During the dry season, when no roof was needed to protect them from rain, families dispersed into open-air camps, or rockshelters. Coastal groups, on the other hand, made camps like those of the Basarwa. They arranged their small shelters in a circle, surrounding an open space used for bonfires or dances (Figure 3, top).

As with many foragers, there was a protocol to be followed. Each woman owned the wild yams she had collected and reserved them for her immediate family. A big fish belonged to the man whose harpoon killed it. A wild pig

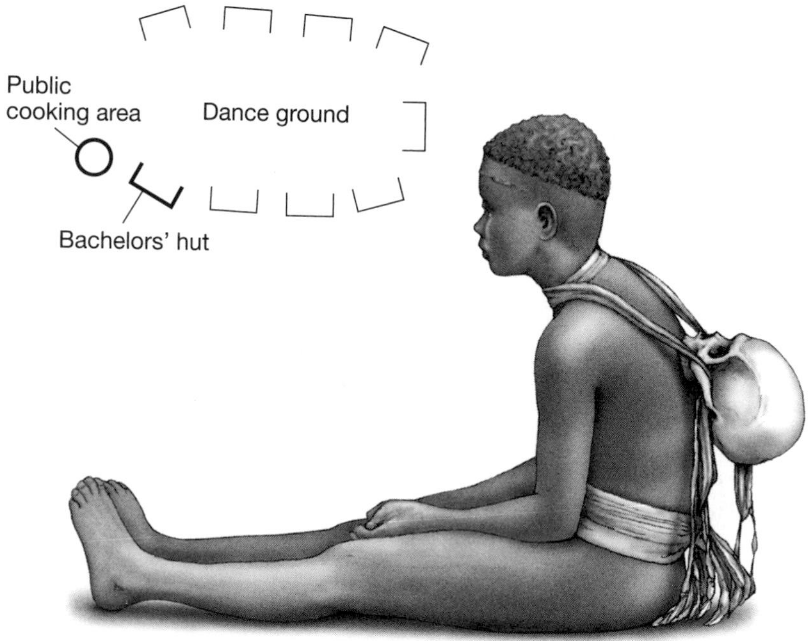

FIGURE 3. Traditional foragers on the Andaman Islands arranged their huts in a circle or an oval, surrounding a space left open for dancing. The settlement might include a bachelors' hut and a public cooking area. Andaman Islanders regularly saved skulls and other skeletal parts from their ancestors, a behavior that can be detected in many of the earliest prehistoric villages in other parts of the world.

belonged to the man whose arrow struck it first, with this exception: when a young, unmarried man killed a pig, its meat would be distributed by an older hunter, with the choice parts going to senior men.

This deference to one's elders was typical of Andaman society. Terms of respect equivalent to "Sir" and "Lady" were used by young people when addressing elders. Seniority, therefore, was a source of unequal treatment, and this inequality extended to reciprocal gift-giving. The Islanders had a system like the hxaro of the !Kung, but when an older man received a gift from a younger man, his reciprocal gift could be of lesser value.

Each of the 13 ethnic groups in the Andamans deferred to a group of elders called *maiola*. Each also had an informal headman called a *maia igla*. Despite being groomed for the position for years, the maia igla had no real authority, only the power of persuasion. His wife had similar influence among the women of her group.

Younger men performed tasks for the headman, waiting their turn to be elders. The position of maia igla did not automatically pass from father to son, but the grown son of a deceased headman would be given preference if he was highly respected. To earn respect he needed to be generous and kind, slow to anger, and skilled at both hunting and combat. Such a man was referred to as *er-kuro*, "big," a term used frequently in larger-scale societies.

Group Violence

During Radcliffe-Brown's stay in the Andamans, he found that two of the 13 ethnic groups—the inland Jarawa and the coastal Aka-Bea—had become mortal enemies. Raymond Kelly, who has restudied the Andaman information, traces their hostilities to the period between 1792 and 1858, when the Aka-Bea took away territory from the Jarawa. After that, whenever foragers of one group came upon their enemies hunting pigs, gathering honey, or collecting shellfish, the larger party would attack the smaller. So frequent were such attacks that the Jarawa, who usually went around naked, took to wearing bark armor to ward off arrows.

Because the Aka-Bea and Jarawa spoke different dialects, it was difficult for them to hold peacemaking ceremonies. In areas where there was no such language barrier, the Islanders did negotiate truces. Any man who had killed an enemy took to the jungle for months of ritual purification; in his absence, the women of both groups arranged for the other men to dance and hunt together, ending hostilities for a time.

Despite having no clans, the Andaman Islanders displayed several behaviors typical of societies with clans. One of these was the painful ordeal of deliberate scarring at initiation, which was thought to ensure toughness. A second behavior was the creation of the bachelors' hut. While its original purpose was probably to make sure teenage boys did not sleep with teenage girls, the hut was also associated with ceremonies; for example, the bachelors often prepared ritual meals for the entire camp. Such bachelors' huts may have been the prototype out of which larger-scale societies created the "men's house."

A third intriguing behavior was the unequal treatment shown certain elders after death. While most Andaman Islanders were buried in the ground, some respected and influential people had their corpses exposed on platforms in trees. Such elders were mourned until all the flesh was gone from their bones. Then—to the accompaniment of dancing—their bones were recovered, washed, painted, saved, made into ornaments, given away as presents, or worn to protect against illness (Figure 3, bottom).

Why do the clanless Andaman foragers show us so many ritual institutions typical of societies with clans? As we mentioned earlier, many anthropologists suspect that it had something to do with the abundant resources of the Andaman Islands. We doubt, however, that this was the whole story. We detect principles of social logic that allowed for differential burial treatments and exceptions to reciprocal gift-giving. This logic was tied not to the Andaman environment but to the premise that elders were more deserving of respect than youths.

AN INTRODUCTION TO AUSTRALIA AND TASMANIA

The eighteenth-century Europeans who arrived in Australia found it already inhabited. The descendants of the hunters and gatherers who entered Australia more than 45,000 years ago had grown to an estimated 300,000, divided among 300 groups called (for want of a better term) "tribes." On the island of Tasmania, southeast of Australia, there may have been another 8,000 foragers.

Every Australian forager belonged to a clan. Many groups had a pair of still-larger social units, each composed of multiple clans. These paired units, called *moieties* (after the French word for "one half"), provided a kind of "loyal opposition" for each other in social and ceremonial situations. Society was thus built of nested units—families within clans, clans within moieties, each unit requiring its own rituals. What we do not know is whether nested units arose from clanless society or were an ever-present alternative, depending on local circumstances.

The Aborigines of Australia (a term they themselves find politically correct) present anthropologists with a number of paradoxes. On the one hand, they once lived in simple windbreaks, and many of their stone tools resembled those used 90,000 years ago in the caves of Mt. Carmel. Isolated for tens of thousands of years, they appeared to many observers to offer insights into life in the Stone Age. On the other hand, as archaeologists and anthropologists soon learned, the Aborigines had been undergoing continuous change ever since they arrived in Australia.

A map published by anthropologist A. P. Elkin documents the spread of several innovations from the northwest part of Australia toward the southeast. Largely restricted to the northern Kimberley region were cave paintings reminiscent of those in Ice Age western Europe. The Aborigines of northern and central Australia used circumcision as an initiation ordeal, but it never reached Queensland and New South Wales. In New South Wales and Tasmania, initiation involved the knocking out of a tooth; evidence from ancient burials suggests that this may have been the older of the two ordeals.

In the north, some Aborigines practiced tree-platform burials like those of the Andaman Islanders, but this behavior never spread to the south. The Australians transformed the throwing stick into the boomerang, but it never reached Tasmania.

No less interesting was the spread of the so-called section and subsection systems, methods of classifying relatives that some Western scholars find difficult to master. These uniquely Australian systems ensured that everyone you ever came into contact with was classed as a kinsman, letting you know how to behave toward him or her. This was done by extending the terms for relatives outward until they embraced the entire tribe. Since this extension was accompanied by incest prohibitions, the subsection system left only about one in eight members of the opposite sex eligible to marry you.

Although the section/subsection system of classification is considered prototypic of Australia, subsections were in fact an innovation, still spreading in the nineteenth century. According to anthropologist M. J. Meggitt, the Walbiri tribe of the Northern Territory adopted sections no longer ago than 1850; it added subsections some 20 or 30 years later. Such classificatory terms spread 100 miles in about 20 years, and by 1896 the Aranda tribe of the Alice Springs area was adopting subsections in imitation of the Walbiri.

The reason for such a rapid spread, Meggitt felt, was that Aborigines did not like it when their neighbors displayed innovations that they lacked. And Meggitt discovered an equally revealing fact: although we know that the Walbiri began using subsections no more than 150 years ago, they had revised their origin myth to allege that subsections were given to them during the fourth stage of Creation.

The lessons of Australia are many. Yes, the nineteenth-century Aborigines provide insights into a long-ago era, but neither they nor anyone else were frozen in time. From the moment they reached Australia, the Aborigines began creating new ways to organize society, some of which were still spreading when Europeans arrived. The Aborigines also show us that, contrary to popular belief, cosmology and religion are not eternal and unchanging. When societies and their situations change, cosmologies get revised as well.

As we have seen, many of Australia's innovations failed to reach Tasmania. Rising sea levels at the end of the Ice Age isolated that island, preventing some ancient behaviors from being superseded. The Tasmanians hunted with javelins and wooden clubs but missed out on the spear-thrower and the boomerang. They initiated youths by scarring them or knocking out a tooth, but evidently they had never learned about circumcision. They had cremations and simple interments but no tree-platform burial. Like the Andaman Islanders, some Tasmanians wore around their necks the bones of deceased relatives. They were comfortable going naked but liked to wear long necklaces of perforated seashells like those of our Ice Age ancestors.

The Australian Aborigines are among the most extensively studied of all hunters and gatherers. Tragically, the same cannot be said for the Tasmanians. The first European colony on the island was established in 1803, and by 1877 virtually every Tasmanian Aborigine had died of disease, neglect, or outright mistreatment. We must piece together the story of the Tasmanians from the accounts of travelers and European colonists. In these accounts the Tasmanians often sound like an amalgam of Olympic athletes and Navy Seals.

Take, for example, the economic partnership of husband and wife. Armed with an 18-foot wooden javelin, which they could reportedly throw 40 yards, Tasmanian men got within range by stalking kangaroos on their hands and knees. At other times, they set fire to the brush and speared the animals as they emerged. They threw kangaroos onto live embers to singe off the fur, and then they cut them into portions that they dipped into ashes "as if into salt." Using a wooden throwing stick called a *waddy* (a precursor of the boomerang), the men also killed birds, which were then placed on embers to singe off their feathers.

Tasmanian women, for their part, are said to have caught opossums by shinnying rapidly up gum trees, cutting toeholds, and using a loop of rope "like a telephone lineman." Considered excellent swimmers, they dived for abalone, prying them from the rocks with wooden chisels. They loved the eggs of black swans, penguins, petrels, and ducks. One Tasmanian woman is alleged to have put Cool Hand Luke to shame by eating, at a single meal, 50 to 60 large eggs of the sooty petrel. Such meals might be washed down with local cider, fermented from the sap of the eucalyptus tree.

Since most Tasmanians vanished before there were anthropologists to interview them, we have only sketchy details of their social organization. We do know, however, that some tribes fought bitterly. Warriors approached their enemies with their hands clasped innocently atop their head, secretly dragging a spear through the brush by gripping it with their toes. To accept the "gift" of a flaming firebrand was to accept a challenge to combat, something unwary colonists learned to their sorrow.

During the 1890s many residents of Australia, appalled by the demise of the Tasmanians, pressured the government to create a post called Protector of the Aborigines, so that similar genocide would not take place in Australia. One of the early Protectors, a magistrate named F. J. Gillen, teamed up with Melbourne biology professor Baldwin Spencer to write two important books on the Aborigines of central Australia. Spencer and Gillen's pioneering work soon inspired anthropologists, including A. R. Radcliffe Brown and W. Lloyd Warner, to record native Australian culture before it vanished. For their part, Spencer and Gillen became so closely associated with the Aranda that they were eventually initiated into the tribe—without, one hopes, having a tooth knocked out.

What the Aborigines have since done for themselves, of course, is more important than anything a magistrate or an anthropologist could do for them. One day, in the 1960s, a man named Bill Kurtzman saw a young Aborigine girl peering through the fence at a tennis court in Barellan, New South Wales. Kurtzman encouraged her to enter the court to see if she liked the game. The young girl turned out to be a member of the Wiradjuri tribe; her name was Evonne Goolagong, and such was her aptitude for tennis that by age 18 she was playing at Wimbledon. Ms. Goolagong went on to win 14 Grand Slam titles, four Australian Opens, a French Open, and Wimbledon Championships in 1971 and 1980.

We would do well to remember that every human being on earth is descended from hunters and gatherers, and we should not underestimate any of them.

CENTRAL AUSTRALIAN SOCIETY

The landscape of central Australia featured broad plains with groves of acacia, dry creek beds lined with gum trees, and occasional mountain ranges rising 2,500 to 5,000 feet above sea level. Each rocky outcrop, hot spring, or water hole had a name and a sacred history. Wild yams and other tubers were there to be dug up with sticks, and there were pools of standing water with ducks, pelicans, spoonbills, and ibis. The Aborigines used oval shelters called *wurleys* and ephemeral lean-tos called *mia-mia*. Night found them sleeping in scooped-out hollows, sometimes covered with a kangaroo hide. Day might find them cooking sedge bulbs in hot ashes, like the foragers of Wadi Kubbaniya.

Many Australian tribes, as mentioned earlier, were divided into two opposing moieties. Each moiety, in turn, was divided into six or more clans, each claiming descent from a different ancestor. Men of a specific clan in one moiety were supposed to marry women of a specific clan in the other moiety. In the case of the Arabana tribe, for example, men of the Dingo clan married women of the Waterhen clan, men of the Cicada clan married women of the Crow clan, and so on. Even if she were of the proper clan, a bride might only be eligible for marriage if she were classified as the groom's "mother's elder brother's daughter."

Needless to say, the section/subsection system greatly reduced one's choice of a wife. To get around this problem, eligible girls were often betrothed to their future husbands early in life, though they continued to live with their parents until their teens. Sometimes a youth's relatives arranged for his circumcision to be done by a man whose future daughters would be *nupas*, or eligible brides, for him. This ritual told the youth whom to seek out when the time came. The section system also affected polygamy; a hunter who decided that he could support a second wife might wind up marrying his first wife's sister, because no one else was eligible.

Each central Australian group owned foraging rights to its territory, but just as with Basarwa territories, the land might be shared with neighbors in good years. Each local band was led by a senior male, called an *alatunja* by the Aranda. Within limits the post was hereditary, passing from father to son or, if no son existed, to a brother or brother's son. The authority of the alatunja did not extend outside his descent group, and he relied on the advice of a small group of male elders.

In large agricultural societies, hereditary leadership is often linked to social inequality. In the case of the alatunja, it had a different origin: passing the role of headman from father to son protected the ritual secrets of the group from outsiders. To understand how important these secrets were, we need to look at central Australian social logic.

At birth, a child's spirit left a state of purity and entered a profane world. Women were thought to remain profane throughout their lives. Men were profane as youths because they were still ignorant of their clan's sacred lore. As a result, elders paid little attention to younger men. Eventually, beginning with his initiation, a youth would embark on an education that would return him to a state of purity. Sons were prized because they could become warriors, bring brides and victories to their clan, and eventually become elders. The sacred knowledge imparted to them was denied to women, but the latter had their own ritual secrets.

Here we see a premise common to many societies with clans: one is not simply born Aranda or Arabana or Murngin or Walbiri; it takes years of effort to become a full member of one's group.

In the process of becoming Aranda, young men were taken to places where sacred paraphernalia were hidden. For example, they might be shown the bull-roarer—a slab of acacia wood on a cord—and see it whirled to produce a roaring sound, thought to be the voice of an ancestor. Called a *churinga,* or "sacred object," each bull-roarer was alleged to have been made early in the world's history. To handle it for the first time was to share in the lives of heroes past.

Conflict, Peacemaking, and Death

In Rousseau's day the existence of religion was routinely attributed to the fact that humans are the only creatures who know that they will eventually die. While this explanation may sound plausible to educated Westerners, it does not work with most hunters and gatherers. To the Australian Aborigines, for example, there was no such thing as inevitable natural death. Death resulted either from homicide or witchcraft. One way to kill an enemy was to point in his direction with a special stick, singing over it to give it magical power. Especially deadly was the sharpened arm bone of a corpse. The pointing of this bone might be accompanied by the curse, "May your heart be rent asunder."

Sometimes a dying man would whisper the name of the person whose magic he believed was killing him. Then an avenger, his body coated with charcoal to make him invisible at night, his footsteps muffled by emu-feather slippers, set out to kill the offender. In the case of group offenses, Raymond Kelly's principle of social substitutability applied. A vengeance party left to kill members of the offending group, traveling as far as 80 miles with spear-throwers, shields, boomerangs, and clubs. Eager to end hostilities, the offending group might agree to surrender two or three of its least popular members as long as the rest were spared. Figure 4 depicts the whole process, from bone-pointing to group revenge.

While many tribes had enemies for neighbors, there was one person who could travel widely without fear of death: this was the sacred messenger. The messenger might, for example, be carrying the red-painted forearm bone of a deceased relative, showing it to normally hostile groups and inviting them to attend his relative's final burial. Owing to his sacred mission, the messenger could not be harmed. However, after the burial ceremony, the bone might be given by the deceased's father to his father's sister's sons. Their task was to avenge the decedent's death, presumed to be caused by witchcraft.

Australian burial ritual varied by tribe. Often it reflected the increased importance of ancestors in societies with clans. The Warramunga exposed corpses on tree platforms and then crushed and buried all the bones except for the one to be carried by the sacred messenger. When that last bone was finally interred, the deceased could be reincarnated. The Luritcha tribe sometimes cannibalized their enemies, adding insult to injury by destroying the forearm bones so that their victims could not be reincarnated.

Burial in Murngin clans consisted of interring the dead, exhuming them later, defleshing the bones, and saving skeletal parts as sacred relics. We call attention to this ritual because archaeologists have detected it in the earliest village societies of the Near East, strongly suggesting that those ancient societies also had clans.

POTENTIAL INEQUALITY IN FORAGING SOCIETIES WITH CLANS

By itself, the formation of clans and moieties did not dramatically increase inequality. To be sure, the men in Aborigine clans did not believe in gender equity, insisting that women could never become as ritually pure as initiated

FIGURE 4. The drawings here, based on photographs more than a century old, illustrate witchcraft and group revenge among the Australian Aborigines. At *a*, two men of Clan A work black magic on a man of Clan B by pointing a bone toward his camp; at *b*, the victim's clansmen examine his tree burial to deduce who caused his death; at *c*, a vengeance party from Clan B sets out to take revenge on Clan A. Such was the origin of small-scale raiding, which set the stage for war.

men; yet to obtain eligible brides, men were prepared to make generous gifts to a girl's family. It is also true that some headmen inherited their position, but it carried limited authority and served mainly to preserve the clan's ritual secrets. Young Aborigines deferred to their elders but fully expected to become elders themselves one day. Perhaps most importantly, there is no evidence that any clan outranked another.

Among the Murngin of northern Arnhem Land, however, one can see the germ of an institution with the potential to create significant differences in prestige. This was an intertribal trading system called *mari-kutatra,* used to obtain much of the paraphernalia used in ritual. Among the items circulating were wooden spears, parrot feathers, beeswax, resin, red ocher, and beads. The farther away a man's trade goods came from, the more highly prized they were and the greater his renown as an entrepreneur. At the time of W. Lloyd Warner's study, the mari-kutatra had yet to turn anyone into a prestigious "Big Man" like those we will meet in later chapters. It would have taken only a few changes in logic, however, to nudge the system in that direction.

FOUR

Why Our Ancestors Had Religion and the Arts

Each hunting-and-gathering society discussed so far had its own distinctive character. All, however, featured a set of common principles, a few of which we list here.

1. Generosity is admirable; selfishness is reprehensible.
2. The social relationship created by a gift is more valuable than the gift itself.
3. All gifts should be reciprocated; however, a reasonable delay before reciprocating is acceptable.
4. Names are magic and should not be casually assigned.
5. Since all humans are reincarnated, ancestors' names should be treated with particular respect.
6. Homicide is unacceptable. A killer's relatives should either execute him or pay reparations to the victim's family.
7. Do not commit incest; get your spouse from outside your immediate kin.
8. In return for a bride, the groom should provide her family with services or gifts.
9. Marriage is a flexible economic partnership; it allows for multiple spouses and variations.

In addition to these principles, which imply no inequality among members of society, we also encountered some premises that allowed for a degree of inequality. They were as follows:

10. Men have the capacity to be more virtuous or ritually pure than women.
11. Youths should defer to seniors.
12. Late arrivals should defer to those who were here first.

In those societies that featured lineages, clans, or ancestor-based descent groups, the following new premises appeared:

13. When lineages grow and divide, the junior lineage should defer to the senior lineage, since the latter was here first.
14. You are born into your family, but you must be initiated into your clan.
15. The bad news is that initiation will be an ordeal. The good news is that you will learn ritual secrets, become more fully a member of your ethnic group, and perhaps gain virtue.
16. Any offense against a member of your lineage or clan, such as murder or serious insult, is an offense against that entire social unit. It requires a group response against some member (or members) of the offending group.
17. Any armed conflict should be followed by rituals of peacemaking.

Many of the aforementioned principles are considered "cultural universals," shared by virtually all societies. It should come as no surprise that another widespread social attitude is ethnocentricity. Each society believes that its behavior is appropriate, while its neighbors do things improperly. Foragers, however, tend to be philosophical about these differences. Convinced that each human group has a different origin and different ancestors, foragers adjust to their neighbors rather than try to change them. Ethnocentricity thus need not lead to intolerance, although in larger-scale societies it sometimes does.

Another widespread principle is that in life there are no accidents; everything happens for a reason. If you fall ill, it is because you have offended a spirit. If you die, it is because someone has worked witchcraft on you. Failed hunts are the outcome of hunting magic done wrong. Failed harvests are the result of rituals incorrectly performed.

The latter premise, of course, did not disappear with the Ice Age. We know that the Power Ball Lottery depends on randomly generated numbers. Yet we often hear the winner, interviewed beside his newly purchased RV, attribute his victory to supernatural intervention. Then he adds, "I believe that everything happens for a reason."

COSMOLOGY AND SOCIAL LOGIC

Cosmology is a universal institution. All societies have a story that explains how the universe and its beings came into existence. Since no humans were present at the origin of the cosmos, the story is of necessity a myth. Anthropologists define *myth* as a folktale believed to be true and regarded as sacred. Myth differs from *legend,* which is also believed to be true but not regarded as sacred.

Most foragers' creation myths begin with a chaotic Earth that is without form or void. Often there is no light until a spirit or creature requests it. The first humans were created from earth or clay, from plants or animals, or from half-formed beasts. The original humans often had superpowers, magic, or the ability to speak directly with animals. They lost these abilities, often as a form of punishment, when they took on their final form.

Creation myths, however, are more than just folktales. Myths serve as charters for social groups. They include instructions from supernatural spirits on how to earn a living and behave toward each other. In the case of the foragers discussed so far, their cosmology generated many principles of their social logic.

That same cosmology supplies yet another universal premise: many beings, objects, and places are sacred.

Questions about the antiquity of the sacred come up frequently. Many Western scientists cannot believe that people as pragmatic as hunters and gatherers would invest their energy in something as irrational as belief in the sacred. A number of biologists and psychologists, whose views are discussed in a recent synthesis by Nicholas Wade, have concluded that religion might have a genetic basis.

For their part, anthropologists are skeptical about the existence of genes for religion. They can think of many ways that a concept of the sacred could emerge from logic alone.

Consider, for example, the Zapotec Indians of southern Mexico, who referred to the wind as *pèe.* No one can see the wind, but the Zapotec were sure it existed because they could feel it on their faces, see it bend trees, and hear it howl during storms. They recognized the similarity between that wind and the equally invisible breath that flowed in and out of their bodies while they were alive. No special gene was required to convince the Zapotec that wind was a sacred force; for them, breath was the difference between life and death. As with all human groups, Zapotec rationality had its limits, and the

Zapotec chose never to say "nobody knows." For them, pèe came to mean "wind," "breath," and "sacred life force."

For hunters and gatherers, as we have seen, the transition from natural to supernatural was seamless. The Netsilik were raised not only to harpoon seals but also to give the dead seal a drink and return its bladder to the sea. Once reincarnated, the seal would remember the hunter's kindness and allow itself to be harpooned again. That seemed entirely logical.

Anthropologist Roy Rappaport, who experienced firsthand the power of the sacred in New Guinea society, has provided us with a framework for the study of religion. Rappaport argues that all religion consists of three components. First are the *ultimate sacred propositions,* beliefs considered irrefutable despite the fact that there is no empirical evidence to support them. These propositions direct the second component, *ritual,* which must be performed repeatedly and correctly in order to achieve its goals. If done correctly, ritual induces the third component, an *awe-inspiring experience.* Because this experience deeply stirs the emotions of the participants, it verifies the sacred propositions in a way that cold, hard logic could not.

For archaeologists, ritual is the key component of this self-validating system. Because ritual requires paraphernalia, costumes, pigments, and musical instruments, and because it must be performed over and over again, it leaves archaeological traces. We have seen them already in our discussion of the Ice Age.

In recent years we have heard several prominent Western scientists argue that religion could be dispensed with. Rappaport, however, points out that any institution as universal as religion must have contributed to the survival of human groups, otherwise it would long since have disappeared or been replaced by something else. Without acknowledgment of the sacred, there would be nothing to give the ultimate propositions the *gravitas* they need to generate the first principles of social logic.

What appears to bother Western scientists the most is that religion so often seems at loggerheads with science and social progress. This situation conflicts with the widespread assumption that humans are rational thinkers.

Part of the problem, we suspect, is that many scientists are wrong about why humans have language and intelligence in the first place. Because those human attributes originated in the context of foraging, they assume that the purpose of language and intelligence was to make us better at hunting and gathering. After all, our ancestors learned to classify hundreds of plants and animals, shout instructions to each other during hunts, and create technologies to convert superficially unappetizing plants into meals.

The trouble with this assumption is that our earliest ancestors shared the African savanna with animals that could hunt game and convert plants into meals more efficiently than humans ever could. So let us suggest an alternative scenario: human language and intelligence evolved not to make us better at foraging but to make us better at social networking.

If our ancestors had been as pragmatic as some scientists believe, there would have been no need for a concept of the sacred. But in addition to being verbal and intelligent, our ancestors were arguably the most emotional, moralistic, superstitious, and (sometimes) irrational creatures on earth.

To be sure, our ancestors had an incredible knowledge of plants and animals, but their most important intelligence was social intelligence. Their classifications often include not only every living human they come into contact with but every ancestor, including some who were supernatural. The result is that foragers can create larger societies, larger networks of sharing and cooperating individuals, than those of any of their primate relatives. To underscore this, let us consider some of the data that have accumulated since Marshall Sahlins wrote his classic comparison of apes and early humans.

What Have You Done with My Dominance Hierarchy?

Chimpanzees, with whom we share 98 percent of our DNA, have strong social inequality. They display a dominance hierarchy or "pecking order" in which alpha individuals dominate all others, beta individuals dominate all but the alphas, and so on down the hierarchy to the lowliest omega.

It is not predetermined who the alphas will be. Chimps live in troops, and their social structure emerges from a series of interactions among individuals in the troop. These interactions, some confrontational, determine who the alphas, betas, and gammas will be. Nor is the hierarchy set in stone; betas have been seen forming alliances to overthrow an alpha by force. One of the victorious betas then takes over the fallen leader's place.

According to primatologists John Mitani, David Watts, and Martin Miller, one of the ways that male chimps learn to create alliances is by hunting colobus monkeys together and sharing the meat. This food sharing could be seen as a precursor to the sharing of meat by human foragers. It does not, however, extend beyond the limits of the troop. No one has ever seen members of two chimpanzee troops meet at the border between their territories and exchange food. In fact, groups of males from Troop A have been observed ambushing and killing isolated males from Troop B. Thus when a troop of chimps has depleted the food in its territory, it cannot appeal for help to a neighboring troop. Chimps cannot do what human foragers do: accumulate social obligations with their neighbors as a hedge against lean times.

It was not the ability to hunt with spears instead of teeth that created the greatest differences between human foragers and apes. By giving humans the capacity for language and culture, natural selection enabled them to reach beyond their local group and make relatives out of strangers. Their use of words to create clan members, section members, gift partners, and namesakes, and to establish mutual obligations and systems of bride exchange, enabled human society to spread to every corner of the earth.

Our Ice Age ancestors temporarily put an end to leadership based on confrontation. As Christopher Boehm reminds us, the headmen of foraging groups were not bullies. They were generous, modest, and diplomatic, because their constituents were too skilled at alliance-building to put up with bullies. The fate of a bully was to be lured into the bush and shot with poisoned arrows.

Those who study apes, however, tell us that their dominance hierarchies provide stability to their societies. Without such a hierarchy, where was the stability in foraging society going to come from?

Some anthropologists argue that in the process of creating the first human beings, natural selection did away with the dominance hierarchies characteristic of our ape ancestors. Proponents of this view suggest that during the centuries since agriculture arose, some societies have done everything in their power to reinstate a social hierarchy.

While we understand why some would hold this view, we would like to play the devil's advocate. We see other ways that the evidence can be interpreted.

When we look at hunters and gatherers, we see a dominance hierarchy as clear as that of chimpanzees. It is, however, a hierarchy in which the alphas are invisible supernatural beings, too powerful to be overthrown by conspiracy or alliance, and capable of causing great misfortune when disobeyed. The betas are invisible ancestors who do the bidding of the alphas and protect their living descendants from harm. The reason human foragers seem, superficially, to have no dominance hierarchy is because no living human can be considered more than a gamma within this system.

Confirmation of this hierarchy will appear later in the book, as we watch inequality emerge in human society. We will see would-be hereditary leaders who attempt to link themselves to revered ancestors or even to supernatural beings. By the time we reach the civilizations of Egypt and the Inca, we will be introduced to kings who actually claimed to be deities. Such strategies for justifying inequality would not have worked if humans did not already consider themselves part of a natural/supernatural dominance hierarchy.

The celestial alphas were the source of the ultimate sacred propositions. Our beta ancestors were the focus of many rituals. The emotions of living gammas made possible the awe-inspiring experience.

Religious conservatives have long argued that secular laws are derived from ultimate sacred propositions. They will be pleased to learn that their view is supported by what we know of foragers. They may be less pleased to learn that ultimate sacred propositions are not eternal and unchanging. In the Aranda view of Creation, humans were once told that initiation required the knocking out of a tooth. They later decided that they had been told to circumcise initiates. Still later, they decided that they had been told to create sections and subsections. Religions transmitted by word of mouth changed constantly to keep up with innovations and altered circumstances.

There is, therefore, nothing wrong with religion per se. Its role in establishing the morals, ethics, values, and stability of early human society is well documented. What bothers some leading scientists is that many of today's huge multinational religions refuse to take significant scientific information into account.

One roadblock preventing these major religions from adjusting to social and scientific progress is the fact that their sacred propositions are now set in type. Several of the world's great monotheistic religions preserve, largely unaltered, the ultimate sacred propositions of Aramaic-speaking societies that lived too long ago to have heard of Copernicus, Galileo, Newton, Darwin, Crick, and Watson. Had those sacred propositions been passed on by word of mouth instead of in printed texts, religious cosmology might very well have changed slowly over the centuries to keep pace with scientific cosmology. What no one could have foreseen was the invention of the printing press and the fossilization of a pre-Copernican view of the world. So if today's multinational religions sometimes seem resistant to social and scientific breakthroughs, Gutenberg will have to share some of the blame.

That Old-Time Religion

During the Ice Age, the ultimate sacred propositions were transmitted not through scripture but through ritual performance. For examples of how this might have happened, we can return to Spencer and Gillen's nineteenth-century accounts of Australia's Aranda and Warramunga people. Their ultimate sacred propositions can be found in their creation myth, which took place in an era the Aranda referred to as the Alcheringa.

During the Alcheringa, Earth was only partly formed. Creation took four stages. During the first stage, two self-created beings called the *ungambikula* discovered rudimentary half-human, half-animal creatures from which true humans could be made. These creatures were limbless, deaf, and blind, living in a featureless world that had just emerged from the sea. The ungambikula used flint knives to release these creatures' limbs, carve fingers and toes, bore nostril openings, make cuts for mouths, and slit open their closed eyelids so that they could see. The half-animals out of which some of the first people were made—dingoes, emus, cicadas, crows—became the totems, or mascots, of later clans.

Early humans wandered Earth, performing rituals that generated spirit people. Each ritual created a sacred landmark, such as a spring or a rock outcrop, to which the spirits were tied. Some spirits eventually became people, but others became part of a reservoir of "spirit children" who lay in wait to be reincarnated. When an unsuspecting Aranda woman passed a sacred landmark, she ran the risk of being impregnated by a spirit. Such spirits, the ultimate source of all babies, returned to their landmarks after the people they inhabited had grown old and died.

In the second stage the Aranda were taught to perform circumcision with flint knives. This ritual replaced their earlier circumcisions, which (all male readers should skip the rest of this sentence) were done with glowing fire sticks. In the third stage they learned ritual subincision, an even more painful mutilation of the male organ. In the fourth stage they learned the section/subsection system. We now know that stages two, three, and four were later additions to an older creation myth. We suspect that stage one of the myth was very old, because ancient rock paintings in northwest Australia depict half-formed humanoids without mouths, like those of the Alcheringa. These cave paintings, like those of Ice Age Europe, were probably visual aids for the teaching of creation myths.

Earth gradually took on its present form as early humans and their animal ancestors traveled, creating landmarks to mark sacred events or places where people died. During their travels, some older men became weak and were given nourishing drinks of blood from the arms of younger men.

This was the cosmology the Aranda conveyed to Spencer and Gillen in the 1890s. Like all cosmologies, it provided the basis for Aranda morality and ethics. It explained why Aranda clans were named for the plants or animals involved in their ancestors' creation. It lessened the trauma of infanticide, which the Aranda considered no more than the returning of a reincarnated spirit to the sacred landmark where it lived. It explained the practice of ritual bloodletting, including the giving of healthy men's blood to sick old men.

Aranda elders knew, of course, that each new generation would have to learn its group's cosmology from scratch. An appropriate time to indoctrinate youths into all this sacred lore would come when they were old enough to be initiated. Young men would learn male lore from older men; young women would learn female lore from older women.

Anyone who has ever tried to deliver a long, complicated lecture to young people knows that they do not always pay attention. Let them watch music videos over and over, however, and they commit every lyric to memory. Combine art, music, and dance, throw in an intoxicating beverage, and they cannot get enough of the awesome experience.

The Aranda held a secret ritual known as *churinga ilpintira,* which integrated art, music, and dance. It was performed at a secret venue in the desert and began with a group of men smoothing an area of bare ground. One or more would provide blood, often as much as a pint, from veins in their arms. This sacred blood was used both to dampen the ground and to serve as a medium for the paint. Impersonating legendary ancestors, the men serving as artists painted their bodies red, white, yellow, and black, adding downy bird feathers glued on with blood. Using a chewed twig as a brush, they slowly painted the earth with white pipe clay, red and yellow ocher, and charcoal. As the painting took shape, the elders sang ballads recounting the mythical exploits of the ancestors; less experienced men watched and learned.

Aranda earth-paintings were geometric, featuring circles, squares, dots, and lines. Each told the story of an ancestor from the Alcheringa. Members of the Emu clan painted yellow, white, and black figures that represented the eggs, intestines, feathers, and droppings of the emu. Members of the Snake clan painted their totemic ancestor slithering through mythological landscapes. Redundancy drove the story home: the paintings, the songs, and the artists' decorated bodies all reinforced the same account. Repetition of the ritual ensured that no one would forget his or her clan's creation myth.

The churinga ilpintira allows us to see why cosmology, religion, and the arts were crucial to hunters and gatherers. Ice Age foragers had language but no writing. The lessons of myth were passed on audiovisually. Performances combining art, music, and dance fixed in memory the myth and its moral lessons. At the same time, some aspects of the story were allowed to change over time.

We doubt that art, music, and dance arose independently. More than likely they evolved as a package that committed sacred lore to memory more effectively than any lecture. If you doubt this, think back to high school and ask yourself which you remember best: your math teacher's lecture on logarithms or the words to the number-one song on the jukebox. Many baby boomers, unable to remember the hypotenuse of a right triangle, will never forget that "Long Tall Sally" was "Built for Speed."

In Western society today we have lost sight of the original purpose of art, music, and dance. We now attribute art to individual "geniuses," born with a "gift" that yearns to "burst out" in an act of "self-expression." Those without talent need not apply.

The truth is that in early human society, everyone was an artist, a singer, and a dancer. What the archaeological data suggest is that the use of the arts increased as larger social units appeared, because each moiety, clan, section, or subsection had its own body of sacred lore to commit to memory. That is not to say that individual talent went unrecognized. Spencer and Gillen reveal that some Aborigine tribes began rewarding good singers or dancers with valuables, encouraging them to perform at other groups' rituals. Even show business, it would seem, began among hunters and gatherers.

The Warramunga, neighbors to the Aranda, had similar ground-painting ceremonies. Spencer and Gillen were invited to watch one of these rituals, which is illustrated in Figure 5. The painting recounted the creation of spirit children by a totemic serpent.

In addition to transmitting secret ritual information, the churinga ilpintira was designed to elicit an awesome emotional response. Rappaport suspects that the precursors to human emotions might have been the deep bonds of love and dependence seen between mother and neonate in the apes. But emotion in humans evolved to be even stronger, strong enough to make intelligent people do irrational things, strong enough to inspire the selfless acts that strengthen society. Dancing, drinking, and singing for days, as some tribes did, opened a window into the spirit world and thereby confirmed its existence.

If all of this sounds mysterious it may be because, as Edward O. Wilson once wrote, the ultimate motivation of religion is probably hidden from our conscious mind, allowing it to be the process by which "individuals are persuaded to subordinate their immediate self-interest to the interests of the group."

Freed from the continual status confrontations of ape society, human foragers created extensive networks of cooperating pseudo-relatives. They transmitted their cosmology and sacred propositions to the next generation with

FIGURE 5. Many hunters and gatherers of central Australia used art, music, costume, and dance to transmit their creation myths to the next generation. In this scene, based on a 100-year-old photograph, men of the Warramunga tribe have painted and sung the story of a mythical serpent. The painting features the undulating body of the serpent, who left spirit children at the places shown as concentric circles. The kneeling men are dressed to represent creatures and places in the myth. Until a young man had learned his clan's sacred lore, he was considered inferior in virtue to his elders.

rituals involving song, dance, and art. Such multimedia performances created highly emotional experiences. Pictures were more memorable than a thousand words, and our ancestors, like modern filmmakers, used music to evoke happiness, sorrow, fear, and tension. "Art for art's sake" is a relatively recent idea; Stone Age art, like the religious art of the Middle Ages, had an agenda.

AFTERTHOUGHTS

Let us close with a few words on the potential for inequality in societies that labor in the shadow of ancestors and celestial spirits. As long as no living human was more than a gamma, the social playing field was level. Emus, cicadas, dingoes, and water hens coexisted peacefully in the Alcheringa, and the living clans named for them were all considered equal.

In later chapters, however, we will encounter societies that revised their cosmology to create inequality. We will see some lineages that claim to have descended from the older of two cosmic brothers, allowing them to outrank the descendants of the younger brother. Others will argue that, in contrast to everyone else's beta ancestors, their lineage descended from a celestial alpha. This closer relationship to the sacred entitled them to be the social unit from which all future leaders would come. Thus the concept of the sacred, which had once strengthened human society by encouraging selflessness and reducing status confrontation, would one day be manipulated to create a hereditary elite.

FIVE

Inequality without Agriculture

Inequality, according to Rousseau, began when self-esteem gave way to self-love. Foragers knew that as long as they suppressed ambition and greed, they would be well thought of. They were obligated to share food and reciprocate all gifts, yet they were discouraged from shaming their partners with gifts too grand to match. Once larger units such as clans had arisen, however, a number of societies witnessed changes in social logic. Such larger units might collaborate to support members who, in Rousseau's words, "desired to be esteemed by others."

Let us consider only two behaviors, gift-giving and marriage. Among clanless societies, bride service and bride-price only passed between individuals and families. Some societies with clans, however, decided to require even larger gifts between the clans of the bride and groom. Sometimes these gifts went on long after the marriage had been consummated.

In some cases the bride's clan considered itself superior because it was giving "the gift of life," that is, future children. That clan might demand such an expensive bride-price from the groom as to drive him into debt, forcing him to borrow from his clanspeople.

Similar changes took place with the custom of reciprocal gift-giving. Most clanless societies took pains to reciprocate with gifts of equal value. No one wanted to be shamed by receiving too valuable a gift from a kinsman. Many clans, however, did not mind shaming a rival clan with expensive gifts. If being generous was good, being more generous than another clan made one superior. Some clans came close to impoverishing themselves while shaming their rivals with lavish gifts and feasts.

Rousseau suspected that self-love became more common after the adoption of agriculture and animal husbandry. He felt that farming could exacerbate natural inequality by allowing smarter, stronger, and more industrious individuals to create the food surplus they needed to outstrip their neighbors. Many of today's anthropologists would agree, to this extent: societies with agriculture and animal husbandry do seem to create more opportunities for inequality.

There are, however, a few exceptions to the preceding paragraph. Most were unknown in Rousseau's day. Some hunting-and-gathering societies did not wait for agriculture to promote self-love. They produced both wealth and inequality using only wild foods.

In this chapter we examine some of the ways in which foragers could wheel and deal their way out of Rousseau's State of Nature. To document the process, we need go no further than the Native societies of western North America. We begin with California Indians who became wealthy middlemen in the movement of seashell ornaments. We end with Alaskan fishermen who were divided into aristocrats and slaves.

In our examination of western North American societies we are not limited to early European eyewitness accounts, because those eyewitnesses were preceded by thousands of years of prehistory. Here is a case where social anthropology and archaeology can work together. Social anthropology gives us, in great detail, the historic tip of the iceberg. Archaeology gives us, albeit in less detail, the mass of prehistoric ice that is hidden from view. Only when we employ both disciplines do we see the whole iceberg.

THE CHUMASH OF THE CALIFORNIA COAST

California once displayed a prodigious diversity of hunting-and-gathering peoples. Its deserts were home to small-scale clanless groups, while its great Central Valley could support societies with lineages and clans. The Yokuts of the San Joaquin Valley, like the natives of Australia, had hereditary leaders and clans with totemic mascots. To the north, in the Sacramento Valley, lived their linguistic relatives the Wintun. Several Wintun leaders not only inherited their positions but also went on to establish regional spheres of influence, expanding their group's territory at the expense of their neighbors. This influence, however, was ephemeral. It depended heavily on the charisma of individual leaders and produced only modest inequality.

A number of archaeologists have committed themselves to explaining the rise of inequality among the Indians of California. One of the best documented cases is that of the Chumash, who once occupied the California coast from San Luis Obispo to Malibu Canyon and colonized the islands of the Santa Barbara Channel. A long-term project, led by archaeologist Jeanne Arnold, has revealed that the historic Chumash were the end product of 7,500 years of social change.

Since the Santa Barbara mainland is now densely covered with modern buildings, Arnold chose to focus her research on Santa Cruz Island. No Spanish missions had ever been built there, so at least five of its historic Chumash villages were relatively undisturbed. These had been villages composed of 125 to 250 people living in very large houses. Each residence was built of poles, thatch, and reed matting, and each housed groups of people related through the male line. Arnold's team located at least 35 deep, circular depressions left by such multifamily dwellings.

Some 7,500 years ago, the Santa Barbara region and the Channel Islands were occupied by nomadic foragers who divided their effort between the inland acorn groves and the marine resources of the Pacific. For the next 5,000 years, as far as one can tell from the archaeological record, their society seems to have been egalitarian. These coastal foragers were limited in their ability to capture large fish by their simple watercraft, which were made of bulrushes and waterproofed with natural asphalt from the California tar pits.

The turning point in Channel Island prehistory seems to have come between A.D. 500 and 700. The key innovation was the creation of a large ocean-going plank canoe. The raw materials for this canoe included redwood logs that had washed up on the Channel Islands as driftwood. One of Arnold's sites contained more than 200 fragments of redwood, as well as asphalt brought from the mainland in abalone-shell containers.

The *tomol*, or Chumash plank canoe, required 500 man-days of labor to make. The result was a vessel 19 to 22 feet long, made of redwood planks sewn together with milkweed cords and caulked with a mixture of asphalt and pine tar. In contrast to the earlier bulrush vessels, which were only eight feet long and held two to three passengers, the tomol could hold either 12 passengers or a ton of cargo. These canoes were capable of going 65 miles out to sea, making the 12–31 mile trip between the coast and the Channel Islands easier.

Between 500 and 1150, the tomol began to alter the archaeological record. First, the ancestors of the historic Chumash began pulling in swordfish and tuna, large fish that would have capsized a bulrush vessel. Second, each plank canoe could carry a ton of asphalt from the mainland for future caulking. Third, the Channel Islanders became producers and middlemen in the shell trade along the California coast.

Between 1150 and 1300 the sources of flint on Santa Cruz Island were increasingly converted into blades and drills for cutting and perforating shell. The Islanders made huge quantities of beads from olive shells, abalone, and Pismo clams. Mainland groups had an insatiable demand for these shell ornaments and were willing to surrender basketloads of acorns, piñon nuts, and edible grass seeds to get them.

According to anthropologist A. L. Kroeber, it is likely that the Chumash furnished the bulk of the shell valuables used in the southern half of California. Not only were the strings of shell beads used for bride payments, they also came to be used as a medium of exchange which, like the wampum of the eastern North American Indians, functioned as currency. By the time Europeans arrived in California, each unit of shell made by the Chumash was worth a third more to the Gabrielino of Catalina Island and four times as much to the Salinans of the California mainland.

Our first eyewitness accounts of the Chumash come from Spanish colonists, many of whom visited the Mission of Santa Barbara during the late 1700s. While they do not seem to have had actual named clans, the Chumash had lineages that reckoned descent in the male line. The Spaniards claim that each large Chumash village had three to four "captains," one of whom outranked the others and was referred to as a *wot* or *wocha* ("head chief"). The role of chief normally passed from father to son, pending village approval. If a suitable male heir was not available, however, the office could be held by the former chief's sister or daughter, allowing her lineage to hold on to leadership until an appropriate male was available.

Most Chumash men painted their bodies with motifs specific to their communities, but they went naked except for a waist-length skin cloak. The chief, on the other hand, was entitled to a special bearskin cape or vest and could wear his skin cloak to his ankles as a way of distinguishing himself from ordinary men. Chiefs were allowed two or three wives, while ordinary men had only one. This was a sign of wealth, because each wife required a bride payment of shell valuables, sea-otter hides, and rabbit-fur blankets.

Chiefs monopolized the ownership of plank canoes. They also controlled access to hunting and seed-collecting territories, served as war leaders during periods of raiding, and presided over ceremonies. The two latter roles were interrelated, since the refusal of a chief's invitation to a ceremony was considered an insult punishable by group violence.

Chiefs received payments of food and shell valuables from their followers. While some chiefs' influence is said to have extended beyond their home villages, Spanish authors stress that their authority was not absolute.

The plank canoe clearly enabled the Chumash to haul in bigger fish. Perhaps of even more importance, however, was its ability to transport massive quantities of shell ornaments at the very time that they were becoming a widespread form of currency. By monopolizing the canoes, Chumash chiefs were able to employ large numbers of lower-ranked craftsmen in the conversion of marine shells to beads. They then used their role as middlemen to increase the shells' value.

Most leaders in societies with lineages or clans call upon their kinsmen when extra labor is needed. Arnold, however, believes that historic Chumash leaders went beyond this, and that they controlled the labor of craftsmen who were not even their kin. The shell trade was so profitable that even members of other descent groups were willing to accept a position of subservience to the chief.

Diversity and Tolerance in Chumash Society

There is one more lesson we can learn from the Chumash. Spanish eyewitnesses observed that a small percentage of Chumash men lived and worked as women, even dressing in the paired, knee-length buckskin skirts of a woman. These men were referred to as *joyas*, the Spanish word for "jewels."

To the Chumash, the fact that some members of their community lived as members of the opposite sex was accepted as part of nature's plan. The Spaniards, on the other hand, were scandalized.

Lieutenant Pedro Fagés was a Spanish soldier who spent the late 1770s at the Mission of San Luis Obispo and traveled among the Chumash of the Santa Barbara coast. He took note of "Indian men who, both here and farther inland, are observed in the dress, clothing, and character of women." Fagés estimated that there were two or three of these men in each village. Some, he said, permitted others "to practice the execrable, unnatural abuse of their bodies. They are called joyas, and are held in great esteem."

In later chapters we will see more examples of transgendered Native American men and women, often referred to by their societies as "two-spirit people."

Almost without exception, two-spirit people were seen as having been supernaturally destined to live life as a member of the opposite sex. They were not merely accepted by their society but considered more attuned to the spirit world than the average individual.

One could hardly imagine a greater contrast than that between the tolerance of the Native Americans and the intolerance of the European colonists. In Lieutenant Fagés's case, we are not sure which fact appalled him the most—that such men existed, or that they were "held in great esteem" by their society.

THE FORAGERS OF VANCOUVER ISLAND

Let us now move north along the Pacific coast to regions where social inequality was hereditary and exceeded that of the Chumash. The social complexity of the Pacific Northwest has often been attributed to its spectacular fish resources, but we believe that there is much more to the story. There were at least two different forms of inequality involved. Sometimes whole kin groups were ranked relative to one another; in such cases a chief was simply the head of a highly ranked kin group. In other cases elite individuals within the same kin group might be ranked relative to one another and the chief. In both systems chiefs displayed their wealth and rank by sponsoring ceremonies at which guests were treated to food and gifts.

There are two conflicting interpretations of these ceremonies. One group of anthropologists considers such "feasts of merit" to have been the mechanism by which chiefs rose to prominence. The chief who threw the most spectacular event, they argue, humiliated his rivals because they could not match his generosity. This scenario is based on the principle that one is shamed by a gift he cannot reciprocate.

Other anthropologists, however, see the feast of merit as a symptom of rank rather than its cause. They point out that when European travelers first reached the Pacific Northwest, the ceremonies in question were relatively modest and did not serve as a major route to prominence. To be sure, guests were given food and gifts, but this was done mainly to repay them for acting as witnesses to an important event, such as the transfer of a chiefly title from father to son.

What gave these feasts such a competitive flavor in later years? We believe that it was the suppression of raiding. It turns out that in the days before European contact, Pacific Northwest chiefs led raiding parties against their enemies, traveling overland or in 60-foot war canoes. Such raids were often over resources, but they sometimes served as punishment for a neighboring group's failure to reciprocate a gift, loan, or act of generosity. In some cases the victors brought back captives and kept them as slaves.

Warfare, however, is one of the first behaviors suppressed by colonial governments. When Euro-American authorities suppressed warfare, the feast of merit became an alternative outlet for the competitive elite. Such feasts evolved into displays at which the host lavished food and gifts on his rivals, flaunting his wealth by destroying valuable possessions and sacrificing slaves.

Among the best known of these displays was the potlatch of the Kwakiutl people of Vancouver Island. According to anthropologist Wayne Suttles, the potlatch was a modest ceremony prior to 1849. After that date, two processes converted it to an instrument of competition. The first was the colonial suppression of warfare. The second was the Euro-American fur trade, which substantially increased the wealth of Kwakiutl leaders.

There is no doubt that late nineteenth-century potlatches were spectacular. They cannot, however, be viewed as the original cause of hereditary inequality in Kwakiutl society. We know of no society, including that of hunter-gatherers, that did not hold feasts. If feasting alone could create hereditary inequality, there would have been no egalitarian societies left for anthropologists to study.

The Historic Nootka

The Japan current warms the west coast of Vancouver Island. One hundred inches of annual rainfall produce dense evergreen forests. Seals, sea lions, porpoises, and whales swim offshore. Salmon swim up the rivers to spawn. The halibut are immense, and an oil-rich fish called the olachen is abundant. On land are elk, deer, and bear, and the inlets teem with ducks and geese. This is the environment in which European explorers encountered the Kwakiutl and Nootka.

The Nootka were the more southern of these two Wakashan-speaking peoples. There were roughly ten groups of Nootka, each occupying its own inlet along the coast. Since "Nootka" is the name of a region and not an ethnic term, in 1978 these Native Americans chose the name Nuu-chah-nulth to cover all local groups.

The nineteenth-century Nootka moved their settlements twice a year. Sheltered locations on the upper part of each inlet accommodated winter villages of big plank houses, 40–100 feet long and 30–40 feet wide. Summer villages were usually on the coast. At the height of each year's salmon run, the Nootka gathered at stations where they could intercept thousands of fish on their single-minded race upstream.

In most years the Nootka caught more salmon than could be eaten immediately. Large quantities were preserved by drying and smoking, and gallons of olachen oil were stored in containers. The forest was a source of wood for planks, shingles, canoes, carved boxes, chests, bark cloth, and blankets. The surplus food, shell valuables, sea-otter pelts, and craft items could be traded for resources from the snow-capped mountains and Fraser River plateau to the east.

The basic unit of Nootka society was a local group led by a hereditary chief called a *ha'wil*. He and his family wore distinctive clothing, elaborate hats, robes trimmed with sea-otter fur, and ornaments of abalone, dentalium (tooth shell), and native copper. The chief himself did no menial labor.

Chiefs practiced polygamy and sought to marry women from other chiefly families, thereby ensuring the high rank of their offspring. Sometimes highly ranked girls were betrothed when they were only eight to ten years old. Usually a chief's oldest son was the highest in rank, his second son next in order, and so on, with nephews in line after sons. Within the extended family, in other words, rank declined as genealogical distance from the chief increased. Senior lineages outranked junior lineages. In later chapters we will see that many agricultural societies on the islands of the South Pacific had a similar system of descending rank.

The children of the chief's more distant relatives had fewer privileges, but they were addressed with terms of honor. They could raise the rank of their offspring by marrying someone of an even higher rank. It was often from among his more distant relatives that the chief selected his war leaders and spokesmen, giving them a way to increase their renown through hard work.

Serving as craftsmen, fishermen, and hunters for the chiefly families were large numbers of Nootka of humble birth. These people were recompensed in various ways for their services, and they endeavored to make sure their expertise was passed from parent to child.

On the bottom rung of the social ladder were the slaves alluded to earlier. Most were women or children captured in raids on enemy villages, and they could be bought, sold, mistreated, or even killed. On the other hand, slaves might also be freed as an act of generosity during feasts of merit, or ransomed by their relatives if the price was right.

During the 1930s, Philip Drucker visited the Nootka and attempted, by interviewing elderly individuals and consulting documents, to reconstruct Nootka life of the era 1870–1900. Drucker's accounts are useful to us because he was at once a social anthropologist, an ethnohistorian, and an archaeologist. He therefore asked many of the questions we would like to ask.

The Nootka of the period 1870–1900 showed a level of inequality that seems surprising compared to foraging societies like the Basarwa and Aranda, or even the Chumash. In Drucker's reconstruction, however, we can see that many principles of Nootka inequality could have been created out of the preexisting principles of egalitarian foraging society. All that would have been required were appropriate changes in social logic.

Many egalitarian foraging societies reckoned descent through both father and mother; so did the historic Nootka. Some individuals in egalitarian foraging societies chose to become spiritual healers or shamans; there were similar individuals in Nootka society. These behaviors, in other words, provided continuity between the historic Nootka and their egalitarian ancestors.

In earlier chapters we saw that among egalitarian foragers, the right to use a resource territory or water hole was usually conceded to the local group that had been using it the longest. "We were here first" seems to have been a first principle. Expanding on this principle, a chiefly Nootka family used prior occupancy to establish its right to a specific inlet and was considered to "own" the associated plank houses, riverine fishing spots, and ocean waters offshore. Chiefs also laid claim to considerable intellectual property, called *tupa'ti,* which included rituals, dances, songs, personal names, and carvings on house posts or totem poles. A chiefly family's rights and privileges were said to have been acquired by its remote ancestors during the course of a supernatural experience.

Let us now consider how such inequality might have been created. We have seen that generosity and reciprocity were important to egalitarian foragers. Such people expected that all gifts would eventually be reciprocated. They fed visitors who were in need but expected that one day their generosity would be returned. They might loan one of their relatives part of his bride payment but expected that loan to be repaid one day. With the passage of time, chronic failure to reciprocate was met first with grumbling and later with anger. Unpaid debts could lead to raiding and confiscation.

Some scholars suspect that in the rich economy of the Pacific Northwest, loans and gifts escalated to a level where defaulting was punishable by raiding, captive-taking, and slavery. An extension of Raymond Kelly's principle of social substitutability meant that the actual debtors did not need to be taken captive; it was enough to enslave women and children from the debtors' village, lineage, or clan.

While reciprocal exchanges of gifts continued to be important among societies like the Nootka, the emergence of inequality led to a new form of wealth transfer. Since the chief was seen as owning all salmon fishing localities, those who fished there were obligated to pay him tribute in foodstuffs. This tribute did not have to be reciprocated by the chief, but it was acknowledged in the following way: the chief used his accumulated surplus to provide periodic feasts for his followers.

These feasts did more than establish the chief's generosity; they also kept his followers loyal. Tribute is a clear symptom of inequality, but the asymmetrical relationship it reflects can be masked by displays of largesse. So important were these displays that followers might abandon stingy chiefs and take up residence with their more generous rivals.

We have seen that even among foragers like the Aranda, leadership could become hereditary as long as everyone else agreed. Nootka chiefs, with their greater inherited privileges, did not have to seek such a consensus. They did, however, pass on their titles in ways that showed a desire for the support of their followers.

When the time came for a Nootka chief to transfer his title and privileges to his children, he began sponsoring a series of feasts. At each of these events, some privilege would be transferred to an heir. In a final ceremony the chief bequeathed his office to his eldest son and gave lesser gifts to his other children. Many of these gifts were heirlooms with a long history of previous owners; this history was chanted to the assembled guests.

The reason these transfers of titles and gifts were done in the context of a feast was because they needed to be performed in the presence of witnesses. The guests at the feast served this role, and the food and gifts they received were considered payment for services rendered.

Such title transfer may have been the original role of the potlatch, long before it escalated under the influence of the Euro-American fur trade and the suppression of raiding. There was indeed competition involved, but according to Drucker it was not among rival chiefs. Each chief's privileges were a legacy from his distant ancestors, and the main pressure he felt was to outperform those ancestors.

To be sure, ancestors were important even to egalitarian foragers. At some point, however, the Nootka had revised their creation myth to include the acquisition of titles and privileges by chiefly ancestors. This revision created the need to meet or exceed their ancestors' displays of wealth.

The Nootka chief also co-opted certain rituals, one of which involved whale hunting. Figure 6 shows a building used in 1904 for hunting magic by the Nootka of Jewitt's Lake, British Columbia. This building contained life-size wooden statues of successful harpooners, wooden carvings of whales, and large beds of deceased ancestors' skulls. Along three of the walls of the house, additional skulls were arranged as if standing guard. The chief of the Yuquot local group visited this shrine and conducted rites of hunting magic to coax the whales closer to shore. During the ritual, his wife lay on one of the beds of ancient whalers' skulls.

FIGURE 6. The ancestors played a crucial role in the traditional hunting magic of the Nootka. In this drawing, based on a 100-year-old photograph from Vancouver Island, we see a ritual building dedicated to successful whale hunters of the past. Included in the building were life-size wooden statues of great harpooners and large beds of deceased ancestors' skulls. As long ago as 9,000 years, Near Eastern village societies were making comparable statues and curating the skulls of clan ancestors.

We have seen that even egalitarian foragers built modest ritual structures, such as a sweat house or bachelors' hut. Some also preserved the skulls of deceased relatives. The Nootka simply saved more skulls and built larger buildings to curate them. Later in the book we will encounter early agricultural societies that also preserved the skulls of their ancestors in special buildings. Such behavior was widespread in the ancient world, whether one lived on wild or domestic foods.

Let us now turn to the topic of intellectual property. Even among egalitarian foragers, names were considered magic. Among the Nootka, certain names and titles became the prerogative of chiefly families. The chief inherited the right to assign these names and titles to others; to display the images of certain supernatural beings; to own certain crests that were analogous to those of medieval heraldry; to erect freestanding figures and totem poles; and to adorn his house with carved beams and paintings. The chief patronized the craftsmen who created these works of art for him, providing a route to prominence for skilled people from families of lesser rank.

Chiefs, as mentioned earlier, owned the large houses in which dozens of people spent the winter. Within these houses, each person's sitting place reflected his or her rank. Individuals of highest rank slept in the rear of the house, an area made private by the erection of a decorated screen. From the perspective of an observer standing in the rear of the house and facing the door, the chief occupied the right rear corner; the person second in rank occupied the left rear corner. The front corners were for the third- and fourth-ranked persons, while the fifth and sixth in rank occupied the middle of the house. Slaves usually slept just inside the front door, the most vulnerable area in case of an enemy raid (Figure 7).

How Might Nootka Inequality Have Been Created?

Archaeologists working in the Nootka region face a daunting task. Vancouver Island's high rainfall can turn into mush plank houses, shingles, carved posts, and beds of ancestral skulls. In spite of these obstacles, dedicated archaeologists are searching for the origins of social inequality on the Northwest coast.

According to archaeologists Gary Coupland, Terence Clark, and Amanda Palmer, the large multifamily houses of the Pacific Northwest have a 2,000-year history. The McNichol Creek site in the land of the Tsimshian Indians, north of Vancouver Island, was occupied between A.D. 1 and 500. Some of its

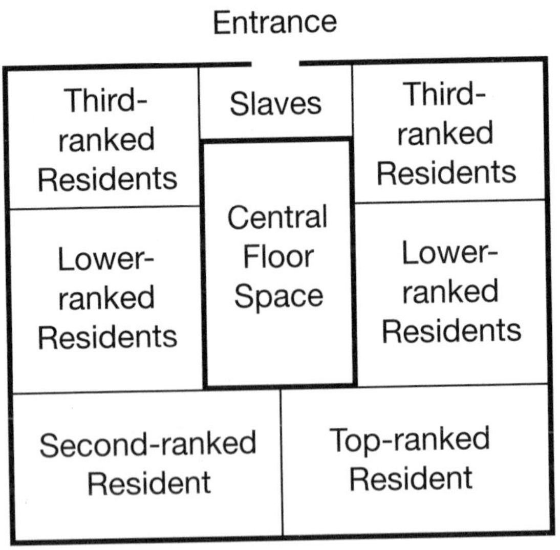

Entrance

Third-ranked Residents	Slaves	Third-ranked Residents
Lower-ranked Residents	Central Floor Space	Lower-ranked Residents
Second-ranked Resident		Top-ranked Resident

FIGURE 7. The Nootka lived in large, multifamily plank houses that could exceed 70 feet in length. In these houses the location of sleeping areas reflected the hereditary rank of each resident. The chief, or top-ranked resident, slept in the right rear, behind a decorated screen. The second-ranked resident slept in the left rear. The remaining residents were distributed as shown in the drawing. Note that slaves were required to sleep near the entrance, the most vulnerable area in case of an enemy raid.

occupants left behind artifacts of polished nephrite, a jade-like stone used for luxury items. More detailed evidence was recovered at the Ozette site on the northwest coast of Washington, a summer village occupied from 60 B.C. to A.D. 1510. The Ozette houses were large—up to 66 by 39 feet—and the quantity of luxury goods varied within and between houses.

House 1 at Ozette had evidently sheltered at least 11 families, each with its own discrete hearth area. From the perspective of an observer standing at the rear of the house and facing the door, it was in the right rear corner that archaeologists found the highest density of luxury items. The second-highest density came from the left rear corner. Fewer luxury items appeared as archaeologists worked toward the door, where the lower-ranking families and slaves presumably slept. The archaeological evidence, in other words, matches Drucker's description of a Nootka chief's plank house.

On the Fraser River plateau to the east of Vancouver Island, rainfall is lower and the preservation of archaeological sites is better. The prehistoric societies of the Fraser River, which had access to their own salmon runs, may have experienced a period of inequality between the years 800 and 1200. While the Fraser plateau evidence does not necessarily explain how the Nootka created inequality, it allows archaeologists to suggest one way that it might have happened.

Two large villages on tributaries of the Fraser River may provide the key. Archaeologist Brian Hayden began excavating the Keatley Creek site in the mid-1980s. Anna Marie Prentiss began excavating the Bridge River site early in the twenty-first century. Each site is covered with depressions left by the collapse of semi-subterranean houses, some small but others exceeding 60 feet in diameter.

Prentiss believes that social inequality may have been present in the region by A.D. 400 but did not become pronounced until 800–1200. Three processes were evident during the latter period. First, the acquisition of luxury items, including polished nephrite, increased. Second, the Bridge River site grew from 17 houses to 29 houses, while the Keatley Creek site may have grown to encompass 40 to 60 houses. The third process was a reduction in the number of small houses, accompanied by an increase in the overall size of the largest houses. One possible implication of the latter process is that small households could no longer amass the resources necessary to be economically viable. As a result, their members were being steadily incorporated into larger households as servants or poor relations.

Prentiss and her collaborators believe that between 800 and 1200, a growing number of impoverished families were willing to accept a subservient role in wealthy households in return for food, shelter, and protection. In turn, more successful families sought to preserve their accumulated wealth by passing on their resources, luxury items, and intellectual property to their offspring. This would represent a significant change in logic from an egalitarian foraging society, where hoarding and refusing to share were anathema.

While she does not phrase the process in such terms, we believe that Prentiss is describing what anthropologists call *debt servitude,* or even *debt slavery.* The first step in such a process is to loan food and valuables to impoverished neighbors. The second step is to foreclose on the loan. Families who accept food and shelter from wealthy neighbors are in a poor position to deny the latter's claims to luxury items and hereditary privileges.

Prentiss reveals that after 1200, the archaeological remains of salmon at the Bridge River site decreased. It is not certain whether this was the result of environmental deterioration, overfishing, or both. Whatever the case, both the Keatley Creek and Bridge River villages were eventually abandoned. These events remind us that even though foraging societies did occasionally develop hereditary differences in rank, there may have been inherent limitations to supporting an aristocracy on wild foods.

THE HISTORIC TLINGIT

Now let us move farther up the Pacific coast, to the panhandle of southeast Alaska. It is a rugged coast with deep fjords, bays, fast-moving rivers, and rocky islands. The Native Americans of this coast spoke a Na-Dené language in which Tlingit was the word for "human beings."

The late Frederica de Laguna, who was both a social anthropologist and an archaeologist, estimated that there may have been 10,000 Tlingit in the year 1740. Unfortunately, by 1838 their numbers had been reduced to less than 5,500. Like the Nootka, the Tlingit built villages on sheltered bays in the winter and lived in hunting-and-fishing camps in the summer. Their winter villages featured plank houses large enough for at least six families and their slaves. Totem poles were erected in front of the houses to honor important ancestors.

Like some Australian foragers, Tlingit society featured two opposing divisions, called the Raven and Wolf moieties. Each moiety was made up of 30 or so clans whose members reckoned descent in the mother's line. Clans were further divided into lineages or "house groups."

Each house group claimed that it could trace its descent from a founding ancestor. In reality, however, the system was fluid. Lineages that grew rapidly might either split in two or become populous enough to declare themselves a new clan. Lineages that shrank below a certain threshold might be absorbed by a more prosperous clan. Recall that Prentiss and her collaborators have suggested a similar scenario for failing households on the Fraser plateau.

The lineages within each clan were ranked, and a Tlingit chief was simply the head of the most highly ranked lineage within his clan. His immediate relatives were a kind of aristocracy, identified by their hats, blankets, crests, ear ornaments, lip piercings, and tattoos. Lower-ranking lineages within each clan were treated as commoners but, as among the Nootka, could achieve renown through craftsmanship or bravery in combat. At the bottom of the social ladder were war captives who were kept as slaves.

The clans and their house groups owned the rights to good localities for winter villages, fishing-and-hunting territories, sources of timber, trade routes to neighboring societies, heirlooms, and a series of personal names. Of all their possessions, however, de Laguna argued that the Tlingit aristocracy favored their heraldic crests. These designs were woven on blankets, carved on canoes and totem poles, and depicted on wooden screens that divided the plank houses into living spaces (Figure 8). Crests could be based on supernatural beings, heavenly bodies, ancestral heroes, or totemic animals such as bears, sea lions, and whales.

FIGURE 8. Among the most prized intellectual property of chiefly Tlingit families were the heraldic crests and symbols bequeathed to them by their ancestors. Such motifs were embroidered onto the chief's robe, carved on his house posts, and painted on the screens that provided privacy for his living quarters. In this drawing, inspired by two photos more than 90 years old, we see a bear and a salmon on the chief's robe and a raven and a salmon carved on his house post. The painting on the cedar screen represents a Rain Deity surrounded by anthropomorphized raindrops.

Each family's crest was said to have been acquired by a remote ancestor, and its owners chanted the history of its acquisition. Tlingit aristocrats displayed their crests at feasts, and guests from the opposite moiety were rewarded for serving as witnesses to the display. Some of the motifs on Tlingit crests were shared by their coastal neighbors, the Haida and Tsimshian. This is not surprising, since Tlingit clans carried on active trade with their neighbors and sometimes absorbed the remnants of shrinking Haida and Tsimshian lineages.

Tlingit houses, like their Nootka counterparts, were divided into multiple sleeping areas. The house's owner lived in the rear behind a decorated wooden screen. In front of this screen was a platform, a place of honor, where the owner and his family sat. After the owner had died he was left on this platform for four days, dressed in ceremonial clothing, his face painted with clan symbols and his valuables displayed beside him.

If the deceased was a chief of the highest lineage, mourners sang continuously and his widows fasted for eight days. Eventually the chief was cremated. Those who built the funeral pyre and the wooden box for his ashes were given gifts. Sometimes a chief's valuables, or even one of his slaves, might be added to the fire. In other cases a slave or two might be freed to symbolize the chief's generosity.

As with Nootka houses, the sleeping area for slaves was just in front of the door. In addition to using their slaves as a first line of defense, the Tlingit kept their doors so small that an intruder would have to enter on hands and knees. Some villages were further defended with a palisade of wooden posts.

The Tlingit threw feasts equivalent to those of the Nootka, during which the host's children had their nobility validated. Boys' and girls' ears were pierced for ornaments, girls' hands were tattooed, and slaves who assisted in the ritual were rewarded with their freedom. Only people who had been honored in this way as children were considered true aristocrats.

De Laguna stressed that nobility came from the title bestowed by one's father, not from the feast itself. As with the Nootka, in other words, the role of the feast of merit was to validate existing rank rather than generate it out of egalitarian life. The more feasts a man sponsored, however, and the more lavish his gifts, the greater his reputation became.

Accumulating wealth for a major feast was so daunting a task that some aristocrats turned to their poor relations, allowing them to work off debts by contributing items. The host's wives also solicited contributions from their clanspeople, who belonged to the opposite moiety from the host. There was tension between rival clans, according to de Laguna, as they competed in gluttony and dancing. On the final day of the feast the host recited his family history and gave away furs, copper valuables, blankets, and even slaves.

Two types of inequality, in other words, were visible during Tlingit feasts of merit. The most important and pervasive type was inherited nobility. Aristocrats inherited titles and privileges from key ancestors and passed them on to their children in front of witnesses. The second type of inequality, less pervasive, was achieved prestige. Highly motivated aristocrats were able to sponsor more feasts and give away more gifts than others. Such displays enhanced a man's reputation during his lifetime, but there was no way to transfer that reputation to his children; they were still too young to have achieved anything.

It is likely that Tlingit feasts, like those of the Nootka, escalated after the colonial suppression of warfare. Once, in the eighteenth century, the Tlingit had made 60-foot war canoes and went on raids with spears, daggers, war clubs, and bows and arrows. They wore body armor made from wooden slats and peeked out through slits in protective helmets. The Tlingit took scalps or heads from male enemies and brought women and children back as slaves.

The distinction between achieved prestige and hereditary nobility is an important one. We will refer to it again in the context of agricultural societies, especially those of Southeast Asia and New Guinea. Achieving renown by hosting a feast builds on egalitarian society's long-standing love of generosity. Turning debtors into servants or slaves builds on society's long-standing dislike of failure to reciprocate gifts or repay loans. Making the master-servant relationship hereditary dilutes the "personal freedom" of Rousseau's hypothetical State of Nature. For a slave, that freedom is erased entirely.

The Impact of the Tlingit on Their Egalitarian Neighbors

The Tlingit had two kinds of neighbors. On the coast to the south were the Haida and Tsimshian, who also had hereditary nobles. To the east, beyond a snowcapped cordillera, were foragers speaking Athapaskan languages.

Trade among the Tlingit, Haida, and Tsimshian began as exchanges of gifts. First came gifts between clans of the same moiety, then gifts between opposing moieties, and eventually exchanges with more distant societies. The Tlingit sought native copper from the interior, dentalium from the south, walrus ivory from the north, decorative porcupine quills from the boreal forest, animal furs from the Athapaskans, and slaves wherever they could get them.

The Tlingit understood exactly how to deal with the Haida and Tsimshian because those groups also had nobles, commoners, and slaves. The farther east the Tlingit traveled, however, the more often they encountered egalitarian hunting-and-fishing societies. Their challenge then became the incorporation of egalitarian trading partners into the hierarchical society of the Tlingit. This incorporation changed some Athapaskans dramatically. In the remainder of this chapter we look at three of those Athapaskan groups: the Tutchone, the Tagish, and the Teslin.

The Tutchone of the interior Yukon were the least affected by the Tlingit, and they probably give us our best glimpse of what an unmodified Athapaskan foraging society might have looked like. Tutchone families belonged to clans who reckoned descent in the mother's line. These clans, in turn, were grouped into opposing moieties. Headmen tended to be skilled hunters and traders who attracted followers but led by example rather than real authority. Despite the fact that sons belonged to their mother's clan, a headman's son could succeed his father if everyone agreed.

Two aspects of Tutchone society strike us as noteworthy. First, their system of clans, moieties, bride service, and polygamous headmen reminds us of other foragers with clans, like the Aborigines of central Australia. Second, the fact that the Tutchone held funeral feasts and reckoned descent in the mother's line suggests that those behaviors may already have been present in the Pacific Northwest before the escalation of inequality.

The Tagish lived in the alpine forests and meadows of the southern Yukon, where they fished in the lakes and rivers, hunted caribou with game drive fences, and trapped fur-bearing animals. Their ancient society, like that of the Tutchone, featured opposing moieties with Athapaskan names, each made up of clans with matrilineal descent. The Tagish were one of the Athapaskan groups to whom the Tlingit sent trading parties.

By the eighteenth century, Euro-American fur traders had reached the coastal Tlingit. One of their first targets was the sea otter, whose fur had long been used to trim the garments of Tlingit nobles. Euro-American trade goods made the Tlingit wealthier, but sea-otter populations were declining by the end of the century. Fortunately, the Tlingit knew that their Tagish and Teslin trading partners had access to the river otter, beaver, mink, fox, marten, and wolverine. It would be key for the Tlingit, however, to prevent the Europeans and Americans from getting at those furs directly.

The Tlingit, therefore, began blocking the trade routes through the cordillera with parties of up to 300 warriors. By the 1850s they controlled all traffic between the Alaskan coast and the Yukon. Tlingit trade partners came into the territory of the Tagish, some taking Tagish wives and others having their daughters marry Tagish men. The bride-price paid to the Tagish included Euro-American trade goods. The bride-price paid to the Tlingit included furs.

Over time, the Tagish changed the names of their moieties to Crow (the inland equivalent of Raven) and Wolf. The most prolific Tagish trappers gave themselves Tlingit names, and their funeral feasts came to resemble potlatches.

Of all their trading partners, however, the people the Tlingit affected the most were the Teslin. Originally the Teslin had lived on the Taku River, a tributary of the Yukon, but they moved to the Yukon headwaters to take advantage of the fur trade. Emulating their trading partners, they learned Tlingit and gave coastal names to their clans and moieties. Some families began claiming high rank as a result of descent from the daughters of Tlingit traders. They fought among themselves over which family had the right to Wolf or Crow/Raven crests. They adopted songs and myths that featured coastal animals never seen in the Yukon. Their funeral feasts became potlatches. They began to keep slaves. Their cosmos became an amalgam of Athapaskan spirits and Tlingit supernaturals. In the words of anthropologist Catherine McClellan, these Athapaskans had become "Inland Tlingit." The Teslin show us that when egalitarian foragers were ready to adopt rank, they just might model that rank on trading partners who already displayed it.

Such transformations from egalitarian to ranked were based on a first principle of social logic: our trading partners are honorary kinsmen. This is the principle that allows them to enter our territory with impunity. The principle is reinforced when I marry my trading partner's daughter. Trade, which began as reciprocal exchanges of gifts, then expands to include bride-price transactions. I am now free to emulate my wealthier relatives and even borrow from them in emergencies.

The Inland Tlingit show us that some forms of social inequality, once established in a region, can spread through emulation. We must not, however, forget that the Inland Tlingit do not represent a pristine case. Their economy was partly a product of the fur trade, which was an intrusion of Western culture into the Native American economy.

Nor should we forget that the Tlingit carried on equally intense trade with their coastal neighbors, the Haida and Tsimshian, who were just as highly ranked as the Tlingit. The Tlingit, Haida, and Tsimshian actually absorbed each other's shrinking lineages into their own clans and borrowed each other's heraldic crests on a regular basis. They created a network of circulating luxury goods, for which we will find analogies in the ancient rank societies of Mexico and Peru.

THE LIMITS OF INEQUALITY AMONG FORAGERS

Clanless foragers usually displayed the "personal and anarchic freedom" of which Rousseau wrote. Some foragers with clans, however, eventually came up with ways to take away the freedom of others.

Such is the nature of wild resources that people in one area may be getting enough, while people in a neighboring area are not. Clanless foragers created meat-sharing partnerships, hxaro partnerships, namesake relatives, and other strategies to forge a safety net. The moieties, clans, and lineages of larger-scale foraging societies provided an even larger network of mutual aid. As they grew and divided, however, these social segments sometimes used the principle "We were here first" to distinguish senior and junior lineages.

Just as senior Aranda men were taken more seriously than junior Aranda men, senior lineages tended to be taken more seriously than junior lineages. Just as senior Andaman Islanders could reciprocate with lesser gifts to junior men, the playing field between senior and junior lineages was not always level.

Archaeologists suspect that on the Santa Barbara coast and the Channel Islands and Vancouver Island and the panhandle of Alaska, lineages who could no longer repay their debts were forced to accept a permanently subordinate position. Two routes they could use to curry favor with their patrons were craft production and bravery in combat.

For their part, superior lineages modified their cosmology to attribute their privileges to their ancestors. In the Pacific Northwest they bequeathed those privileges to their children at public feasts, lavishing presents on the guests who witnessed the event. They had, in Rousseau's words, forced their poor relations to sign a contract accepting inequality in return for food, shelter, and occasional gifts.

How often might hereditary inequality have arisen among prehistoric hunters and gatherers? To answer this question, archaeologists must be able to distinguish between achieved prestige and inherited nobility. As we saw among the Tlingit, both forms of inequality were sometimes on display at the same feast. Prestige accrues to the generous host. Nobility belongs to the child who inherits his father's titles, crests, and sumptuary heirlooms. As a result, archaeologists pay special attention to children buried with what appear to be symbols of nobility.

Rousseau would have been interested to learn that not all foragers had to adopt agriculture in order to emerge from the State of Nature. Because of a desire to be thought of as superior, some hunter-gatherers manipulated cosmologies, reciprocal exchange, social obligations, wealth transfer, and the subservience of junior lineages to create societies based on hereditary rank.

II

Balancing Prestige and Equality

SIX

Agriculture and Achieved Renown

We can excuse Rousseau for not knowing that some foragers found ways to create hereditary inequality. After all, societies like the Tlingit and Nootka were largely unknown to Europeans in 1753. It is also the case that for most parts of the world, Rousseau was right: not until people had begun to raise crops or animals do we see signs of emerging inequality.

To be sure, even successful agriculture does not always lead to inequality. Many societies remained egalitarian even after thousands of years of farming. Others, as we see in this chapter, allowed modest amounts of achieved renown but still resisted hereditary rank.

Even after rank began to appear, it could not always overcome the widespread desire for a level playing field. There were, as we will see later in this book, societies that oscillated between equality and hereditary rank for decades. To be sure, some of those societies eventually made inequality permanent. They were in the minority when they arose but often, like the Tlingit, had a dramatic impact on their egalitarian neighbors.

WHICH FORAGING SOCIETIES WERE GOOD CANDIDATES FOR AGRICULTURE?

Agriculture is a delayed-return activity, and we suspect that it most often arose among foraging societies with delayed-return economies. By the end of the Ice Age a number of hunter-gatherers were burning wild vegetation to increase its productivity, replanting the excess tubers they had harvested, broadcast-sowing excess seeds, or building fences to drive wild animals into temporary enclosures. Most farming and herding probably began as extensions of those practices. For many foragers, in other words, the first attempts at horticulture may not have felt like a dramatic behavioral change.

On other occasions the introduction of agriculture has been known to trigger significant changes in behavior. In the 1960s the //Gana of the Kalahari region—foragers who lived next door to the !Kung—began raising domestic beans, melons, and goats like their Bantu-speaking neighbors. By the late 1970s anthropologist Elizabeth Cashdan had noted the following changes:

1. The //Gana stopped moving their camps and became virtually sedentary during the rainy season.
2. Families began to preserve rather than share their meat.
3. People began to tolerate successful families' accumulation of food.
4. Families were allowed to acquire, store, and trade valuables without criticism.
5. People began to purchase cattle from their Bantu-speaking neighbors.
6. Polygamous marriage among //Gana men increased to 25 percent, while it remained at 5 percent among their !Kung neighbors.
7. A man who wanted to marry might have to pay the bride's family as many as ten goats.
8. Older men with growing wealth, many of whom had two to three wives, began passing themselves off as "headmen" who spoke for the whole group. Their behavior was tolerated because the former prohibition against accumulating property was beginning to fade.

These behavioral changes show us agriculture's potential to overcome the egalitarian logic of foragers. We must keep in mind, however, that the //Gana changes took place in the politically altered world of the twentieth century. We must therefore ask whether similar changes followed the adoption of agriculture in the pre-industrial world. For this we turn to the island of New Guinea.

EARLY AGRICULTURE IN NEW GUINEA

We have seen that many foragers of Australia and the Andaman Islands collected wild yams. We now believe that by the end of the Ice Age, some societies in Southeast Asia and the Pacific Islands had begun to encourage yams and other native plants by tending and managing them. Eventually they began to plant them in gardens. This did not happen among the Aborigines of central Australia, who remained hunters and gatherers. It did, however, happen in New Guinea, which had once been connected to Australia by the lowered sea levels of the Ice Age.

New Guinea is the world's second largest island, covering more than 300,000 square miles. The spine of the island is a snowcapped mountain range flanked by dissected high plateaus. To the north and south of this spine are swampy lowlands covered with tropical forest.

Archaeologists do not agree on the date when the transition from tending to gardening was complete. Nor is it always clear which of the domestic plants were native to New Guinea and which were introduced from mainland Asia by watercraft. Among the key plants were the sago palm, whose pith can be made into flour; taro, a relative of the calla lily, whose starchy root is edible; the Asian yam, another plant with a starchy root; the pandanus or kara nut plant; plantains and bananas; and the familiar coconut palm.

Archaeologists suspect that the active tending of these plants began as early as 6,000 years ago. Perhaps 2,000 years later, some New Guinea highlanders were digging drainage ditches to encourage them. This activity suggests a delayed-return labor investment, and perhaps even active horticulture. Sago would have done well in the coastal swamps and yams on the drier plateaus. Eventually two more foods were introduced by watercraft. These were the domestic pig (possibly brought from Indonesia 3,000 years ago) and the sweet potato (introduced by ship from the New World after it had been discovered). The introduction of the sweet potato, which is superior to the yam when it comes to gardening above an elevation of 6,000 feet, led to population increases in the mountains of New Guinea.

Gardening required more than land clearance, planting, and drainage ditches. It also required magic. Some wet crops, like taro, were considered female; some dry crops, like yams, were considered male. In the plateau country a magical plant called cordyline—considered male—had to be planted on the female side of each garden to neutralize unwanted femininity.

Horticulture alone did not lead to inequality in New Guinea society. It is likely, however, that it led to the following changes in cosmology and behavior:

1. The creation myth was revised to claim that spirit ancestors (among their other teachings) showed humans how to garden.
2. Even those tribes with a history of immediate-return economy converted to a delayed-return economy, justifying the investment of labor in clearing and planting gardens.
3. Prohibitions against hoarding were relaxed so that gardeners could begin storing plants such as yams and sweet potatoes.
4. Previous behaviors in which men shared meat with everyone and women collected plants only for their family were modified. Now men pressured their wives to produce surplus plants for lavish feasts to which guests were invited.
5. Bride-price escalated.

AN INTRODUCTION TO NEW GUINEA SOCIETIES

Like the //Gana people described earlier, New Guinea gardeners stopped moving from camp to camp during the growing season. Once having cleared land and established gardens, they began to spend longer periods of time in substantial houses of pole and thatch. They now lived in *autonomous villages*, meaning that each settlement was politically independent even though it had economic ties to other villages.

Some tribes lived in large communal houses that sheltered multiple families. In other tribes each family built its own house. New Guinea societies had their own versions of moieties, clans, and subclans or lineages, and they often created cycles of marriage exchange that were longer and less symmetrical than those of the Australian Aborigines.

New Guinea men believed in the same gender inequity we saw in Australia: women could never be as virtuous as men. Among the Etoro tribe, anthropologist Raymond Kelly discovered a "hierarchy of virtue" whose premises followed from Etoro cosmology. The steps in logic were as follows:

1. Generosity is a highly virtuous behavior.
2. Men provide society with both meat and semen; that is, they contribute life force.
3. Women accept meat and semen, receiving life force rather than giving it.
4. Hence, men are more generous and, by logical extension, more virtuous than women.

In addition to beginning life in a more virtuous state, Etoro men could enhance their prestige by achieving one of two statuses: (1) *tafadilo* (one of the respected senior males who shaped community decisions, directed raids on enemies, resolved witchcraft accusations, and authorized executions); or (2) spirit medium (an individual who could make prophecies, cure illness, conduct séances, and preside at rituals).

Women began life as less virtuous and had few ways to increase their virtue. They were forgiven for gradually depleting a man's semen (which led to his senescence) as long as they bore him children; barren women were lower in the hierarchy of virtue than mothers.

Etoro society did not produce "Big Men" like those of some New Guinea tribes. Their hierarchy of respect went no higher than tafadilos and spirit mediums, followed by ordinary men, then mothers, and then barren women. Lowest of all in respect were witches, people accused of having stolen life force in order to cause illness or death. If found guilty of witchcraft, people might be banished or killed, unless they paid compensation to their victims' kin.

Other New Guinea groups had even more extensive hierarchies of respect. The Chimbu tribe was noted for its high population density, which by the 1960s had reached 500 persons per square mile. When we consider that many hunter-gatherers lived at densities of less than one person per square mile, and that the overall density of the Etoro averaged three per square mile, the density of the Chimbu was impressive. To be sure, some of this density resulted from crowding together for defense from hostile neighbors.

According to anthropologist Paula Brown, Chimbu men fell into at least four prestige categories, as follows:

1. Men who raised few crops, failed to accumulate the bride-price for a wife, and played only a small role in regional exchange were called *yogo,* "rubbish men" or "nothing men."
2. The majority of married men, who provided adequately for their families and met their obligations in regional exchange, were the average citizens of Chimbu society.
3. Twenty percent of Chimbu men were more active than average in exchange and speech making and more successful at gardening and pig raising. They often had two or more wives and supported more dependents. It often took these men until age 30 to become truly "prominent," as Brown calls them. Those over 50 had usually created a following of

sons, sons-in-law, and brothers-in-law who contributed food and labor to their quest for renown.

4. At the top of the prestige ladder were *yomba pondo,* "Big Men." They represented no more than 5 percent of the Chimbu men; usually there were only one or two in each lineage or subclan. While most prominent men made speeches, Big Men were the ones chosen to speak when outsiders were present. They could initiate or veto group activities and were major participants in regional exchange. Often Big Men were responsible for directing the construction of a ritual men's house, about which we will hear more later. Big Men also supported a certain number of "rubbish men" who ran errands for them. Perhaps, therefore, early Big Men should be credited with inventing the entourage.

As impressive as this ladder of prestige seems, even Big Men possessed no more than a strong influence. They did not occupy an office that came with any real authority. Their renown resulted entirely from their accomplishments. Moreover, in order to retain the prestige they had acquired, yomba pondo had to fight off constant challenges from ambitious younger men. Chimbu society, according to Brown, was in constant flux, with individual Big Men, lineages, and clans competing to see who could grow the most sweet potatoes, raise the most pigs, give the most spectacular feasts, and accumulate the most trade goods.

As fierce as the competition may have been when Brown visited the Chimbu, it had once been fiercer. Previous Big Men had been daring war leaders who assassinated enemies and led raids against other tribes for pigs, valuables, and revenge. "Chimbu men used to be strong fighters," Brown was told; "now they are like women and children." The suppression of warfare, head-hunting, and cannibalism in New Guinea was, of course, the work of the Dutch and Australian authorities who controlled parts of the island.

In retrospect it appears that men in precolonial New Guinea detected at least three routes to prominence and that they used all three. The first route—based on the premise that displays of generosity were good—was to use the labor of one's wives, lineage mates, and clansmen to produce surplus yams, sweet potatoes, pigs, and other goods to give away at impressive feasts. The second route—which probably developed out of the vengeance attacks we saw among some foragers—was to lead head-hunting and pig-stealing raids against enemy groups. The third route—which probably developed out of an exchange network like the one described in Chapter 3 for the Murngin of

Australia—was to acquire impressive amounts of mother-of-pearl shells, cowrie shells, parrot feathers, bird of paradise plumes, and other exotic trade goods. Many anthropologists now believe that when colonial authorities denied New Guinea men the war leadership route, the latter redoubled their competition for trade goods. This activity raised their entrepreneurial skills to levels that fascinate us.

The combinations of strategies used by New Guinea leaders resulted in an amazing diversity of societies. At the same time, this diversity was built on an underlying set of shared principles. By raising their own plants and pigs, New Guinea tribes were able to overcome foraging society's insistence that leaders remain humble and slow to anger, work to suppress violence, and give away everything they accumulate. Horticultural society took the desire to be well thought of by one's peers, which until now had suppressed ambition, and surrendered it to the sin of pride.

The Era of Good Old-Fashioned Tribal Warfare

Even pacified New Guinea tribes expressed nostalgia for the days when "men used to be strong fighters." Among the Chimbu, fighting was both a source of personal prestige and a reason some clans became rivals rather than allies. The causes of fighting included murders, the theft of food or valuables, the failure to reciprocate a gift, or insults of various kinds (what today's youth would call "dissing"). Raiding parties carried spears, bows and arrows, stone axes, clubs, and large shields. During most Chimbu battles, fewer than 10 men out of 200 fought hand-to-hand, while 60 to 70 shot arrows from a distance, and the remaining men waited in the wings to see if they would be needed. After a few casualties there would be a ceremonial truce, with reparations paid for the victims.

For their part, the Etoro told Raymond Kelly of a raid in which tribe members allied themselves with their neighbors, the Petamini and Onabasulu, to burn a longhouse occupied by their traditional enemies, the Kaluli. Only two Etoro died, while many Kaluli were killed as they fled the blazing residence. The Etoro later paid the Kaluli 54 strings of cowrie shells and three stone axes to compensate for the Kaluli's higher level of casualties.

The most legendary raids, of course, were carried out before colonial pacification. But anthropologist Bruce Knauft, who lived on the south coast of the island with the Marind tribe, found that their postcolonial cosmology still justified head-hunting. The Marind, according to Knauft, had been enthusiastic headhunters in precolonial days. As many as six large villages would join in a foray to *kui-mirav*, "the head-hunting grounds." So exciting was this endeavor that whole villages were deserted during the head-hunting, or nonagricultural, season. About the only thing this activity lacked was tailgating spectators.

Raiders traveled over special trails in the forest, or by canoe over inland waterways, carrying sago flour as a provision. In 1884 a British ship came upon 1,200 Marind warriors, in 30 to 40 canoes, nearly 170 miles from their homes. Outnumbering their unsuspecting enemies, the warriors separated into platoons made up of men from the same village. They surrounded an enemy settlement by night and attacked at dawn. A warrior would crack a man's skull with a club and then behead him; the victim's last cry was assumed to be his name, and that name would be given to a future Marind child. The warriors spared children and young women because, in a polygamous world, young women were always in high demand.

Marind warriors often returned with canoe loads of heads, an estimated 150 a year. The heads were kept as trophies, while other body parts might be cannibalized. This head-hunting was not a random act of aggression but a death-defying ritual, believed to bring good fortune and abundant harvests to the takers of the heads.

CHIMBU KINSMEN AS BROKERS

Most horticultural tribes still valued the generosity and reciprocal gift-giving we saw in hunting-and-gathering societies. Indeed, these behaviors escalated in societies with lineages, clans, and moieties, because now each of these larger units had reciprocal relations with others. We will use the Chimbu as an example.

The Chimbu reckoned descent through the father's line. Above the family was a lineage of 15 to 60 related men and their families. These lineages, in turn, were grouped into clans numbering 600 to 700 people each.

Sometimes the largest descent group in the area claimed to have been founded by a specific male ancestor, and each of its subgroups claimed to have been created by one of the founder's sons. These men were all supposedly related through a common ancestor and had to marry outside their own clan. This practice required large marriage payments to the bride's clan, involving mother-of-pearl shells, headbands covered with cowrie shells, parrot feather headdresses, bird of paradise plumes, special bridal axes, and pigs. The groom's lineage mates and clansmen contributed much of this bride wealth, because few young men could afford it. When the groom was older and owned more valuables, he would be expected to reciprocate.

Chimbu clans, and even whole tribes, invited their neighbors to feasts that were supposed to impress them (and eventually to be reciprocated). Paula Brown describes piles of surplus vegetables 20 to 50 yards in diameter, proudly displayed before being given away. Every six to ten years, Chimbu villagers would invite several hundred guests to watch them sacrifice pigs to their deceased ancestors; later, the meat would be cooked and distributed to the guests. Each host group expected to receive a reciprocal feast one day, and no one ever forgot; inadequate payment of pork debt could provoke armed retribution. To be sure, after a successful raid the victorious group was expected to pay war reparations, once again involving trade goods and pigs.

Big Men also used lavish feasts to gain prestige. They quickly learned that the more wives a man had, the more yams and sweet potatoes his family could grow, and the more pigs they could raise. As we will learn later on, amassing an unrivaled quantity of food allowed a Big Man to humiliate his rivals by giving them more than they could repay. To achieve this, Big Men not only led pig-stealing raids against neighbors but also put pressure on their entire clan to contribute. There was risk involved, for if an ambitious man "maxed out his credit"—borrowing more than he could ever repay—he could lose his prestige and spend years in servitude to his creditors. Some archaeologists, as we saw earlier, suspect that this kind of debt servitude led to inequality on the Fraser plateau and in the Pacific Northwest.

THE ENGA TRIBE: "ONE PIG AND ONE PIG ONLY"

One of the longest and most complex histories of competitive trading and feasting is that described by anthropologist Pauline Wiessner and her colleague Akii Tumu, a member of the Enga tribe. The story began 9–12 human generations ago when the sweet potato was introduced to the Enga, and the tribe's population increased from 20,000 to 100,000 in roughly 220 years.

Like so many New Guinea tribes, the Enga were organized into clans that reckoned descent from a common male ancestor. Some men were described as *kamongo,* Big Men, having achieved renown by mediating disputes, giving speeches, and manipulating trade in shells, feathers, aromatic oils, ceremonial drums, and pigs.

Sweet potato gardening increased the women's workload but also raised their value in the eyes of ambitious men. Now the pigs that created prestige for the men were fattened on sweet potatoes grown by the women. As populations grew, new ceremonies and exchange systems were spun off.

An ancient initiation ritual called the *kepele* eventually evolved into a cult honoring the ancestors, and then into a major ceremonial exchange system. Because one of the kepele's goals was to unify the Enga tribe and integrate it with its neighbors, great pains were taken to prevent anyone from giving gifts too large to be reciprocated. The equality of all participants in kepele was expressed in the obligation to contribute "one pig and one pig only." Kepele, in other words, counterbalanced some of the intense competition that took place at the Big Man level.

Another institution, slowly evolving, was the *tee* cycle. The tee originally began as a mechanism for wealth accumulation, financing bride-price and funerary gifts, by tapping into valuables that circulated beyond the immediate kin group. But as the population grew and the level of tribal warfare increased, the tee came to be used as a way to accumulate valuables for war reparations. An unanticipated consequence of the latter move was that some groups began engaging in war just to avail themselves of the generous reparations.

Since the goal of these new battles was wealth rather than revenge, no one really wanted to die in them. Wars became increasingly ritual, and the reparations increasingly involved pigs rather than land and crops. From 1915 through 1945 the Enga came to realize that killing pigs was better than killing people, and the tee gradually became an exchange cycle.

At its peak the cycle involved more than 375 clans. The first would provide its trading partners to the west with pigs, pearl shells, axes, salt, oil, and edible flightless birds called cassowaries. These items would be passed westward to a second clan, then to a third, and so on. The last clan to receive the items launched a west-to-east countercycle. Half the pigs they received would be butchered, and the cooked pork was then sent east to the very people from whom they had received the original pigs.

Australian authorities were naturally happy to see warfare decline. The Enga tribe, for its part, had managed to preserve the dual values of equality and delayed reciprocity, even as its Big Men were creating differences in prestige.

THE MT. HAGEN TRIBES: "I'LL SEE YOUR ONE PIG AND RAISE YOU TWO"

In the highlands of New Guinea, just east of the Enga and west of the Chimbu, live the tribes of the Mt. Hagen region. In the 1960s roughly 60,000 Melpa tribespeople occupied 530 square miles. Their neighbors, the Gawigl, numbered 30,000. The women of Mt. Hagen, described in an important study by anthropologist Marilyn Strathern, tended gardens and raised the pigs that constituted the family's major source of wealth. Their men cleared the forest, dug drainage ditches, and made war on their enemies. Men also participated in a spectacular exchange system called *moka,* which was studied by anthropologist Andrew Strathern.

Like the tee cycle of the Enga, moka almost certainly began as a way of creating wealth for war reparations. With the suppression of war by the Australian government, moka surged in importance as a way for men to compete without human casualties. Moka, however, did not admonish men to give one pig and one pig only. It dared men to say, "I'll see your one pig and raise you two." It turned exchange into a high-stakes game, at the end of which someone would fold and someone would emerge with renown.

Like the Chimbu, men in the Mt. Hagen region climbed a ladder of prestige. About 15 percent of the population wound up as unproductive "rubbish men" who attached themselves to Big Men as errand boys. Many rubbish men never married, because their kinsmen considered them such losers that they would not contribute to their purchase of a wife.

Ordinary men, usually monogamous and at least minimally involved in trade, made up 70 percent of the population. The remaining 15 percent were Big Men, who averaged between two and three wives and were very successful at trade, controlling almost all the valuables referred to colloquially as "shell money."

At roughly ten years of age, all boys left their families and went to live in a communal men's house. This circular, dormitory-like structure was almost certainly an outgrowth of the bachelors' huts of earlier foraging societies. Once they had joined the men's house, boys' ritual lives would be very different from that of their sisters. From that point on they would spend less time with women, because such contact could deplete their life force. For their part, the women lived in oval longhouses with two doorways, one for women and children and one for the pigs they nurtured as a source of wealth.

In the good old days, Big Men led raids against enemy villages. During the subsequent phase of peacemaking and reparations, the killers presented the victims' group with live pigs, cooked pork, fruits, cassowary eggs, and shell valuables. The recipients of these reparations would then return gifts of equivalent value, plus an additional amount, like the "vigorish" of modern loan sharks. It was this additional amount that was referred to as moka. The victors would accept the moka and return it with yet more vigorish. This back-and-forth "I'll see you and raise you" was called "building a road of pigs between us," and it was intended to prevent counterattacks.

With the suppression of warfare, ceremonial trade in shell valuables, feathers, salt, stone axes, animal fur, and red ocher escalated, always including moka. A man might give his trading partner two mother-of-pearl shells and a pig; the recipient gave back eight to ten pearl shells and shared the cooked meat of the pig with others. A man who received baby pigs might give back fully grown pigs or sides of cooked pork (a kind of "value added" pig). One gained prestige by finally giving a gift so big that it could not be returned with moka.

Eventually the Big Men of Mt. Hagen began wearing one bamboo "tally stick" for every event at which they had given a moka of eight to ten shells. Figure 9 shows a Big Man wearing long strings of tally sticks, called *omak*, communicating to everyone that he is a man of renown. Big men were skilled at timing their gifts and very persuasive at convincing their kinsmen to help them accumulate resources. However, they had no authority to give commands and no way to enforce them. Nor could the sons of Big Men inherit their fathers' prestige; they had to earn it on their own.

Andrew Strathern discovered, however, that while Big Man was not a hereditary position, it certainly helped if one's father was a role model. Out of his sample of 88 Big Men, 49 (or 56 percent) had fathers who were Big Men. Even more significantly, of the 32 who were considered truly major Big Men, 23 (or 72 percent) had fathers who were Big Men.

After proudly wearing his omak in life, a Big Man received special treatment at death. His kinsmen flexed his corpse by tying his hands and legs, placing him on a platform for a day. They then buried him underground with an offering of shells and feathers. Later on they exhumed his skull and erected a shrine pole to him as an important ancestor. The shrine where his skull was kept became a place where would-be Big Men performed rituals. A village whose most prominent Big Man had died was temporarily directionless. No one doubted that he had been killed by witchcraft.

FIGURE 9. A Big Man of the Melpa tribe, New Guinea, stares down a rival entrepreneur. Before him is a display of pearl shell valuables, arranged on presentation disks of hardened resin. In the 1960s a man who gave away eight to ten of these shells, trumping his rival's gift of two shells and a pig, was entitled to a bamboo tally stick called an *omak.* This Big Man has more than 50 omak hanging from his neck, identifying him as a person of great renown.

Ritual Buildings

One of a Mt. Hagen Big Man's greatest contributions was to organize his kinsmen to build a *moka pene*, or grassy plaza, for ceremonial exchange. The moka pene was a cleared area bordered with casuarina trees and magical cordyline plants, and it often served as the beginning of a long row of ritual buildings and spaces. At the head of the moka pene the Big Man directed the construction of a ceremonial mound, and beyond that a ritual men's house, different from the dormitory where young men slept. Behind this ritual men's house was a hut where pigs could be sacrificed. Finally, at the opposite end of the ceremonial alignment from the moka pene lay a clan cemetery. Magical stones were buried in this cemetery, and its periphery was planted with trees that symbolized ancestors.

Let us consider the implications of the alignment consisting of a moka pene, a ritual mound, a ceremonial men's house, a sacrificial hut, and a cemetery with sacred trees and magical stones. Almost any archaeologist coming upon it would consider it a "ceremonial complex" and assume that its construction had been directed by the hereditary leaders of a society with high levels of inequality. In reality it was the creation of a society whose leaders had achieved renown but possessed no real authority. That society received crucial support from its ancestors, invisible betas who were now in a position to lobby the celestial alphas on behalf of their descendants. Prevented from collecting enemy heads to bring their village good fortune, the Big Men of Mt. Hagen were at least able to earn renown through moka and, in some cases, to have their own skulls become objects of veneration.

ACHIEVING RENOWN IN ASSAM: THE ANGAMI NAGA

Assam is India's easternmost province. Its borderland with Burma (modern Myanmar) is an area of forested mountains where a variety of tribes, mostly speaking Tibeto-Burman languages, once supported themselves by raising rice, millet, cattle, and pigs. Identifying themselves with names such as Angami, Lhota, Ao, Rengma, Sema, and Konyak, these hill people were referred to generically as the Naga. Today they belong to a separate nation called Nagaland.

While they shared no history with the people of New Guinea, Naga societies displayed many behaviors similar to those we have just discussed. Many lived in autonomous villages, were divided into clans, built dormitory-style men's houses, ambushed enemies and brought back their heads, and ascended a ladder of renown by funding a series of increasingly prestigious rituals. Some of the most impressive rituals involved the hauling of multi-ton stones to villages, where they would be set up as permanent monuments.

It was through head-hunting that life force was accumulated by Naga villages. Life force was something that passed from humans into the rice and millet they grew, then into the humans and animals who ate the crops. That life force could be recovered later, when animals were sacrificed or enemies beheaded. Warriors who returned to their village with enemy heads underwent rituals of purification and then were allowed to wear insignia of prestige. The Sema Naga awarded the head-taker a boar-tusk collar. The Rengma and Lhota let him wear clothes of special cloth. The Konyak allowed him to wear special tattoos and a pendant in the shape of a trophy head.

The final resting place of an enemy head varied by tribe. The Angami buried the heads in the earth with their faces downward; the Konyak and Ao kept the heads in their men's houses; the Lhota, Rengma, and Sema displayed the heads in a tree near the edge of the village. A man who had beheaded no one was considered such a wimp that he had trouble getting a wife.

We will focus here on the Angami Naga, who lived near the Burmese border, south of the Rengma, Lhota, and Sema. Angami villages were defended with walls and ditches. Each clan had its own men's house, and often its own fortifications. The clans, which reckoned descent in the male line, sometimes bickered until they split in two. In spite of losses to warfare and fissioning, however, clans worked hard to keep from shrinking or going extinct, even adopting outsiders if necessary. Though they quarreled with each other, the clans in a village would come together to fight other villages or ethnic groups.

The Angami went to battle with five-foot spears of sago palm, shields made from elephant hide, and a machete-like knife called a *dao.* Each warrior also carried a bag filled with *panjis,* or sharpened bamboo spikes. Retreating after a raid, he would mine the trail behind him with panjis to impale the feet of pursuers.

Prisoners were often beheaded, but they were rarely tortured. An exception was made for a Lhota warrior named Chakarimo, who was captured by the Angami after he had speared 30 of them. The Angami war leader had Chakarimo tied to a tree and let young boys cut small pieces off him. He reportedly died after piece 312.

A pioneering account of the Angami was written by John H. Hutton, who lived among them prior to World War I. Hutton discovered that the Angami cosmos was filled with spirits, the most important of whom was Kepenopfü. The creator of all living beings, Kepenopfü dwelt in the sky, where most successful people would go after death. There were spirits of fertility, spirits who delivered game to the hunter, and malevolent spirits who brought death. Individual humans were protected by lesser spirits.

The Angami believed that they were descended from two brothers who had emerged from the earth. These brothers gave rise to two groups of human descendants, who sound like opposing moieties. Because Thevo, the older brother, emerged first, his descendants were allowed to begin eating before the descendants of Thekrono, the younger brother. This myth provided justification for a principle with which we are already familiar: whenever a clan or lineage splits in two, one division is considered senior, the other junior.

Angami men, like their New Guinea counterparts, had two pathways to renown. Both paths were related to the acquisition of life force. One could become a *pehuma,* or war leader, by supplying one's village with life force in the form of enemy heads. Before colonial rule, this path produced legendary pehumas whose names were widely remembered, and some of these leaders managed to pass on their positions to their first-born sons.

An alternative pathway was to become a *kemovo,* a kind of "holy man" or "ritual sponsor," of which there might be two or more in each village. One became a kemovo by first amassing wealth and then distributing it through a series of lavish ritual feasts, each more impressive than the one before it. The message sent by these feasts was that the would-be kemovo had acquired a surplus of life force that he was willing to share.

Each ritual feast had its own Angami name, but all could be lumped under the borrowed Assamese term *genna.* The keys to a genna were (1) dances and music that created an emotional response; (2) abundant meat from the host's sacrificial pigs and cattle; and (3) unlimited quantities of *zu,* or rice beer, which enhanced the awe-inspiring ritual experience.

The entire sequence of gennas could take years. First and easiest was the *kreghagi* feast, which could be held by anyone who had a surplus of rice. The host was blessed by a ritual expert, then fed his guests by sacrificing a cow. Afterward the host was entitled to adopt a special hairstyle.

More challenging were later feasts like the *thesa,* at which hundreds of pounds of rice were converted to beer, and the guests were fed the meat of four bulls and two pigs. This ceremony entitled the host to decorate his house in a special way. At the even grander *lesü* feast, the host needed to amass three times as much rice as for the thesa; in addition, he had to sacrifice ten bulls and five pigs. Afterward he could decorate his house with special wooden horns and assume the title of "house-horn-bearer."

Only a man who had performed lesü was entitled to move on to *chisü*, or "stone pulling," the month-long feast that would make him a kemovo. For this ritual he provided twelve bulls, eight pigs, and four times the quantity of rice, most of it converted to beer. The climax of the chisü involved hauling a huge stone to the host's village and setting it up as a monolith to commemorate his climactic genna.

For this spectacle the host turned to all the young men of his clan, all the alumni of his men's house, and perhaps even his entire village. It was not unusual for 50 clansmen to turn out for this task, and when an entire village was involved, the crew could grow to several hundred men. All stone pullers had to be in ceremonial dress, which could include dyed cotton kilts, huge necklaces of conch shell beads, armbands of brass and elephant ivory, and the plumes of a tropical bird called the hornbill.

To be of value the monument had to come from a distant quarry. The Lhota Naga placed the stone on a heavy litter of wooden poles that could be carried by six rows of men, 12 per row. The Angami, who hauled even larger stones, levered the monument onto a sledge of heavy logs. Wooden rollers were placed in the path of the sledge and hundreds of men, using strong ropes made of tropical vines, pulled it along jungle trails for hours while singing aloud. All knew that at the end of their journey they would be welcomed with gallons of rice beer.

At the host's village, participants dug a hole and tilted the sledge until the base of the stone slid into it. Once the monument was upright, it became the abode of important spirits and would keep alive the memory of the host's chisü (see Figure 10). He could now call himself a holy man.

Instead of receiving an ordinary citizen's burial, the kemovo would now be buried in his village's *tehuba*, or sitting circle. This was a waist-high platform 30 to 45 feet in diameter, built over the grave of the village's first kemovo and including the burials of most later kemovos.

The chisü ritual took an earlier premise—that rocky landmarks were the abode of spirits—and added the premise that a host who created such a landmark within his village had demonstrated exceptional life force. Through the circularity of Naga logic, a host's good relationship with the spirit world also explained his ability to accumulate the food necessary for a chisü.

Like the linear ritual complexes of Mt. Hagen, the stone monuments of the Naga provide a cautionary tale for archaeologists. Many of the chisü stones

FIGURE 10. One hundred years ago the Angami Naga of Assam had several routes to achieving renown. For example, one could become a *kemovo,* or "holy man," by amassing the surplus necessary to sponsor a series of increasingly lavish rituals. The final ritual in the sequence was a month-long celebration culminating in *chisü,* or "stone pulling." As many as 50 to 100 men were rewarded for hauling a multiton stone from a distant quarry to the host's village, where it served both as the abode of important spirits and a monument to the host's achievements. Here we see a chisü stone erected by the village of Maram.

overlap in weight with the carved monuments of Mexico's Olmec, Colombia's San Agustín culture, and even the ancient Maya—three societies thought to have very powerful hereditary elites. Naga monuments thus warn us not to underestimate societies whose leadership was based solely on achievement.

THE GOALS AND LIMITS OF ACHIEVED RENOWN

The societies discussed in this chapter allowed ambitious men to achieve inequality in prestige but limited their actual authority. War leaders commanded respect until enemies took *their* heads, increasing the life force of a rival village. A Big Man commanded respect until someone gave him a gift he could not match. The kemovo became a holy man, but sooner or later someone would come along to haul a bigger stone, sacrifice more bulls, and provide more beer.

It is significant that leaders in achievement-based societies wanted very much for their sons to succeed them. As role models, of course, they had shown their sons how to achieve renown. They could not, however, do what Tlingit and Nootka chiefs did regularly—confer titles and privileges on their offspring. Moreover, surrounding every man of achievement were ambitious young rivals who were determined to become Big Men, war leaders, or holy men themselves.

Achievement-based societies became very common once agriculture had arisen. To demonstrate this we will focus our attention on one of achievement-based society's most widespread institutions: the men's house, or clan house. We will look first at the ritual houses of several living societies. We will then show that similar houses were built by ancient societies in Mexico, Peru, and the Near East. By so doing, we are following a principle we spoke of earlier: when one sees people doing the same thing at 8000 B.C. and A.D. 1900, one probably has identified a behavior that arose repeatedly in world history.

SEVEN

The Ritual Buildings of Achievement-Based Societies

Foragers often create ritual space by arranging their shelters in an oval. The enclosed area can then be used for feasting or dancing, sometimes around a communal hearth.

Farming villages, for their part, often formalize ritual space by creating a building to house it. In aboriginal North America that building could be a sweat house, a kiva, or a ceremonial lodge. We will see examples of those buildings in the chapters that follow. In other regions the ritual building might be a men's house.

The ground plans of men's houses vary considerably. Some are circular, and others are rectangular. Some have benches on which men can sit and some have beds or platforms on which they can sleep. Architectural diversity can be present even within the same ethnic group. Figure 11 shows us the ground plans of Rengma Naga men's houses from early twentieth-century Assam. Note that the men's house of the eastern Rengma had large communal sleeping platforms, while that of the western Rengma had rows of beds. As we will see later, many prehistoric men's houses had sitting or sleeping benches.

In this chapter we look at three different types of men's houses used by Old World societies. Each of these buildings reflects a slightly different route to achieved inequality. Each type of men's house is also potentially identifiable in the archaeological record. Its presence can therefore provide a date for some of the world's first achievement-based societies.

The three societies we look at are representative of, but do not exhaust, the variation in men's houses. One society, the Ao Naga of Assam, had a dormitory-style men's house in which every young man slept while he learned the rules of Ao society. Once he embarked on his campaign of self-promotion, a prominent Ao man could count on the support of the residents and alumni of his men's house.

The second society we examine is that of the Mountain Ok of New Guinea. Rather than one large men's house that was open to all, the Ok built a cluster

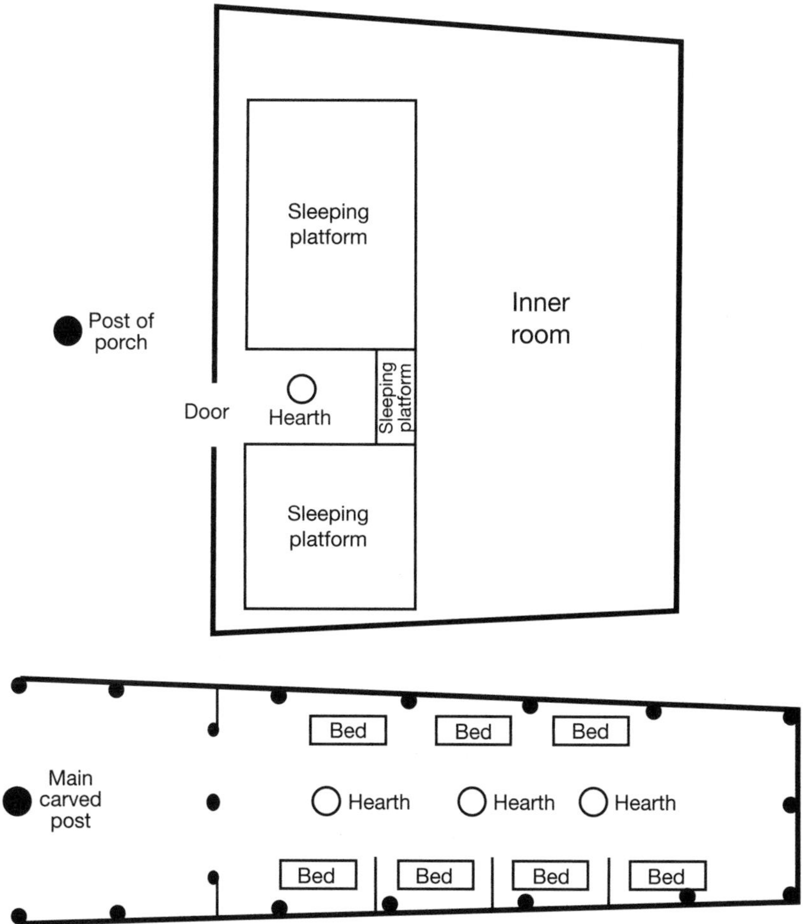

FIGURE 11. One hundred years ago, the Rengma Naga of Assam built men's houses in which all the village's young men slept during their formative years. The men's house of the eastern Rengma, seen above, was 14 by 20 feet (excluding the porch) and had large sleeping platforms. The men's house of the western Rengma, seen below, was 30 feet long (excluding the porch) and had rows of beds.

of smaller houses accessible only to initiated men. One had to earn his right to enter, a process that weeded out "rubbish men." In our opinion societies with exclusionary men's houses had the potential to create greater differences in social inequality. They had, after all, already allowed a small number of men to monopolize key ritual information.

Finally we look at the Siuai of Bougainville, the largest of the Solomon Islands. Among the Siuai the men's house was built by a Big Man who had allegedly been chosen for prominence by a demon. The Big Man provided blood to the demon, who both protected him and was nourished by him. Such a men's house, in our opinion, had the greatest potential for being converted to an actual temple. Unlike the typical men's house, where initiates sat around on benches and celebrated their ancestors, the Siuai men's house was considered the favored venue of a powerful supernatural being.

What makes the latter possibility exciting to archaeologists is that in many regions of the world—the Near East, Mexico, and Peru, to name only three—there came a moment when the men's house gave way to the temple. And that moment was often accompanied by evidence for hereditary inequality.

THE ARICHU OF THE AO NAGA

Like their Angami neighbors, the Ao Naga had an achievement-based society. An ambitious man could rise to prominence by sponsoring a series of increasingly expensive ritual feasts. His clan, as well as the alumni of his men's house, contributed to these feasts and basked in the reflected glory. For weeks preceding each ritual event, the women of his clan ground the rice that would be turned into beer.

The men's house, called *arichu* in Ao, or *morung* in Assamese, was built inside the village's defensive palisade. It was a magnificent building, 50 feet long and 20 feet wide, with its front gable soaring 30 feet above the ground. Since the arichu was a dormitory-style men's house, it had sleeping benches around the walls. To protect sleeping youths from the enemy spears thrust through the walls during raids, the eaves of the house extended to the ground. The largest vertical posts were carved with human figures, tigers, hornbills, or elephants. The arichu was rebuilt every six years, at which time animals were sacrificed and a neighboring village was raided to obtain an enemy head for good luck.

A pioneering account of the arichu was provided by James P. Mills, who lived in Assam during the early twentieth century. Mills reports that the boys of each village were divided into three-year cohorts based on age, and that they remained within their cohorts for life. Boys between 12 and 14 years of age, referred to as "unripe," entered the men's house for the first time. Between 15 and 17, when they were "ripening," they were joined by a new cohort of boys, 12 to 14. They became "arichu leaders" between 18 and 20 and could accompany older kinsmen on head-hunting raids. After two more three-year stages, most of them were married men and considered "clan leaders." When they were between 27 and 29, they were "councilors" and got the biggest share of meat at ritual feasts. Finally, when they were between 36 and 38, they were declared "priests." This term implies that a gradual increase in virtue was involved.

Well into middle age, each man still considered himself part of his age cohort and an alumnus of his men's house. Every Ao boy in an arichu, according to Mills, had undergone socialization equivalent to that received by boys at British schools such as Eton and Harrow. In other words, among Ao males, it "took a men's house to raise a child."

THE RITUAL MEN'S HOUSES OF THE MOUNTAIN OK

Let us look next at the western highlands of New Guinea. Here 15,000 people, referred to as the Mountain Ok, made a living in roughly 4,000 square miles of forest by growing taro, raising pigs, and hunting and fishing. Many Ok considered themselves the descendants of Afek, a female creator and ancestor, who gave birth to people and designed their ritual life. According to anthropologist Maureen Anne MacKenzie, the Ok venerated their ancestors and beseeched them for success in hunting, warfare, gardening, and pig raising. To maintain contact with those ancestors, the Ok kept skulls and skeletal parts of the deceased in their ritual houses.

Among the Ok, men's and women's ceremonies were separate; each gender was therefore excluded from the other's ritual houses. The village of Telefol, where MacKenzie lived, had five kinds of ritual buildings: three for the men and two for the women (Figure 12). Examples of such buildings were as follows:

The *kabeel am,* or "hornbill house," for new male initiates
The *yolam,* or *ogen am,* "ancestor house," for previously initiated men

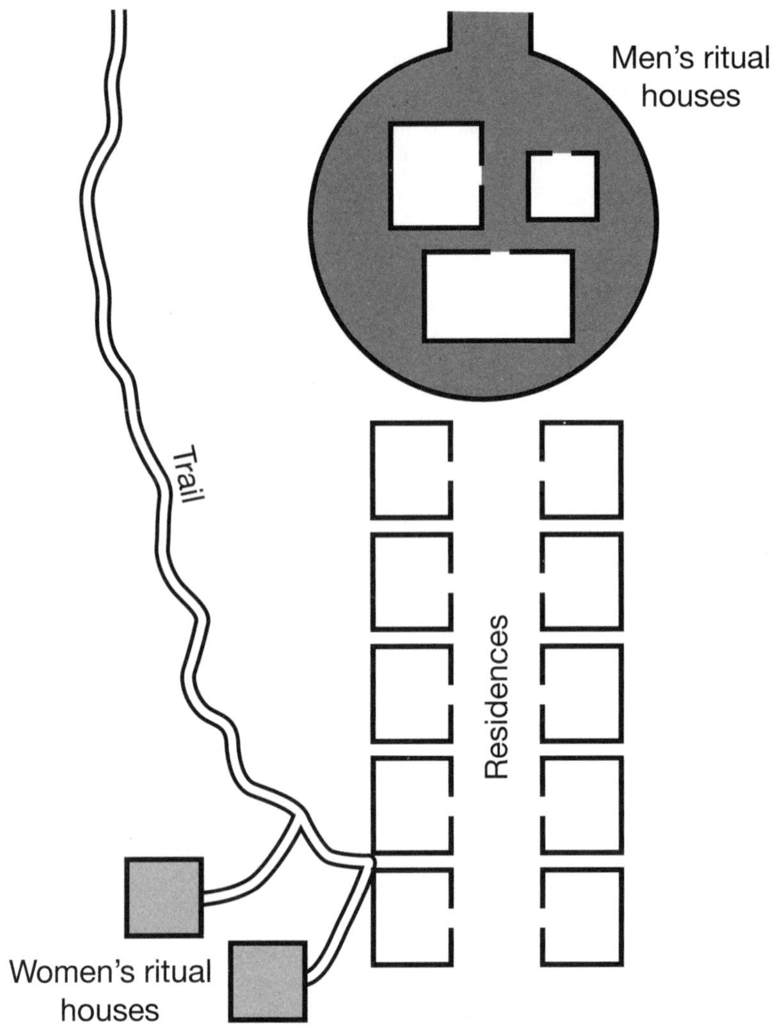

FIGURE 12. The traditional Mountain Ok of New Guinea built multiple men's houses, allowing relics from logically contradictory parts of their cosmology to be kept separate. The village of Telefol, for example, had three small ritual houses for male initiates and two small ritual houses used by women.

The *katibam,* or "old men's house," also for previously initiated men
The *dungam,* or *am katib,* a hut where women were secluded while giving birth
The *unan am,* a ritual venue for women

Not far from Telefol lay Baktaman, another Mountain Ok village. This village had four ritual men's houses, as follows:

The *katiam* was a building for curating hunting trophies and making sacrifices to increase the yields of gardens. Miscellaneous bones of the ancestors might be kept there.
The *yolam,* or "ancestor house," was dedicated to rituals of warfare and crop increase. It often contained two sacred fires and the skulls of ancestors from several different clans.
The "taro house" was a residence for senior men.
The "house of the mothers" was, in spite of its name, a place for the veneration of male ancestors.

Anthropologist Fredrik Barth, who lived for a time in Baktaman, explains why the Ok needed four men's houses with different ritual themes. All cosmologies have logical contradictions, and Ok cosmology was no exception. Having four different ritual houses enabled the Ok to keep incompatible relics from different parts of their cosmology in separate buildings.

Because Ok religion was focused on the ancestors, the bones of deceased kin were often used in rituals. Many of the skulls kept in the yolam were decorated for this purpose. Barth says, however, that the specific identities of deceased individuals were not important; the skulls simply represented generic ancestors. This situation fits the findings of anthropologist Igor Kopytoff, who discovered that among African societies comparable to the Ok, "ancestor" was also a generic category. Once deceased, the ancestors lost their individual characteristics and became another class of village elder.

Achieved inequality can be subtle. It becomes visible in Ok society when we consider who had access to the men's house. Ok men were initiated into the secrets of ritual one step at a time, and the whole process could take 10–20 years. "Rubbish men" never made it. A small core of prominent men monopolized the bulk of the mythical and cosmological knowledge, setting themselves apart as a low-level religious elite.

Barth found 11 initiated men actually living in the katiam at Baktaman. Fifteen partially initiated men had attended rituals in the katiam but were not allowed to live there. Some 128 uninitiated men had never been allowed past the door. Thus 80 percent of the Ok men were just as excluded from the katiam as the women were.

It is also the case that only fully initiated men could enter the yolam to perform ceremonies. It was in the yolam that village elders planned raids against enemy villages and carried out rituals to guarantee success. The enemies of the Ok knew this, of course; as a result, the men's house was often the first building burned when an Ok village was raided.

As Barth points out, having so much ritual information concentrated in the minds of so few men could have consequences. Sometimes senior men died without passing on the full cosmology. Sometimes the training took so many years that a young man's mentors forgot the details. All such lapses, Barth learned, provided opportunities for revising cosmologies and modifying rituals. Religions transmitted orally, as mentioned earlier, can change in ways that religions with printed texts cannot.

THE KAPOSO OF THE SOLOMON ISLANDS

Six hundred miles out in the Pacific, east of New Guinea, lie the Solomon Islands. The largest of the Solomons is Bougainville, whose societies share with New Guinea the raising of pigs, sago, taro, yams, and sweet potatoes.

In 1938 there were 4,700 members of an ethnic group called the Siuai living on 250 square miles of Bougainville. Prior to coming under Australian administration, the Siuai (pronounced "See-ooh-eye") had tended to live in small, scattered hamlets of one to nine households. These hamlets were linked by a series of trails through the forest.

Siuai society paid more attention to the mother's descent line, because it was the corporate custodian of garden land. As a result, a mother's genealogy extended back four or five generations, while the father's genealogy tended to be shallower.

Men were seen as inherently superior in virtue, strength, and ambition, yet their wives raised the pigs that facilitated a man's social advancement. A further irony lay in the fact that although the immediate source of a man's prestige was his accumulation of pigs and shell valuables, he could not afford the latter without the capital created by the garden land of his mother's lineage.

There were four types of gifts among the Siuai: the usual reciprocal gifts between relatives and friends; bride-price and dowry exchanges between the relatives of a bride and groom; coercive gifts, used to enforce social obligations; and competitive gifts, used to humiliate rivals. The latter were given by men with *haokom,* or ambition, who desired *potu,* or renown. If such a person succeeded, he became a Big Man.

Like the Chimbu of New Guinea, the Siuai had many levels of prestige. At the bottom were lowly men, referred to as "legs" because they performed menial tasks for men of renown. In the middle were the modestly successful men who constituted the bulk of society. At the top were men of prestige, a few of whom stood out as *mumis.* Mumis were the biggest of all Big Men, individuals of intelligence and industriousness, charisma and diplomacy, and generosity and executive ability. Just as in the Mt. Hagen region of New Guinea, a man whose father was a mumi, or whose mother was a mumi's daughter, had role models who increased his chances of making it to the top.

When anthropologist Douglas Oliver lived among the Siuai during the period 1938–1939, he overheard the same nostalgic comments that Paula Brown had heard among the Chimbu. There had been "real mumis" in the good old days, the Siuai said, before the suppression of warfare. Long ago there were war leaders who "lined our men's houses with the skulls of people slain." Today the Siuai were reduced to "fighting verbally," humiliating rivals with gifts and feasts.

By 1938 the most common path to renown was to accumulate *manunu* ("wealth") while attracting supporters, messengers, and "legs." The more wives a man had, the more gardens he had access to, and the more pigs his family could raise. He used his entrepreneurial skills to trade for shell valuables that could be exchanged for additional pigs. As his prestige grew, he would be welcomed into a men's ritual society, giving him access to a *kaposo,* or men's house. His hope was that one day he would have the resources to build his own.

Every man of renown, of course, had rivals. Putting pressure on his relatives and in-laws, he accumulated the plant foods, pigs, and shells to eliminate the competition. His strategy was to plan a huge pig feast called a *muminae* and then, at the last minute, announce that one of his rivals would be the guest of honor. That news would be spread by the throbbing beat of the wooden slit-gongs in his men's house, and the message would be defiant: "So-and-so many dozens of pigs will be butchered, and my honored rival will be rendered near death by humiliation, since there is no way that he can reciprocate."

In his rise to the top a mumi had supernatural help. He had been befriended by a *kapuna,* a *horomorun,* or both. The kapuna was a humanlike supernatural being who was an ancestor of the line of women leading to his mother. The horomorun was a malevolent spirit, "the mumi of all demons," so powerful that it could cause death. This demon spent time in the men's house, identified potential leaders, and then announced its desire to ally itself with one of them by making the man temporarily ill.

Once the future mumi had recovered from his illness, he bonded with the horomorun, and the man was protected by the demon's magic. In return the mumi would supply the demon with its favorite beverage: pig's blood. At each feast the horomorun would drink its fill before the human guests were served. In this Faustian relationship the well-nourished demon grew stronger as the mumi grew in prestige. No one could be jealous of the mumi's success because he had been chosen by the demon itself.

Somewhere along a major trail between settlements, the mumi built a men's house where the demon could enjoy itself. Using accumulated shell

FIGURE 13. In the traditional Siuai society of Bougainville, the *kaposo,* or men's house, was built by a Big Man called a *mumi.* The mumi's extraordinary achievements were attributed to the supernatural aid of a demon. Some mumis paid to have huge tree trunks hauled to the men's house and carved into slit-gongs. This illustration, based on a 70-year-old photo, shows the hauling of a future slit-gong across a river.

valuables, the mumi paid men from other hamlets to help his relatives complete the task.

Oliver watched a Big Man named Kuiaka create his own kaposo in the late 1930s. It took 85 men 3,600 man-hours to build it, and Kuiaka paid for it with 18 pigs, masses of steamed taro, and the milk of 2,000 coconuts. In olden days such a men's house would have been consecrated by beheading a man from an enemy community and keeping his skull in the clubhouse.

Critical features of the Siuai men's house were its nine to ten wooden slit-gongs, ranging from three feet long and one foot in diameter to 15 feet long and five feet in diameter. Each gong, hollowed out of a tree trunk, was given a different name and produced a different tone. The largest gongs weighed many tons, and their procurement was a feat equivalent to the stone-pulling of the Naga.

Oliver witnessed the delivery of one immense tree trunk, a task in which 200 men participated (Figure 13). They cut a 25-foot-wide trail through the forest, destroying valuable coconut palms for which they would have to pay. Men from several hamlets, using ropes and sledges, struggled for days to transport the giant log across rivers and swamps. They were paid with pork and coconut milk, but only after the demon had drunk deeply.

The death of any mumi was attributed to witchcraft. For him there was no burial in a place of honor, as there was for an Angami Naga holy man. Because of his close association with the demon, the mumi was seen as brimming with black magic. He could pass on neither his wealth nor his prestige to his son. He and his accumulated shell valuables would be cremated, out of fear that contact with them would cause illness.

ACHIEVED INEQUALITY AND THE MEN'S HOUSE

In each of the three societies we have examined, the relationship of the men's house to the sources of achieved inequality was different. The arichu of the Ao Naga was open to every boy; sleeping there conferred no prestige. A man began his pursuit of renown only later, after he had married and moved out and could use his wives' labor to accumulate wealth for a feast. At this point, however, he could count on the support of current and former members of his men's house.

Among the Mountain Ok, on the other hand, simply being allowed into the katiam was a source of prestige. It meant that one had been initiated into a small group of ritual leaders, an honor afforded to only one in five men. Ok men's houses were smaller than those of the Naga, and owing to their different ritual functions there could be as many as four in use at a time.

Finally, among the Siuai, the kaposo was strongly associated with the Big Man who paid to have it built, just as an Angami Naga stone monument was associated with the man who paid to have it set up. Inside the kaposo lived a demon that drank pig's blood and protected its favorite Big Man with black magic, magic so powerful that even a deceased mumi's shell valuables had to be burned. The mumi's reputation reflected a premise that we also see in societies with hereditary nobles: Our leader has a closer relationship to the supernatural world than the rest of us.

Let us close with the limitations of leadership in this chapter's three societies. Their leaders had prestige but no actual political power. They could pay people to build men's houses but not order them to do so. Most importantly, they could not pass on their prestige to their sons. The latter were forced to earn it on their own.

EIGHT

The Prehistory of the Ritual House

At the start of the twentieth century, village societies with achievement-based leadership were among the most common in the world. They were remarkably stable societies, made up of descent groups that exchanged brides and gifts, honored their ancestors, considered everyone equal at birth, yet threw their support behind gifted kinsmen who sought to achieve renown.

Such societies were also widespread in prehistory; we probably all have ancestors who lived in one. Once you know what to look for, you can identify them in the archaeological records of the Near East, Egypt, Central and South America, North America, and Africa. Achievement-based societies became common as soon as each of those regions had adopted agriculture and village life.

When did the first achievement-based village societies appear? Perhaps 10,000 years ago in the Near East, 4,500 years ago in the Andes, and 3,500 years ago in Mexico. No two of these regions were exactly alike, but all three had a series of recognizable behaviors in common. One of those behaviors was the building of ritual venues, some of which were almost certainly men's houses.

As we have seen, men's houses came in many varieties. They could be built by Big Men, clans, or entire villages. They could be inclusive dormitories or places for an exclusive few. Ancestors' remains might be kept in them, as well as enemies' heads, since men's houses often played a role in turning youths into warriors; this could make the men's house the target of an enemy attack. Such ritual buildings were typical of societies where prestige was based on leadership in raiding, head-hunting, trade, and exchange, or the underwriting of ritual feasting, stone monument pulling, and public construction.

In this chapter we consider three regions, some of whose early ritual buildings displayed many features of men's houses: benches for sitting or sleeping, curated skulls and skeletal parts, sunken floors, white plastered surfaces, or other attributes not shared with residences. We begin in the Near East because it provides our oldest examples.

FROM FORAGING TO ACHIEVEMENT-BASED VILLAGE SOCIETY IN THE NEAR EAST

During the peak cold of the Ice Age, 20,000 years ago, many of the higher mountains of the Near East were covered by treeless steppe. The valley of the Jordan River, however, was a warm refuge. One of the landmarks of this valley was the brackish lake called the Sea of Galilee, which lies more than 600 feet below sea level. The slopes surrounding the lake supported a Mediterranean parkland of oak, pistachio, almond, fig, and olive trees.

On the southwest shore of the lake lay a prehistoric camp that Israeli archaeologists have called Ohalo II. The foragers of Ohalo lived in shelters made of branches and thatch, not unlike the conical huts of the Hadza and !Kung. Some shelters included beds of grass, and there were hearths outdoors.

The foragers of 20,000 years ago were harvesting more than 140 varieties of wild plants. In contrast to the bulb and tuber collectors of the Nile embayments, the occupants of this Mediterranean parkland concentrated on high-calorie nuts and sugary fruits. They harvested acorns, almonds, and pistachios, wild olives, and the fruits of the wild fig, grape, and raspberry. At least 20 percent of the plants in their archaeological refuse, however, were the seeds of wild grasses that other hunter-gatherers might have ignored. Half of these seeds were of bromegrass, not the most appetizing of foods. The remaining seeds were of alkali grass, creeping foxtail, and four varieties of wild cereal grasses.

Two of the cereal grasses—wild barley and emmer wheat—were the ancestors of Mesopotamia's most important future crops. The stage was therefore set for a process analogous to the raising of sago, taro, yams, and plantains in ancient New Guinea: first a period of intense exploitation of wild cereals, and then the cultivation of wheat and barley.

When the Ice Age ended 10,000 years ago, conditions for the growth of cereals were improving over the entire Near East. Rising temperatures allowed oaks, pistachios, and large-seeded grasses to spread to higher altitudes.

Glaciers melted, sea levels rose, and the amount of carbon dioxide in the world's atmosphere increased from 180 to 280 parts per million, a 50 percent increase in just a few thousand years. Since plants grow better at this higher carbon dioxide level, it was a good time to experiment with agriculture.

At least some of the early seed collectors arranged their huts in a circle or oval, like the Basarwa or the Andaman Islanders. One of the clearest examples comes from the site of M'lefaat in northern Iraq. M'lefaat lay 950 feet above sea level on a tributary of the Tigris River, 20 miles from the city of Mosul. Ten thousand years ago, the foragers of M'lefaat cleared an area roughly 300 feet long and 200 feet wide and covered it with hard-packed clay. On this surface they laid out ten huts, surrounding an oval area that could have been used for dances or other ritual activities (Figure 14, top). Huts varied in size from 16 to 26 feet in diameter, and there were differences among them in stone tools. These differences suggest that the occupants of each hut—male, female, married, unmarried—varied as widely as those of the Basarwa huts shown in Figure 2. The foragers at M'lefaat relied on a wild cereal called goat-face grass, along with smaller amounts of wild barley, wheat, and rye. They also collected pistachios and lentils and hunted gazelles.

Let us now look at the well-studied Natufian people of Israel, Jordan, Lebanon, and Syria, who from 12,000 to 10,000 years ago made the collection of wheat and barley the centerpiece of their economy. To harvest grain, the Natufians made sickles by inserting flint blades into bone or wooden handles. To store it, they wove baskets and created pits waterproofed with lime plaster. To turn the grain into porridge, they ground it in mortars of stone.

The Natufians soon learned that wild cereals ripen in different months at different altitudes. At sea level they ripen in late April; at 2,000 to 2,500 feet they mature in mid-May; at 4,500 feet they ripen in early June. By starting their harvest at sea level and moving steadily higher, foragers could lengthen the season of availability. The most permanent-looking Natufian settlements were at lower elevations, where their refuse included the remains of waterfowl that winter in the Near East. In the higher mountains they sometimes occupied caves. Let us follow them through a series of camps.

In the cliffs of Mt. Carmel, only two miles from the Mediterranean Sea, lies the Cave of el-Wad. Here the Natufians built shelters in the mouth of the cave and terraced the slope below it so that more could be added. In addition to having portable grinding stones, they hollowed seed-grinding basins into the living rock of the cliff. They harvested grain with sickles, used bows and arrows to hunt gazelles and deer, made bone harpoons for fishing, and whittled

Ten circular huts at M’lefaat, Iraq

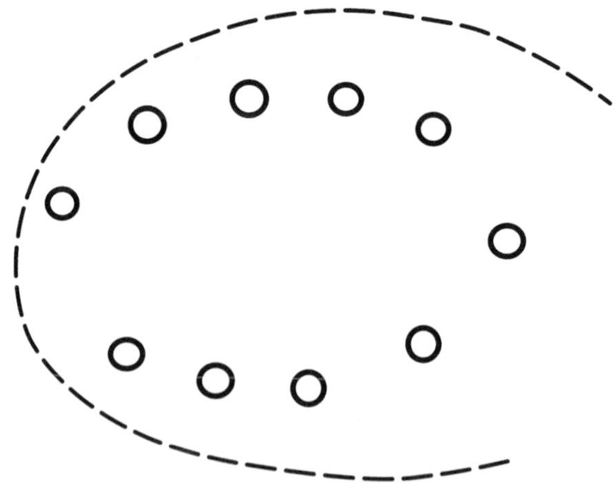

Men’s house with carved monolith at Wadi Hammeh 27, Jordan

FIGURE 14. Caption on facing page

gorgets to snare unsuspecting waterfowl. Like the mammoth hunters of the Ice Age Ukraine, the Natufians also kept domestic dogs.

The foragers of el-Wad were fond of ornamentation, carving pendants from gazelle bones and producing strands of cowrie shells. Their favorite ornament, however, was dentalium, a tubular seashell that they wore in such quantity that it brings to mind the strings of bride-price beads worn by the Hadza.

There were more than 50 Natufians buried at el-Wad, many of them wearing dentalium. Inside the mouth of the cave were men, women, and children, buried fully extended. Outside on the terraced slope were more men, women, and children, some with their limbs so tightly bent that their kinsmen must have tied them or wrapped their bodies in bundles. Some of these bodies were complete, but others were the partial skeletons of people who had been buried earlier, exhumed, and added to their relatives' graves.

Under the traces of a poorly preserved shelter were two reburied individuals, one with a head covering, bracelet, and garter of dentalium. Not far away was an adult, likely male, with seven rows of shells around his head; two exhumed and reburied individuals accompanied him. Another man had an elaborate head covering and a garter of dentalium, as well as a necklace of bone and shell beads; he was accompanied by another adult and a child, apparently reburials.

Now let us move higher, up to 800 feet above sea level in the wooded hills of Galilee. Here another group of Natufians camped in the Cave of Hayonim, just eight miles from the sea. Much as they had done at the Cave of el-Wad, the Natufians modified both the chamber of the cave and the slope below.

In the case of Hayonim, the Natufians built more substantial shelters. The floors were sunk into the earth, the lower walls were of stones fitted together without mortar, and the roofs were framed with branches. Five of

---

FIGURE 14. As the Ice Age drew to a close, permanent-looking settlements of circular huts began to spread over the whole area from the Mediterranean Sea to the headwaters of the Tigris River. In many cases the huts were arranged around an oval or circular area devoted to ritual activities. Some settlements had buildings that appear to be men's houses with sitting benches, lime plaster, or carved monoliths. Above we see the archaeological site of M'lefaat, Iraq, where ten huts were laid out on an oval clay floor 300 feet long (dashed line). Below is a probable men's house from the site of Wadi Hammeh 27, Jordan, which featured a bench and a carved monolith; it was 46 feet in diameter.

these circular huts, most of them with hearths, formed a cluster inside the cave. On the terraced slope outside the cave was a structure that may have been used for the storage of grain.

Three generations of Israeli archaeologists, including Ofer Bar-Yosef and Anna Belfer-Cohen, have excavated burials inside and outside the cave. The remains tell an interesting story. An analysis of 17 of the skeletons by anthropologist Patricia Smith revealed that eight, or nearly half, were congenitally lacking the third molar, or wisdom tooth. Such a high frequency of this genetic anomaly suggests that the people of Hayonim were mating within a small gene pool.

There are several possible explanations for such inbreeding. We have seen a case where one prominent male in a foraging society contributed his genes to 20 percent of his local band. We have also seen that the wives of polygamous hunters were sometimes sisters. Finally, we have seen that the marriage rules of some Australian societies reduced the number of eligible mates to one-eighth of the population. In a small population of foragers, any or all of these processes might have increased the frequency of a genetic anomaly.

The Natufians of Hayonim also decorated themselves with dentalium, and herein lies another story. Some of the dentalium belonged to a species from the Mediterranean Sea, only eight miles to the west. Other dentalium, however, belonged to a species from the Red Sea, some 400 miles to the south. The latter almost certainly reached Galilee as the result of exchanges with other groups. Why would one trade for a shell that one could get by taking an eight-mile walk? The answer is, it was not the intrinsic value of the shell that mattered but the social relations generated by the exchange.

Let us continue our ascent to higher elevations, this time to a stream canyon 25 miles northwest of Jerusalem. Here, at an altitude of 1,000 feet in the mountains of Judea, lay the Cave of Shukbah. Archaeologist Dorothy Garrod found abundant sickles and hearths at Shukbah but surprisingly few grinding stones and storage pits. This raises the possibility that the Natufians stayed at Shukbah only long enough to harvest the available wheat and barley, after which they carried as much grain as possible to more permanent camps at lower elevations.

A good example of a longer-term, lower-elevation camp would be Ain Mallaha in the Jordan River Valley. Here the Natufians lived on the slopes above Lake Huleh, a reed-bordered expanse of fresh water north of the Sea of Galilee. The lake itself lay only 200 feet above sea level, flanked by mountains with cereal grasses and oak, pistachio, and almond trees.

Several generations of archaeologists, including Jean Perrot and François R. Valla, have uncovered a sequence of repeated encampments at Mallaha. The Natufians lived in circular or crescent-shaped huts whose lower halves were sunk below ground. The posts supporting the roof were set into the dry-laid masonry walls that lined the subterranean part of each structure. The huts varied from 13 to 26 feet in diameter. Because so many have been excavated, we can see that some huts were large enough for a nuclear family, while others would have been appropriate only for an individual—perhaps a second wife, a widow, or a widower—and still others were big enough to be bachelors' huts.

One building discovered by Perrot stood out from the rest. Its outer diameter was roughly 21 feet, and it had a sitting bench more than two feet wide running around the interior. Unlike a typical residence, this structure and its bench were coated with lime plaster. The floor was carefully paved with flagstones. A small fireplace was set against the wall, and a single human skull rested nearby. We believe that this building may be one of our oldest examples of a bachelors' hut or men's house.

Was the Natufian men's house a dormitory for all youths or a ritual house for the initiated few? The building at Mallaha seems too small to have held more than the initiated. The encampment of Wadi Hammeh 27 in Jordan, however, appears to have had a larger Natufian ritual house. This one was 46 feet in diameter and had an apparent bench, ten feet wide, running around the wall. On the sunken circular floor of the building lay a broken monolith, carved with geometric motifs (Figure 14, bottom). The Wadi Hammeh structure is twice as large as the lime-plastered building at Mallaha. It is possible, therefore, that both inclusive and exclusive men's houses were present in the Near East at this time.

In addition to possible men's houses, multigenerational cemeteries, and exchanges of shell valuables, the Natufians provide us with evidence of feasts at which large quantities of meat were shared. One of the earliest examples comes from Hilazon Tachtit Cave in Lower Galilee.

In the center of the cave, the Natufians created two small subterranean structures and three burial pits. At least three wild cattle and more than 70 tortoises were cooked and eaten at some point in the burial ceremony. This is far more meat than the occupants of a typical Natufian camp would have consumed at a funeral and hints at substantial social networking.

The Spread of Achievement-Based Villages and Ancestor Ritual

Ten thousand years ago, from the Bay of Haifa to the Tigris River in Iraq, at least three major processes were under way. One process was the emergence of domestic races of wheat and barley, mutant strains that left the cereals with no seed-dispersal system but made them easier for humans to harvest and thresh.

This first process led to a second: the gradual conversion of long-term camps into permanent, multigenerational villages. In the course of this transformation, circular huts were replaced by larger rectangular houses. Some of the latter had their own storage rooms and, if the walls could bear the weight, even a second story.

As life became sedentary, it facilitated a third process: the hunting of herd animals with drive fences and corrals, followed by the penning, imprinting, and raising of their young. Small numbers of goats, sheep, and pigs gradually became residents of the village. Wild cattle were bigger and more dangerous animals, but under domestication even they became smaller and more docile.

During the course of all three processes, ancestor ritual escalated, and men's houses became increasingly well built and decorated. Some villages seem to have had more than one ritual house, either because each clan built its own (like the Ao Naga) or because the society wanted to keep contradictory parts of its cosmology separate (like the Mountain Ok).

Some of the most spectacular ritual houses are those excavated by Klaus Schmidt at Göbekli Tepe, a site east of the upper Euphrates River in Turkey. Göbekli sits on a high limestone ridge that makes it visible from more than ten miles away. This location was apparently chosen so that huge limestone blocks could be quarried from the ridgetop and hewn into upright posts for the ritual houses.

The earliest ritual houses at Göbekli Tepe, dating to perhaps 10,000 years ago, were round or oval. The lower half of each building was subterranean, with walls of dry-laid stone masonry. A sitting bench ran the entire length of the interior, except for the doorway.

The most distinctive attributes of these buildings were the immense, T-shaped limestone pillars that supported the roof. Usually the two largest pillars were set in the center of the floor, while a ring of slightly smaller pillars was set into the walls (Figure 15, top). Many of the pillars were carved in low relief with realistic images of animals; depending on the building, they might feature foxes, lions,

Circular ritual house at Göbekli Tepe

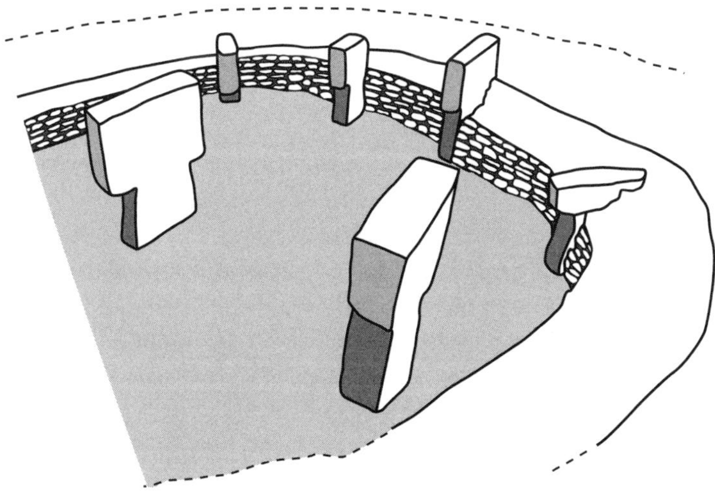

Rectangular ritual house at Nevali Çori

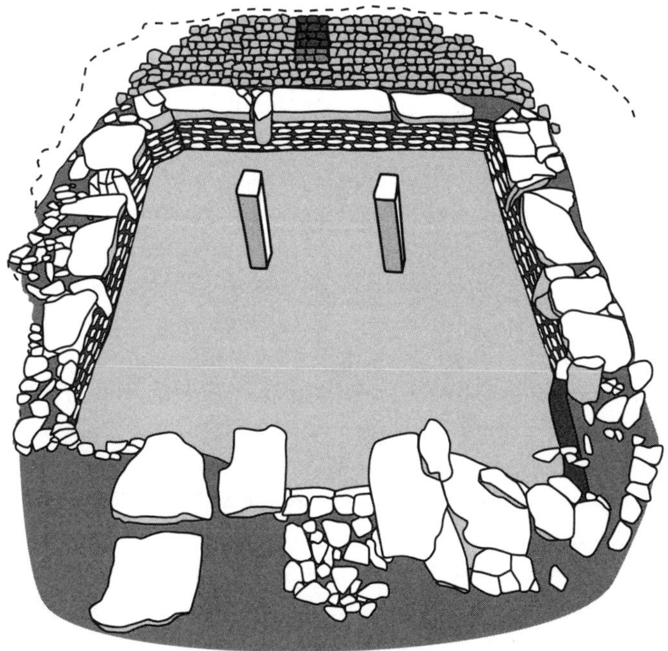

FIGURE 15. Some of the most spectacular men's houses (or clan houses) in the early Near East can be found in the Euphrates headwaters of Turkey. They made use of dry-laid stone masonry and monolithic stone pillars. Most pillars used at Göbekli Tepe were T-shaped; most pillars at Nevali Çori were straight. Note the flagstone sitting or sleeping benches in the Nevali Çori ritual house.

cattle, boars, herons, ducks, scorpions, or snakes. Some pillars showed signs of having had earlier images ground off, so that new ones could be carved. One immediately thinks of the carved posts of the Ao Naga men's house, with its depictions of tigers, hornbills, or elephants.

Once their period of use was over, the villagers deliberately buried these ritual houses with earth and domestic refuse. This act may reflect an unwillingness to leave the village's most important ritual venues accessible to outsiders.

Roughly 9,500 years ago, the occupants of Göbekli Tepe changed their architecture: they built a ritual house that was rectangular. Nicknamed the Lion Pillar Building, this new structure was partly subterranean. Its floor space was roughly 20 by 17 feet, and because of its thick stone masonry walls, its outer dimensions were greater than 33 feet. The roof had been supported by T-shaped pillars, one of which was carved with the lion that gave the building its name. From this point on, as we shall see, most ritual houses in the Euphrates headwaters would be rectangular.

The remarkable complex of ritual structures at Göbekli Tepe was built by a society with virtually no evidence of domestic plants and animals. The people harvested wild cereals with sickles, collected almonds and pistachios, and hunted gazelle, boar, and wild cattle.

We should consider the possibility that the hilltop ritual complex at Göbekli Tepe was maintained by multiple clans or descent groups, each of which built and decorated its own ritual house. Perhaps each descent group competed with the others to make its ritual house the most elegant. The carvings of animals may relate to a common cosmology or set of origin myths. Instead of curating ancestors' skeletal parts, the people of Göbekli Tepe carved limestone statues of what may have been mythical ancestors.

We have no doubt that quarrying huge T-shaped pillars and lowering them into place brought great renown to the leaders who organized the labor. One still-unfinished T-shaped pillar, left in the ridgetop quarry, weighed an estimated 50 tons. This is heavier than any stone monument erected by the Angami Naga or the Olmec of Mexico, although it did not have to be hauled as far.

Now let us move a short distance to Nevali Çori, another site in the upper Euphrates drainage of Turkey. Nevali Çori was roughly contemporaneous with the Lion Pillar Building at Göbekli Tepe and had several ritual buildings that remind us even more of Naga men's houses.

Nevali Çori was founded on the slope overlooking a stream in the rolling foothills of the Taurus Mountains, 1,600 feet above sea level. There its inhabitants supported themselves on wheat in the early stages of domestication. The limestone bedrock provided a ready source of stone blocks for construction.

The people of Nevali Çori both saved skulls like the Natufians and carved limestone statues like the people of Göbekli Tepe. Some of the statues depict human skulls, while others show people with their arms folded on their chests. Still other statues were of animals, and there were even statues that combined human and animal attributes. One shows two humans dancing with an unidentified animal, perhaps a scene from a mythological era such as the Alcheringa of the Australian Aborigines.

Archaeologist Harald Hauptmann discovered two rectangular ritual houses of dry-laid stone masonry at Nevali Çori. Each was semi-subterranean, its lime-plastered floor reached by descending stone steps. The earlier of these buildings, called Structure II, measured 45 feet on a side. Its floor area, however, had been reduced by the addition of sitting or sleeping benches. The roof would have been supported by limestone pillars, two of which were set up in the center of the floor and the rest of which ran along the periphery.

The later ritual house, Structure III, was 44 feet on a side. It had very clear sitting or sleeping benches covered with flagstones, flanking the entire floor except for the area of the stone steps. As at Göbekli Tepe, the roof would have been supported by limestone pillars (Figure 15, bottom).

If Structure III was for initiates only, its benches could have accommodated 40 to 50 sitting men. If, on the other hand, it was a Naga-style dormitory, it could have accommodated perhaps 15 sleeping youths. We do not know which was the case.

The streams passing Göbekli Tepe and Nevali Çori carry water to the Euphrates. That river flows south out of its Turkish headwaters and enters its Great Bend in northern Syria. On a bluff overlooking the Euphrates, some 990 feet above sea level, lay the archaeological site of Abu Hureyra. Here excavations by Andrew M. T. Moore revealed the slow transformation of a two-acre encampment of circular huts into a 28-acre village of large, rectangular houses.

The first occupants of Abu Hureyra, who lived there 10,000 years ago, harvested wild rye, barley, and two species of wild wheat. They stalked herd animals with bow and arrow, relying most heavily on gazelle. The unit of residence was the extended family, but instead of sharing one large building, they lived in clusters of five to seven small, circular houses grouped around an open area.

Over the next 2,000 years Abu Hureyra became a village of rectangular houses. By this time Abu Hureyra's economy was based on domestic wheat, domestic barley, wild rye, lentils, field peas, faba beans, and flax or linseed, from whose fibers linen could be woven. The villagers harvested the bulbs of sedges and rushes and collected pistachios and wild caper fruits. They kept early domestic sheep and goats and harvested wild clover and alfalfa, perhaps as fodder for their flocks.

One building, partly exposed by Moore, was a likely ritual venue. Its full dimensions are unknown, but its walls were up to four mud-bricks wide (twice the width of a typical house wall) and its corners were aimed at the cardinal points. Basically rectangular, it had a crescent-shaped room at one end that resembled the apse of a church (Figure 16, above). This apsidal space appears to have been a charnel room, that is, a place where remains of the dead were kept. A large pit inside the apse contained the skeletons of 25 to 30 individuals—men, women, adolescents, and children—many with their skulls missing. Elsewhere in the room were additional burials and isolated skulls. This building, constructed 8,000 years ago, evidently served a purpose similar to the "ancestor houses" of New Guinea.

One young man buried at Abu Hureyra had an arrow point in his chest cavity. Given the raiding we have seen in societies of this type, our only surprise is that this young man was shot once rather than pincushioned. His untimely death makes us wonder why palisades or other defensive works have not been more often discovered at early Near Eastern villages. One answer is that archaeologists do not always excavate the outskirts of the village, where such defensive works were usually built. When they do investigate the village periphery, they sometimes find ditches or walls.

Consider, for example, the village of Tell Maghzaliyah in northern Iraq. Some 8,500 years ago its occupants were planting wheat and barley on a rolling plain west of the Tigris River. Archaeologist Nikolai Bader discovered that Maghzaliyah had once been surrounded by a defensive wall, which he was able to trace for roughly 200 feet. The lower part of the wall was made of upright blocks of stone, standing five feet tall in places, while the upper wall was built of hard-packed earth.

Equally impressive defenses have been found at the site of Tell es-Sultan, known to the Israelis as Jericho. This large village lay in a hot, arid region, 900 feet below sea level, near the point where the Jordan River enters the Dead Sea. The source of its water, both for drinking and irrigation, was an oasis created by the spring of Ain es-Sultan.

FIGURE 16. Some early Near Eastern ritual houses had apse-shaped rooms where the ancestors' skeletal remains were curated. Above we see an apsidal building from Abu Hureyra, Syria, where the remains of more than 30 people were kept. Below we see the apsidal Skull Building from Çayönü, Turkey, where the remains of more than 400 individuals were kept. The Skull Building featured sitting benches and a rectangular stone altar, on whose surface was found hemoglobin from both human and cattle blood. Owing to later damage, the full dimensions of these buildings are not known.

Roughly 8,000 years ago, Jericho was defended by a ditch and a 14-foot-high stone masonry wall. It also seems to have had a lookout tower, in this case preserved more than 25 feet high. The tower was flanked by storage facilities, a few still holding charred cereals.

Some archaeologists, reluctant to accept the fact that village societies often engaged in raiding, have offered nondefensive explanations for Jericho's wall. Such attempts to pacify prehistory underestimate the worldwide evidence for raiding in achievement-based societies, as well as the fact that contemporaneous sites such as Maghzaliyah had clear defensive works.

Jericho, then a village of mud-brick houses, had a multistage burial program similar to that of some Australian Aborigines. People were buried first as individuals. At a later date their graves were reopened so that their skulls, and perhaps some limb bones, could be reburied in a charnel room. Archaeologist Kathleen Kenyon found many partly decayed bodies that had, in her words, been "searched through for the particular purpose of removing the skulls."

One cluster of seven skulls stood out from the others. Their facial features had been reconstructed by coating the skull with lime plaster and placing seashells in the orbits to represent eyes (Figure 17, left). According to the biological anthropologists who have examined them, all these plastered skulls came from adult men. This should not surprise us, given the tendency of Near Eastern societies to reckon descent through male ancestors.

Archaeologists have found similar skulls at Ain Ghazal, an early village on the outskirts of the Jordanian city of Amman. Ain Ghazal was founded 9,000 years ago on an embayment of the Wadi Zarqa, a spring-fed river traversing the arid region. Some 1,500 years later the site had grown to cover more than 30 acres, making it one of the largest villages of its time. The villagers of Ain Ghazal lived in extended families whose houses averaged 18 by 26 feet. They grew barley, wheat, peas, chickpeas, and lentils; raised goats, cattle, and pigs; and hunted gazelles with bow and arrow.

One ritual building at Ain Ghazal, partially destroyed by the growth of Amman, was almost 50 feet long. It had four rooms of differing sizes, and its walls bore unusually thick lime plaster, painted red with ocher. This building was as large as the dormitory-style arichu of the Ao Naga, but owing to its destruction we cannot confirm what kind of building it was.

Archaeologist Gary Rollefson found that Ain Ghazal had a multistage burial program similar to that of Jericho. Many people were buried originally below the floor of a residence, only to be exhumed later so that their skulls

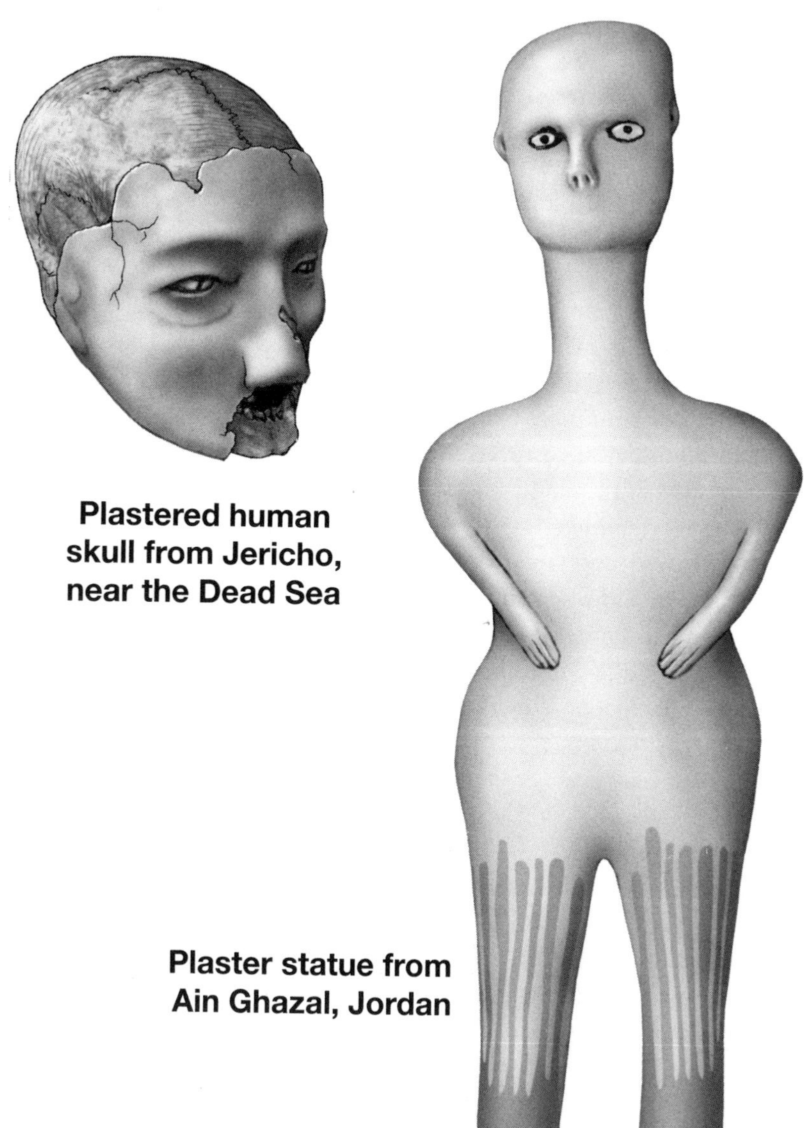

FIGURE 17. In the early agricultural villages of the Near East, the ancestors were a major focus of ritual. On the left we see the skull of a presumed male ancestor from Jericho, near the Dead Sea; his facial features have been reconstructed in lime plaster, and his eyes have been replaced with seashells. On the right we see a three-foot-tall plaster statue of an ancestor from Ain Ghazal, Jordan. This statue should be compared to the wooden ancestor figures made by the Nootka (Figure 6).

could be removed. Groups of skulls were then buried elsewhere. One pit at Ain Ghazal contained four skulls whose features had been reconstructed with lime plaster.

Ain Ghazal had kilns for converting limestone into plaster, much of which was undoubtedly used to cover walls. At least some of it, however, was used to make plaster busts and statues of the ancestors, some as tall as three feet. The statues were given a pink coating (perhaps diluted ocher), and the black pupils of their eyes were drawn in tar from an asphalt seep (Figure 17, right). Gaunt and spooky-looking, the statues of Ain Ghazal remind us of the Nootka ancestor statues shown in Figure 6.

Finally, let us travel east to the Boğazçay River, a tributary of the Tigris, where it flows out of the Taurus Mountains near the city of Ergani, Turkey. Nine thousand years ago the Ergani region was a savanna-woodland with oak, pistachio, almond, and tamarisk trees overlooking a grassy valley floor. Here, some 2,700 feet above sea level, lay the ancient village of Çayönü, which has produced one of the most detailed series of ritual buildings in the early Near East. These buildings have been investigated by several generations of archaeologists, including Robert J. Braidwood, Halet Çambel, and Mehmet Özdoğan.

The earliest houses at Çayönü, built nearly 10,000 years ago, were oval and semi-subterranean. The lower walls were made of stone slabs and the domed roofs were made from reeds daubed with clay. Çayönü at that time subsisted on a combination of wild and domestic wheats, field peas, chickpeas, and lentils; its meat came from hunting wild pigs, sheep, goats, cattle, and other animals. Çayönü later went over to large rectangular houses with mud or mudbrick walls, set on grill-like foundations that kept the houses dry by allowing air to circulate beneath them.

The oldest identifiable ritual house at Çayönü, nicknamed the Bench Building, resembles a typical men's house: a single room 14 to 16 feet on a side, with a floor of clean sand and massive stone benches running along three of its four walls. The second ritual structure, called the Flagstone Building, was reminiscent of the men's houses at Nevali Çori. Its floor was paved with flagstones, except where two large stone monoliths rose vertically to support the flat roof. Instead of a series of pillars around the periphery, however, the builders of the Flagstone Building had added buttresses to its dry-laid stone masonry walls. This allowed the 35-by-18-foot building to be spanned by wooden roof beams.

The third ritual building at Çayönü resembled the apsidal men's house from Abu Hureyra: a rectangular building with a crescent-shaped room at one end, serving as a charnel room (Figure 16, bottom). Stored in a pit in the apse were the skulls of 70 people, giving the structure its nickname, the Skull Building.

The rectangular part of the building had a large central room with sitting benches on two sides. In front of one bench a large stone table or altar rested on the lime-plastered floor. Between this room and the apse were three small rooms described as "cellars," some of which were stacked high with additional skulls and human bones. In all, the remains of some 400 deceased individuals had been curated in the Skull Building.

Two other details of the Skull Building are noteworthy. First, a forensic analysis of residue on the surface of the stone table/altar revealed crystals of both human and cattle hemoglobin. This suggests rituals involving the shedding of human and cattle blood. The second noteworthy detail is that this ritual building shows signs of massive burning. We have already noted that in many societies, men's houses were the prime targets of enemy raids.

Perhaps 8,700 years ago Çayönü underwent a number of transformations. Its economy now included both cereal cultivation and the herding of sheep and goats. Archaeologists suspect that the occupants had added second stories to their residences, requiring a sturdy foundation of stones set in clay mortar. The lower floor was often divided into eight to 12 small storage rooms. These storage units were connected to each other by crawl holes, and to the upper story by hatches. An extended family could live in the upper story and produce as much surplus grain as it chose to, secure in the knowledge that its neighbors could not see how much was being stored in the cells of the lower floor. The stage was set, in other words, for an ambitious household to outstrip its fellow farmers. Then, like the high achievers of Assam and New Guinea, the household head could use his surplus (and that of his relatives) to build renown.

There are signs that such differences in surplus and renown were indeed emerging at this point in Çayönü's history. The village now had a cleared plaza, 165 by 83 feet in extent, not far from its ritual house. The residences north of the plaza were among the largest in the village and contained the most prestigious valuables. The residences to the west of the plaza were smaller and showed little evidence of luxury items.

During this period the leaders of Çayönü directed the building of a new structure called the Terrazzo Building, roughly 40 feet long and 30 feet wide (Figure 18). Its corners were aimed at the cardinal points. The thick stone walls were given decorative buttresses. The floor that gave the building its name was made of thousands of red stone chips, set in clay and polished. Forensic analysis once again revealed crystals of human hemoglobin, this time on the rim of a heavy circular basin.

The Terrazzo Building had no sitting bench, no charnel room, and no curated remains of ancestors. In our opinion it may signal an important transition in the development of Near Eastern society: the shift from societies with men's houses to societies with actual temples. The implications of this transition require so much discussion that we must defer them to a later chapter.

FROM FORAGING TO ACHIEVEMENT-BASED SOCIETY IN THE MEXICAN HIGHLANDS

During the late Ice Age the climate of the central Mexican highlands was cooler and drier than today's. Temperatures began to rise 10,000 years ago, covering many mountain valleys with a thorny forest of organ cactus, acacia, and mesquite trees. The understory of this forest was a wonderland of edible plants.

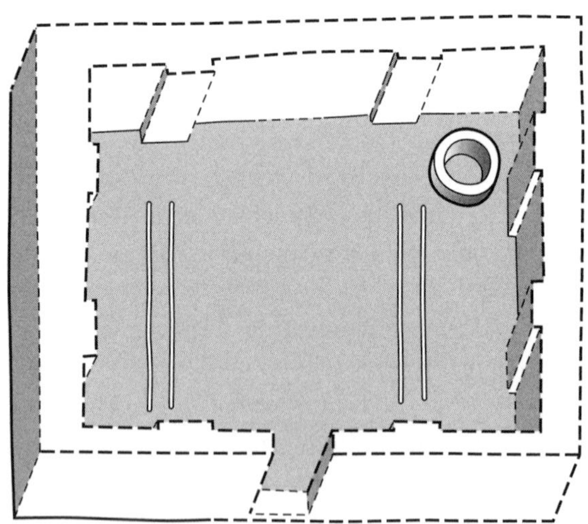

FIGURE 18. The Terrazzo Building at Çayönü, Turkey, measured 40 by 30 feet. It was rectangular, and its corners faced the cardinal directions. This ritual building, where traces of human hemoglobin were found on a heavy circular basin, dates to 8,700 years ago. It may fall near the transition from men's houses to temples.

Foragers in highland Mexico had an immediate-return strategy. Rainy summers and dry winters forced them to move their camps often. During lean seasons they dispersed into family-size bands of four to six persons. In seasons of abundance families came together to form camps of 25 to 30 persons.

Since pottery had yet to be invented, Mexican foragers used bottle gourds to transport water. Their familiarity with gourds led them to recognize their relatives, the wild squashes, as potentially cultivable plants. Soon they were growing squash for its protein-rich seeds and its useful, gourd-like rind. (Wild squashes have no edible flesh, and some species even have a bad odor.)

It is at this point that we can see major differences between the Natufians of the Near East and the so-called Archaic hunter-gatherers of highland Mexico. Wild wheat, barley, and goat-face grass grew so densely, and provided so much carbohydrate, that they could sometimes support life in villages. Squash and gourds could not support such a sedentary lifestyle.

Little by little, however, the nomadic foragers of highland Mexico began increasing their cultivation of plants. Some 8,000 years ago they added beans, tomatoes, and chile peppers. Most of their domesticates were weedy, resilient plants that could be grown in floodplains or humid canyons. We believe that they were also tending wild fruit trees such as the avocado, whose seeds appear in Archaic archaeological sites.

Two sites in the Valley of Oaxaca, 250 miles south of Mexico City, provide us with evidence that Archaic rituals were held mainly during seasons of abundance, when numerous families camped together. Gheo-Shih was a camp made by an estimated 25 to 30 persons during the summer rainy season, when gourds and squash could be planted and mesquite pods and hackberry fruits could be harvested. It lay on a river floodplain at an elevation of 5,400 feet above sea level. At the base of a cliff a mile and a half to the north, at an elevation of 6,400 feet, lay the small cave of Guilá Naquitz. A family of four to six people camped there during the early dry season, when acorns and piñons were ready for harvest. Both sites were occupied 8,600 years ago.

No evidence of ritual showed up in the cave. Gheo-Shih, however, appears to have been one more example of a camp where foragers arranged their shelters around an open area devoted to ritual. In this open area excavator Frank Hole found a rectangular space 65 feet long and 23 feet wide, delimited by two parallel rows of boulders (Figure 19, top). This ritual feature, resembling a small dance ground, had been kept virtually clean; to either side lay abundant debris from Archaic shelters. It is therefore likely that certain rituals were held on an ad hoc basis, whenever enough people lived together to make it worthwhile.

Archaic foragers may already have been performing some of the blood sacrifice for which later societies in Mexico became famous. Our best evidence for such sacrifice comes not from Oaxaca but from Coxcatlán Cave in the Tehuacán Valley, roughly 100 miles northwest of Gheo-Shih. In Level XIV of

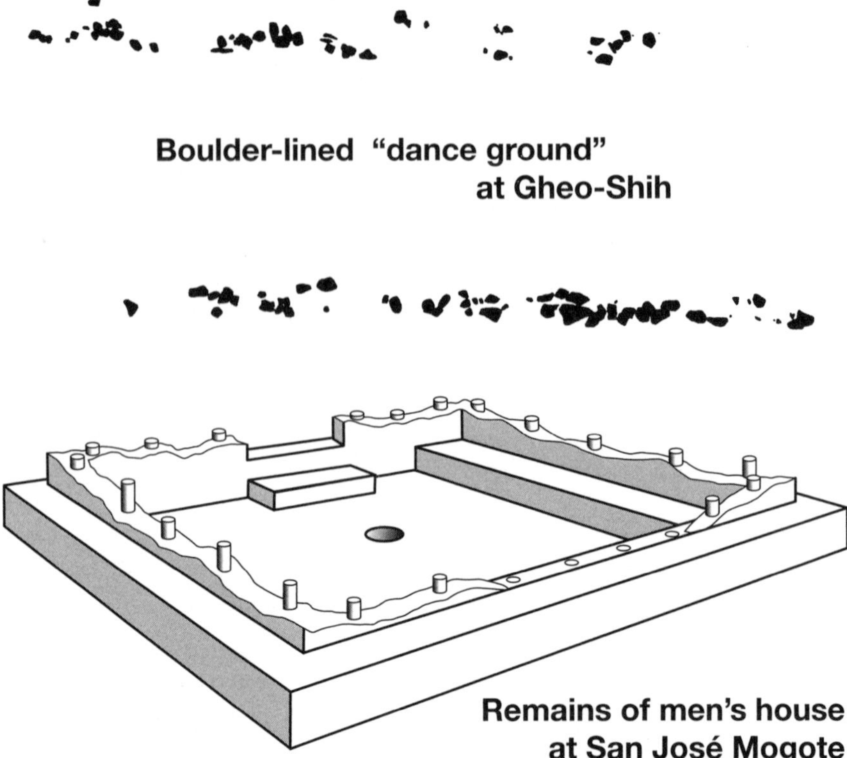

FIGURE 19. During the long transition from foraging to agriculture, societies of highland Mexico displayed a number of behaviors in common with societies of the early Near East. One was the practice of arranging huts or shelters around an open area devoted to ritual. Another was the building of men's houses. Above we see traces of a 65-foot-long ritual area from Gheo-Shih, Mexico, defined by two parallel lines of boulders. Below we see the ruins of a men's house from San José Mogote, Mexico, with a sitting bench and a pit holding powdered lime. Such buildings averaged 20 by 13 feet.

that cave, the remains of a camp occupied 7,000 years ago, archaeologist Richard MacNeish found two sacrificed and cannibalized children. One decapitated child was wrapped in a blanket and net, its skull lying in a basket nearby. A second decapitated child, also wrapped in a blanket and net, had its skull placed beside it. This skull had been burned or roasted, scraped to remove the flesh, and broken open so that the brains could be eaten. These children were accompanied by nine to ten baskets containing the desiccated remains of plants. This hints that child sacrifice may have accompanied rituals of thanks for a good harvest.

None of Mexico's earliest domestic plants were sufficiently productive to modify the nomadic lifestyle of Archaic society or support social units larger than the extended family. That situation would change with the domestication of an unassuming wild grass called teosinte. Teosinte looks superficially like maize, or Indian corn, except it has no cob. It has instead a single row of kernels in very hard shells called fruitcases. Anyone attempting to eat teosinte would either have to crush the fruitcases in a mortar or explode the kernels like popcorn.

Geneticists believe that the race of teosinte first cultivated by Archaic foragers was native to the Balsas River drainage in the Mexican state of Guerrero. The founding population of domestic teosinte may have been no more than 600 plants. Its cultivators took note of a mutation that softened the fruitcase, making it easier to get at the kernels. Another mutation doubled the kernel rows, creating a tiny cob the size of a cigarette filter. Still other mutations improved the taste by increasing the starch and protein content of the kernels. By selecting plants with favorable mutations, early cultivators from Guerrero in the west to Oaxaca in the east developed corn out of teosinte.

Some 6,250 years ago a small group of visitors to Guilá Naquitz Cave discarded tiny corncobs bearing two rows of kernels. By saving the best of each harvest to plant, the Indians of central Mexico encouraged the cobs to grow in length and the kernel rows to double, first to four and then to eight. Archaic gardeners at last had a carbohydrate source that could be stored for months, which allowed them to stay longer in their camps.

As corn became increasingly productive, large rainy-season camps in Mexico began to look more like Natufian settlements. On a stream terrace in Atexcala Canyon, which enters the Tehuacán Valley from the west, Archaic campers created a settlement of oval, semi-subterranean houses. One house, reported by Richard MacNeish and Angel García Cook, had been excavated two feet into the terrace and measured 17 by 13 feet. The roof featured a central ridge pole, supported by two vertical posts set in the floor and a series of smaller posts slanting in from the sides. The builders used stone slabs to reinforce the subterranean part of the house and created a shallow hearth in the floor. Scattered around the house were stone mortars and grinding slabs for reducing kernels to cornmeal. The Atexcala Canyon camp was occupied about 4,500 years ago.

As long-term camps grew into villages, families became less likely to disperse during the dry season. While women, children, and the elderly stayed at home, small groups of men went on hunting trips and returned with deer. One such all-male camp was made in Cueva Blanca, a cave in the Oaxaca mountains not far from Guilá Naquitz.

Despite the changes brought about by agriculture, we can see at Cueva Blanca the same weapon-sharing behavior we saw among the Basarwa hunters of the Kalahari Desert. Each hunter at Cueva Blanca seems to have made his own stylistically distinct flint points for the darts launched by his atlatl, or spear-thrower. He then evidently exchanged darts with his hunting partners, much the way the Basarwa exchanged arrows. Within the cave, a variety of distinctive points was found in each hunter's work space. There were even dart points left behind in Cueva Blanca that appear to have been made in the Tehuacán Valley, which was a journey of four or five days to the north. This evidence suggests that some system of reciprocal exchange, analogous to hxaro among the !Kung of Botswana, existed in the Mexican Archaic.

Some 3,600 years ago highland Mexico experienced a transition similar to the one we saw earlier at Çayönü and Abu Hureyra. Encampments of semisubterranean oval huts gave way to permanent villages of rectangular houses. These early Mexican houses had a framework of pine posts, a thatched roof, and walls of cane bundles plastered over with clay. Each house measured 10 by 17 feet, sufficient for a nuclear family, and each was surrounded by an outdoor work area with storage pits and earth ovens. Highland village life was supported by a combination of agriculture, wild plant collecting, the hunting of deer and rabbits, and the raising of dogs as an additional meat source. Villagers now made pottery and were active in the circulation of valuables such as mother-of-pearl.

Like many New Guinea societies, these early villages kept men's and women's rituals separate. The household was the woman's ritual venue. There she made small ceramic figurines of the ancestors that could be arranged in ritual scenes. These figurines probably provided a physical body to which the spirits of the deceased could return, while they were ritually offered sustenance and petitioned for favors. The venue for the men's ritual was separate from the residences and in some villages included what appear to be men's houses.

San José Mogote, on the Atoyac River in the Valley of Oaxaca, was a village estimated at 150 to 200 persons. There was a palisade of pine posts on the western edge of the village, and several buildings showed signs of having been burned. This evidence for intervillage raiding suggests a society featuring clans or descent groups and Kelly's principle of social substitutability.

The men's houses at San José Mogote averaged 20 by 13 feet, seemingly too small to serve as dormitories; they were more likely to have been restricted to the fully initiated. These ritual venues typically contained two to three times as many posts as ordinary residences. Each was built on a low platform into which its floor was recessed. While the walls were built of cane bundles daubed with clay, the builders had coated the floors and walls with lime plaster. In cases where the walls were preserved to a sufficient height, one could see that some had sitting benches running along them (Figure 19, bottom). Each men's house was given the same orientation, eight degrees north of east, undoubtedly an alignment with ritual significance. Fragments of ceramic masks, presumably parts of ritual costumes, were found in and around the buildings.

Built into the center of the floor was a storage pit filled with finely powdered lime. Based on what we know about later Indian societies in Oaxaca, we suspect that this powdered lime was for mixing with a ritual plant such as wild tobacco, jimson weed, or morning glory. Wild tobacco, finely ground and mixed with powdered lime, was believed to increase men's physical strength, making it an appropriate drug to use before raids.

During this period most people were buried in the fully extended position. Three burials, however, received different treatment. All were middle-aged men buried in a seated position, their limbs so tightly bent as to suggest that the corpse may have been placed in a bundle and kept around for a time before burial. Two of these seated men were buried near a men's house. This pattern suggests that, just as in so many achievement-based societies, certain men had earned the right to be treated differently in death.

Because it required so many genetic changes to convert teosinte into truly productive maize, highland Mexico took longer than the Near East to produce sedentary, achievement-based villages with clans or descent groups. Once that type of society had arisen in Mexico, however, it displayed many of the same social institutions as the Near East: men's houses, ancestor ritual, intervillage raiding, interregional exchanges of shell valuables, and ways of recognizing prominent individuals after death.

Achievement-based societies characterized highland Mexico until roughly 3,150 years ago. At that point the archaeological record begins to show signs of hereditary social inequality. Not long after that, men's houses were replaced by temples with evidence of blood sacrifice. The implications of this social transformation will be discussed in a later chapter.

FROM FORAGING TO ACHIEVEMENT-BASED VILLAGE SOCIETY IN THE CENTRAL ANDES

Over the years archaeologists have come up with countless theories that use the natural environment to explain the rise and fall of ancient societies. Peru is the graveyard for all those theories.

Peru's desert coast, where half an inch of rainfall would be considered a wet year, gave birth to precociously complex societies. The same is true of the frozen tundra of Peru's altiplano, 12,500 feet above sea level. Spectacular sites can be found in canyons so narrow that when visitors stretch out their arms, they fully expect to touch the cliffs on either side. They can also be found on the tropical eastern slopes of the Andes, where rivers carrying the water from melting glaciers descend to the Amazon jungle.

By the time the Ice Age ended, Peru's Archaic foragers had created several alternative lifeways. Societies on the desert coast took advantage of the Humboldt current, an upwelling of nutrient-rich water that supports one of the world's great fisheries. So abundant were fish and shellfish that some parts of the coast could support encampments as impressive as those of the Natufians. Many of the campers left behind enormous heaps of mollusk shells, fish and sea lion bones, crab claws, and sea urchins.

One large shell heap near the southern Peruvian city of Ilo formed a ring 85 feet in diameter. This strongly suggests that the Archaic people had arranged their shelters in a circle, just as some Mexican and Near Eastern foragers did. First occupied more than 7,500 years ago, the Ring site took on its distinctive form 5,000 years before the present.

Foragers in the Andean highlands encountered a different set of resources. Armed with spear-throwers, they stalked game through wooded valleys and high-altitude meadows of bunch grass. At elevations of 8,000 to 10,000 feet they pursued the white-tailed deer and the guanaco, a wild member of the camel family. At 12,000 to 14,000 feet they stalked the *taruca,* or huemul deer, and the vicuña, a smaller relative of the camel. These highland hunters lived sometimes in open-air camps, sometimes in caves and rockshelters. They slow-cooked their prey in *pachamancas,* or earth ovens, like those of Ice Age Europe.

At the start of the Archaic period Peru's hunter-gatherers seem to have had an immediate-return strategy. There is much, however, that we still do not know about them. In the Near East we can show that foragers harvested the wild ancestors of wheat and barley for thousands of years before they produced domestic varieties. In the Andes we know what the earliest domestic plants were, but we have less information on the period when their wild ancestors were being harvested. Part of the problem is that plant preservation is poor in the highlands, where many of the wild ancestors lived. An equally significant problem is that the area involved was huge, and Andean peoples had eclectic tastes. Some of the domestic plants they adopted were native to the coast; some were native to intermontane valleys; some were native to high-altitude tundra; and some came from as far away as Brazil and Paraguay.

The Pacific coast of Peru is one of the world's most extreme deserts. Thousands of square miles of the coast could not possibly have been farmed in the Archaic. At various points, however, rivers carrying water from the snow-capped peaks of the Andes descended to the sea. Each of these rivers was a green, linear oasis in the beige desert, not only creating an alluvial floodplain but even supporting marshes, seeps, and canebrakes. Once domestic plants had reached the coast, the Archaic fishermen could turn these localized patches of humid soil into gardens.

We suspect that, just as in Mexico, the first Peruvian domestic plant was the bottle gourd, followed not long afterward by squash. Early Peruvian squash did not, however, belong to the same species as the earliest Mexican squash; it belonged to a species whose wild ancestor lived in Colombia. That squash was brought from its native habitat and grown in Ecuador and Peru almost 10,000 years ago.

Cotton grew wild on the coastal plains of southwest Ecuador and northwest Peru and was soon domesticated there. The combination of bottle gourds and cotton meant that Archaic people were no longer limited to fishing with hooks and spears. They could now use nets of cotton cordage, with net floats made from gourds. Nets were particularly useful for harvesting anchovies and sardines, two small fish that were available in enormous numbers.

Many of the key Andean domestic plants were grown for their roots, bulbs, tubers, or other underground parts. Wild manioc, or cassava, is native to the Brazilian states of Mato Grosso, Rondônia, and Acre, and it was probably first domesticated there. That same region may have produced both the jack bean *(Canavalia)* and a species of chile pepper, different from the peppers domesticated in Mexico. Peru welcomed those foreign plants. The peanut grows wild in forest and savanna environments between southeast Bolivia and northern Paraguay. It may have reached the Peruvian coast 7,000 years ago.

The high altiplano of Peru is a land rich in edible tubers, protected from frost by their below-ground development. The best known of these is the wild potato, probably domesticated somewhere in the southern highlands. Potatoes and manioc gave Andean farmers the carbohydrate source they needed for sedentary life. Another product of the altiplano was a weedy plant called quinoa, whose tiny seeds could be ground into flour. It reached the Pacific coast during the Archaic period.

Some plants from the Amazon jungle and the high tundra of the Andes reached the north coast of Peru even before they had undergone sufficient genetic change to be recognizable as domestic varieties. The only reason we consider them cultivated plants is because they had clearly been brought hundreds of miles from their native habitats. The spread of such crops suggests that Archaic foragers traveled long distances and had trading partners with whom they eagerly exchanged products.

While all these domestic plants were spreading to the coast, another domestication process was taking place in the highlands. Hunters who for thousands of years had pursued, surrounded, and driven the guanaco and vicuña into cul-de-sacs were beginning to pen up captive animals and tame their young. Archaic highlanders built corrals in caves the way Natufians had built circular huts in caves.

One of the oldest corrals was created more than 4,000 years ago by building a stone wall across the mouth of Inca Cueva 7, a small cave in the *puna,* or high-altitude tundra, of the Argentinian Andes. The penned-up animals had left pellets of their dung on the floor of the cave. Tukumachay, a cave in the puna near Ayacucho, Peru, had a corral roughly as old as the one at Inca Cueva 7.

Corrals were not limited to caves, of course. Archaeologist Mark Aldenderfer excavated an Archaic campsite on a terrace of the Osmore River, 50 miles inland from the aforementioned Ring site. Asana, located almost 9,500 feet above sea level, had been repeatedly used as a campsite for thousands of years by the hunters of guanaco and huemul deer. As long as 6,800 years ago, the hunters at Asana were building small oval structures that may have been ritual sweat houses. The best preserved of these structures had a clay floor, a basin to hold water, and a mass of fire-cracked rock. In such a sweat house, water was poured over heated rocks to create a sauna for rituals of purification.

In a later camp at Asana, occupied at roughly the same period as Inca Cueva 7, the campers built an apparent corral of wooden posts. Forensic analysis of the soil inside the corral revealed the breakdown products of animal dung, although individual dung pellets could not be recognized.

Some of the most compelling evidence for keeping such animals captive was found by archaeologist Jane Wheeler at a cave called Telarmachay. This cave lay in the tundra near Lake Junín in central Peru. The evidence consists of very high levels of infant mortality, typical of domestic herds in the centuries before modern veterinary medicine.

Eventually, two domestic members of the camel family appeared in Peru. One, the llama, was a beast of burden derived from the guanaco. The other, the alpaca, was raised primarily for its fine wool, which according to recent DNA evidence reflects some vicuña genes. Both animals may have assumed their present form more than 3,600 years ago.

The emergence of the llama as a pack animal, in particular, made possible a dramatic escalation in the movement of products from region to region. One could load a llama caravan with highland potatoes and descend to the Pacific coast, where the potatoes would be exchanged for dried anchovies. Such exchanges made every product available to every region.

While all of this was taking place, maize was slowly spreading from group to group on its way south from Mexico to Peru. By the time it reached the north coast of Peru, perhaps 4,000 years ago, it was simply one more plant added to the potatoes, manioc, squash, beans, peanuts, and other crops. The Archaic Peruvians did not see corn as a staple but as a plant whose sugary kernels could be turned into *chicha,* or maize beer. From that point on, chicha took on the ritual importance that rice beer did among the Naga.

From Camps to Achievement-Based Villages with Ritual Houses

Coastal Peruvian societies underwent a transformation similar to that seen in Mexico and the Near East. From seminomadic societies living in circular, semi-subterranean huts, they developed into achievement-based societies living in autonomous villages of rectangular houses.

Two and a half miles inland from San Bartolo Bay on the central Peruvian coast lay the Archaic site of Paloma. Here Robert Benfer and Jeffrey Quilter discovered a semipermanent settlement of 30 to 40 people living in shelters like those at Ohalo. The people of Paloma tended gardens of squash, beans, and gourds during the time period of 6,500 to 5,000 years ago, but the real staples of their diet were anchovies and sardines, augmented by larger fish and shellfish.

The occupants of Paloma salted the corpses of their ancestors to slow down decay. They buried their dead with limbs tightly flexed, wrapped in reed mats. An analysis of 200 skeletons revealed a disproportionately high frequency of men, suggesting that female infanticide may have been practiced. The men had a high incidence of inner and middle ear damage, probably caused by a lifetime of diving for mollusks in the cold ocean water.

Nine miles from Paloma, the Chilca River empties into the Pacific Ocean. Two miles inland lay another Archaic camp, this one occupied 5,500 to 4,500 years ago. At the Chilca camp, archaeologist Christopher Donnan discovered a number of conical huts. While most were made of canes, the upright posts of at least one hut were rib bones salvaged from a beached whale. The families living at Chilca caught their fish with cotton nets or with hooks carved from mollusk shells. They also took advantage of the high water table near the river, creating small vegetable gardens by digging down to the humid soil just below the surface.

By this time the rugged canyons through which many Peruvian rivers approached the coast had been recognized as optimal for small-scale irrigation. In each canyon there were places where farmers could divert water into small canals that descended in elevation more slowly than the river itself, allowing them to irrigate most of a river terrace. The village of La Galgada lay in just such a location.

One of the mightiest watercourses on the north coast of Peru was the Santa River. Fifty miles from the ocean its flow was augmented by a tributary called the Tablachaca River, which carved its way through a dusty canyon 3,300 feet above sea level. Roughly 4,400 years ago the village of La Galgada was founded on a terrace of the Tablachaca, and its families began irrigating gourds, squashes, chile peppers, common beans, jack beans, lima beans, peanuts, cotton, and orchard crops such as guavas and avocados. Corn had arrived but was still too rare to be considered a staple. The villagers also hunted deer and traded for dried fish and shellfish from the coast.

While most crops grown at La Galgada were foods, the wealth of the village was based on cotton. According to C. Earle Smith Jr., the botanist who analyzed its plant remains, "the volume of cotton recovered at La Galgada indicates that it must have been a crop produced for export." There was reason to believe that La Galgada had also added value to those exports by converting the cotton into textiles before sending it on.

According to archaeologists Terence Grieder and Alberto Bueno Mendoza, the villagers of La Galgada lived in circular houses large enough to accommodate a nuclear family. The walls were made of stones, set in mud mortar, and the peaked roofs were thatched with grass. The fact that most residences preserved the circular plan of earlier times made it easy to recognize ritual houses at La Galgada: the latter were rectangular, with slightly rounded corners at first (Figure 20, top), which then became fully rectangular over time.

The La Galgada ritual houses were small, and it appears that several might have been in use at any one time. Their accumulated remains created two earthen mounds, called the North Mound and the South Mound. The North Mound began growing first; the South Mound began growing later, and from that point on the two were occupied simultaneously. In both mounds there were ritual houses painted pearly white with a mineral called talc.

We see at least three possible scenarios for La Galgada's multiple ritual buildings. In one scenario La Galgada might have maintained multiple men's houses for the same reason the Mountain Ok did, to provide venues for potentially contradictory parts of their cosmology. In a second scenario there might have been multiple lineages or descent groups at La Galgada, each of which built its own men's house. In a third scenario men and women might have maintained separate ritual venues as the Ok did. The data do not permit us to decide among these alternatives.

The earliest ritual houses at La Galgada were only about seven by nine feet in size. The walls, which in one case had survived to a height of five feet, were of broken fieldstones set in mud mortar, plastered over with clay and painted white. Some walls were decorated with rows of niches, and the roofs were of acacia poles plastered with clay. Each floor consisted of a rectangular sunken area, surrounded by a bench wide enough for sitting or sleeping. A crucial feature of the sunken area was a circular hearth, connected to the outside world by a vent below the floor. This vent would have provided oxygen to keep the hearth burning, even if the building's door was closed to maintain ritual secrecy. Archaeologist Michael Moseley suspects that the sunken floors and hearths reflect a widespread Andean cosmology, one in which the first humans ascended to the earth's surface through caves, springs, and holes in the ground.

Later ritual houses in the North and South Mounds, built perhaps 4,000 years ago, continued to feature benches, sunken floors, and central hearths similar to the earlier versions. These buildings, however, had small variations

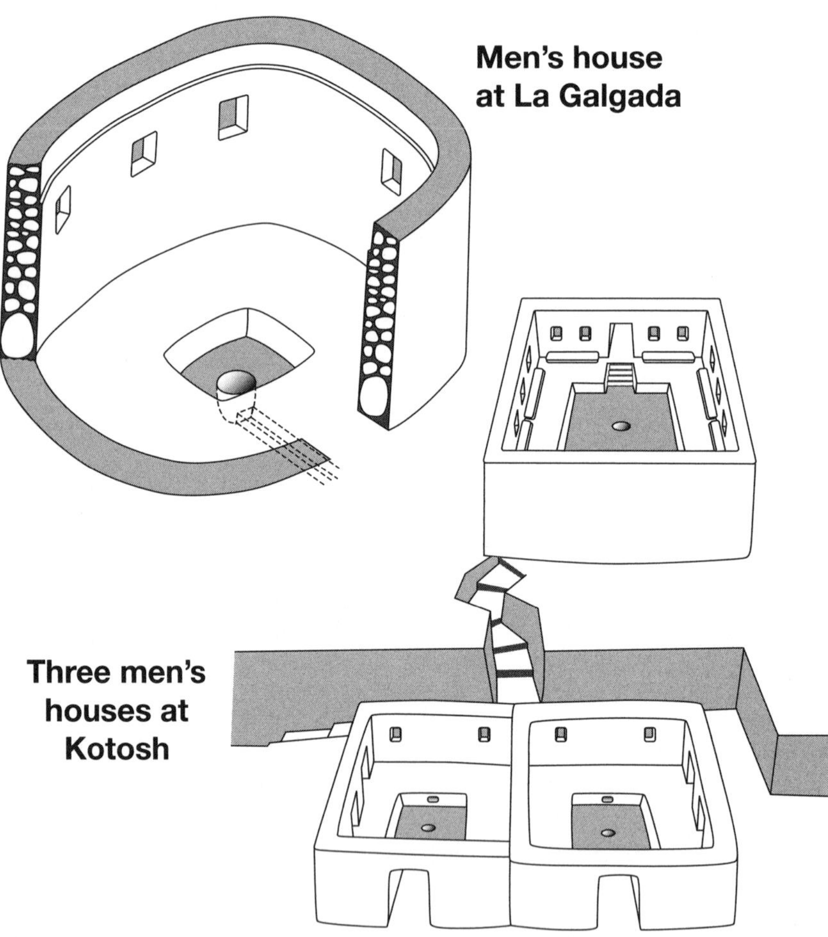

FIGURE 20. Early men's houses in Peru were often painted white, had decorative wall niches, and featured sitting or sleeping benches around a sunken floor with a hearth. Above we see a seven-by-nine-foot men's house from La Galgada. Below we see three men's houses from Kotosh, the largest measuring 30 feet on a side.

in wall decoration, either because they were built by different social units or because they addressed different ritual needs.

Left behind on the floors of some ritual houses were the downy feathers of white, green, and orange tropical birds, obtained in trade and probably used in costumes or body decoration. Among the burnt offerings found in the central hearths were carbonized chile peppers. This discovery leads us to hope that those attending the ritual did not inhale.

Once each ritual house fell into disuse it was burned and then used as the final resting place for the bundled remains of men, women, and children. Many of the corpses were wrapped in cotton textiles or sleeping mats, and a few were supplied with cotton bags bearing the designs of birds and snakes. Some burials were provided with gourd bowls or stone cups, and others were ornamented with bone pins inlaid with turquoise. Each abandoned ritual house was then deliberately filled with earth to the height of its surviving walls.

Do the diverse burials in the ritual houses at La Galgada imply that both men and women used these buildings? Might they have been analogous to the kivas of Pueblo Indian societies? Not necessarily. For all we know, the women and children buried there might simply have been family members of male lineage heads. Many features of the La Galgada buildings—the benches, the sunken floors, and the pearly white plaster—strike us as being similar to those of men's houses elsewhere. The deliberate post-abandonment filling with earth reminds us of the ritual buildings at Göbekli Tepe, and the bundled human remains remind us of the charnel rooms at Abu Hureyra and Çayönü.

Ritual houses similar to those of La Galgada have been found 9,000 feet above sea level at Huaricoto, on a different tributary of the Santa River. At Huaricoto, archaeologists Richard Burger and Lucy Salazar-Burger found small ritual chambers that also had sitting or sleeping benches, central fire pits, and ventilator shafts below the floor to provide oxygen for burnt offerings.

Traveling east from Huaricoto, one would cross the crest of the Andes and begin a long descent toward the Amazon basin. Some 6,500 feet above sea level one would reach the Higueras River, still in the highlands but only 25 miles from the tropical slopes of the eastern Andes. Three miles from the Peruvian city of Huánuco, on a terrace of the Higueras, lies an early village with ritual buildings that could be men's houses.

The site of Kotosh, excavated by Seiichi Izumi and Toshihiko Sono, was founded 4,000 years ago. Like La Galgada, it had more than one ritual house in use at a time. The ritual buildings at Kotosh were rectangular and somewhat larger than those at La Galgada. Built of stone masonry plastered over with clay, they had the same central fireplace with an underground ventilator shaft, the same sunken floor surrounded by sitting or sleeping benches, and similar niches decorating the walls.

The oldest ritual house at Kotosh, which we will call the White Building, had an earlier and later version sitting side by side (Figure 20, bottom). Upslope from the White Building and connected to it by a narrow, twisting stairway was a second ritual house. This cobble masonry building was square with rounded corners and roughly 30 feet on a side; it was large enough to have been a dormitory-style men's house. Its wall decoration included rows of ornamental niches, as well as a clay frieze depicting a pair of crossed human forearms. This unusual frieze gave the structure its nickname, the Building of the Crossed Hands.

Now consider the wide altitude range covered by the societies to which these three villages belonged. La Galgada grew cotton and received shipments of fish from the nearby coast. The region of Huaricoto was too high and cold for cotton production, but it could have received cotton textiles from lower-altitude villages analogous to La Galgada. Villages founded near the tropical forest, such as Kotosh, could have provided villages like Huaricoto and La Galgada with the feathers of tropical birds. Such was the network of interactions among early village societies in the Andes.

Roughly 3,500 years ago, the ritual houses at Kotosh and La Galgada were replaced by actual temples. The main temple atop the North Mound at La Galgada, reached by a long, narrow stairway, took the form of a giant U. Much larger than La Galgada's earlier ritual houses, it could have accommodated 50 people. This architectural change, to paraphrase Michael Moseley, reflects a shift from small, private rituals to larger, more public performances. It is a change seen also in Mexico and the Near East, and we will consider its more universal implications later in this book.

NINE

Prestige and Equality in Four Native American Societies

The early village societies of Mexico, Peru, and the Near East went on to develop hereditary rank and never looked back.

Not every society with achievement-based leadership, however, underwent such a transformation. Many agricultural village societies resisted every attempt to increase inequality. They found a way to let talented people rise to positions of prominence while still preventing the establishment of a hereditary elite. The balance they struck between personal ambition and the public good allowed their way of life to endure for centuries.

Some of the best known of these societies were the Tewa, Hopi, Mandan, and Hidatsa of North America. In this chapter we look at their prehistoric origins and reflect on the balance of prestige and equality they were able to achieve.

AGRICULTURE AND VILLAGE LIFE IN THE SOUTHWEST UNITED STATES

We have seen that maize, or Indian corn, passed from group to group for thousands of years as it made its way south from Mexico to Peru. Maize also passed from group to group on its way north through the sierras of western Mexico, accompanied on its journey by squash and beans.

The Mogollon highlands on the Arizona-New Mexico border proved receptive to these Mexican plants. This is a mountainous region 4,500 to 6,500 feet above sea level, where rocky canyons alternate with woodlands of juniper and piñon. The Native Americans who lived there had a long tradition of harvesting seeds and nuts, a lifeway into which Mexican seed crops were quickly accommodated.

Roughly 2,800 years ago a group of foragers camped at Bat Cave, an opening in a volcanic cliff above New Mexico's San Augustine Plains. The group harvested piñons, walnuts, juniper berries, prickly pear cactus fruits, and dozens of other local plants. Mixed in among the wild foods were squash seeds, beans, and fragments of corncobs. According to archaeologist W. H. ("Chip") Wills, these Mexican plants were more than a thousand miles from their native habitats and must, therefore, have been locally grown.

Corn, beans, and squash had reached the Southwest, but owing to the region's aridity it would take centuries for them to bring about sedentary life. At first, cultivated plants were only a supplement to traditional wild resources such as piñons and jackrabbits. Rockshelters remained popular places to spend the night and store food. Little by little, however, the natives of the Southwest began to create encampments similar to those of the Natufians. They built circular semi-subterranean shelters and created storage pits lined with grass or basketry.

In time these encampments gave way to more permanent villages, featuring semi-subterranean houses lined with stone slabs. Two archaeological sites in western New Mexico, both occupied 1,500 years ago, show us some of the regional diversity. Shabik'eschee Village near Nageezi, New Mexico, had more than 60 semi-subterranean houses. They ranged from circular to rectangular, suggesting that the Southwest was going through a change in house shape similar to that seen in Mexico, Peru, and the Near East. The fact that the storage pits lay outside the houses suggests that harvests had not yet been privatized.

It is significant that Shabik'eschee also had a largely subterranean, one-room building with a bench running around the wall, the Southwest version of an early ritual house. This is circumstantial evidence for the emergence of social units larger than extended families.

The SU (pronounced "Shoe") site near Reserve, New Mexico, provides a contrast. SU had roughly 40 semi-subterranean houses, some with more than 800 square feet of floor space. These larger houses could easily have accommodated whole families. It also appears that SU families had privatized their harvests by constructing storage pits inside the house. The pits were larger than those at Shabik'eschee and could, on average, have held more than 500 pounds of corn. Wills has calculated that this amount would supply a family of five with enough corn for three months.

What we may see at SU is the same behavior we saw at Çayönü late in its history: families building larger houses, keeping quiet about how much food they were storing, and gearing up to outproduce their less industrious neighbors. SU, like Shabik'eschee, had a ritual building that was larger than most houses.

At this point it appears that the Southwest had developed politically autonomous villages with clans or descent groups, analogous to those we saw in Mexico, Peru, and the Near East. It is no surprise, therefore, that some villages engaged in raiding. Archaeologist Steven LeBlanc describes warfare as "endemic" in the Southwest 1,500 to 1,000 years ago. Some villages relocated to steep defensible ridges or mesas. Others surrounded themselves with palisades of wooden posts.

Warfare in the Southwest seems to have involved both ambushes and direct confrontations. Some groups of burials, according to LeBlanc, suggest that male victims had their skulls crushed with clubs; the killers may have spared young women, however, much as the Marind of New Guinea did. In other cases, probably ambushes, the women were killed along with the men. In several dry caves occupied 1,500 years ago, archaeologists found men buried with trophies of the enemies they had slain: human scalps, preserved by desiccation in the desert environment.

Sometimes ritual cannibalism was added to the skull cracking and scalping. A 900-year-old village near Mancos, Colorado, provides us with an example. There, biological anthropologist Tim White discovered that nearly 30 men, women, and children had been butchered and cooked, presumably following a massacre. In other words, the evidence from the Southwest is consistent with what we know of achievement-based societies elsewhere in the world, with the exception that scalps were more often collected than heads.

Over time the kind of rectangular, above-ground architecture that we associate with the historic Pueblo villages of the Southwest began to appear in the archaeological record. Between 1,240 and 1,140 years ago, in the region of Dolores, Colorado, semi-subterranean houses had given way to apartment-like blocks of above-ground rooms. In some of these blocks, large residential rooms were lined up in front of even more substantial storage rooms. Virtually the only remaining circular structures appear to have been for ritual; they were forerunners of the kivas built by historic Pueblo clans.

According to archaeologist Stephen Plog, the Southwest reached its maximum prehistoric population about 900 years ago, and then it began to decline. This period of decline, with many villages being abandoned and others accepting refugees, has provided archaeologists with insights into the origins of the historic Pueblo societies. For example, the region of Black Mesa, Arizona, was abandoned 900 years ago and repopulated 250 years later by people who may have been the ancestors of the historic Hopi. Later in this chapter we look at the Hopi village of Old Oraibi.

Old Oraibi was considered one of the Western Pueblos. The Eastern Pueblos are believed to have had separate origins. For example, many archaeologists suspect that when the spectacular cliff dwellings of Mesa Verde, Colorado, were abandoned, their former occupants moved south and east toward the headwaters of the Río Grande in northern New Mexico. There, along with other immigrants, they contributed to the creation of the Eastern Pueblos. It is probably no accident that the legendary histories of many Pueblo villages describe the arrival of groups from diverse regions. The order in which various groups arrived often determined their rights and responsibilities.

One of the forerunners of today's Eastern Pueblos was Arroyo Hondo, near Santa Fe, a site occupied 700 to 600 years ago. Arroyo Hondo grew to 1,200 rooms, arranged in 24 room blocks two stories tall, and had 13 open ritual plazas. After it was destroyed by fire, its population is believed to have helped create one or more of the surviving Río Grande Pueblos. Later in this chapter we look at one of these pueblos, the village of San Juan.

WHAT LEVEL OF INEQUALITY WAS REACHED IN THE SOUTHWEST?

The Pueblo societies visited by anthropologists during the nineteenth and twentieth centuries were considered both egalitarian (because everyone started out equal at birth) and achievement-based (because certain individuals achieved positions of prominence through initiation into increasingly exclusive ritual societies).

Many archaeologists, however, suspect that there was a time in the prehistory of the Southwest when Native American society experienced greater levels of inequality than those seen in historic Pueblo communities. The period in question was from 1,150 to 880 years ago, and the evidence consists of archaeological sites whose size, burial ritual, and accumulation of valuables stand out from those of their contemporaries.

One of those atypical sites is Pueblo Bonito in Chaco Canyon, not far from Shabik'eschee Village. To say that opinions on Pueblo Bonito differ would be putting it mildly. To some archaeologists, familiar mainly with sites in the Southwest, Pueblo Bonito seems as spectacular as the ancient cities of Mexico and Peru. To most archaeologists familiar with Mexico and Peru, Pueblo Bonito just looks like a big village. In this chapter we follow the thoughtful middle ground established by two experts on the region, Stephen Plog and Linda Cordell.

Chaco Canyon is 20 miles long. Within its watershed, it had large expanses of alluvial soil that could be irrigated with canals. There were pine forests on the nearby mountains for fuel and construction materials, and the canyon had unlimited sandstone for masonry walls. Archaeologist Gwinn Vivian once reported masonry dams up to 120 feet long in the region. There were at least ten ancient canals that could be traced for distances of anywhere from 2,000 feet to more than three miles. Many irrigation systems collected and channeled rain runoff, watering fields that received only nine inches of rainfall in a normal year.

The settlements in Chaco Canyon included many small hamlets that probably housed no more than one clan each. In addition, there were nine large, multiclan villages. Each hamlet had only one kiva, while the larger villages had many, reinforcing the likelihood that each clan maintained its own ritual building(s). Some of the larger villages had two clear divisions, suggesting that clans may have been grouped into opposing moieties. Large villages with this dual organization occasionally had ritual buildings 34 to 63 feet in diameter. These so-called Great Kivas may have been the scene of rituals that united all the clans within a moiety.

Let us now take an "outsiders' view" of Pueblo Bonito, which covered several acres and was four stories high, consisting of 650 stone masonry rooms with different functions. The village was shaped like a half moon, with its convex side presenting blank walls to the outside world. In the interior of the half moon lay a great open plaza divided into two sections, each with its Great Kiva. Scattered throughout the site were 36 smaller kivas, perhaps one or two per clan.

Archaeologists estimate that the walls of each room at Pueblo Bonito required 44 tons of sandstone blocks. Taken together, the roof beams and wooden floors of the nine largest villages in Chaco Canyon reflected the felling of 200,000 pine trees, sometimes from forests 70 miles away.

There are hints that the corn grown in the canyon itself would have been insufficient to feed the labor force that had built Pueblo Bonito. Chemical isotope analysis of ancient corn from the site shows that some of it was grown in regions where the groundwater had different chemicals from those of Chaco Canyon. For example, some corn from Pueblo Bonito came from the Chuska Mountains, 50 miles downstream, while other cobs came from the floodplain of the San Juan and the Animas Rivers, 55 miles to the north.

Because the large villages of Chaco Canyon were drawing on the resources of an area greater than 35,000 square miles, archaeologists were not surprised when they began finding ancient roads. Not only did these roads connect villages within the canyon, some even extended more than 60 miles outside. As archaeologists began to trace these roads, however, they found that many led nowhere. Some followed absolutely straight sight lines, even if it meant cutting steps in stone cliffs rather than following natural contours. Many archaeologists now believe that the roads were ritual, part of a sacred landscape in which cosmological landmarks were connected to centers of human occupation.

Roads and earthworks, even large ones, are well within the capacity of societies where leadership is based on achievement. But there are hints that the society of Pueblo Bonito might have had a higher degree of social inequality than the historic Southwestern Pueblos. Unfortunately, some of the evidence was recovered more than 100 years ago by avocational archaeologists who lacked many of today's excavation skills. Plog is currently compiling and reanalyzing the data from a century of work at Pueblo Bonito, allowing us to draw on his insights.

In 1896 George H. Pepper excavated burials in several rooms at Pueblo Bonito. He found 14 burials in Room 33, two of which had been placed below an unusual wooden floor. These two burials were accompanied by hundreds of turquoise pendants, thousands of turquoise beads, a conch shell trumpet, more than 40 shell bracelets, and a cylindrical basket covered with a turquoise mosaic. The 12 burials above the wooden floor were accompanied by turquoise and shell beads, bracelets, pendants, seven large wooden flutes, and dozens of wooden ceremonial staffs. In nearby rooms Pepper found burials wrapped in colored feather robes.

While the burials found by Pepper were unique, smaller amounts of valuables were found elsewhere at Pueblo Bonito. Archaeologist James Judge describes the craftsmen of Chaco Canyon as having played "an increasingly dominant role" in the processing of turquoise into ornaments for the region. The turquoise came from the Cerrillos mines near Guadalupe, New Mexico, 60 miles to the southeast. Only a mile from the mines were small villages with Chaco-style pottery, possibly places supplying the Pueblo Bonito craftsmen with raw material. Pueblo Bonito also had access to unusual amounts of other valuable items, such as copper bells, chocolate, and scarlet macaw feathers from Mexico, shell trumpets from the Gulf of California, obsidian from Jemez, New Mexico, and quantities of mica and selenite ore.

What level of social inequality was required to produce a village like Pueblo Bonito? Let us refer back to the early villages of Peru and the Near East. Pueblo Bonito's architecture was no more impressive than that of Ain Ghazal or Jericho, and Ain Ghazal covered at least ten times the area of Pueblo Bonito. The kivas of Chaco Canyon were no more spectacular than the ritual buildings of Göbekli Tepe, Nevali Çori, or Çayönü, and the irrigation canals were no longer than those of Archaic Peru.

Pueblo Bonito has evidence for two opposing moieties, each with its Great Kiva, and for the division of each moiety into a number of clans, each maintaining one or two smaller kivas. Such evidence alone, as we will see in later chapters, does not imply that leadership had become hereditary. Later in this chapter we will examine a Hopi society where one clan "owned" the most important rituals and monopolized the office of village leader. Such ritual preeminence, under the right circumstances, might enable one segment of society to accumulate shell valuables and macaw feathers.

Plog points to the luxury goods found by Pepper as potential evidence for the emergence of a Chacoan elite. This seems plausible for the huge quantities of turquoise, but the shell trumpets, flutes, and ceremonial staffs look more like items of ritual authority. Whatever the case, any emerging social inequality in Chaco Canyon was in remission 900 years ago.

Some archaeologists have attributed the decline of Chacoan society to a deteriorating climate. An analysis of growth rings in tree trunks from archaeological sites suggests that the period A.D. 1050 to 1130, when Pueblo Bonito peaked in importance, was rainier than average. The period A.D. 1130 to 1180, when Pueblo Bonito declined, was drier.

We do not dispute the climatic data. We are simply unwilling to put all the burden of explanation on the environment. Later in this book we will examine several Asian societies that created hereditary inequality, only to overthrow it periodically and return to a more egalitarian way of life. In none of these

Asian cases was a drought to blame. What happened was that a long-standing desire for equal treatment, found in most of the societies we have examined so far, periodically overcame hereditary privilege. It is possible that similar processes were at work in the U.S. Southwest.

AN OVERVIEW OF THE EASTERN AND WESTERN PUEBLOS

When Spanish colonists arrived in the Southwest, they found Pueblo communities from the Upper Río Grande in the east to the Colorado River in the west. Some villages greeted the Spaniards with a hail of rocks and arrows. In other cases people abandoned their homes and took refuge in remote areas.

Despite the outward similarity of many Pueblo communities, these societies were the product of very different ethnic groups and language families. The Hopi spoke a language of the Uto-Aztecan family, making them distant relatives of both the Ute and the Aztec. The Tewa spoke a language of the Kiowa-Tanoan family, making them linguistic relatives of some Plains Indian groups. The people of Acoma and Cochiti spoke Keresan languages. The language of the Zuni was like no other.

All Pueblo societies combined (1) a system by which talented individuals could rise to positions of respect, and (2) a series of built-in safeguards that prevented hereditary inequality from developing. Anthropologists, however, have called attention to some basic differences between the Eastern Pueblos (central and eastern New Mexico) and the Western Pueblos (Arizona and western New Mexico). The Western Pueblos, where the clan was the key unit, did not display much centralized social control of labor. The Eastern Pueblos, where a system of opposing moieties provided a lot of the social structure, had a stronger centralized control of labor.

Anthropologist Edward Dozier (himself a Tewa speaker from the village of Santa Clara, New Mexico) offered an explanation for this difference. He pointed out that most Western Pueblos relied on rainfall or floodwater farming, labor for which could be handled at the level of the extended family, lineage, or clan. The Eastern Pueblos relied on systems of canal irrigation, whose creation and maintenance might have required a more formal control of labor.

Many Western Pueblo peoples, including the Hopi, Zuni, and Acoma, claimed descent through the mother's line (matriliny). Many Eastern Pueblo peoples, including the Tewa, Tiwa, and Keres, either claimed descent through the father's line (patriliny) or considered both lines equal. Other differences between the Eastern and Western Pueblos, pointed out by anthropologist Fred Eggan, were as follows.

Like the Angami Naga, the Tewa could achieve respect by working their way up through a series of increasingly prestigious ritual societies. In order to prevent one group from becoming a permanent ritual elite, the Tewa broke the village into two divisions named for their culture heroes, the Summer People and the Winter People. Each division had its own headman and ritual assistants, and each was allowed to run the village for half a year. The Tewa did not reckon descent rigidly through either parent's line, and people were not required to marry outside their own clan. Instead, everyone had a major group of relatives called the *matu'i,* radiating out from all four sets of great-grandparents.

For most Western Pueblos, on the other hand, mother-daughter and sister-sister ties formed the core of social groupings. Often the bonds of brother and sister were so strong that a husband felt like the proverbial third wheel. Divorce was not uncommon. The woman owned the house, and the crops stored there were in her care. For the Zuni the matrilineal family was the most important unit; for the Hopi that role was assumed by the clan, which held all ritual knowledge in trust.

The lineages within each Western Pueblo clan varied in prestige, and within each village one clan tended to be ritually prominent. The specific clan that stood out, however, varied from village to village. What prevented a permanent elite from emerging was the fact that the clans, the secret ritual societies, and the managers of the kivas had such independent constituencies that, in effect, power was shared. It is also worth noting that until Euro-American colonists began to suppress raiding, each village had its traditional enemies, and one path to renown was to become a war leader.

All Pueblo villages had kivas, that distinctly Southwestern venue for ritual performance. Kivas remained semi-subterranean not only because they were modeled on an ancient house type but also because of the cosmological premise that human beings first reached the surface of the earth by emerging from the underworld. While the kivas of Acoma and many Eastern Pueblos, like those of Pueblo Bonito, were circular, the kivas of Western Pueblos, like those of the Hopi and Zuni, were rectangular. Kivas shared many of the features of men's houses, such as sitting benches, sunken floors, and sacred hearths. Pueblo society, however, did not display as much separation of men's and women's ritual as we saw in New Guinea.

The walls of some Zuni kivas were painted with murals of deer, birds, and other creatures, the two-dimensional counterparts of Göbekli Tepe's carved pillars. The superimposed levels of some Hopi kivas recapitulated the stages through which early humans ascended from the underworld to the surface of the earth. The Acoma people envisioned their ancestors using a primordial kiva whose components were the sun, moon, Milky Way, and rainbow. Many Western Pueblo kivas were entered by dancers disguised as ancestors or supernaturals, called *shalako* by the Zuni and *katcina* by the Hopi.

In the pages that follow we look at two of the best-known Pueblo societies, one from the east and one from the west. The lesson they teach us is that regardless of how many immigrants it absorbed, which gender its clans emphasized, and what its ancestors were called, an autonomous village society, run by ritual specialists, could limit inequality but still provide gifted people with a path to leadership and respect.

MADE PEOPLE AND DRY FOOD PEOPLE: THE TEWA OF SAN JUAN PUEBLO

Long, long ago, at a time when death was still unknown, humans and animals and supernatural beings all lived together. Their home was in the underworld beneath the waters of Sandy Place Lake, far to the north of Santa Fe.

Among the supernatural beings were two mothers of the future Tewa: their summer mother, Blue Corn Woman, and their winter mother, White Corn Woman. These two mothers sent a man out to find a route to the surface of the earth, but he found only haze and mist because the world was still moist and unripe. Finally, after searching in all four of the great World Directions, the man came to a place where some animals gave him a bow, arrows, and clothing. He returned to the underworld as Mountain Lion, the hunt leader. Mountain Lion appointed two men to be Summer and Winter leaders of the Tewa. He and they were now the first three *patowa,* or Made People, those who had become fully Tewa.

Next the people of the underworld sent out six pairs of brothers to explore the earth. Blue Man went to the north, Yellow Man to the west, Red Man to the south, White Man to the east, Dark Man to the zenith, and All-Colors Man to the rainbow. They all returned to the lake, but from that point on, each world direction would be associated with the color of the man who had traveled to it.

The Tewa then tried again to emerge from the lake but were still not fully formed. They returned from their attempt, however, transformed into Warm Clowns, Cold Clowns, Scalp Leaders, and members of the *kwiyoh*, or Women's Ritual Society. Fully Tewa at last, they emerged from the lake and traveled south. Along the way the Winter People survived by hunting, and the Summer People lived by growing corn and harvesting wild plants. They eventually founded the six Tewa-speaking villages of northern New Mexico.

This cosmology of the Tewa explained why each village was divided into Summer and Winter People, and why those two major divisions were crosscut by ritual societies featuring Warm Clowns, Cold Clowns, Scalp Leaders, Medicine Leaders, and Women's Ritual Leaders. The story of creation also revealed that no one was born fully Tewa; one had to become Tewa gradually, through ritual achievement. In the course of initiation into successive ritual societies, the Tewa would recapitulate the stages through which their ancestors had worked their way to the surface of the earth.

This story of Tewa creation was recorded by anthropologist Alfonso Ortiz in San Juan Pueblo in the 1960s. Ortiz, born and raised in San Juan, did not have to travel far to collect the story. He sat down with the older members of his own community.

In an earlier chapter we mentioned that foragers often had ad hoc, unscheduled rituals, held whenever resources permitted a large group to live together. In contrast, sedentary agriculturalists were able to hold scheduled, calendrical rituals at the same time every year. The Tewa had between 40 and 50 rituals of this type. They relied on the vernal and autumnal equinoxes and the summer and winter solstices to set the ritual calendar. Most ritual activity was concentrated between the fall and spring equinoxes, because that was a time of reduced agricultural activity.

The Tewa planted corn, chile peppers, beans, melons, squash, and gourds after the spring equinox. Later in the spring they collected wild plants. Vegetable crops were harvested in early fall, after which everyone helped clean the irrigation canals. After the fall equinox it was time to harvest piñons. In the early winter the village authorities began to redistribute stored food. After the winter solstice, in an act of public welfare, the village's poorest families were given food.

Like the Etoro, the Chimbu, and the tribes of Mt. Hagen, the Tewa had a hierarchy of virtue. This hierarchy involved three levels of "being" for living humans and three levels of "becoming" for the spirit world.

At the bottom of the hierarchy were ordinary Tewa who, over the course of their lives, rarely qualified for ritual or political roles in the village. They were called Dry Food People, a reference to those late-arriving ancestors who had walked on the world only after it had dried out, and had themselves become hard and dry. At the top of the hierarchy were the Made People, or patowa (literally, "completed"), those who had become fully Tewa by ascending to the top of the ritual/religious hierarchy. The patowa were considered the most virtuous people in society. They were at the core of Tewa ritual organization, controlling and directing all public ceremony, a task for which they were helped by lay assistants.

Intermediate in virtue, between the Made People and the Dry Food People, were the *Towa é* ("persons"), who symbolized the six courageous pairs of brothers sent out to explore the world while it was still moist. They lay at the core of Tewa society, mediating between the Made People and Dry Food People.

The path to becoming a Made Person led through at least eight ritual societies. It began with the Women's Society, followed by the Scalp Society. In the past there was a "scalp leader" whose expertise ensured success in raiding traditional enemies like the Navajo. One role of the Women's Society was to curate the scalps for him.

Next on one's ritual ascent was the Hunt Society, followed by the Warm Clown Society and the Cold Clown Society. The terms "warm" and "cold" referred not to actual temperatures but to the Summer/Winter dichotomy. Then came the Bear Medicine Society, so named because its members, like bears, were thought to be able to heal themselves.

Finally came the climactic stages of one's ritual ascent, the Summer Society and Winter Society. Those who made it to the top were now patowa. In the 1960s these Made People and their lay assistants numbered 52 people, roughly 6 percent of a village of 800. This was small by past standards; according to Ortiz, a much higher percentage of Tewa became patowa prior to 1900.

Described by some anthropologists as "part-time priests," the Made People endeavored to keep the seasons moving normally while maintaining peace and harmony in their village. They were the representatives on earth of the most respected supernatural beings, Blue Corn Woman and White Corn Woman. After death their souls would be honored; they would come to symbolize those supernatural beings who had remained below the lake and, therefore, never dried out.

For their part, the Towa é had their own souls and, as we shall see, their own final resting place in the ritual landscape. The Dry Food People became lesser spirits, known as the Dry Food Who Are No Longer.

In olden days one kiva in the center of San Juan village had been the Navel of the Earth. From there, four lines went out to sacred mountains at the four corners of the Tewa world: north, south, east, and west. These were the first mountains seen by the pairs of mythical brothers who left the underworld. Each mountain had a lake or pond where the spirits of Made People went to live with the supernatural beings. On each mountaintop the spirits of past Towa é stood watch. Those Towa é had created a mesa on the way to each mountain. Between San Juan village and each of these mesas was a shrine to

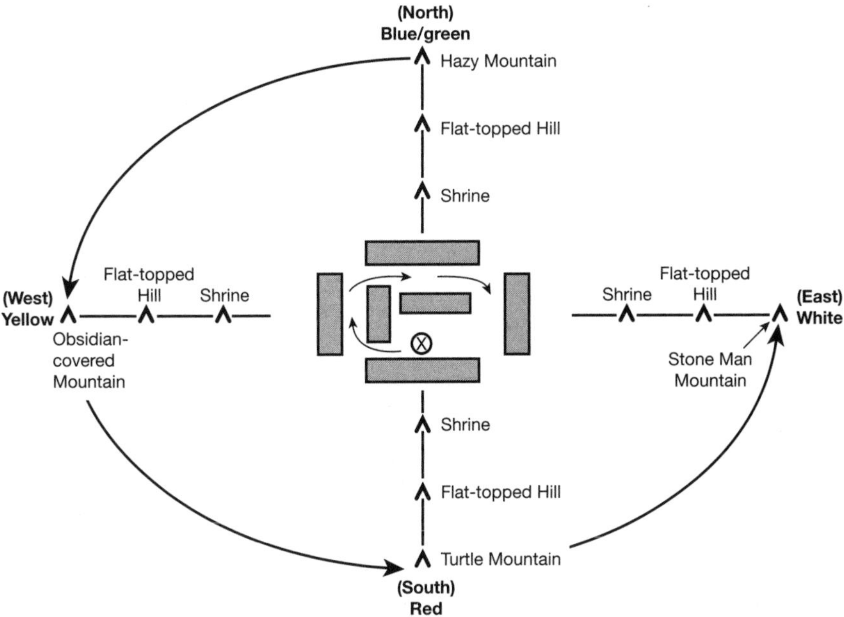

FIGURE 21. The Tewa of San Juan Pueblo lived at the heart of a sacred landscape. Its center was the Navel of the Earth (marked with an "X"), which was surrounded by dance plazas (gray rectangles). From here, pathways radiated out past shrines and flat-topped hills to a series of sacred mountains up to 80 miles distant. These mountains stood at the four great World Directions, each associated with a color. Whether one's soul went to a shrine, a mountaintop, or a mountain lake was determined by how far one had ascended the ladder of ritual/religious achievement. Arrows indicate the direction of ritual movement.

one of the four Great World Directions. It was to these shrines that the souls of Dry Food People went.

In other words the Tewa world was a carefully laid out, sacred, quadripartite landscape composed of mountains, mesas, shrines, and kivas, all connected by roads or sight lines (Figure 21). The existence of such a landscape lends credibility to those archaeologists who have reconstructed Pueblo Bonito as the center of a road system leading to an even older and even grander sacred landscape.

ETERNAL BICKERING AMONG CLANS: THE HOPI OF OLD ORAIBI

Black Mesa rises 6,000 to 7,000 feet in northern Arizona, just east of the Grand Canyon. The hills are dotted with piñon and juniper trees, but the growing season is barely long enough for agriculture. Hopi Third Mesa is a peninsula-like extension of Black Mesa, and it is here that the village of Old Oraibi was founded.

Like the Tewa the Hopi once resided in the underworld. One day they heard footsteps above them and investigated but found that the surface of the earth was still cold and dark. Eventually they discovered that the footsteps were those of a supernatural being named *Masau'u* ("Skeleton"). Seeing the light of a fire in the distance, they approached and came upon a garden of corn, beans, squash, and other cultivated plants. Skeleton met them, fed them, and warmed them by his fire.

Now fortified for their journey, the Hopi began to wander. They were already divided into clans that reckoned descent in the mother's line, and their leader was called Matcito or Machito, head of the Bear clan. Matcito led them to Old Oraibi, where Skeleton allotted them land. Soon after, other clans began to arrive, each offering to perform a beneficial ceremony if allowed to settle there.

Members of the Bear clan picked the best land for themselves and erected a stone boundary marker on which they carved a bear claw. Matcito was named *mongwi,* or village headman, and he allowed other clans to cultivate land at Oraibi on the conditions of good secular behavior and proper ritual.

So large was the plot of land set aside for Matcito and his Bear clan that they could use it to support a war leader. Members of the Kokop clan, late arrivals at Oraibi, were allowed to settle there because they had helped the war leader fend off an enemy raid. Henceforth the Kokop clan would consider the defense of Old Oraibi to be one of its main responsibilities.

This Hopi account of creation was recorded in the early 1930s by anthropologist Mischa Titiev. Titiev was the son of Russian immigrants rather than a Native American like Alfonso Ortiz. He became such a friend to the people of Oraibi, however, that he was eventually adopted into the Sun clan.

Hopi cosmology provided justification for the leadership role played by the Bear clan. It drew on that widespread principle of social logic, "We were here first." The Oraibi origin myth also supported a scenario long advanced by Southwest archaeologists, namely, that many historic Pueblo villages were multiethnic in origin, that lands were allocated on a first-come, first-served basis, and that late arrivals had to be on their best behavior.

In the matrilineal society of the Hopi, daughters remained for life in their mothers' households; their husbands joined them there. A typical extended family included a woman's maternal grandparents, her parents, her mother's sisters and their husbands, and her unmarried brothers and sisters. Such a family occupied a block of contiguous rooms in the village.

Families at Oraibi were grouped into lineages, about 39 in all, and the lineages were grouped into 21 clans. These clans were in turn grouped into nine larger units. Had there been only two of these larger units, each representing half the village, they would have been considered moieties like the Summer and Winter divisions of San Juan Pueblo. Since there were nine of these larger units, however, anthropologists refer to them as *phratries,* borrowing the ancient Greek word for a group of clans.

Titiev discovered a number of obsolete clan names at Oraibi, suggesting that some former clans had become extinct. Many clans shared kivas peacefully with each other. Others, however, bickered continuously, validating prophecies that they were destined to quarrel.

Oraibi had both clan houses and kivas. There were 31 clan houses in which the women kept *tiponi,* or clan fetishes, the equivalent of the sacred bundles of the Plains Indians discussed later in this chapter. There were also 13 kivas, eight of which could host major rituals and five of which could host only minor events. At least nine of the kivas had mountain shrines associated with them, creating a sacred landscape similar to the Tewa. Titiev believed that in the past each clan had "owned" a specific ceremony, usually held in one of the kivas it controlled.

Hopi kivas were rectangular and subterranean, entered through a hatch in the roof by a ladder (Figure 22). In the middle of the floor was a *sipapu,* a cavity representing the hole through which humans had emerged from the underworld. The kiva also had a fireplace and a hollow bench, used to conceal sacred objects from view. The kiva was for ritual performances; the clan house was for private meetings and the curation of ritual paraphernalia.

Like the Tewa, the Hopi scheduled many of their rituals to coincide with solar or lunar events. One of the most important rituals commemorated the departure of the Hopi from the underworld. According to Hopi cosmology, when their ancestors left the underworld they brought with them a number of

FIGURE 22. Instead of building men's houses, the achievement-based village societies of the American Southwest built subterranean kivas that were entered through the roof. This drawing, based on a 100-year-old photograph, shows three Hopi men exiting a kiva at Old Oraibi Pueblo, Arizona.

spirit beings called *katcinas,* anglicized "kachina." The kachinas had accompanied the Hopi during their wandering but were killed in an enemy attack and returned to the underworld. Each year they were allowed to return to earth for a period extending from the winter solstice to the summer solstice, during which time they could mediate between the Hopi and the spirit world.

At the proper time of the year, dancers from two clans impersonated kachinas by dressing in sacred masks and costumes that the kachinas had allegedly left behind. Like the bull-roarers of the Australian Aborigines, the costumes of the kachinas were so old that no one could remember when they had been made. Prior to initiation, children watched the kachinas in awe, believing that they were supernatural beings; after initiation, they knew them to be Hopi men in costume.

The Hopi of Third Mesa tended gardens of corn, beans, squash, gourds, and melons, always trying to produce a surplus; the ideal was to have a year's supply stored to ward off drought. They relied on underground seepage of water and, because of the slope of the mesa, were able to capture most available rain runoff. Once the harvest of carbohydrate sources was in, a hunt leader added protein to the diet by organizing a rabbit drive.

Another post-harvest activity was raiding. The Hopi people's official position was that they fought only in self-defense. In days of old, however, after praying to ancestral warriors for help, they set off enthusiastically with bows and arrows, tomahawks, spears, and throwing sticks. Oraibi's traditional enemies included the Apache and Ute, whose scalps would once have been brought back, displayed on poles, and ritually "fed." These scalps were treated as "sons" of the *nina,* or slayer, who had taken them. They would be buried with him when he died, just as scalps had been left with burials in dry caves 1,500 years earlier. Among the Hopi, the scalp taker had to seclude himself in a kiva for four days and nights, fasting and undergoing rituals of purification so that the spirits of his victims could not take revenge.

Inequality and Conflict

Let us look now at the sources of inequality at Old Oraibi. They lay principally in the field of ritual leadership and were based on the alleged sequence of the arrival of various clans. Because the culture hero Matcito had reached Oraibi first, his Bear clan was preeminent in ritual. At the time of Titiev's stay, the Pikyas clan was second only to the Bears.

Ritual authority, however, did not necessarily translate into secular authority. The head of the village, always drawn from the Bear clan, could urge proper behavior. The heads of the 21 clans, who served as his advisers, could agree. The war leader could threaten the disobedient with punishment. But in the final analysis there was no monopoly of force, no power to carry out commands. Since the highest authorities were ritual leaders, the ultimate punishment for wrongdoing would be supernatural.

Even within a phratry, the ties between clans were so weak that bickering was endemic. In 1934, for example, the Pikyas and Patki clans (both part of Phratry VIII) began quarreling. The Patki argued that the ceremonies they owned entitled them to be ritually superior to the Pikyas. In fact, they claimed that the ancestors of the Pikyas were late arrivals, Tewa speakers from the village of Hano, and therefore not "true Hopi." This quarrel confirmed an ancient prophecy that warned the Pikyas to beware of the Patki.

The most famous Oraibi conflict took place in 1906 and is still the subject of heated debates. It involved the Bear and Spider clans, both part of Phratry II. According to Titiev, the Spiders argued that they were equal to the Bears in ritual authority but were never allowed to provide Oraibi with its headman. The Bears sought justification for their ritual preeminence in the legend of Matcito. The Kokop clan of Phratry VI sided with the Spiders. More and more clans began to choose sides in the dispute, and eventually half of Oraibi's population picked up and moved to nearby Hotevilla.

The Bear-Spider conflict had been simmering for decades before Oraibi split, and many scholars believe that the dispute had multiple underlying causes. Anthropologist Jerrold Levy has examined many potential causes, which include the destabilizing effects of population growth, drought, the erosion of farmland, and interference in Hopi life by everyone from Anglo-American ranchers and missionaries to the U.S. Cavalry.

We have no doubt that one can find multiple causes for every social upheaval, but we would like to focus on a few widespread principles. Achievement-based societies fissioned, or gave rise to daughter communities, all the time. The archaeological record is full of periods when a handful of villages appeared in a region, grew, split, and sent junior segments off to found new villages, while senior segments remained at the parent community.

Although certain clans were treated as the ritual leaders of the village, that did not give them the secular power to prevent fissioning. A hierarchy of ritual authority was little more than a hierarchy of virtue, and the larger an achievement-based society became, the harder it was for such a hierarchy to hold it together—especially if a sizable group considered its position in the hierarchy unfairly subordinate.

THE ORIGINS OF ACHIEVEMENT-BASED VILLAGE SOCIETY ON THE PLAINS OF NORTH AMERICA

Corn, as we have seen, reached the U.S. Southwest from Mexico. In the Southwest the major obstacle to its success was drought, which Native Americans overcame with irrigation. The story was different along the Missouri River, from North Dakota to St. Louis. In this case the major obstacle to corn's success was the threat of frost.

The first types of corn to enter the Midwest and the Plains were Mexican varieties that needed 180 to 220 days to mature. They succeeded in the warm Southwest but were too cold-sensitive for the shorter growing season of the central United States. Corn reached the Midwest at least 2,000 years ago but did not grow well enough at first to become a staple.

What probably opened up the Plains and the Midwest was a new type of corn called Northern Flint, which matures in only 160 days. Archaeologist David Brose suggests that this newly evolved corn was given one of its first serious tryouts by the Indians of northern Ohio and western Ontario, a region whose climate is ameliorated by the waters of Lake Erie. The lake was surrounded by humid soil and provided Native people with the fish needed to sustain life while they experimented with the growing of 160-day corn. Once Flint corn, with its eight to 12 rows of kernels, had established itself, 1,000 to 900 years ago, virtually all Native American groups in the Midwest made a greater commitment to corn agriculture.

For millennia the Plains, the land of the bison or American buffalo, had been home to hunters and gatherers. Now the Missouri River became a corridor leading horticulturalists north to the Dakotas. A prehistoric society known as the Middle Missouri tradition, using hoes made from bison shoulder blades, began cultivating the floodplain of the Missouri and its major tributaries. They planted gardens of corn, beans, squash, gourds, sunflowers, and tobacco, fished and collected mussels from the river, and hunted buffalo on the prairies beyond the river.

The earliest Middle Missouri villages, occupied perhaps 1,000 years ago, were at the mercy of marauding foragers who came off the Plains to raid their food supplies. In response the villagers surrounded themselves with defensive ditches and palisades. Villages typically had anywhere from 15 to 100 rectangular houses, each big enough for an extended family. The houses had fireplaces and storage pits, and the doors were kept narrow for defense. The ritual buildings in these villages, called ceremonial lodges, were framed with cedar posts and covered with an insulating layer of prairie sod.

In spite of the distances involved, the villagers of the Middle Missouri tradition carried out active exchanges of valuables with other regions. They acquired native copper from Lake Superior, conch shells from the Gulf Coast, dentalium from the Pacific, and a stone called catlinite from which tobacco pipes could be carved.

In 1541 Spanish explorers under Francisco Vásquez de Coronado reached what is now Kansas. Over the next century, many of the Spaniards' horses escaped into the Plains and changed the lifeways of Native American groups. Some tribes abandoned horticulture and escalated their hunting of buffalo by making use of captured horses. Other tribes chose to continue farming the floodplain of the Missouri. It is the latter societies that we examine in this chapter.

During the late 1700s fur traders, following the Missouri River into what is now North Dakota, encountered two Native American societies called the Mandan and Hidatsa. In 1804 the Lewis and Clark expedition met the same two groups. Some 29 years later, Prince Maximilian of Wied-Neuwied, a German explorer, reached Fort Clarke on the upper Missouri. He was fascinated by the Mandan, Hidatsa, and Crow, three allied tribes who spoke languages of the Siouan family. While the Mandan and Hidatsa lived in horticultural villages on the Missouri, the Crow had become equestrian buffalo hunters. Despite their different lifestyles, all three societies were allies, defending themselves against mobile horsemen such as the Cheyenne, Blackfoot, and Lakota.

Our first detailed look at the Mandan and Hidatsa comes from a French trapper named Charbonneau, who had lived among the Hidatsa for 37 years when Prince Maximilian interviewed him. By 1907 anthropologists such as Robert H. Lowie had begun to visit the societies of the upper Missouri, combining their personal observations with the historic accounts of Maximilian and Charbonneau.

Based on these pioneering descriptions, the precolonial Mandan and Hidatsa seem to have been typical achievement-based societies. They were composed of clans who reckoned descent in the mother's line. Leadership was based on achieving respected elder status, for which raiding and ritual sponsorship were alternative routes.

Life in these Siouan-speaking villages was an endless search for *xo'pini*, a supernatural essence or life force that lay at the heart of success and renown. Some anthropologists have translated xo'pini as "power," but it reminds us more of the magical, electric life force that the ancient Polynesians called *mana*. Xo'pini could be acquired either from a supernatural being or from a person of renown.

According to the age-old principle of reciprocity, one could not acquire xo'pini without paying a price. Often the price was self-inflicted suffering, such as the cutting off of the final joint of one's own finger or the suspension of oneself by skewers through the flesh. Suffering could lead to visions in which a spirit or sacred animal revealed one's destiny.

The man or woman who received a vision put together a sacred bundle, a tightly wrapped collection of objects associated with the supernatural encounter. Some bundles remained personal; a warrior, for example, might take his bundle along on a raid in hopes that its xo'pini would protect him. A bundle kept around so long that its origins were lost, however, became a clan or tribal bundle, one whose life force was all the greater because it went back to mythical time. The clan created a story about its origins, sang songs to it, and performed rituals over it. Some bundles were curated by clan elders for generations, making them the Plains equivalent of the fetishes curated by the Hopi.

Sacred bundles were by no means small. One Mandan bundle is reported to have contained the following items: two rings and a crescent made from native copper; one gourd rattle; six magpie tail feathers; 12 owl tail feathers; the scalp of a Cheyenne warrior killed by a respected Mandan ancestor; the skull and left foreleg of a grizzly bear; one tuft of chin whiskers from a buffalo; one skull and one horn from a buffalo; the hide from the head of a buffalo calf; and a stuffed jackrabbit of the type used to bait eagle traps.

In comparison, one well-known Hidatsa bundle contained two human skulls, one buffalo skull, a tobacco pipe used in ritual, a turtle shell, and a fan made from the wing of an eagle. The two human skulls were said to be from enormous eagles that had assumed human form.

The Hidatsa bundle just described had an interesting history. It was originally in the possession of a man named Small Ankle, a member of the Water Buster clan. When Small Ankle died suddenly, his son Wolf Chief was persuaded to sell the bundle to Christian missionaries. In 1907 it made it to the Heye Foundation's Museum of the American Indian in New York.

The Dust Bowl drought of the 1930s convinced many members of the Water Buster clan that they were being supernaturally punished for selling

FIGURE 23. The sacred bundles of Mandan and Hidatsa clans contained *xo'pini,* a powerful life force. In 1938 the Water Buster clan of the Hidatsa recovered a sacred bundle that had been lost to them for more than 30 years. Among other items, this bundle contained two human skulls which, according to legend, came from enormous eagles that had assumed human form. In this drawing, inspired by a 70-year-old photo, clan elders Foolish Bear and Drags Wolf have partially unwrapped the bundle to reveal the skulls.

the bundle. A delegation asked President Franklin Delano Roosevelt to intercede on their behalf, and in 1938 the Heye Foundation returned the bundle to the Hidatsa in exchange for a buffalo medicine horn. Lo and behold, the drought ended with the return of the bundle to Fort Berthold (Figure 23).

While personal sacred bundles were sometimes buried with their owners, others could be "purchased" in the context of a ritual. A father, for example, might host a ceremony at which his son was allowed to purchase his bundle. This would only happen, however, if his son had already had a vision of himself making the purchase.

Once his vision had been reported, the son might need a year to accumulate sufficient resources for the purchase. He borrowed from his relatives and the members of his age cohort, much as New Guinea tribesmen borrowed from their relatives to pull off an important act of moka. The seller of the bundle was then expected to distribute to others all the property he had received. Accumulating wealth was frowned on in Plains society.

Many Plains feasts and ceremonies centered on the intergenerational transfers of bundles. In theory each bundle could be sold four times. The first three purchases, however, involved a duplicate bundle. Since four was a sacred number, it was only during the fourth purchase that the original bundle was used. Taking possession of a sacred bundle increased the buyer's xo'pini, and the more bundles one acquired, the more life force one possessed.

Bundles could be curated either by men or women. It was widely believed, however, that because women did not participate in raiding, they did not need to accumulate as much xo'pini as men. Women also had an alternative strategy: by occasionally sleeping with a renowned elder, a woman could acquire some of his life force. This act, of course, depleted the elder's reservoir of xo'pini, and that—added to the life force he had already expended during a lifetime of raiding—hastened his senescence. This strikes us as the Plains equivalent of a premise we saw in New Guinea: men grow old and feeble because women drain their life force.

Let us now consider some of the social logic involved in the pursuit of xo'pini, which can be summarized in the following steps:

1. Suffering leads to a vision.
2. The vision allows one to create a sacred bundle.
3. If kept long enough, the bundle acquires a mythical origin and becomes a source of life force for one's entire clan.

4. The objects in a bundle are memory aids for the recounting of myths and legends.
5. Bundles stimulate rituals and lead to the building of new ceremonial lodges.
6. Each new ritual inspires the self-inflicted suffering of younger individuals.
7. Step 6 leads to Step 1, continuing the cycle.

While each Plains horticultural society was unique, all shared a set of general principles with other achievement-based societies. Men who sought renown made sacrifices, borrowed food and property from kinsmen to sponsor ceremonies, and directed the building of ritual lodges. To prevent successful men from becoming a permanent elite, however, society discouraged them from accumulating property and encouraged them to lead by example rather than giving them actual political power.

RISING TO PROMINENCE IN MANDAN SOCIETY

There were three levels of the Mandan world: the earth on which humans lived, the world above it, and the world below it. Beings of one kind or another lived on all three.

The earth on which the Mandan lived was the product of First Creator and Lone Man. It floated on water from the lower world, a place from which springs continue to bubble up. Above the earth lay the upper world, an earth lodge whose four huge posts held up the sky. The Sun traveled the roof of this lodge. He had three sisters—Sunrise Woman, Above Woman, and Sunset Woman—who lived at places where the Sun paused to have a smoke.

This Mandan cosmology was recorded by anthropologist Alfred Bowers. His information came from elders who remembered Mandan life in the 1870s at Like-a-Fishhook Village, North Dakota.

It is estimated that the Mandan may have numbered 9,000 before European traders and trappers introduced smallpox. Tragically, by 1910 the Mandan had been reduced to fewer than 200. Bowers visited the survivors during the period 1929–1931 and combined their firsthand descriptions with the historic accounts of Lewis and Clark, Maximilian, Charbonneau, and others.

In days of old the epicenter of each Mandan village had been a big cedar post, driven into the center of a ceremonial circle 150 feet in diameter. This post was as important to the Mandan as the Earth Navel kiva was to the Tewa. Outside the ceremonial circle, families built residential lodges by placing layers of prairie sod over a framework of poles. Once, long ago, Mandan houses had been rectangular; by the time Europeans reached the area, they had become round. The old rectangular shape, however, had been retained for the ceremonial lodge.

Most Mandan villages were defended by a ditch and a palisade of cedar posts. Inside the defensive works were scaffolds for the drying of corn or the exposure of burials to the elements. The Mandan also modified the landscape outside the village. They created catfish traps in the shallows of the river, built game-drive fences for hunting pronghorn antelope, and dug traps on the prairie where eagles could be captured for their feathers.

Mandan villages moved either when their gardens were no longer fertile, or when all available firewood had been used up. Even in wooded riverside locations, where settlements endured for years, villages might be temporarily abandoned for the summer buffalo hunt. Loading their possessions onto a *travois*—a sled pulled by a team of dogs—families moved out onto the Plains, where they lived in tepees until the hunt was over. Once they returned to their village for the winter, they set about building corrals in the cottonwoods along the river, where unsuspecting bison could be driven when they sought shelter from the cold.

Bowers found that surviving Mandan families belonged to at least 16 clans, all claiming descent in the mother's line. Nine of these clans were grouped into a West moiety and seven into an East moiety, a dual division like that of the Summer and Winter moieties of the Tewa. Each clan owned its own sacred bundles, which were ritually transferred from one generation to the next. Possession of certain bundles conferred the right to perform a specific ceremony, and each clan felt that it held a copyright to the songs, chants, dances, and costumes used in that ritual.

Like the Nootka and Tlingit, the Mandan had a strong concept of intellectual property. The difference is that while Nootka and Tlingit nobles could simply bequeath that property to their offspring, a Mandan youth was required to purchase it. This is a significant difference between achievement-based and hereditary leadership.

Some sacred bundles gave their owners the rights to rituals that guaranteed success in eagle or catfish trapping or the driving of bison into corrals. Anyone who wanted to carry out those activities had to purchase the rights from the clan owning the bundle. There were also personal bundles that could be purchased for their life force; the price might be garden produce, bison hide robes, or even horses, once the latter had arrived in the Plains. A good horse could cost a man three war bonnets, the end product of 108 black-tipped feathers from nine eagles. This made eagle trapping a crucial activity, one requiring payment to the owner of the eagle trapping bundle.

In theory the fact that all clans owned sacred bundles kept the playing field level. In practice, however, not all ceremonies were of equal importance. The most crucial Mandan ceremony was the *Okipa,* a four-day ritual recapitulating the creation of the earth. Because the Waxi' Ena clan of the West moiety owned the sacred Okipa bundle, it held a position of ritual importance like that of the Bear clan at Old Oraibi.

One became a Mandan gradually, moving up one step at a time through a system that anthropologists call age-grades. Such a system had both age-cohort aspects, like those of the Ao Naga, and grades of accomplishment, like the ritual societies of the Tewa. W. Raymond Wood and Lee Irwin list eleven age-grade societies for Mandan males and seven for women.

Several of the age-grades deserve special mention. One age-grade for older men was the Black Mouth society, whose members had proven themselves to be implacable warriors. One of the most important age-grades for women was the Goose society, whose members possessed exceptional ritual knowledge. Additionally, women who had passed menopause became eligible for the prestigious White Buffalo Cow society.

All Mandan boys and girls who were able to do so had their parents or grandparents purchase their membership in the next age-grade society. This purchase was carried out so that youths could work their way up to the status of elder. The latter was everyone's goal because all villages were run by elders. The elders, however, led by consensus and bent over backward to avoid offending any faction.

From among the male elders, one brave warrior was chosen to be the community's War Leader, and one consummate ritual expert was chosen to be Peace Leader. To avoid resentment, these two leaders were drawn from opposing moieties. The Peace Leader outranked the War Leader until the village suffered an enemy attack, at which time their relationship would be reversed.

Both boys and girls began fasting at age eight or nine, hoping to induce a vision of their destiny. They also performed self-torture, cutting off finger joints or suspending themselves by skewers inserted through the skin of their backs or chests. Pieces of skin or fingertips might be offered in sacrifice to the spirit world. As youths grew older (and with the help of their family), they began accumulating food and valuables to pay for the right to host a ceremony like the Okipa.

While women most often chose a ritual route to prominence, men could gain prestige by stealing rival tribes' horses, killing and scalping enemies, or "counting coup" by touching an enemy in battle and living to tell about it. A man who had scalped an enemy was allowed to paint one of his buckskin leggings black and the other yellow or white. He could also wear a coyote tail at each ankle or an eagle feather in his hair.

Many lineages wanted their young women to marry successful warriors, and they were willing to buy a sacred bundle for the groom in order to increase his life force. Advancing through warfare, however, was a high-risk pathway. Each time a warrior risked his life, he expended some of his accumulated xo'pini, and if any of the men he led into battle were killed, he lost respect. In fact so many Mandan men were killed in raids that some unmarried women had no choice but to become a man's second wife.

The highlight of any year was the Okipa, an elaborate four-day ceremony held in a special ritual lodge. The ritual depicted the creation of the Mandan world, and the lodge in which it was performed symbolized Dog Den Butte, the mythical hill where Speckled Eagle had once kept all living things prisoner. Costumed dancers impersonated Speckled Eagle, the culture hero Lone Man, The First Day of Creation, Night, and important animals such as bison, bears, beavers, swans, and snakes. Permission to perform the songs and dances was purchased from the Waxi' Ena clan, which held the permanent intellectual rights. Many young Mandan saw the Okipa as an opportunity to suspend themselves from the lodge's roof by ropes attached to skin-piercing skewers, paying for life force through suffering.

Balancing Prestige and Equality

Mandan life allowed for social advancement without the emergence of a hereditary elite. On the individual level, one could accumulate xo'pini through fasting, self-torture, offering flesh, purchasing sacred bundles, sponsoring ceremonies, accumulating ritual knowledge, or displaying bravery in scalping and coup counting. On the community level, however, War Leaders and Peace Leaders were chosen from opposing moieties—each clan owned the rights to its sacred bundles and rituals, and the elders led by consensus. To be sure, one clan held intellectual rights to the Okipa, but anyone could sponsor the ritual as long as he or she paid for it.

RISING TO PROMINENCE IN HIDATSA SOCIETY

The Hidatsa lived just to the north of the Mandan, along the Missouri and its Knife River tributary. In the past, before their decimation by smallpox, the Hidatsa numbered more than 4,000.

Even after losing part of their population to smallpox, the Hidatsa retained much of their traditional organization. Seven clans claiming descent in the mother's line were grouped into two opposing moieties, four clans in one and three in the other. Men advanced toward elder status through 12 age-grade societies, each of which owned the intellectual rights to a series of songs, dances, and costumes. One difference from Mandan society was that members of the Black Mouth society were not warriors but senior men who occupied the final age-grade before becoming village elders.

For their part, Hidatsa women had four or five age-grades. Two of the most senior were the Goose society and the White Buffalo Calf society. The most prestigious, however, was the Holy Women society, whose members were supreme in their knowledge of religious lore.

The Hidatsa had an important ceremony called the *Naxpike,* which, like the Mandan Okipa, lasted four days. Rather than recounting the creation of the world, however, this ritual was a dramatization of the sacred Naxpike bundle's origin myth. The bundle was said to represent the Hidatsa culture hero Long Arm, leader of the People Above, who had directed the self-torture of a mythical figure named Spring Boy. During the four-day ceremony the Naxpike bundle was transferred to the next generation, with the seller of the bundle impersonating Long Arm and the buyer impersonating Spring Boy. As with the Okipa, the Naxpike was seen as an opportunity for young men to cut off finger joints, suspend themselves with skewers, or endure branding with hot irons in pursuit of xo'pini.

In the Hidatsa system each ascending age-grade purchased the rights to the next grade from the group ahead of it and sold the rights to its grade to the group coming along behind it. As with the Mandan, senior Hidatsa tried to be fair and democratic elders. They sometimes chose multiple Peace Leaders and War Leaders from opposing moieties. Peace Leaders curated the village's most important sacred bundles and, if they could, kept their leadership positions in the family by allowing their sons to purchase the bundles at the appropriate time.

Hidatsa babies were thought to have an origin like that of some Australian Aborigine babies. They began as spirits who lived in certain sacred hills, waiting for the chance to enter an unsuspecting woman's body. Virtually from birth, the Hidatsa prepared children for their adult roles, encouraging them to fast, endure pain, accumulate life force, and seek the vision that would determine their destiny. Hidatsa men received extensive tattoos on their bodies and were sent out in war parties to take scalps and steal horses. Each man paid for his wife with a gift of horses, and each man was also expected to hunt for his wife's household, an obligation similar to the one we saw earlier among the Hadza.

The potential for achieved inequality in Hidatsa society lay in the fact that large families could work hard, grow more crops, trap more eagles, trade more war bonnets for horses, and acquire more life force through the purchase of sacred bundles. There was a limit, however, to how much wealth or status one could accumulate. For example, there was constant pressure to give away your possessions to others, and if you began to act superior, you would be ridiculed even by your own relatives. Bravery and ritual expertise were admired, but in the end your task was to live out your destiny, following the vision that the spirits had allowed you to see.

Two-Spirit People

In Plains society, as we have seen, children as young as eight years old were encouraged to seek a personal vision through fasting and pain. For most young people, a window opened into the spirit world and set them on a gender-specific course. Men in buckskin leggings stole enemy horses, killed buffalo, and counted coup. Women in buckskin skirts hoed gardens, made buffalo hides into robes, and cooked their family's meals.

Once in a while a Hidatsa youth began dreaming of a spirit called "Village-Old-Woman." This dream was considered his vision, and it meant that his destiny was to be a "two-spirit person." For the rest of his life he would dress in women's clothing and perform women's work. He might set up housekeeping with a man and even adopt children.

Anthropologist Raymond DeMallie estimates that there might have been a dozen two-spirit people in a Hidatsa village of 100. Such individuals entered the women's age-grade system and often became respected members of the Holy Women society. They were, in fact, the only Hidatsa allowed to participate in every ceremony. In the social logic of the Plains, two-spirit people had an air of mystery about them and were thought to have a closer relationship with the supernatural world.

Two-spirit people were, of course, not unique to the Hidatsa. They were a widespread feature of Plains society, from the Blackfoot of Alberta and the Assiniboine of Saskatchewan to the Mandan of North Dakota, the Ponca and Lakota of South Dakota, and the Arapaho of Colorado. In fact, it has been estimated that more than 100 Native American societies included men who dressed and lived as women. Perhaps a third as many societies are estimated to have had women who dressed and lived as men.

In the Blackfoot language the word for two-spirit men meant simply "acts-like-a-woman." The Blackfoot believed that such a man was possessed by a unique spiritual force. Far from being shunned, DeMallie reveals, two-spirit men "were in demand as wives because of their physical strength in carrying out womanly duties and for their artistic abilities." Sometimes a married man, believing that he could support a second wife, added a two-spirit person to his household.

We have already seen that marriage among the Eskimo was an economic partnership, expressed in four varieties: a man and a woman, a man and two women, a woman and two men, and two men sharing two wives. To these we can now add (1) one man and a two-spirit person and (2) one man, one woman, and one two-spirit person, giving us at least six varieties of marriage among the indigenous people of the New World.

Feeling that two-spirit people were better suited for women's work, the Mandan did not ask them to go along on raids. The Arapaho believed them to have their own special type of life force. Ponca youths had dreams in which the moon asked them to choose between a bow and arrow and a woman's pack strap. Those who chose the pack strap were destined to dress and live as women.

For their part, some Plains women had visions of themselves as men. Among the Blackfoot, for example, there were "manly hearted women" who joined men's raiding parties, stole their enemies' horses, dressed in men's clothing, and even took wives. This behavior gives us a seventh type of marriage: two women, one of whom was manly hearted.

It is significant that in no Plains society were transgendered individuals looked down upon or ostracized. Their destiny had been predetermined by the spirit world, and Plains society created a place for them. They were often prized for their hard work, respected for their deeper understanding of the sacred, or admired for their craftsmanship.

All this changed, of course, when Euro-American missionaries reached the Plains. They branded the two-spirit people "berdaches," a corruption of the Spanish term *berdajes*, "male prostitutes." Countless two-spirit people were persecuted and driven into hiding.

The Plains Indians' view of two-spirit people was that they owed their way of life to forces beyond their control rather than to human choice. This view is closer to that of today's social scientists than was the pejorative attitude of the missionaries. No Plains society argued that in order to preserve the institution of marriage it had to be restricted to one man and one woman.

BALANCE AND LONG-TERM STABILITY

There are lessons to be learned from the traditional communities of the Tewa, Hopi, Mandan, and Hidatsa. All four groups struck a balance between personal ambition and community spirit. These ethnic groups created a socially accepted way for talented individuals to rise to positions of respect while working to prevent the development of a hereditary nobility.

In many parts of the ancient world, archaeologists can point to periods when society remained remarkably stable for hundreds upon hundreds of years. Often, following further investigation, that stability turns out to have been the product of achievement-based, politically autonomous village societies.

A group's initial attempts to create hereditary nobility, on the other hand, could bring on great instability. The contradictions in social logic between privilege and equality could result in years of oscillation and even bloodshed, as we will see in the next chapter.

III

Societies That Made Inequality Hereditary

TEN

The Rise and Fall of Hereditary Inequality in Farming Societies

Leadership in New Guinea and the Solomon Islands and in the Southwestern pueblos and Plains villages of North America was traditionally based on achievement. Those societies had no hereditary aristocracy. Mandan leaders could sell their sacred bundles to their sons, but they could not present them with noble titles in the manner of Nootka chiefs.

The archaeological record tells us that at various times in the past, a number of achievement-based societies must have altered their social logic to allow for hereditary privilege. Unfortunately, archaeology shows us the results but not the logic itself; to reconstruct the latter, we must turn to studies of living societies. In this chapter we begin with a village that shows us how one ambitious subclan tried to become its society's hereditary elite. To do so, it had to convince rival subclans of its right to intellectual property that had previously been shared.

AUTHORITY IN THE VILLAGE OF AVATIP

The Sepik River flows east through Papua New Guinea and then turns north and empties into the Bismarck Sea. Along the lower Sepik and its tributaries, clearings open to reveal the gardens, hamlets, and villages of the yam-growing Abelam, the farmers and fishermen of the Iatmul tribe, and their neighbors the Manambu, just to mention a few.

The Abelam had a classic Big Man society. The Iatmul worked feverishly to accumulate shell valuables. The Manambu, however, were different. Their level of surplus production was so low that even with the help of kinsmen, one could rarely amass enough shells and pigs to become a Big Man.

In the 1970s anthropologist Simon Harrison came to live in Avatip, a dispersed community of roughly 1,600 Manambu. He discovered a society with several paths to leadership. First, there were secular political leaders who had achieved respect for their work ethic, hunting prowess, strength, generosity, and debating skills. Second, there were ritual leaders who had been initiated through successively higher levels of sacred knowledge. In the past there had been a third path to renown through head-hunting and warfare, activities predictably suppressed by colonial authorities.

Individuals in Avatip were grouped into lineages, lineages were grouped into subclans, and subclans were grouped into clans. Political leaders were drawn from the ranks of clan elders and possessed only secular power; they rose to prominence by winning debates against rivals. Ritual authority, on the other hand, lay in the hands of men called *simbuks,* each of whom was the head of a *laki* (one ceremonial division of a ritual cult).

Each simbuk desired to pass on the office to his oldest son in order to preserve all his cult's ritual secrets. Such a move was predicated on the son's passing all levels of initiation. If those hurdles had not been passed, the order of succession was to the simbuk's younger brother or, if none existed, to his sister's son. Even when they were successful at keeping ritual office within the family, simbuks had little secular power.

Subclans, of which there were 16, were the most dynamic units in Avatip society. Each subclan built its own men's house, featuring big wooden posts, beams, spires, and slit-gongs. Every object stored in the men's house embodied an ancestral spirit. Inside the building, men sat at one of three hearths, depending on whether they had been initiated into the first, second, or third level of sacred knowledge.

One of the greatest sources of tension in Avatip society was the rivalry between secular and ritual leaders. Subclans struggled to grow in population, because if their numbers declined it became harder to win debates. Secular leaders, who battled for years to achieve fame as debaters, envied the simbuks who inherited their office while still in their 20s, an age when most debaters were still nonentities.

Occasionally a simbuk would pick up debating skills, becoming a leader in both the secular and ritual spheres. Such men were so envied that they ran the risk of being murdered. For their part, some secular debaters took advantage of disputes in ritual succession, usurping a simbuk office to which they were not entitled.

Soon after doing this, one such usurper announced that he had just discovered three previously unknown ancestors in his genealogy. These ancestors were used as justification for declaring that the junior section of his lineage (to which he belonged) should become a separate lineage under a new name. This strategy, combined with population growth, was one way for a junior lineage to achieve parity with a senior lineage.

Any attempt by a lineage or subclan to improve its position provoked a debate. Such contests were held on formal debating grounds near the men's house (Figure 24). There, men from two rival subclans faced each other across a vine boundary. Each subclan used an overturned canoe as a drum; each also erected a series of sticks, spears, and arrows representing important ancestors. The women of each subclan danced and prepared food for the participants.

Each debater held a bundle of magical cordyline leaves while he spoke, throwing one leaf to the ground to dramatize each point he was making. As long as he held the leaves, he could not be interrupted; when the leaves were all on the ground, the debater could be heckled. When tempers flared, onlookers used humor to prevent violence.

While secular leadership required oratorical skills, ritual leadership required a prodigious memory. It also employed a principle with which we are already familiar: Names are magic.

All the men and women of an Avatip subclan considered themselves the namesakes of mythical ancestors. The names themselves were not secret, but the myths they referred to were. Each subclan "owned" between 1,000 and 2,000 names, the total of its past, present, and future members. For the entire community of Avatip, that could mean an estimated 32,000 names. Each subclan jealously guarded its names and tried to grow in numbers so that it could own more and more names over time.

During Harrison's stay at Avatip, the largest and most powerful subclan was the Maliyaw. This subclan comprised 246 members, or 15 percent of the community, the result of three generations of deliberate population growth. The Maliyaw had six highly acclaimed orators, ranging from 40 to 70 years old. For four decades they had been aggressively debating the ownership of disputed names, gradually winning by superiority of numbers.

The goal of the Maliyaw was nothing short of the monopolization of all names, and hence all ritual authority. They usurped several positions of genealogical seniority. They then attempted to revise the genealogical record to legitimize their usurpation. They co-opted the ancestors of a mythical village once claimed by their rivals, the Nanggwundaw subclan. The Nanggwundaw objected, so the

Maliyaw debated them and won. When one Nanggwundaw orator collapsed and died, the Maliyaw claimed to have killed him by sorcery. In their own words, their aim was to "tread other subclans underfoot."

The Maliyaw were out to eliminate the traditional Avatip separation of secular and ritual authority. Their goal was to unite the roles of political and

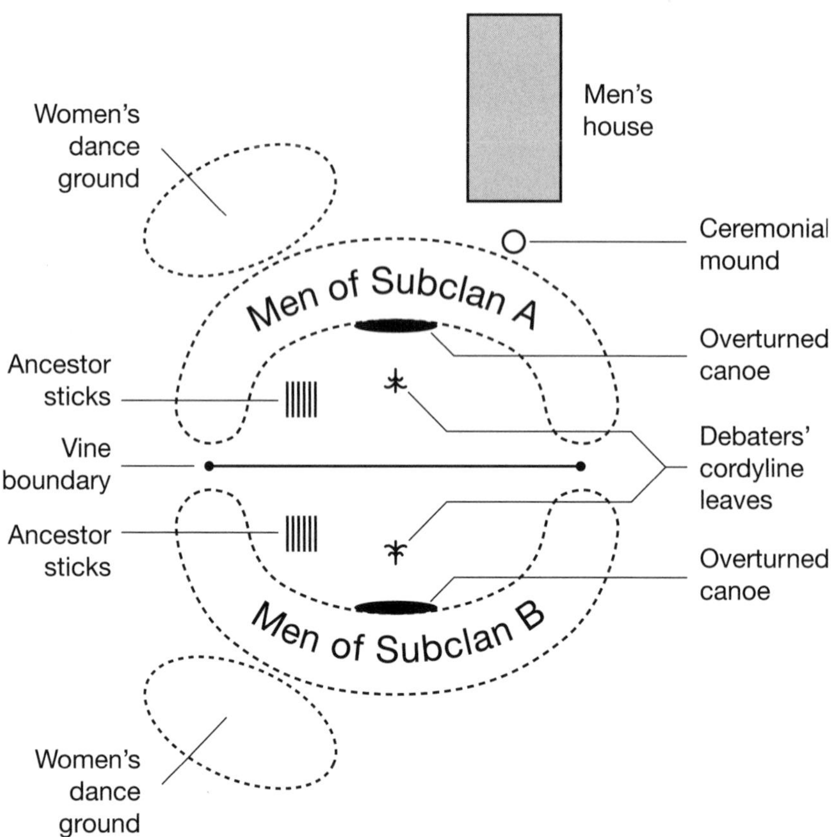

FIGURE 24. The formal debating ground of the village of Avatip, New Guinea. Here, during the 1970s, rival subclans debated the ownership of sacred ancestral names. The men from two subclans faced each other across a vine boundary. Each subclan used an overturned canoe as a drum; each erected a series of sticks, spears, and arrows to represent totemic ancestors. The debaters held bundles of magical cordyline leaves, throwing one leaf to the ground to drive home each point. The Maliyaw subclan hoped to monopolize all sacred names, making it Avatip's de facto elite.

ritual leader and create an office for which only men born into the Maliyaw subclan would be eligible. All the other 15 subclans, for their part, were trying to keep the Maliyaw from succeeding.

Owing to interference by colonial authorities, it seemed unlikely that this attempt to create a permanent elite would bear fruit. Had the Maliyaw succeeded, however, they would have become even more envied than the Bear clan of Old Oraibi. Even the Bear clan, while providing most of Oraibi's leaders, was not able to co-opt all the other clans' rituals.

Hereditary Rank and Social Logic

Harrison's study of Avatip reinforces one of Rousseau's most important conclusions: inequality results from people's efforts to be thought of and treated as superior. Whatever the supporting role of factors such as population growth, intensive agriculture, and a beneficent environment, hereditary inequality does not occur without active manipulation of social logic by human agents. The privileges the Maliyaw wanted would have to be taken away from their fellow subclans. To endure, they would eventually have to be justified by changes in cosmology—attributing them, for example, to legendary ancestors or supernatural spirits.

We do not believe that Avatip was an isolated case. We suspect that prehistory is full of cases where one segment of society manipulated itself into a position of superiority; the problem for archaeologists is finding a way to document the process. We also suspect that debates such as those of the Avatip are the preindustrial forerunner of the political campaign.

In the pages that follow we look at an Asian society that saw elite privileges created, overthrown, and periodically reinstated. The anthropologist who witnessed this repetitive cycle has identified some of the changes in social logic for us.

THE KACHIN OF HIGHLAND BURMA

We looked in previous chapters at the traditional Ao and Angami Naga of India's Assam province. Across the border in Burma (modern Myanmar) lived their neighbors, called the Kachin. All three groups spoke languages of the Tibeto-Burman family.

As it happens, Naga and Kachin are generic terms for diverse groups of societies, some of which had hereditary rank and some of which did not. To complicate matters further, some Kachin societies had a history of shifting back and forth between hereditary privilege and equality. Archaeologists refer to such repeated shifts as "cycling."

The world first learned of Kachin cycling from anthropologist Edmund Leach, who spent time in the northern Burmese district of Hpalang during the early 1940s. The Kachin themselves used the term *gumlao* to refer to societies in which all social units were considered equal. When such units became ranked relative to one another, they used the term *gumsa.*

The key unit involved was one that reckoned descent in the father's line. The Kachin themselves called this unit a *htinggaw,* meaning "of one household." Leach refers to it as a lineage.

There may once have been more than 300,000 Kachin living in the hills of northern Burma. Hpalang lay 5,800 feet above sea level in forested hills receiving 120 to 150 inches of rain a year. The Kachin cleared patches in the forest, growing rice, millet, buckwheat, yams, and taro by *taungya,* or slash-and-burn agriculture. Taungya is called a long-fallow system because new land must be constantly cleared, while old fields are given 12 to 15 years to regain their fertility. The Kachin also raised zebu (humped) cattle, water buffalo, pigs, and chickens. The meat of the larger animals, however, was eaten only after the latter had been sacrificed during ritual, and at such times many guests shared in the feasting. These ritual feasts resembled the ones we saw among the Angami Naga.

In the cosmology of Hpalang, the world had been created by a bisexual deity named Chyanun-Woishun. This creator was reincarnated in spirit form in Shadip, the most powerful of all the *nats,* or supernatural spirits. Shadip was both the chief of the "earth spirits" *(ga nats)* and the parent of all "sky spirits" *(mu nats).*

The youngest of the sky spirits was Madai. Because they themselves practiced *ultimogeniture,* a system in which the youngest son inherits all property, the Kachin knew that his youth made Madai the most important of Shadip's offspring. In their logic, this cosmological premise was used to justify chiefly ultimogeniture when the Kachin were in the rank, or gumsa, mode of their cycle.

Madai's daughter, the spirit Hpraw Nga, married a human being. This made her husband the ancestor of the first Kachin chief. When the Kachin were in their rank mode, this cosmological premise validated the lofty position of the chiefly lineage. It allowed chiefs to sacrifice animals directly to Madai, and through him to the supreme earth spirit Shadip.

Such sacrifices were considered the ongoing payment of a bride-price to Madai's celestial lineage, since he had given his daughter in marriage to humankind. This relationship between Madai (a highly ranked nat) and his son-in-law (a human) supported another principle of Kachin social logic: The lineage giving the bride was seen as superior to the lineage receiving the bride. Bride-givers were called *mayu,* and bride-takers were called *dama.*

When the Kachin were in their gumlao, or egalitarian, mode, they kept the marital playing field level in the following way. Men of lineage A married women of lineage B. Men of lineage B married women of lineage C. Men of lineage C married women of lineage A, and so on. Thus no lineage was ever left in a permanently inferior position.

Another part of Hpalang cosmology, however, justified rank society: Storm, the daughter of the sky nat Thunder, married an orphan human of lowly status. Her husband then became the ancestor of all low-ranking Kachin lineages. As a result, all members of those lineages had to make preliminary offerings to Storm before they could even think of sacrificing animals to Thunder. And they could not sacrifice directly to Madai or Shadip at all.

Human ancestors, of course, also played a role in this cosmology. The ancestors of every lineage became *masha nats,* "ancestor spirits," and every household had shrines to them. Ancestor spirits were thought to intercede with the celestial nats on behalf of their descendants. When the Kachin were in rank mode, their chief had two household shrines, one for his human ancestors and one for Madai. Lower-ranked households, on the other hand, had only one shrine, at which they supplicated or scolded their human ancestors before making sacrifices to less-powerful nats.

The animal sacrifices of the Kachin, called *nat galaw,* or "spirit making," were built on the age-old principle of reciprocal gift-giving. One sacrificed to a nat to put him in one's debt, expecting him to return the favor. The nat took only the *nsa,* "breath or essence," from the sacrificed animal, leaving the meat to be shared by humans at a feast. When the animal was the size of a water buffalo or zebu bull, it could feed a large crowd of guests and bring prestige to the host for his generosity.

When the Kachin were in rank mode, the ritual required an additional step: one hind leg from every animal sacrificed was given to the hereditary chief. This act was a form of tribute, justified by the chief's genealogical relationship to Madai. The high nat partook of the essence of the animal, while the chief's family ate the meat. As some Kachin expressed it, they were ruled by "thigh-eating chiefs."

The chief often used his house to entertain distinguished visitors. This justified calling upon his followers to help build and repair his house, much the way clans repaired their men's houses in egalitarian societies.

When a respected man sacrificed animals and used the meat for a feast, it was commemorated in ways reminiscent of the Angami Naga. The host might create a circular dance ground 45 to 60 feet in diameter. In front of his house he set up a sacrificial post. This post was decorated with symbols of the nat being honored by the sacrifice, and the skulls of the sacrificed animals were hung there. When the chief himself hosted a major celebration called a *manau,* he sometimes commemorated it by erecting a stone monument. The chief was not seeking to achieve renown, since he had already been born to

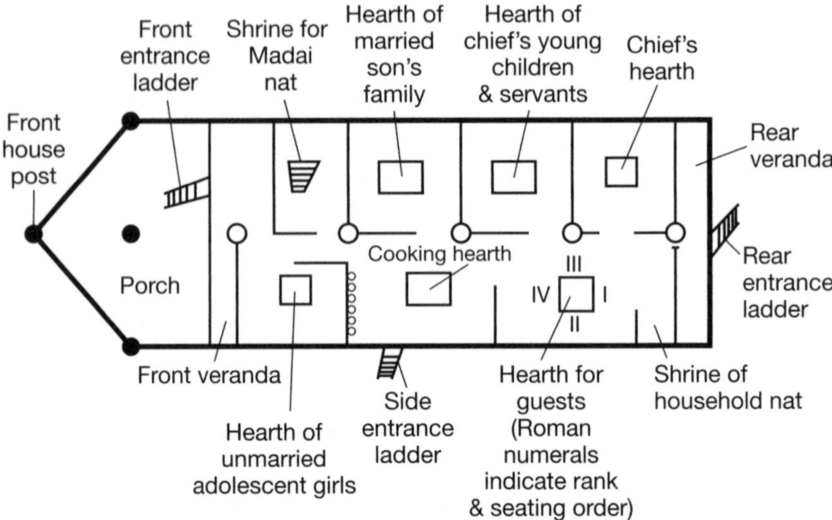

FIGURE 25. In the early twentieth century, the Kachin of highland Burma oscillated between (1) egalitarian society with achievement-based leadership and (2) rank society with hereditary chiefs. When a village was in its *gumsa,* or rank mode, the Kachin built their chief a large house like the one shown here. The porch lay at ground level, while the rest of the house was elevated for protection. All Kachin families could pray to their household *nats,* or ancestral spirits; only chiefs, because of their noble ancestry, could pray directly to the supreme nat Madai. The chief maintained a room for distinguished guests, whose sitting places reflected their relative ranks.

privilege. Instead, like the Nootka and Tlingit chiefs, he was seeking to reconfirm the high rank of his lineage by meeting everyone's lofty expectations.

Along with monuments to his greatness and lavish hospitality, a major Kachin chief might live in a house called a *htingnu* (Figure 25). Built of bamboo and thatch, the htingnu could be up to 100 feet long, large enough to accommodate the chief's multiple wives and children, loyal followers from lower-ranked lineages, servants, and slaves. A special hearth was set aside for entertaining guests, whose seating positions were ranked according to their social standing. We call attention to how similar this sounds to the sleeping positions within the plank houses of the Nootka.

Cycling between Egalitarian and Rank Society

The contrast between gumlao and gumsa leaders was great. Under gumlao, each village was autonomous. Some gumsa chiefs, on the other hand, oversaw more than 60 villages at a time. They could ill afford to forget, however, that it was the chief's entire lineage that enjoyed high rank, not the chief alone. This led to a complex dynamic among brothers.

Some anthropologists suspect that ultimogeniture was a social adaptation to long-fallow agriculture. Slash-and-burn agriculture requires so much land clearance that older sons were encouraged to emigrate to new patches of forest, leaving the youngest son behind to support his elderly parents.

The older brother who decided to emigrate took a group of followers with him to carve out a clearing elsewhere. He then purchased from his younger brother the right to make sacrifices to the ancestral nats of the chiefly lineage, becoming the "thigh-eating chief" of his own junior lineage. An older brother who declined to emigrate had only two choices: he could become a ritual specialist, or he could remain subservient to his younger brother.

An older brother who remained subservient could become so disgruntled as to become a political rival. In 1940, according to Leach, five villages in the Hpalang district recognized one man as their chief, while four other villages recognized a rival. No marriages were allowed across this factional divide, and Leach suspected that the ultimate outcome would be the collapse of hereditary rank and a return to achievement-based society.

Armed with these data, let us now consider the logical premises of gumlao and gumsa society. Our goal will be to determine the ways in which an egalitarian, achievement-based society had to change in order to produce a society with ranked lineages.

The premises of gumlao society, according to Leach, were as follows:

1. All lineages are considered equal.
2. All villages in a territory are politically autonomous.
3. Each village has a headman, to whom no tribute is owed.
4. Debts require modest repayment, with what we would call interest. (We discuss this in detail later.)
5. The price for all brides is the same.
6. Men of lineage A marry women of lineage B. Men of lineage B marry women of lineage C. Men of lineage C marry women of lineage A.
7. All siblings are equal. It makes no difference whether one is born first or last.
8. When a lineage grows and divides, there is no senior or junior division; both are equal.
9. One's loyalty is to the place where one lives.
10. Each headman is to be advised by a council of elders.
11. Land is controlled by all the lineages that originally entered the region. Late arrivals must negotiate for land.
12. Everyone makes sacrifices to his or her household ancestors, to one of the lesser sky spirits, and to one of the lesser earth spirits.
13. The head of each lineage does the above and also makes sacrifices to a regional spirit, to a sky spirit other than the supreme spirit Madai, and to an earth spirit other than the supreme spirit Shadip.

In contrast, the premises of gumsa society were as follows:

1. All lineages are ranked relative to one another.
2. Villages are no longer autonomous; all settlements within a territory are controlled by a single chief.
3. Everyone who does not belong to the chief's lineage must pay him tribute, usually in the form of a thigh from every animal sacrificed.
4. Individuals of high hereditary rank must pay more compensation (interest) for their debts.
5. Families of elite brides can request a higher bride-price.
6. The giver of the bride is considered superior to the recipient.
7. To encourage older sons to leave home and found a new lineage elsewhere, all property is left to the youngest son.
8. Any lineage that grows and splits results in senior and junior lineages, with the former dominant.

9. One's loyalty is to one's lineage rather than to a place.
10. The hereditary chief is to be advised by a council of lineage heads.
11. All land is controlled by the chief's lineage.
12. Lower-ranking people continue to make sacrifices to their household ancestors, and to lesser sky and earth spirits. Chiefs alone make sacrifices to the regional spirit of their lineage, as well as to the supreme sky spirit Madai, his daughter Hpraw Nga, and the supreme earth spirit Shadip. Chiefs are allowed to sacrifice to the highest spirits of earth and sky, because those spirits are now considered remote ancestors of the chief's lineage.

Explaining the Shift from Achievement-Based Leadership to Hereditary Rank

We know which premises of social logic had to change in order for Kachin lineages to become ranked. We now consider three alternative scenarios for how it might have happened.

One scenario, proposed by Leach, includes interactions with a more complex neighboring society, called the Shan. The Shan differed from the Kachin in significant ways. Instead of practicing long-fallow, slash-and-burn agriculture in the highlands, the Shan were supported by permanent wet-rice paddies in the riverine lowlands. Shan agriculture was so productive that it could support princely states with lineages of aristocrats, commoners, and slaves. While the Kachin sacrificed to spirits of the earth and sky, Shan rulers had been converted to Buddhism.

Hereditary aristocrats sought to communicate their rank through displays of valuables called sumptuary goods. The sumptuary goods sought by the Shan included jade, amber, tortoise shell, gold, and silver. The resources of the Kachin hill country included all these items. Significantly, the Kachin were chronically short of rice, while the Shan produced a surplus.

For several generations the family of the *saohpa,* or Shan prince, of a district called Möng Hkawm sent noble Shan women to marry the Kachin leaders who controlled the jade mines of the hill region. Sometimes a dowry of wet-rice land accompanied the bride. The Kachin chief reciprocated with raw materials for sumptuary goods.

One effect of this intermarriage, according to Leach, was that it encouraged the shift from gumlao to gumsa. Having a Shan wife raised the prestige of a Kachin leader and encouraged him to model his behavior on that of a Shan prince. Incipient Kachin chiefs might convert to Buddhism, dress like a Shan, and adopt Shan ritual and symbolism. They did so in spite of a serious contradiction in social logic: the mayu-dama relationship of the Kachin, in which the recipient of the bride was inferior, was incompatible with Shan logic. Shan princes all had multiple wives, and it would be unthinkable for any of their marriages to make them someone else's dama.

While ambitious Kachin leaders considered Shan-like behavior a mark of prestige, it only increased their followers' resentment and hastened their overthrow. The result was an inherently unstable situation in which hereditary inequality was repeatedly created, lasted for a few generations, and then collapsed.

The strength of Leach's scenario is its grounding in historical fact. While it accounts for the imitation of Shan behavior by Kachin chiefs, however, Leach's scenario relies on intervention by Shan princes. We would prefer a scenario like Harrison's for Avatip, which shows us one subclan trying to steal all ritual authority from its rivals. The strength of the Avatip scenario is that it does not assume intervention by princely neighbors.

Anthropologist Jonathan Friedman has proposed such a scenario, based mainly on Kachin behavior but augmented by what we know of various Naga groups. Friedman's scenario begins with a society whose lineages are equal in rank, like the gumlao version of Kachin society. Each local lineage has its own set of ancestor spirits, arranged in short genealogies of three or four generations. There is also a village nat whose domain is the local territory. On a higher plane lie the earth nats and sky nats which, in the egalitarian mode of Kachin society, can receive sacrifices from any lineage through the intervention of its ancestral spirits.

In Friedman's scenario the creation of hereditary rank takes place when one lineage convinces all the others that the village nat is its ancestor. That move converts one Kachin social unit into a chiefly lineage, descended from the nat who rules the whole territory. At this point the Kachin revise their cosmology to allege that their most highly ranked lineage is descended from Madai, while their lower-ranking lineages are descended from the lesser nat Thunder.

Friedman was aware that the most difficult task facing a would-be chiefly lineage was making its privileges palatable to others. In his scenario that palatability was based on a familiar premise, one we saw earlier among achievement-based societies like the Siuai of Bougainville: if one was extraordinarily successful, it meant that one had a special relationship with a supernatural being.

In Kachin society the lineages that worked the hardest and produced the greatest surplus could sponsor the most prestigious sacrifices and feed the most visitors. Their fellow Kachin, however, did not attribute such success to hard work; they believed that one only obtained good harvests through proper sacrifices to the nats. Wealth was seen not so much as the product of labor (and control over others' labor) as the result of pleasing the appropriate celestial spirits. The key shift in social logic was therefore from "They must have pleased the nats" to "They must be descended from higher nats than we are."

Once one lineage was seen as having descended from the nats that ruled a region, it made sense that that lineage should control the region's lands. It was also entitled to receive tribute from other lineages, because it alone could intercede on society's behalf with the highest nats.

As we shall see later in this book, Friedman's scenario resonates with the archaeological record in highland Mexico, where the earliest evidence for hereditary rank was accompanied by depictions of what appear to be the spirits of Earth and Sky.

Unfortunately, some archaeologists have oversimplified Friedman's scenario to the point of implausibility. What they have argued is that hereditary inequality was generated by competitive feasting. There are several problems with this oversimplification. As we have seen in previous chapters, competitive feasting in achievement-based societies usually escalated only after warfare had ceased to be a path to prominence. Instead of creating hereditary rank, it produced individual Big Men who had no way of bequeathing renown to their offspring. Let us repeat what we said in an earlier chapter: if feasting were all it took to produce hereditary inequality, there would have been no achievement-based societies left for anthropologists to study.

A Third Scenario: Debt Slavery

There is a third possible scenario for the establishment of rank society among Tibeto-Burman speakers, including both the Naga and the Kachin. Its premises are to be found in Leach's description of Kachin marriage and the mayu-dama system.

In the 1940s a moderately well-to-do Kachin groom might have to give his bride's lineage four head of cattle, plus valuables such as slit-gongs, swords and spears, coats and blankets, and pottery vessels. In many cases the haggling over bride-price went on for a long time, with negotiators using tally sticks to represent cattle and valuables.

Often a groom had to go into debt to pay for a bride. This was as true for wealthy grooms as for ordinary grooms, since bride-price was set higher for the former. One of the contradictions of Kachin logic was that bride-price was supposed to reflect the prominence of the bride's family, while in practice it reflected what the bride's family believed the groom could pay. A prominent groom could thus go even further into debt than a man of modest means.

It is no accident that the Kachin word *hka* meant both "debt" and "feud." Although debts might be left unpaid for long periods, thereby allowing social relations to continue, failure to pay could eventually have repercussions.

In Charles Dickens's England there were debtors' prisons for those who failed to repay their creditors. The Kachin punishment was just as grim: debt slavery. Many Kachin, unable to repay their loans, had to sell themselves into bondage to work off such debts. Leach estimates that in days of old, up to 50 percent of the Kachin may have been *mayam*, or slaves, nearly all owned by the chiefs or village headmen who extended the loans. A rule similar to Raymond Kelly's principle of social substitutability held a debtor's whole lineage accountable for his failure to pay. This swelled the ranks of the mayam.

Slaves in societies such as the Kachin, to be sure, do not fit our stereotype of chattel slavery in the pre-1860 United States. Mayam status was more like that of an illegitimate child, or a poor son-in-law working off his bride service. Debt slaves were considered Kachins, but of a particularly low lineage. Some eventually worked off their debts or married into nonslave lineages.

We consider debt slavery a third scenario that might have brought about the inequality of lineages, both among the Kachin and (as we saw earlier) the Native Americans of the Pacific Northwest. It is a scenario supporting Rousseau's conclusion that the most unpleasant inequalities of human society were the result not of nature but of developments in society itself.

One of the strengths of the debt slave scenario is that it is based on events Leach actually witnessed. One question, however, remains: What would happen to slaves when society cycled from ranked to unranked? Leach reveals that some Kachin chiefs came to rely more on the loyalty of their personal slaves than on the loyalty of families from lower-ranked lineages. As gumsa society broke down, therefore, the most loyal slaves may have been assimilated into their owners' extended kinsmen so that this close relationship could continue.

CYCLING AMONG THE KONYAK NAGA

Let us now go west from the Kachin hills and cross the border into Assam, north of the territory of the Ao and Lotha Naga. Here one encounters a land of forested hills, 4,000 feet above sea level, drenched with 160 inches of monsoonal rain between April and September. This part of Nagaland was the realm of the Konyak Naga.

The Konyak grew rice, millet, and taro and raised water buffalo, pigs, and a burly species of cattle called the mithan. They shared a number of institutions with their Ao and Angami neighbors: the morung, or men's house, the taking of enemy heads, the building of prestige through the hosting of ritual feasts, and so on. They differed from the Ao and Angami in that, like the Kachin, their society had a long history of cycling between rank and egalitarian.

Such was the nature of cycling that some Konyak villages had a mixture of institutions from achievement-based and rank society. A visit to such a village was like seeing a snapshot taken during the transition from one social system to another.

Like many speakers of Tibeto-Burman languages, the Konyak lived in a cosmos peopled by spirits—some friendly, some malevolent—who were approachable through proper ritual. Their highest supernatural being, whose various regional names meant "Earth/Sky," was equivalent to the highest nat of the Kachin. Portrayed as an immense Naga, this supreme spirit created Thunder and Lightning, Earth, humans, and rice. He was the ultimate guardian of the moral order, rewarding the virtuous and punishing the evil.

The Konyak were visited in the 1920s by John H. Hutton and James P. Mills. Anthropologist Christoph von Fürer-Haimendorf lived among them during the period 1936–1937 and returned in 1962 to see how their life had changed. Fürer-Haimendorf spent most of his time in Wakching, a village of 1,300 Naga, but he also visited 20 other villages, including Niaunu. These travels enabled him to see every stage in the transition from achievement-based to rank society.

The Konyak were divided into descent groups called *li* ("clans"), each led by an elder. All reckoned descent in the father's line, and people had to marry outside their li. Villages were divided into residential wards, each of which built its own men's house. Wakching was divided into five wards, ranging in size from 40 to 82 houses. Clans were spread among different wards, and some clans occupying the same ward shared a men's house. The heads of the men's houses inherited their position within each ward, and they collectively made up the village council.

The Konyak morung was easily the most impressive of all Naga men's houses. It could be 84 feet long and 36 feet wide, with a porch extending out another 24 feet. The lintels and doorjambs were carved with elephants and tigers. Inside were bamboo bunks for the young men, all of whom slept there once they had reached a certain age. Three sides of the porch had benches that allowed older alumni to gather, without having to enter the boys' dormitory. The morung built by the chief's clan was particularly elegant, featuring slit-gongs, baskets of enemy heads, and a marimba whose notes summoned members to ritual.

Like the Kachin, the Konyak had two different modes of social organization. They used the term *thenkoh* for villages where, as Fürer-Haimendorf puts it, "one could live for a considerable time without being conscious of distinctions of rank." They used the term *thendu* for villages where hereditarily ranked clans were clearly evident. Of the villages Fürer-Haimendorf knew best, Wakching was thenkoh and Niaunu was thendu.

Let us begin with Niaunu, where there were at least four levels of hereditary rank. People of the highest category were known as *wangham* ("great Angs"); those of the second rank were *wangsa* ("small Angs"). People of intermediate rank were called *wangsu*. At the bottom of the hierarchy were *wangpeng* ("commoners"), debt slaves, and captives. Because Niaunu was a thendu village, people of high rank could be recognized immediately by their dress and sumptuary goods. Only elite women could wear their hair long and use red-and-white-striped skirts. Brass earrings, bracelets, and bangles were widespread. Highly ranked men received elaborate facial tattoos, while those in thenkoh villages did not.

It was important for chiefs to preserve the nobility of their line through strategic marriage. Only the union of two people of great Ang rank could produce a future chief. When a great Ang took a bride of lower rank, their son became a small Ang, suitable only to be the subchief at a minor, tribute-paying village. A great Ang named Nyekpong, who ruled Niaunu village in 1962, had two wives of Ang rank and 24 wives of commoner status.

The chief's roles were to administer a district with the help of his small Ang subchiefs; to direct the affairs of his own village; to receive tribute in rice, pigs, fish, and water buffalo; to punish criminals; to resolve disputes; and to lead raids against enemies. Chiefs were proud and dignified figures who traveled with a large entourage of bodyguards, followers, and servants. Commoners approached the chief bowing, never looking directly at his face. They cleared his rice fields for him and helped him build his house, which could be 360 feet long. Inside the house were halls as elegant as those of the men's house and a carved bench on which the chief alone could sit.

When thendu, or chiefly villages, cycled into thenkoh mode, it was usually not because of an egalitarian overthrow like those of the Kachin. A chronic problem for the Konyak was that population growth among commoners outpaced that of Angs. Sometimes the chiefly family of a village literally died out. When this happened, the village either had to settle for a lower-ranked leader or switch its allegiance to a thendu community that could supply it with a new Ang. If no Ang family was available, the rank structure might collapse, returning the village to a more democratic, achievement-based society.

Even when a village obtained a new Ang, that leader might fail to gain the loyalty of his new subjects. In such cases, the unpopular Ang might be ousted with the aid of a rival village. It was prohibited to kill a member of one's own chiefly lineage, but "hit men" from another group could be paid to do it. This was dangerous work, because the Konyak believed that great Angs possessed magical power.

Fürer-Haimendorf noted that even when a village cycled into thenkoh mode, the rank differences among clans did not completely vanish. They were simply not emphasized by dress or behavior, because now the village had to be run by the leaders of the men's houses. These men's houses continued to collect the same tribute from satellite villages that had once been paid to the chief. These ongoing tribute payments were justified on the grounds that the satellite villages were using land that had once belonged to the parent village.

Thenkoh villages also revived an institution of autonomous village society: the prestige-building "feast of merit." Ambitious men sacrificed mithan cattle or water buffalo and supplied their guests with quantities of rice beer. Hosts also paid gangs of men to haul massive slit-gongs to their village, where they were kept in a special shed.

Villages with a great Ang chief did not need feasts of merit, because everyone's rank was determined at birth. The chief himself did host spectacular feasts, but their purpose was simply to reaffirm his greatness. An Ang might arrange for the hauling of a massive slit-gong, but it was delivered to his house rather than to a public shed.

What we see in the thendu-thenkoh cycle, therefore, is an interesting mix of institutions from achievement-based and rank societies: chiefs, sumptuary goods, and tribute payments sometimes coexisted with men's houses and feasts of merit. Slit-gongs would be hauled to a public shed or to a chief's house, depending on a village's stage in the cycle. Villages without a great Ang might ask their neighbors for one, only to hire hit men if he did not work out. As we will see in subsequent chapters, the archaeological record provides evidence for similar mixtures of institutions during periods of transition.

Head-Hunting and Territorial Expansion

Like all the other Naga we have examined, the Konyak went on head-hunting trips. Every man owned a machete-like *dao* for clearing brush, slaughtering cattle, and beheading enemies. The Konyak went to war with five-foot spears, skull-cracking clubs, and shields of water buffalo hide. Human heads had powerful magic that enhanced the fertility of crops and the prosperity of the village. Men who had taken heads received special tattoos and were allowed to wear a brass pendant in the form of a trophy head (Figure 26).

Unlike the Marind headhunters of New Guinea, who spared young women, the Konyak had only one rule: As long as the victim was old enough to possess teeth, his or her head was fair game. People were thus beheaded regardless of sex, age, or rank. Warriors returning to the village with enemy heads were greeted enthusiastically, bathed, and honored with dances. The heads were put in a basket near the slit-gong, so that they could be "fed" rice and beer at all feasts. Eventually they were hung from posts in the men's house.

Because of the frequency of raids, the Konyak often placed their villages on defensible ridges with a reliable water supply. Some were enclosed by a palisade with a gate and had a bamboo lookout tower. In hilltop settings, residential wards might be separated by deep ravines. Panjis, or sharpened bamboo spikes, were used to mine the approaches to the village.

The villages, which varied from 50 to 250 houses in size, had carefully locked granaries of rice and millet. The villagers built circular dance platforms and erected stone monuments near the men's houses. Outside the village was the cemetery, where rotting corpses were exposed on platforms and bleached skulls were piled in sandstone basins. These skulls had been wrenched from the body after six days of decomposition in the tropical heat, cleaned and emptied, and kept where they could be given their usual share of rice and beer for three more years. Konyak treatment of ancestors' skulls recalls the behavior of villagers at ancient Near Eastern sites such as Jericho, Ain Ghazal, and Çayönü.

When a great Ang died, his corpse was allowed to decompose in a wooden coffin rather than the usual bamboo bier. After cleaning, his skull had its orbits filled with white tree pith to resemble eyes. A craftsman then painted the skull with the same tattoos the great Ang had worn in life and glued some of his hair to the skull with resin. The remains of mighty warrior chiefs were always buried with their weapons. They would need them because while en

FIGURE 26. During the 1930s the Konyak Naga of Assam cycled between (1) egalitarian society with achievement-based leadership and (2) rank society with hereditary chiefs. When the village of Longkhai was in its *thendu*, or rank mode, it featured both inherited and achieved prestige. On the left we see a Longkhai hereditary chief dancing in ceremonial costume, brandishing a spear and a machete-like *dao*. On the right we see a Longkhai commoner whose prowess in battle had won him the right to wear special tattoos and a pendant in the form of a trophy head.

route to Yimbu, the land of the dead, they would have to confront the spirits of the men they had killed.

Konyak warfare owed its origins to head-hunting. Hereditary rank, however, provided raiding with another goal: territorial expansion. Fürer-Haimendorf reports that 12 generations before his 1962 visit, the thendu village of Niaunu was founded by a great Ang chief named Maipupa. Maipupa then declared Niaunu the "parent" of four other villages, each of which would in the future be ruled by men of his chiefly lineage.

One of the four villages that Maipupa claimed as a satellite was Mintong, a community ruled by a rival Ang lineage. Maipupa sent a force of Niaunu warriors to raid Mintong. They wiped out the chief's family (including the children who were his heirs), slaughtered most other members of highly ranked clans, and killed at least half the commoners. Maipupa then installed a new chief, drawn from the junior branch of his own lineage.

The use of raiding for the increase of chiefly territory became a strategy of societies with hereditary rank, virtually whenever and wherever they arose. Once seen only as a means to obtain enemy heads and settle scores, raiding had been turned into a tool of political expansion. Warfare would never be the same again.

THE CREATION OF HEREDITARY RANK

There is no substitute for eyewitness accounts of inequality in the making. The struggles of the Manambu, Kachin, and Konyak Naga put a human face on the creation of rank.

Those struggles show us that hereditary inequality is not something that appears spontaneously once population has increased, or agriculture has produced a surplus, or people have accumulated lots of shells and pigs. Inequality is orchestrated. At the same time, it is not enough for one segment of society to demand privileges for itself and its heirs. Would-be nobles need leverage, an advantage of some kind, or their privileges will be taken back by the rest of society. That is presumably why so many societies remained achievement-based for thousands of years.

When did evidence of hereditary rank first appear in prehistoric farming societies? This is a difficult question for archaeologists to answer, since they rely on inference rather than direct observation. Making their task more difficult is the fact that many prehistoric societies combined both inherited and achieved inequality. This fact forces archaeologists to ask whether the unequal treatment they detect could have resulted from a lifetime of accomplishment or was more likely someone's birthright.

That said, we believe that we can see signs of hereditary rank in Mesopotamia between 7,300 and 7,000 years ago, and in Peru and Mexico between 3,200 and 3,000 years ago. We shall present the evidence later in this book.

We have used the building of men's houses as an indicator of village societies where leadership was based on achievement. This enables us to use the decline of the men's house and the rise of the temple as an indicator of societies with some degree of hereditary leadership. In the societies we have examined, the transition from the men's house to the temple seems to have been associated with the decreasing importance of ordinary people's ancestors and the increasing importance of the celestial spirits in the chief's genealogy.

ELEVEN

Three Sources of Power in Chiefly Societies

We have seen that agricultural villagers do not surrender their equality without a fight. No sooner does one social segment achieve elite status than its privilege is challenged, forcing it to resume its quest for supremacy. Cycling between ranked and unranked was probably common in the preindustrial world. Eventually, however, the leadership roles in some societies became hereditary in perpetuity.

One part of the world where hereditary rank flourished was the South Pacific. To be sure, most Polynesian islands were colonized by people from places that already featured some degree of inequality. On a number of archipelagoes, however, the level of inequality continued to escalate after the first canoes arrived.

Anthropologist Irving Goldman once took a close look at 18 Polynesian societies. He succeeded in identifying three widely shared sources of chiefly power. All Goldman intended to do was break down hereditary Polynesian leadership into its component parts. Afterward, by recombining those parts in different ways, he hoped to account for the variety in Polynesian societies. As it turned out, however, Goldman gave us a way of comparing rank societies worldwide.

The central concept of chiefly power was a life force the Polynesians called *mana*. Goldman defines mana as an odorless, colorless, invisible, supernatural energy that pervades people and things. To be sure, all the societies we have examined so far believed in a life force and had ways of accumulating or losing it. In Polynesia, however, people of high rank were automatically born with more mana.

The person with the largest supply of mana was the chief. He had so much life force that he was described as *tapu,* a term from which we get the English word "taboo." Anyone or anything tapu was approached with extreme caution. Some Polynesian chiefs had so much mana that by touching them inappropriately, one could receive a jolt akin to being Tasered.

A second source of power in Polynesia was *tohunga,* a term usually translated as "expertise." Tohunga could refer to administrative or diplomatic skill, ritual skill, or craftsmanship. While innate talent was certainly involved, individuals could increase their expertise through education, training, or apprenticeship. Sometimes a chief would provide incentives to the craftsmen who produced his sumptuary goods.

The third of Goldman's sources of power was *toa.* While toa referred to a durable tree known as "ironwood," it was also a metaphor for bravery and toughness. Toa was applied to warriors in general, and especially to those who distinguished themselves in battle. A key aspect of toa was that it allowed for a certain degree of social mobility. A warrior of humble birth could rise in prominence to the point where he had to be taken seriously, even by chiefly individuals. For his part, a chief who fought bravely became a legend.

All chiefly Polynesian societies relied on a combination of mana, toa, and tohunga. The emphasis, however, was different from island to island. In the case of the Maori and Tikopians, chiefs relied on a combination of sacred authority and genealogical seniority. On Samoa and on Easter Island, chiefs relied more heavily on political expertise and military force. In Tonga and Hawaii, which had the highest levels of social inequality, chiefly families utilized the entire playbook: sacred authority, genealogical seniority, military force, and political and economic expertise.

Polynesian societies did not oscillate between ranked and unranked, as the Kachin and Konyak Naga did. The island societies, however, had their own form of cycling: status rivalry. Polygamous chiefly families produced brothers, half brothers, and first cousins who were almost equal in rank. Sometimes the heir to a chiefly office did not control as many warriors as his ambitious junior rival. In such cases assassination, overthrow, and usurpation could cause one chiefly lineage to collapse while another rose.

All three of Goldman's principles, of course, had antecedents in earlier, achievement-based societies. They had been transformed by changes in social logic, as follows:

1. Achievement-based groups pursued their own versions of life force. The Naga obtained it from the heads of their enemies. The Mandan obtained it from self-induced suffering. Chiefly Polynesians, however, possessed it from birth and could increase it or lose it depending on their own behavior.
2. Leaders in achievement-based societies had expertise of various kinds. They could memorize thousands of sacred names, like the villagers of Avatip, or develop skills at moka, like the people of Mt. Hagen. They could master ivory carving or eagle trapping. In the chiefly societies of Polynesia, however, certain craftsmen were more respected than others, for example, the makers of war canoes, purveyors of sumptuary goods, or carvers of giant statues such as those on Easter Island.
3. In achievement-based societies, bravery in war was already a path to renown. Chiefly societies converted war to a strategy of territorial expansion. Tired of negotiating for the products of a neighboring region, chiefs might just subjugate the region and demand its products as tribute. This enhanced the value of military prowess.

In this chapter we look at three societies with hereditary rank. In the first, sacred authority was paramount. In the second, war was endemic and a chief's patronage of the crafts enhanced his prestige. In the third, a chief's mana, toa, and tohunga made him almost as powerful as a king.

We also take note of a change that accompanied the rise of many rank societies: men's houses were replaced by temples. This change reflects an important social and political transition. Men's houses were built by clans or Big Men and tended to be places where men sat around communing with their ancestors. Temples tended to be places where actual deities lived on a full-time or part-time basis. Temples were staffed not by initiated clansmen but by people trained as priests. Often the construction of a temple was directed by the chief because, after all, there were supernatural spirits in his ancestry.

THE PREHISTORY OF TIKOPIA

The small island of Tikopia ("Tee-ko-PEE-a") lies at the western margins of Polynesia. When anthropologist Raymond Firth arrived in 1929, the island had 20-odd lineages grouped into four clans. One lineage in each clan was considered chiefly and provided the *ariki,* or hereditary chief, for its clan. In addition, all four clans and their ariki were ranked relative to one another. The authority of the chiefs was primarily moral and religious; it represented, in other words, an enhanced version of the hierarchy of virtue seen in achievement-based societies.

Tikopia is only three miles long and a mile and a half wide. In 1929 it was occupied by 1,200 people who supported themselves by fishing and cultivating taro, yams, bananas, coconuts, and breadfruit (*Artocarpus* sp.). The breadfruit was stored in pits in the form of a paste.

The people of Tikopia told Firth that over the centuries they had been visited by canoeloads of visitors from island groups such as Samoa, Tonga, and Pukapuka. Almost anyone arriving by canoe had been accepted and protected by one of the chiefs. Not all overseas visitors, however, had been friendly. One group of Tongans in particular had been "fierce, ruthless and even cannibalistic." Fortunately the Tikopians were able to repel them.

On one occasion an amicable Tongan noble named Te Atafu arrived just when the Tikopian clan called Taumako was facing extinction. Te Atafu was adopted by the Taumako chief, married a highly ranked woman, and helped rejuvenate the clan. Centuries later the Taumako clan still talked about its Tongan connections. Firth came to suspect that Tikopian society "was the result of a fusion of a number of elements from a variety of islands," but he realized that he had no way to confirm his suspicions without archaeological research.

Fortunately archaeology came to the rescue. In 1977, almost half a century after Firth's first visit, archaeologist Patrick Kirch and botanist Douglas Yen arrived in Tikopia. Their excavations added nearly 3,000 years to the island's history.

Kirch and Yen discovered that Tikopia had been colonized 2,900 years ago. Occupation began in the southwestern lowlands and gradually spread to the shores of Te Roto, a saltwater bay. The native plants and animals of Tikopia were in pristine condition at that time. The refuse heaps left by the colonists contained fish, large mollusks, abundant sea turtles, and a wild fowl called the megapode.

The colonists brought with them domestic pigs, chickens, and dogs. Their tools suggest that they had brought cultivated plants as well, although remains of the actual crops had not survived the tropical climate. From the shell of the giant *Tridacna* clam, the colonists made adzes to carve seafaring canoes. They made "peelers" of cowrie shell for use on taro, yams, and breadfruit. They cleared land with fire, practiced shifting cultivation, cooked in earth ovens, and fished with hooks and nets. Their skill at long-distance canoe travel, which had gotten them safely to Tikopia in the first place, kept them supplied with obsidian and stone axes from distant islands.

Over the next 800 years, the colonists had a serious impact on Tikopia's environment. The island's original forests were depleted, the megapode was driven to extinction, sea turtles were reduced in number, and mollusk resources were shrinking. As the island's population grew, the Tikopians increased their pig raising to keep pace with the loss of wild resources.

Beginning 1,000 to 800 years ago, further changes could be seen in the archaeological record of Tikopia. Some of these changes likely reflected the arrival of people from other islands. Agriculture was intensified, with the cultivation of permanent plots replacing shifting cultivation and land clearance by fire. *Tonna*-shell peelers for taro and breadfruit replaced the earlier cowrie-shell peelers. The islanders began digging the type of pit in which breadfruit paste could be stored. Coconuts and *Canarium* nuts showed up in the refuse. With the intensification of agriculture, pigs gradually disappeared from Tikopia. Possibly the hungry pigs had become a threat to the garden plots.

One significant innovation, which occurred roughly 600 years ago, was the introduction of the Tongan-style temple. Such temples were founded on platforms of cut-and-dressed stone masonry, using material quarried from old coral beds. Along with these temples came Tongan-style burial mounds with rectangular, stone-slab-faced foundations. The Tikopians even borrowed the Tongan term *fa'itoka* for such mounds. In other words, Kirch and Yen's archaeological discoveries support the accounts of the Taumako clan, whose members told Firth that immigrants from Tonga had reached Tikopia and helped rejuvenate them. The archaeological data also supported Firth's suspicion that Tikopian society, while unique in its own right, had incorporated behaviors from a variety of islands.

Tikopia in 1929: Four Chiefs Are Better Than One

The Tikopians who talked to Firth, of course, told a story of their origins that differed from Kirch and Yen's. They believed that the first inhabitants of Tikopia were the *atua,* or spirit beings who could assume human form. Not long after, humans appeared. Next came the births of lesser, or tutelary, deities, followed by the *Pu Ma,* or principal twin deities, Tafaki and Karisi. These twin gods became the patrons of the chiefly lineage called Kafika. Tikopian cosmology thus explained why the Kafika lineage was the most highly ranked.

Recall that Tikopian clans were ranked relative to each other, and lineages were ranked within each clan. The chiefs of Clan A came from the Kafika lineage; this clan also had six commoner lineages, distributed through 18 villages. The chiefs of Clan B came from the Tafua lineage; this clan also had five commoner lineages, distributed through 14 villages. The chiefs of Clan C came from the Taumako lineage, famed for its infusion of elite Tongan visitors; Clan C also had seven commoner lineages, distributed through 16 villages. The chief of Clan D, the smallest, came from the Fangarere lineage; this clan had only one commoner lineage, distributed through four villages. Commoner lineages were led by ritual elders, who served as advisers to the chief of their clan.

Because there were four chiefs in place at any one time, Tikopia had no unified central authority. Firth described Tikopian society as a "loosely structured oligarchy," using the Greek term for rule by a privileged few. The ariki ruled as aristocrats—by virtue and mana rather than by wealth—and one clan's chief could not impose his will on the other three clans.

Under ordinary circumstances it would have been difficult for an outsider to know that the *ariki Kafika,* or chief of Clan A, was "first among equals." However, on a second visit to the island in 1952, Firth witnessed a ceremony at which the relative ranks of the four chiefs were on display. At this ceremony colonial officials had come to distribute gifts to all four chiefs, while crowds of their clanspeople watched. The meeting was held on a ritual yard in front of a sacred canoe shed. Each chief sat on a coconut-grating stool so that the seated commoners' heads would be lower than his.

Firth noted that the placement of each chief's stool on the yard reflected his rank relative to other chiefs, based on the importance of his clan's tutelary deities and the length of his genealogy. The ariki Kafika, for example, claimed 19 generations of glorious ancestors, while the ariki Fangarere could claim only eight or nine.

Hereditary chiefs were shown great deference. A commoner delivering a gift to an ariki would touch his nose to the chief's knee and say, "I eat your excrement ten times." The chief, for his part, might humble himself to a spirit ancestor by saying, "I eat your excrement." This act of humility was appropriate because the spirit world was the ultimate source of the chief's mana. Through his relationship with the spirit ancestors, the chief controlled the natural fertility of gardens, the weather, the health of his human subjects, and the abundance of fish. When a chief called to the mackerel, it approached.

When he spoke to the breadfruit tree, it bore fruit. When a chief pointed angrily at a man, the man sickened and died.

Each chief's authority extended beyond his home village to all the settlements of the district in which members of his clan lived. Anthropologists sometimes refer to the territory controlled by a chief as his "chiefdom." Note that this term refers to a territory and not to a type of society; Tikopian society should simply be referred to as a rank society. It was, in fact, only one variety of rank society—one in which corporate segments, such as lineages and clans, were ranked relative to each other. In later chapters we will see rank societies in which there was an even more complex ranking of individuals within chiefly lineages.

The ritual buildings on Tikopia reflected its society's position along the continuum from egalitarian to chiefly. We have already seen that Tongan-style temples were introduced about 600 years ago. These were buildings in which deities were propitiated and carefully memorized chants were recited. However, because the clan was still a very important unit on Tikopia, the island had also retained the bachelors' house, an institution surviving from earlier and more achievement-based times.

Another link to earlier, achievement-based society was the *pora,* a large feast thrown by each new ariki of the Taumako clan. Each pora involved a huge food outlay, with taro pudding being especially favored. A pora might also be held when the Taumako clan rebuilt its temple. Thus, despite the hereditary authority and mana possessed by the chief, he was expected to put on displays of generosity as great as those of leaders in achievement-based societies.

One other Tikopian institution, this one involving the dead, should be mentioned. In the large nuclear family houses on the island many dead were buried below the house or its eaves, always on the side of the house reserved for ritual. The family would place a coconut-leaf mat over the grave, after which it was taboo to walk there. After a certain number of burials had accumulated, the dwelling might be declared a "house of the dead," a Tikopian version of the charnel houses found in some ancient villages. The family would then build itself a new home.

Tikopia also provides us with examples of the way bride-price could escalate in rank society. The groom's family presented gifts to the bride's parents because they were losing her; to her mother's brother, who was usually a member of a different clan; and to others who might be classified as the bride's kin, even though they were not blood relatives. As many as 15 different transactions might be involved at the family, lineage, and clan levels. Most significant, from our perspective, is the fact that gifts were also sent to the chief of the bride's clan, allegedly to compensate him for the loss of her labor. In effect, the ariki—like the "thigh-eating chiefs" of the Kachin—skimmed off a share of the food and valuables.

The Limits of Inequality in Tikopia

Goldman considered Tikopia a "traditional" or fairly modest rank society. Leadership was based largely on religious authority and genealogical credentials, with little or no use of force. In addition to keeping a series of commoner lineages happy, each chief had three other arikis' opinions to consider. He resorted to war only to fight off invaders from other islands.

Although it had clear hereditary chiefs, Tikopian society still preserved many of the institutions of achievement-based societies. Bachelors' houses existed side by side with temples. Some individuals might be interred in a Tongan-style burial mound, but others were buried beneath a residence that eventually turned into a charnel house. Commoners had to sit with their heads lower than the ariki, but the ariki was supposed to provide them with generous feasts. Chiefs controlled entire districts and their garden land, but there was no central authority for the island as a whole. The simultaneous presence of four chiefs acted as a system of checks and balances, preventing one ambitious leader from taking over all of Tikopia.

Of Goldman's three sources of inequality, mana was by far the most important in Tikopia. Toa, or military prowess, came to the fore mainly in response to hostile immigrants.

We have not singled out Tikopia for its craftsmanship, or tohunga. To be sure, the carving of canoes with shell adzes was a highly respected profession, and expertise was needed to maintain permanent garden plots. We see no craft in Tikopia, however, that rose to the level of the goldwork and polychrome pottery produced by the Central American societies we consider next.

We must end on a cautionary note. As we have seen, Kirch and Yen did a wonderful job of adding 3,000 years to Tikopia's history. At no point in the archaeological sequence, however, could one see exactly when the pattern of four chiefs, arranged in rank order, first appeared. This pattern was apparent in the 1952 ceremony that Firth witnessed but would have been archaeologically invisible. Clearly, some kinds of inequality cannot be detected by archaeology alone.

THE RANK SOCIETIES OF PANAMA AND COLOMBIA

When the Spaniards explored Panama and Colombia in the early 1500s they encountered hundreds of Native American societies led by hereditary chiefs. Many of these explorers kept diaries and sent reports back to Spain. While the authors of these manuscripts were not trained social scientists, they were eyewitnesses to societies never before seen by Western observers. Their writings have therefore become as important to Latin American archaeologists as Julius Caesar's accounts of the Gauls are to Classical archaeologists.

Panama and Colombia can be discussed in tandem because their chiefly families paid long-distance visits to each other. Colombia's goldwork and Panama's polychrome pottery both required high levels of the expertise called tohunga by Polynesians.

The Cauca Valley of western Colombia is 300 miles long and 35 miles wide. It was formed by the Cauca River, which flows north to the Caribbean between rugged mountain ranges. At least 80 different rank societies occupied the valley before the arrival of Europeans. The earliest documents describing these societies began arriving in Spain around 1535 and have since been studied by ethnohistorian Hermann Trimborn and anthropologist Robert Carneiro.

In recent years the Cauca Valley has become famous as the heartland of the Medellín drug cartel, but in ancient times its economy was supported by the growing of corn, manioc, and cotton. The population of the valley on the eve of Spanish contact has been estimated at 500,000 to 750,000.

One important revelation of the Cauca documents is the diversity one finds in a sample of 80 rank societies. The two largest were the Guaca in the north (downstream) and the Popayán in the south (upstream). The Quimbaya of the mid-valley region were intermediate in size. The Catío in the north were unimpressive compared to their Guaca neighbors. They spent most of their time in autonomous villages, uniting under a "war chief" only when threatened. If we knew more about the Catío, we might find that they periodically cycled between egalitarian and ranked modes.

The Quimbaya comprised 80 villages, organized into five separate rank societies. In other words, each Quimbaya chief controlled a district, or chiefdom, averaging 16 villages. The largest villages had more than 1,000 occupants.

At the large end of the scale were the Guaca and Popayán, both of whom were expanding when the Spaniards arrived. The Guaca numbered between 48,000 and 60,000 people, united under one paramount chief. Below the paramount were his subchiefs (often his brothers, half brothers, or cousins), who controlled villages subordinate to his. Still lower in the hierarchy were hamlets, which had no members of the chiefly lineage and were subservient to the subchiefs.

The Spaniards described the most powerful Cauca societies as displaying the following ranks. First came the hereditary chief, who was succeeded by his son, or by his sister's son if he had none of his own. Below him were "nobles by blood," that is, others of his chiefly lineage. One member of this lineage, often the chief's younger brother, was named "war chief." There were also "nobles by command," men from commoner lineages who had been rewarded for distinguishing themselves in battle. Still lower in the continuum of rank were "nobles by wealth," essentially commoners who had done well at accumulating food and valuables. It is likely that a certain number of these "nobles by wealth" had risen by expertise at trade or craftsmanship.

On the bottom rungs of the ladder were free commoners and slaves. The commoners were mostly farmers. The slaves were mostly captives, taken during incessant raids by Cauca Valley warriors.

Important members of the larger rank societies included priests, who served in the temples, directed important public rituals, and provided sacred justification for the chief's authority. Priests did not interfere directly in political decisions, but they did memorize thousands of ritual procedures and direct sacrifices. They also officiated at chiefly funerals.

Let us now look at the relative roles played by war and craftsmanship in the rank societies of the Cauca Valley. Their wars had many underlying causes, two of which were a desire for political expansion and an insatiable need for slaves. Several societies, including upstream groups such as the Jamundí and the Lile, had fought wars of expansion around the time the Spaniards arrived. For example, a Lile chief named Petecuy had defeated five other Lile chiefs and consolidated all their territories into a single chiefdom.

Even in cases where political unification was beyond the reach of a chief, enemy villages were raided to obtain slaves. Some of these slaves were used as forced labor in the gold mines, recovering the ore that Cauca goldworkers would turn into prestige goods. Other slaves became sacrificial victims at chiefly funerals.

The number of warriors a chief could muster, of course, varied with the number of villages he controlled. Even minor battles could involve 200 to 400 warriors, many carrying special cords to bind the wrists of prisoners. The powerful Guaca could produce a force of 12,000 warriors, ordered to spare no one. Villages were burned; men, women and children were killed or taken captive; warriors took trophy heads; and cannibalism was so common that human flesh became a trade commodity. The Lile chiefs, mentioned earlier, were said to have owned 680 drums made from the skin of war captives. Many of these behaviors were considered terror tactics, intended both to show the disdain of chiefs for their enemies and to demoralize anyone who considered resisting.

Many Cauca chiefs owned either gold mines or the streams in which gold could be panned. They had patron-client relationships with the craftsmen who turned the gold into crowns, headbands, ear and nose ornaments, pendants, and scepters for them. It is not surprising that Cauca chiefs adorned themselves with gold; the surprise is that the wearing of gold was not restricted to the chief's lineage. Nobles by blood, nobles by command, and nobles by wealth were allowed to acquire as much as they could afford, although there were certain items made only for the chief.

The most famous gold-producing societies were those controlled by the five Quimbaya chiefs, whose territories were modest compared to those of the Guaca. The expertise of Quimbaya craftsmen was a source of prestige for the chiefs who supported them.

Ambitious Cauca chiefs were not held in check by a council of elders, although they did pay attention to their military advisers when it came to war tactics. The most powerful chiefs were carried from place to place on litters or hammocks. They lived in wooden houses with thatched roofs, surrounded by servants, slaves, messengers, and interpreters who helped them deal with foreigners. Large crews of commoners tilled their fields.

In addition to his principal residence a chief might have a second house for war trophies, such as the skulls or dried heads of his enemies. At the chief's death, some Cauca societies preserved his body by smoking or mummifying it, after which it would be kept in the house of his successor. In other societies the chief was buried in a cist or shaft tomb, accompanied by fabulous offerings and sacrificed prisoners and servants.

In his discussion of the sixteenth-century documents, Carneiro points to the following differences in strategy between the rank societies of the Cauca Valley and the achievement-based societies in the neighboring regions of South America:

1. Like the Siuai of Bougainville, many achievement-based societies destroyed a prominent man's property at his death. Cauca societies let

the son inherit his father's property, allowing it to grow generation by generation.

2. Achievement-based societies brought captives home to torture or kill. Cauca societies considered prisoners a commodity, to be kept (or traded) as slaves.
3. Achievement-based societies (and even modest rank societies like Tikopia) tended to expel criminals. The rank societies of the Cauca Valley added criminals to the slave population, increasing their labor force.

To Carneiro's points we can add a fourth. While Cauca chiefs were not without aristocratic and moral authority, they made greater use of war and crafts than did the chiefs of Tikopia. They aggressively increased the number of villages under their control and thought nothing of making mincemeat out of neighboring chiefs. They sponsored and protected craftsmen such as the goldworkers of Quimbaya, using their products as a source of tribute, adornment, inheritable wealth, and chiefly gift-giving.

Rank Societies in Panama

When the Spaniards arrived in Panama, they discovered some three dozen districts under the control of hereditary chiefs. None of the Panamanian chiefs, however, were as powerful as those of the Guaca and Popayán.

Panamanian commoners practiced slash-and-burn agriculture, which required them to move their hamlets as fields were left fallow and new areas of tropical forest were cleared. Men slashed the trees and burned them, but the actual planting and harvesting was done by women. Panamanian agriculture relied on a mixture of Mexican plants such as corn and South American plants such as manioc and sweet potatoes. Some corn was converted to beer, another South American custom. While the women farmed, the men hunted deer and peccaries, which were smoked and salted to preserve the flesh. Villagers ate fish, sea turtles, manatees, crabs, and shellfish, along with iguanas and many species of tropical birds.

While the shifting settlements of the commoners featured circular houses with cane walls, clay daub, and conical thatched roofs, the houses of the chiefs were more permanent. Since they did not have to move around, clearing new land the way commoners did, chiefs could build substantial residential compounds called *bohíos.* One chief named Comogre lived with his servants and bodyguards in a bohío measuring 150 by 80 yards. Its various buildings had timber beams and were strengthened with stone walls. The compound had impressive carved ceilings, decorated floors, storehouses, cellars, and rooms for the burials or mummies of previous chiefs.

In her reading of the eyewitness accounts of the sixteenth-century Spaniards, anthropologist Mary Helms detects at least five social ranks in Panamanian society. The principles on which these ranks were based seem remarkably similar to those used by Cauca societies (with whom the Panamanians were in contact) and even those used by Pacific Island societies (who came up with the principles independently).

The highest Panamanian rank was *queví,* the category to which both the chief and his principal wife belonged. Below this category came lower-ranking members of the chiefly lineage, known as *sacos,* the equivalent of the Cauca Valley's "nobles by blood." Still lower in rank were *çabras,* commoners whose prowess in war had given them a status equivalent to the Cauca Valley's "nobles by command." These prominent warriors were rewarded with women and slaves, and their sons could inherit their titles if they fought ferociously for their chief. Lacking the genealogical credentials, however, çabras could not rise to be sacos.

Even commoners who did not become çabras could achieve a measure of renown through expertise at goldworking and the production of polychrome pottery. Archaeologist Richard Cooke and his colleagues have argued that Panama had too much gold of its own to need inputs from Colombia. Despite their self-sufficiency, however, Panamanian chiefs-in-training were encouraged to lead expeditions to Colombia to set up trading partnerships before they took office. Almost certainly this was done because the foreign relations themselves were more important than the commodities traded.

At the bottom of the social hierarchy were slaves called *pacos,* who were mostly war captives. One front tooth of each slave would be knocked out (to indicate his slave status), and his face would be tattooed in such a way as to identify his owner. This was not the only context in which the Panamanians used body markings. All the warriors who fought for a particular chief painted their faces and bodies in a way that identified their overlord.

When we examine the underlying principles of rank society in Panama, we see many parallels to the concepts of mana, toa, and tohunga in Polynesia. For example, the Panamanians believed in *purba,* an individual's invisible, immortal essence. Purba was inherited by chiefs, who nurtured it through their monopoly of esoteric knowledge. Purba was reflected in a special chiefly language that the queví used during ceremonies.

Then there was *niga,* an aura of power that was generated by acts of bravery in battle and public works in the interests of society. Chiefs had more niga than commoners, but all had enough niga to improve their position if they worked to do so.

Finally there was *kurgin,* the innate talent for a craft. Kurgin varied from individual to individual and could be enhanced with training. We are struck by how similar these Panamanian principles of life force, bravery, and expertise were to those displayed by rank societies elsewhere in the world.

Panama did not have priests as powerful as those of the Cauca Valley, but it did have ritual specialists called *tequinas* who, through the use of strong tobacco or corn beer, could enter into trances. While in these trances they could communicate with the spirit world, predict the future, control the weather, and cause crops to flourish.

Let us now consider a significant difference between chiefly behavior in Tikopia and Panama. We have already suggested that many aspects of rank in Tikopia would be hard for archaeologists to detect. The behavior of Tikopian chiefs was not flamboyant. The chiefs did not make lavish use of sumptuary goods; they expected to be shown deference but were not carried around on litters; and no slaves or servants were sacrificed at their funerals.

The flamboyance of Panamanian chiefs, in contrast, was an archaeologist's dream. The chief's lineage had a distinct emblem that was carved on both houses and tools, much like a Tlingit chief's heraldic crest. Chiefs accumulated gold jewelry and were buried with it. Their funerals were elaborate because chiefs were thought to have eternal life, while dead commoners were merely converted to air. In addition, any wife, concubine, servant, or slave buried with the chief shared in his eternal life. The result was that many individuals volunteered to be sacrificed at the chief's funeral, making his grave easier for archaeologists to identify.

In the Darién region of eastern Panama, the chief's desiccated corpse was preserved in a special house where his remains, and those of his chiefly ancestors, were set in chronological order along the wall. In the realm of chief Comogre on the Caribbean coast, the dried and richly attired corpses of past chiefs were kept in the innermost chamber of the bohío, suspended by ropes in order of rank.

Parita, a deceased paramount chief of the Azuero Peninsula on the Pacific coast of Panama, was discovered in 1519 by the Spaniard Gaspar de Espinosa. Parita's corpse had been dried by the heat of a fire and dressed in a gold helmet, four or five gold necklaces, gold tubes for his arms and legs, gold breastplates, and a gold belt with bells that tinkled when he walked. With him in the charnel room were the corpses of three previous chiefs, each wrapped in a cloth bundle and suspended in a hammock. At Parita's head was one sacrificed woman; at his feet lay another. Both were adorned with gold. In an adjacent room the Spaniards found 20 living warriors from enemy chiefdoms, bound and awaiting sacrifice at Parita's funeral.

A conquistador named Gonzalo Fernández de Oviedo witnessed the funeral of another Panamanian chief. Preparations began while the man was still alive, but fading fast. As his death approached, his relatives and allies dug a rectangular pit 12 to 15 feet on a side and more than 10 feet deep. They set up a bench in this pit and covered it with a brightly colored blanket.

The well-dressed chief, now dead at last, was propped up on the bench and supplied with water, corn, fruit, and flowers for the afterlife. Women who had volunteered to be buried with him took their seats on the bench beside the corpse; they, too, were dressed in finery and adorned with gold.

Then came a two-day funeral feast with dancing and chanting. There were songs praising both the chief and the women, who presumably included some of his wives and concubines. Large quantities of corn beer were consumed. As soon as the women on the bench were completely drunk, workmen rapidly filled the pit with earth and timbers, smothering the women. Trees were then planted on the grave, perhaps to conceal its location.

Fernández de Oviedo was told that up to 40 or 50 wives and servants might be entombed with a chief. In some cases the sacrificial victims were given an herbal concoction that brought on rapid loss of consciousness. Many of these sixteenth-century accounts of Panama's chiefly burials have now been confirmed by excavation, thanks to archaeologists such as Samuel Kirkland Lothrop and Olga Linares.

A key archaeological site on Panama's Pacific coast was Sitio Conte, a chiefly burial center whose heyday occurred 1,500 to 1,100 years ago. There Lothrop found 60 graves with elegant gold items, polychrome pottery, and human sacrifice. Grave 26, for example, appears to have been the burial of a chief with 21 sacrificed subordinates. The subordinates were buried at the bottom of the grave in fully extended positions. Some of the sacrificial victims had gold items with them, suggesting that they may have been relatively important individuals. For his part the chief was seated atop the layer of subordinates in such a way as to suggest that he may have been placed on a wooden bench or stool, long since disintegrated. Many Central American chiefs had servants carry their stools wherever they went, much the way

Tikopian chiefs used coconut-grating stools to keep their heads above everyone else's.

THE BEMBA OF ZAMBIA: MALE CHIEFS IN A MATRILINEAL SOCIETY

Agriculture and animal husbandry had a long history in the Nile Valley. From the Nile headwaters south, however, Africa was for thousands of years occupied by hunters and gatherers analogous to the Basarwa and Hadza. Impeding the spread of agriculture were regions of poor tropical soils. Impeding the spread of herding was the tsetse fly, which transmits sleeping sickness to cattle.

Roughly 2,400 years ago the craft of ironworking reached western and central Africa, probably by means of trade routes crossing the Sahara Desert. Iron tools made it possible to till tropical soils and grow crops such as sorghum and millet. Sheep and goats provided meat, and cattle were added wherever the tsetse fly allowed.

Ironworking spread rapidly over the next few centuries, often through the dispersal of Bantu-speaking peoples. The homeland of the Bantu language may have been north of the forests of the Congo, but 1,700 years ago, speakers of Bantu had spread east to Lake Victoria in Tanzania and south to the Limpopo River, on the border between Zimbabwe and South Africa. During the next 500 years they crossed the Limpopo and, finding no tsetse flies, turned much of South Africa's steppe land into cattle-herding country.

Bantu societies took millions of acres away from hunters and gatherers, driving them into refuge areas such as the Kalahari Desert. One reason for their success was that Bantu speakers not only lived in societies with clans but were also hierarchically organized under hereditary chiefs, making it nearly impossible for foragers to compete with them.

Some of the earliest hints of hereditary rank come from a twelfth-century cemetery at Sanga, near Lake Kisale in the Democratic Republic of the Congo. Among the artifacts in the cemetery were fragments of iron bells, which, according to historian Andrew Roberts, have long been symbols of chiefly rank. Similar bells have been found at the fourteenth- or fifteenth-century chiefly center of Great Zimbabwe, an archaeological site that has lent its name to an entire nation.

The Sanga cemetery lies in the territory of a Bantu-speaking group called the Luba. This is significant because the historical legends of several other

Bantu groups claim that they emigrated from Luba country to their present location.

One such group was the Bemba, a Bantu-speaking group occupying what is now Zambia. In 1933, when first visited by anthropologist Audrey Richards, the Bemba lived at 4,000 feet above sea level on the Tanganyika plateau, surrounded by Lakes Malawi, Mweru, Bangweolu, and Tanganyika. They grew finger millet *(Eleusine coracana),* sorghum or kaffir corn *(Sorghum bicolor),* a variety of legumes, and two crops imported from the New World: corn and pumpkins. Because Zambia has tsetse flies the Bemba were unable to raise cattle; however, the Bemba added protein to their diet by hunting and fishing. Bemba agriculture involved slash-and-burn cultivation on relatively poor soils, requiring villages to move every four to seven years while the old fields recovered. As a result, 140,000 Bemba needed a territory of more than 22,000 square miles.

The Bemba of that era were as impressive as the largest of the sixteenth-century Cauca Valley societies: all 140,000 citizens were ruled by a single paramount chief, or *chitimukulu.* The paramount lived at the *umusumba,* or chiefly village, which was both a secular capital and a ritual center. Boasting an estimated 150 to 400 households, it was the largest single settlement in Bemba territory and was supported by tribute so that the chief need not move.

The chitimukulu was at the apex of a political hierarchy. Below him were a series of *mfumu,* or district chiefs, each of whom was in charge of a district called an *icalo.* The district surrounding the paramount chief's village needed no mfumu because it was administered by the paramount himself. Finally, at a level below the mfumu, there were subchiefs who controlled individual villages or small tracts of land. These subchiefs moved around within each district whenever old fields were left fallow.

As we have seen, rank societies that reckoned descent in the father's line (or in both parents' lines) had a variety of rules for the succession of chiefs. These rules could range from primogeniture (the firstborn son inherits the office) to ultimogeniture (the last-born son inherits). The Bemba were different: their chiefs were male, but they were chosen from a clan that reckoned descent in the mother's line. Let us examine their system.

Each Bemba was born into his mother's lineage, or "house." Several such houses were, in turn, grouped into one of 30 matrilineal clans. In Bemba social logic a child was formed from the blood of a woman. Since a man could not pass on his blood, there was no continuity between father and child.

The 30 Bemba clans were ranked relative to one another, and a woman's position within a lineage determined her rank. Like many societies described earlier in this book, the Bemba used the order in which each clan allegedly arrived in the area as justification for its rank. The Bemba told Richards that they had emigrated to Zambia from Luba country in what is now the Democratic Republic of the Congo. Such an origin, according to Richards, was supported by the fact that a number of words used by the Bemba chief during religious ritual were Luban words. This knowledge of Luban confirmed the priority of the chiefly clan's arrival. Here, once again, we encounter the principle "We were here first."

The Bemba in 1933

Each of the 30 Bemba clans studied by Richards took its name from an animal, a plant, or a natural phenomenon such as rain. The Crocodile clan was first in rank, providing both the paramount chief and the mfumus. The paramount chief was chosen from among the sons of the highest-ranking women of the Crocodile clan. His father played little role in the decision, since he was a member of a different clan and had simply had the good fortune to marry a Crocodile woman.

The paramount chief's mother, or *candamukulu,* served on the high council of the Bemba and controlled several villages of her own. For their part, the paramount chief's sisters were given their own villages and allowed to have sexual relations with as many men as they chose—even men of low rank, since any offspring that resulted would be credited to the elite woman's blood.

The chitimukulu administered all of Bemba territory; he was not only a secular authority and a high judge but also a ritual expert. Many paramount chiefs-in-training served for years as mfumus, both to hone their administrative skills and to allow their performance to be evaluated. The chitimukulu was supposed to observe ritual taboos and remain virtuous and pure because the agricultural success of the Bemba depended upon his reservoir of life force. The chief was supported by tribute and could call upon his citizens for the kind of obligatory labor the French call *corvée.* He had the right to mutilate anyone who offended him. As a result, commoners were sometimes heard to say, "The Crocodile clan tears common people apart with their teeth."

Chiefs were the offspring of elite mothers, but the Bemba also created fictional ties between a new chief and his male predecessors. Upon his inauguration the new chief inherited his predecessor's title, insignia, rights, and duties. Eventually he became so identified with past chiefs—to the point of assuming their names and histories—that it became difficult to tell whether he was referring to events in his own life or the lives of his predecessors. He was aided in this process by assistants who memorized the oral histories of 25 to 30 former chitimukulus.

In the Bemba system elite brothers, half brothers, and cousins did not compete for leadership; rather, the women of the Crocodile clan competed to have their sons chosen, and there were cases of usurpation by district chiefs with ambitious mothers. Usurpers, of course, then had to rewrite history, claiming that they were somehow part of a long line whose privileges extended back to the first chitimukulu.

Upon his death the Bemba chief was treated in ways that remind us of funerals in Panama and Colombia. His corpse was desiccated for a year and then buried in a sacred grove with the bodies of sacrificial victims. The chief's burial was supervised by a council of 30 to 40 elite elders known as the *babakilo.*

The institution of the babakilo is worth examining because it shows us that even the most despotic chief was subject to power sharing. Because the babakilo inherited their position, it was impossible for the chief to replace them with his cronies. Members of this council advised the chief on matters of policy. They wore special feather headdresses, paid no tribute, and received elegant funerals. The fact that the chief was merely chosen from among eligible men while his counselors inherited their posts made it easier for the council to steer him away from chiefly abuse.

Another hereditary office was that of military leader. War was very important to the Bemba, because the tsetse fly prevented them from making cattle a source of wealth. They therefore relied on slaves and tribute obtained through conquest. Other sources of wealth were ivory (mainly in the form of elephant tusks), grain, iron, and salt. Richards reveals that, as with so many societies we have examined, war was endemic until stifled by the colonial government.

The Bemba built no impressive temples, but throughout their land they maintained shrines at sacred places. Each shrine was assigned its priests and guardians, who used Luban words in their ceremonies and kept Bemba ritual secret and elitist. This secrecy reinforced the gulf between the Crocodile clan and the lineages of commoners.

The Crocodile clan displayed both a powerful life force that required the maintenance of ritual purity and a commitment to conquest that supported hereditary war leaders. As for the kind of expertise the Polynesians called tohunga, it took several forms among the Bemba. The memorization of Luban words and long lists of past chiefly accomplishments was one skill. Crafts were another, with wood carving, basket weaving, and ironworking being the most important.

Ironworkers had a special role in society, not only because they were the suppliers of weapons and agricultural tools but also because they took iron from the sacred Earth and subjected it to temperatures that no other craftsman could produce. There was therefore something magical about this particular craft.

The Nature of Bemba Inequality

The territory controlled by the chitimukulu would be described by anthropologists as a paramount chiefdom. This term refers to the territory of a rank society with a three-level political hierarchy. At the top level was the paramount chief, who lived in a large permanent village. At the second level of the hierarchy were the mfumus, who commanded entire icalos. At the third level were the subchiefs, who commanded the smaller villages and hamlets that shifted around within each district. Tribute flowed upward from the subchief, to the district chief, to the paramount chief. Orders and policies flowed down the same chain of command.

All paramounts were chosen from a Crocodile clan that could "tear common people apart with their teeth." The paramount, nevertheless, had to share power with a council of elders, which he could not replace because the elders' offices were hereditary. Society's main war leaders also inherited their offices, meaning that they could not be coerced into taking the paramount's side against the council.

The chief's authority flowed ultimately from his highly ranked mother. He was expected, however, to demonstrate administrative skill as a district chief before moving up to paramount chief. Although every chitimukulu was a potential despot with the power to mutilate his subjects, even he had to operate within a system of constraints.

Let us consider the differences between the Bemba and the Tikopians. The most highly ranked Tikopian ariki was still only one of four chiefs. Commoners allegorically offered to eat his excrement, but no one was sacrificed at his funeral. He took advice from the elders of commoner lineages and did his best to keep them happy. He could point angrily at a subject or banish a criminal, but he would have been criticized had he used force to mutilate his subjects merely for annoying him.

The paramount chief of the Bemba was the supreme leader of 140,000 people. His potential for the abuse of power was vastly greater than that of a Tikopian ariki. No elder from a commoner clan could have held him in check. Almost certainly for this reason, his society had taken the most important political advisers and war leaders out from under his thumb by making their offices hereditary.

Despite its institutions of power sharing, Bemba society seems to have reached an important threshold. It had so many levels of political hierarchy, and so many categories of personnel who sound like bureaucrats, that we do not believe it would have taken many changes in social logic to transform its chief into a king and its territory from a chiefdom to a kingdom. Later in the book we will see how the Zulu of South Africa, another group of Bantu speakers, did exactly that.

TWELVE

From Ritual House to Temple in the Americas

Some 3,500 years ago, achievement-based societies spread over the highlands of Mexico and Peru. Many of these societies built small ritual structures that resemble the familiar men's house of the preindustrial world.

Parts of the Third World continued to build men's houses well into the twentieth century—but not so in Mexico and Peru. In those two countries there came a time when achievement-based society gave way to hereditary rank. Once that happened, society's leaders began to have temples built. Temples and small ritual houses coexisted for a while, but the latter eventually disappeared.

To be sure, the temples of early rank societies were not like the standardized churches, mosques, and synagogues built by today's industrialized nations. Some early temples were so nonstandardized that it can be difficult to identify them. In the pages that follow we look at some archaeological examples.

FROM MEN'S HOUSE TO TEMPLE IN OAXACA, MEXICO

The period from 3,200 to 2,900 years ago was one of significant change in Mexico's Oaxaca Valley. More than a dozen lines of evidence indicate the emergence of hereditary rank, comparable to that among the Konyak Naga or the Kachin. This change took place during a period of striking population growth. The number of villages in the valley nearly doubled, from 19 to 40, and the estimated population nearly tripled, from 700 to 2000. In addition, 50 percent of the valley's population lived in one large village, San José Mogote, which consisted of multiple residential wards scattered over 150 acres.

Privileged families at San José Mogote engaged in several behaviors seen in rank societies elsewhere. They artificially deformed their children's heads to make their aristocratic ancestry clear and buried sumptuary goods with elite children. Among those goods were pottery vessels with carved motifs of Earth or Sky. Recall that chiefly Kachin families claimed a special relationship with the supreme earth spirit Shadip and the supreme sky spirit Madai. Oaxaca villagers seem to have had a comparable Earth/Sky dichotomy. They depicted both of these celestial spirits in their most dramatic or "angry" forms—Earth as Earthquake and Sky as Lightning. Such carved vessels were also buried with elite adult men, but not with adult women. This fact makes it likely that the children buried with Earth or Sky vessels were the sons of highly ranked parents.

Sumptuary goods included mirrors of polished iron ore. Most iron came from outcrops within a day's walk of San José Mogote, and the bulk of its conversion into mirrors was done by craftspeople living in one residential ward. Although most mirrors were worn by (or buried with) local men and women, iron objects from Oaxaca were also traded to chiefly centers in neighboring regions.

While iron-ore mirrors seem to have been restricted to chiefly families, other valuables such as jadeite/serpentinite, mica, and marine shell were not. This situation may be analogous to the pattern in Colombia's Cauca Valley, where gold was accumulated not only by "nobles by blood" but also by "nobles by command" and "nobles by wealth." The elite at San José Mogote wore more jadeite/serpentinite and mother-of-pearl ornaments than anyone else, but smaller amounts were worn by people of lesser rank.

A cemetery discovered at Santo Domingo Tomaltepec, a three-acre village southeast of San José Mogote, reveals the way elite and nonelite men were treated at death. There were more than 60 graves in the cemetery, and because some held more than one person, the number of skeletons was close to 80. There were many paired burials of men and women, indicating that some couples were laid to rest as husband and wife rather than as individuals. In the case of one couple, the man had a bowl carved with Lightning motifs and the woman had an iron-ore mirror. There were also complete skeletons accompanied by the incomplete remains of others, suggesting that some of the deceased had been exhumed and reburied with a spouse or relative.

A group of six middle-aged men in the cemetery stood out as different. All were so tightly flexed that we suspect they had been wrapped in bundles. These men represented only 12.7 percent of the cemetery but were buried with 50 percent of the pottery bearing Lightning motifs and 88 percent of the jadeite/serpentinite beads. Most of the exhumed and reburied skeletal parts had been added to the graves of these six men, suggesting that several of them may have had more than one wife. Almost certainly these were the leaders of the village, whose tightly bundled remains had been kept around for a time, honored, and perhaps even dried or smoked in some way before their burial.

During this period the people of San José Mogote were building both men's houses and temples. This was, in other words, a period of transition from one type of ritual building to another, reminiscent of the situation we saw on Tikopia.

Men's houses remained simple one-room buildings, plastered with lime. Because San José Mogote had grown so large and was so clearly divided into residential wards, it was easier than before to see that each ward built its own men's house.

The main locus for women's ritual remained the household. There, women performed a traditional Zapotec form of divination, casting beans or corn kernels into shallow, water-filled basins and reading the floating seeds the way fortune-tellers read tea leaves. Women also made hundreds of small ceramic figurines of the ancestors, providing a tangible venue to which ancestral spirits could return. These figurines were sometimes arranged in scenes, presumably to be addressed, ritually "fed," and asked to intercede with higher spirits on behalf of their descendants.

The figurines also seem to reflect social rank. A minority depict men seated in what appear to be positions of authority. These seated men were sometimes shown with cranial deformation, filed teeth, jadeite ear spools, and other ornaments. Archaeologists have also recovered miniature versions of the kinds of four-legged stools on which men of chiefly descent probably sat. A much larger number of figurines, however, showed people in postures of deference or obeisance, standing with arms folded across their chests. In one figurine scene, a seated male authority figure was placed atop three obeisant figures. The scene reminds us of some Panamanian graves, where the chief was buried in a seated position above a layer of subordinates.

One of the innovations of this time period was the temple. This building was raised above the level of the village on a pyramidal earthen platform, unlike anything associated with a men's house. The outer casing of each platform was a masonry wall of stones, carefully fitted together without mortar. The internal structure consisted of earthen fill reinforced with adobe walls. The temple itself was a perishable building of pine posts, cane walls coated with clay daub, a thatched roof, and a burnished clay floor. The first temples were nonstandardized, and the stairs ascending the platforms were narrow and single-file.

An unexpected clue to the size of the territory controlled by the chiefs of San José Mogote was found among the stones of one temple platform. Bedrock at the site was volcanic tuff, and most of the stones were of that raw material. A number of stones, however, were boulders of limestone or travertine, types of rock available near villages three to five miles away. It would seem that the leaders of San José Mogote could now call upon the manpower of neighboring villages for the construction of their temples. This was not true of any of the men's houses, which were likely built by smaller units such as descent groups.

How big a territory might the chief of San José Mogote have controlled? An archaeological survey shows that 12 to 14 smaller villages surrounded San José Mogote, like three-to-five-acre satellites caught in the gravitational pull of a 150-acre sun. The distances involved suggest that the chief's authority extended outward at least a half-day's travel.

Chiefly Cycling in Oaxaca

As happened in so many parts of the world, chiefly societies in the Valley of Oaxaca went through cycles of expansion and contraction. Roughly 2,900 years ago, no other village in the valley rivaled San José Mogote in population, number of satellite communities, or size of temples. Over the next two centuries, however, San José Mogote began to experience competition from rival villages. One of these was Huitzo, a community ten miles to the north, which had built its own impressive temple. Another was San Martín Tilcajete, some 20 miles to the south.

It is difficult to assess the extent to which these emerging chiefly centers affected San José Mogote's control of its resources and satellite villages. One clue can be found in the dwindling amounts of iron ore that were now reaching San José Mogote; it appears that Huitzo denied access to the northernmost iron-ore source, while Tilcajete denied access to the southernmost.

The temple built at Huitzo some 2,800 years ago was as impressive as any at San José Mogote (Figure 27). Its underlying pyramidal mound was six feet high and more than 50 feet on a side. Its outer face was of boulders or cobbles set in hard clay, and its interior consisted of earthen fill strengthened with adobe walls. Above this was the temple itself, a building whose thick cane and clay walls rested on an adobe platform. The latter was four feet high and at least 37 feet long, and the temple was reached by a staircase 25 feet wide. This broad staircase would have permitted greater public access than the narrow, single-file stairways seen on earlier buildings.

It is significant that by this time men's houses had disappeared from Oaxaca's archaeological record. The transition to rank society was now complete.

With rival chiefs challenging San José Mogote, its elite implemented various strategies to hold onto their satellite villages. It appears that one strategy was to send women of high rank to marry the leaders of those satellite communities. This strategy, called *hypogamy* (a Greek word used when a woman marries a man of lower rank), raised the status of the satellite village's leader, much the way a Shan bride raised the status of a Kachin chief.

Our best evidence for hypogamy comes from Fábrica San José, a five-acre village some three miles east of San José Mogote. Fábrica San José was an

FIGURE 27. This temple at Huitzo, Mexico, built 2,800 years ago, was a one-room building without the usual sitting bench or lime-filled pit of earlier men's houses. It was elevated on an earthen platform, given a 25-foot-wide staircase, and provided with incense burners.

economically important satellite for San José Mogote because its saline springs supplied the entire region with salt.

Between 2,800 and 2,600 years ago, the richest burials at Fábrica San José were those of women. Some of those women displayed the same type of cranial deformation seen among high-status families at San José Mogote and had likely come from the latter village. Burial 39 was interred with a pendant, 53 beads of jadeite/serpentinite, and a drinking vessel of elegant white pottery imported from outside the valley. Burial 54 was laid to rest with six fine gray pottery vessels, offerings of seashells, and a large, hollow ceramic sculpture representing an ancestor.

Between 2,700 and 2,500 years ago, San José Mogote rebounded from the challenges of the previous era and solidified its place as the most influential chiefly center in the Valley of Oaxaca. There were now between 70 and 85 villages in the 810-square-mile valley, which is shaped like a three-pointed star. San José Mogote occupied the northern arm of the star and was surrounded by 18 to 23 smaller satellite villages. In the eastern arm of the star, the most important chiefly center was Yegüih, which may have had eight to ten satellites of its own. In the southern arm, San Martín Tilcajete was still the largest chiefly center and may have had eight to ten satellite communities.

So great was the rivalry among these three chiefly societies that the center of the valley, where all three arms converged, was left as a virtually unoccupied buffer zone or "no-man's-land." This buffer zone, covering 30 square miles, was evidently considered a dangerous place to live.

During this period, San José Mogote built the largest and most impressive temple in the valley. To ensure that this temple would be visible from a great distance, the builders chose a 40-foot hill that was the village's most prominent landmark. The new temple, built about 2,600 years ago, was placed above the ruins of an abandoned men's house.

The temple itself had extra-thick walls of cane bundles, daubed with clay and whitewashed. Its floor was recessed into an adobe brick platform measuring 46 by 43 feet. Buried beneath each of the temple's four corners were large serving vessels—brown vessels below the northeast and southwest corners and gray vessels below the northwest and southeast. These bowls, which might have held food for celebrants during the temple's inaugural ceremonies, had been buried as dedicatory offerings. Lying broken on the floor of the temple was an imitation stingray spine for ritual bloodletting, chipped from a large blade of imported volcanic glass.

The adobe platform mentioned earlier was in turn supported by an even larger stone masonry platform, whose earliest construction stage was 55 feet on a side and more than six feet high. This platform was enlarged at least twice. Its final stage measured 93 by 70 feet and was built of limestone blocks weighing up to half a ton. These blocks had been brought from a quarry three miles away; they would have to have been rafted across a river and hauled to the top of the hill.

Unfortunately for its builders, even this impressive temple was not immune from raiding. Late in its history, it was the scene of an intense fire that destroyed the building and converted much of its clay daub to glassy cinders. To vitrify clay in that way, the fire must have been deliberately set.

The villagers of San José Mogote responded to the desecration of their temple in several ways. First they built a new temple only a short distance to the north. The new temple was constructed of sturdier adobe walls over stone masonry foundations. The platform beneath it was made of the same kinds of limestone blocks used for the earlier temple but stacked even higher. A radiocarbon date from one of its wooden posts shows the temple to have been built 2,590 years ago.

A second response to the burning of the earlier temple seems to have been violent retaliation. This retaliation was commemorated on a carved stone slab, installed as the threshold for a narrow corridor flanking the new temple. Anyone entering or leaving this corridor would have stepped on the figure carved on the upper surface of the slab. The carving depicts a naked man sprawled awkwardly on his back, with mouth open and eyes closed (Figure 28). A complex motif shows where his chest had been opened to remove the heart during sacrifice. A ribbonlike stream of blood extends from this motif to the border of the monument, ending in two stylized drops of blood. Between the feet of this sacrificial victim is his Zapotec hieroglyphic name. The fact that his name was added to the monument indicates that he was an enemy of some importance, that is, a member of a rival chiefly family.

We do not know from which enemy village the victim came. We do know that over the next century or so, the rival chiefly center of San Martín Tilcajete erected its own important temple (Figure 29). Built 2,500 years ago, this temple was excavated by archaeologists Charles Spencer and Elsa Redmond. The architects of Tilcajete first built a platform with stone masonry retaining walls and earthen fill, three and a half feet high and measuring 40 by 25 feet. Atop this platform they built a one-room temple measuring 22 by 9 feet, distinguished by its two built-in basins for burnt offerings. Associated with this

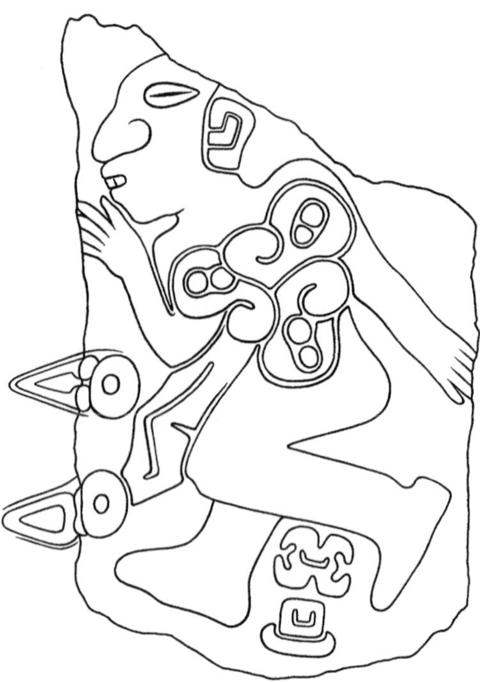

FIGURE 28. This carved stone at San José Mogote, Mexico, formed the threshold for a corridor between the old and new versions of a 2,600-year-old temple. It gives the hieroglyphic name of an elite enemy whose heart has been removed, leaving a stream of blood that runs off the edge of the stone and down the side. The monument was just under five feet long.

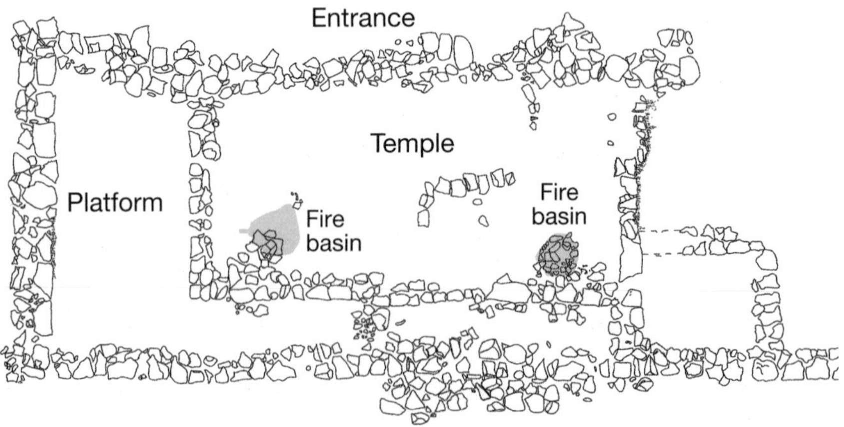

FIGURE 29. The stone foundation for a temple built 2,500 years ago at Tilcajete, Mexico. This temple, measuring 9 by 22 feet, had two built-in fire basins and a series of incense burners.

temple were fragments of braziers for burning incense, a common activity in later Zapotec temples.

Oaxaca's Rank Societies in Regional Perspective

Multiple lines of evidence, as we have seen, suggest that rank societies arose in the Valley of Oaxaca between 3,200 and 3,100 years ago. Of Goldman's three sources of chiefly power, we believe that religious authority was uppermost. For one thing, the most elite-looking households and burials had more vessels carved with references to Earth and Sky than did their lower-ranked neighbors. For another thing, the temple gradually replaced the men's house. This suggests that the worship of the supernatural spirit (or spirits) in the genealogy of the chiefly line had become a community concern.

What happened in Oaxaca, however, did not happen in a vacuum. Chiefly societies were arising over all of central and southern Mexico, an area many times the size of Colombia's Cauca Valley. From the Basin of Mexico in the north (the region of present-day Mexico City) to the Pacific coast of Chiapas in the south, social inequality was growing.

One indication that all these emerging rank societies were in contact with each other is that vessels carved with Earth and Sky motifs were actively exchanged by their elites. An analysis of minerals in the clay of the vessels, for example, suggests that villages in the Basin of Mexico, the Puebla and Oaxaca valleys, and the Gulf coast of Veracruz all exchanged carved pottery.

This exchange system probably resembled the circulation of heraldic crests among the Tlingit, Haida, and Tsimshian and the movement of pottery and goldwork among chiefly families in Panama and Colombia. It is the kind of exchange that leads chiefly families to become the patrons of skilled artisans. In the Mexican case our impression is that although the Basin of Mexico was not home to the largest and most powerful rank societies of this era, its craftsmen produced the most elegant pottery and large, hollow ceramic sculptures. That would make its potters the Mexican equivalent of the Quimbaya goldsmiths.

It is likely that all the rank societies of this era went through cycles like those of the Konyak Naga and Kachin. From 2,850 to 2,700 years ago, as we have seen, rival chiefs arose to challenge the supremacy of Oaxaca's largest rank society. That society regained its preeminent position, but only through the multiple strategies of building impressive temples, attracting new satellite communities, sending elite women to marry the leaders of those satellites, and using war parties to capture and sacrifice rival elites. This increased use of raiding transformed highland Mexican society from Irving Goldman's "traditional" type, based mainly on religious authority, to a more powerful type that combined religious authority, military expansion, and chiefly support of craftsmen.

Now the Valley of Oaxaca had reached an important threshold. Three rival chiefly societies glared at each other across a sparsely occupied buffer zone. The northernmost of those societies was the largest, but its size advantage was insufficient to eliminate the others. What happened next was without precedent for the region and led, after several centuries of struggle, to the creation of one of Mexico's first kingdoms. That remarkable process will be described in a later chapter.

FROM RITUAL HOUSE TO TEMPLE IN CENTRAL PERU

When we last discussed the Peruvian site of La Galgada, its talc-plastered ritual houses had just been replaced by a large U-shaped temple. The time was roughly 3,500 years ago, half a millennium before the first temples were built in Oaxaca.

In fact, some societies on the central coast of Peru were building temples even before La Galgada did. Let us turn first to the valley of the Río Supe, 90 miles north of Lima. The Supe River begins high in the Andes, descends through dusty canyons like the one supporting La Galgada, and breaks out onto the coastal plain some 15 to 20 miles from the sea. The final 15 miles of its floodplain supported a high density of settlements during the Late Archaic period, and some of those settlements built what appear to be temples.

An archaeological team led by Ruth Shady surveyed ten square miles of the Supe Valley and located 18 sites of the Late Archaic period (4,500 to 3,800 years ago). The three largest sites were Era de Pando (197 acres), Caral (143 acres), and Pueblo Nuevo (135 acres). Near the mouth of the river lay Áspero, a 37-acre site with monumental architecture.

The Late Archaic societies of the Supe Valley were supported by a mix of agriculture, fishing, and foraging. There are suggestions that inland horticulturalists sent cotton and gourds to the coastal fishermen, who reciprocated by sending dried anchovies and sardines to the horticulturalists.

The community of Áspero occupied a shallow basin, framed by rocky hills that jut into the sea. Between 15 and 20 artificial mounds still rise above its blackened refuse heaps. Two of the largest mounds were excavated by archaeologist Robert Feldman. Each had an impressive multiroom building on its summit.

One of those buildings, reached by a stairway, had a large entry court with a centrally placed fire pit. Two human sacrifices were associated with this court. One was an adult with a gourd vessel, buried tightly flexed inside a bundle. The other, an infant, had been wrapped in a cotton textile and placed inside a basket. The infant wore a cap covered with 500 beads of seashell, clay, or plant material, and its burial basket had been covered with a four-legged stone basin. Possibly the sacrificed child belonged to a family of some importance.

The second building, reached by a ramp, measured more than 90 by 60 feet and featured a high-walled entry flanked by smaller rooms and courts. Its mud-brick walls were decorated with rectangular niches and brick friezes, and its contents included a cache of 13 unbaked clay figurines. Feldman hesitated to declare this building a temple, in part because the ritual architecture of 4,500 years ago was still relatively nonstandardized. The building's elevated position, complex mud-brick decoration, and lack of domestic refuse, however, all suggest that it was a temple of some kind.

Our sample of early Peruvian temples is enlarged by the Late Archaic site of Caral, 13 miles inland near the left bank of the Río Supe. The sprawling settlement covers a landscape where rocky hills alternate with sandy plains and depressions. Perhaps a dozen prominent buildings encircle the low central area of the site.

The layout of Caral is unlike that of any village we have looked at so far. Each of its public buildings seems to have been accompanied by a nearby multiroom residential complex. These buildings have different styles and orientations and do not give the impression of components integrated into a master plan. Instead, the impression is one of multiple social units, each of which built its own ritual complex. Some of the major buildings were temples atop pyramidal mounds. Others were linear complexes whose varied elements suggest the stages in a ritual procession.

Some of the most impressive buildings were built around natural rock outcrops. This labor-saving use of bedrock was most evident in the case of the so-called Quarry Pyramid, where the stairs and terraces of the building were clearly quarried from a natural hill that formed the core of the building.

Where no such outcrop was available, Caral's architects relied on a technique that Peruvian archaeologists call *shicra.* Shicra refers to coarse net bags filled with stones. These bags could be stacked one upon another to form a wall or terrace, much the way we use sandbags to build levees along flooding rivers.

The so-called Amphitheater Building, built on the south edge of the site, combined rough stone masonry and shicra fill. This building's layout seems designed to regulate the movement of a procession (Figure 30). The linear complex extends more than 500 feet. Worshippers would have entered through a corridor flanked by rows of storage rooms. Then, using a series of narrow stone stairways, they would step down into a sunken circular court more than 90 feet in diameter. Ascending the stairs to the opposite side of this amphitheater-like court, they then would have passed through a narrow doorway into a rectangular entry room. From this point on, the procession would have continued to rise in elevation as it passed through three or four more narrow doorways and stairways. Finally, it would have reached the inner sanctum of the complex, a room elevated by a platform more than 260 feet wide. Every door and stairway in the linear complex seems to have been designed to

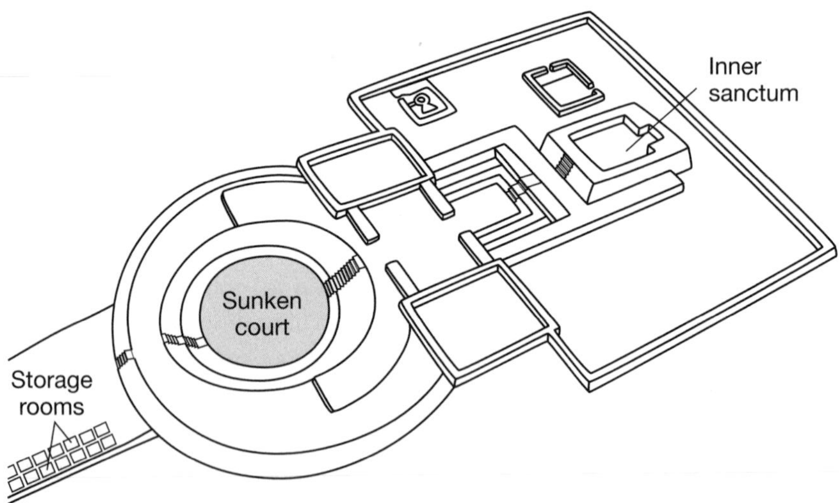

FIGURE 30. The so-called Amphitheater Building at Caral, Peru, appears to have been designed to direct processions of worshippers along a linear, 500-foot route from one ritual venue to the next, eventually reaching the highest and most restricted inner sanctum.

force ritual participants to move deliberately through six or seven levels before reaching the final, and least accessible, room.

This complex was not the only building at Caral to feature a sunken circular court. A smaller court, 60 feet in diameter, had been built at the base of the stairway to the site's largest pyramid, which bore on its summit a temple complex. Our suspicion is that such courts were the scene of preliminary rituals that worshippers were required to perform before being allowed to proceed to the inner sanctum. In many parts of the ancient world, the life force housed in a temple was considered so great that one could not leave the secular world and enter the temple without a preliminary ceremony.

Both the pyramidal structures and the linear complexes at Caral seem to have culminated in temples. Caral, however, includes some ritual structures reminiscent of earlier, achievement-based societies. These structures were circular, white-plastered rooms with centrally placed fire pits, to which oxygen was supplied by subfloor ducts; they look, in other words, like circular versions of the ritual houses at La Galgada. Caral thus seems to lie at the transition from (1) a society with small-scale, private rituals to (2) a society with large-scale, public-performance rituals.

One of the most interesting aspects of Caral is the apparent association of multiroom residential complexes with ritual buildings. Many residences had cobblestone foundations, upright posts of acacia or willow wood, and walls of cane daubed with clay. Shady believes that the residential complexes nearest to the temple pyramids were occupied by extended families of somewhat higher rank.

The number of temples, as well as the attention devoted to ritual processions, makes it clear that spiritual authority was a major source of power at Caral. As at Áspero, human sacrifice was practiced. During a period of architectural renovation in Caral's largest pyramidal mound, the builders incorporated into the fill an adult male victim with his hands tied behind his back.

Music also played a role in the rituals at Caral. Excavations in the Amphitheater Building produced more than 30 flutes made of bone. Each flute was carved in such a way that the sound hole appeared to be the mouth of a spirit or mythical ancestor (Figure 31).

The largest of the residential complexes at Caral, called Sector A, lay almost equidistant from the Amphitheater Building and the Quarry Pyramid. Refuse from the cane-and-clay rooms of this residential complex was a major source of information on Caral's crops. By drawing off water with irrigation canals, the villagers had turned the Río Supe floodplain into gardens of squash, beans, sweet potatoes, and chile peppers. Their orchards featured avocados, guavas, and a series of fruit trees for which we have only indigenous Peruvian names. They grew cotton for textiles and fishing nets, as well as gourds for containers and net floats.

Despite its distance from the ocean, Caral was well supplied with fish and shellfish. Residential Sector A alone produced more than 20,000 specimens of anchovies and another 7,000 of sardines. These small fish were probably dried by communities such as Áspero and transported to Caral in baskets. Occasional larger fish—drums of several species, catfish, bonito, and even the *tollo,* or smoothhound shark—also showed up in Caral's refuse.

From roughly 4,300 years ago onward, Caral was in contact with regions producing corn. As with so many early Andean villages we have examined, however, corn was too rare at Caral to have been a staple. Our suspicion is that Caral obtained enough from its highland trading partners to make *chi-*

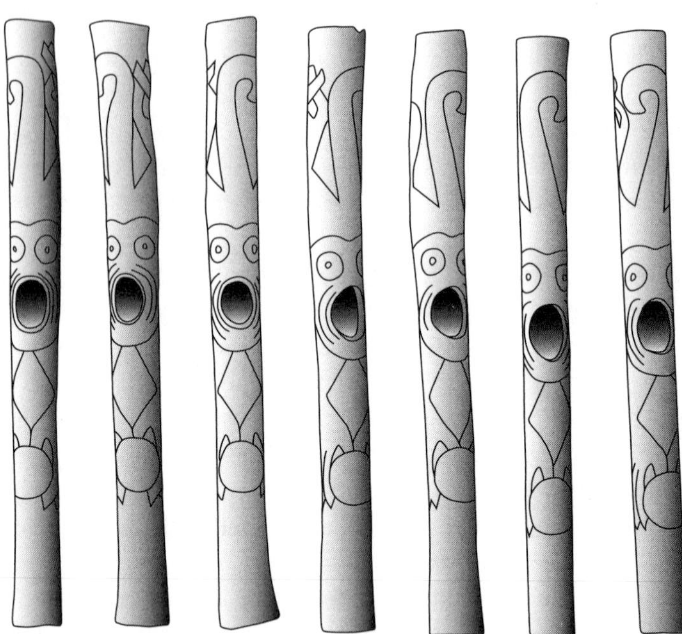

FIGURE 31. These bone flutes from Caral, roughly six and a half inches in length, were carved in such a way that the sound emerged from the mouth of a spirit or mythical ancestor.

cha, or corn beer, for ceremonial use. There are hints that those same highland partners supplied Caral with *charki,* or dried portions of llama meat. Charki (from which we get the English word "jerky") frequently included bone segments that made the rehumidified meat juicier, and such segments are recognizable in Caral's refuse.

Evaluating Inequality at Caral

Several factors make it difficult to evaluate the degree of inequality at Caral. On the one hand, the scale of its ritual architecture makes it precocious within the New World; Mexico did not produce comparable buildings until more than a millennium later. On the other hand, Caral was not even the largest village in its own valley, and there is no evidence that it stood at the apex of a hierarchy of satellite communities. Caral moved large stones around, but so did egalitarian societies such as the Angami and Lotha Naga. So few burials have been reported from Caral that we are not sure whether elite children received sumptuary goods.

To us, one of the most significant lines of evidence at Caral is the apparent association of each major temple complex with a multifamily residential compound. What this suggests is that Caral society consisted of relatively large descent groups of some kind, each of which designed, built, and maintained its own temple. Social units with access to rocky outcrops turned the latter into pyramids. Those without access to outcrops used shicra to construct processional complexes. Both groups of builders placed sunken courts at the interface between the secular and sacred worlds.

The Rise of Chiefly Warfare on the Peruvian Coast

In his analysis of Polynesian societies, Irving Goldman concluded that the most powerful rank societies arise when military force is combined with religious authority. Such a combination seems to have emerged on the Peruvian coast in the centuries following the occupation of Caral.

Head-hunting already had a long history on the Peruvian coast. Consider the evidence from Asia, a Late Archaic site some 65 miles south of Lima. The villagers of Asia lived in multiroom compounds built of stones and clay, irrigating many of the same crops grown at Caral. While excavating an area of burials at Asia, archaeologists came across something unexpected: four severed human heads, all carefully wrapped in a bundle. One of the skulls had incisions left by the removal of the facial skin.

Even the trophy heads from Asia, however, do not fully prepare us for the escalation of raiding depicted at Cerro Sechín. That ancient chiefly center lay in the Casma Valley, perhaps 180 miles north of Lima.

Some 3,500 years ago, villages in the Casma Valley were irrigating squash, beans, cotton, and root crops and producing simple pottery. Cerro Sechín

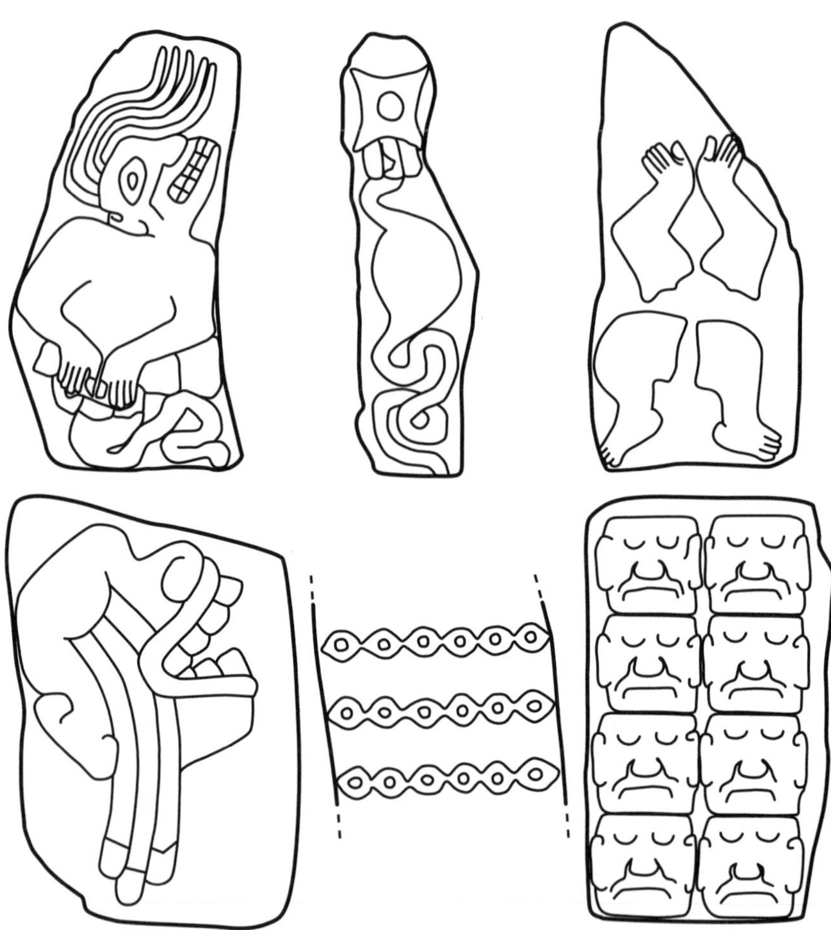

FIGURE 32. The temple at Cerro Sechín, Peru, was surrounded by a wall made of carved stones depicting a massacre. In the top row we see a severed torso, loose intestines, and severed limbs. In the lower row we see blood pouring from plucked-out eyes, a collection of eyeballs, and a stack of trophy heads.

covered more than 12 acres, not far from the confluence of the Sechín and Moxeke Rivers.

The leaders of Cerro Sechín created a multiroom temple with an entry court and an inner sanctum. The temple sat on a three-tiered stone masonry platform more than 170 feet on a side. Access to the temple was controlled by a single doorway in a massive wall, constructed of upright stones quarried from the hill behind it. Because of their irregular shapes the upright stones had gaps between them, and those gaps had been filled with smaller stones stacked one above another. Both the larger and smaller stones were carved with details of a massacre (Figure 32).

Some of the large upright stones at Cerro Sechín depict warriors holding clubs. Others show enemies cut in half, sometimes with their intestines dangling. Severed arms and legs are featured on some stones; vertebral columns decorate others. Trophy heads were a particularly favored theme. Heads stacked many layers high appear on the taller stones; individual heads are portrayed on smaller ones. Some heads have blood streaming from below their eyelids, indicating that the eyes had been gouged out; not surprisingly, one carved stone depicts a collection of eyeballs. In all, there may once have been 700 gruesome carvings on the enclosure wall of the temple.

Such use of art, of course, has a propaganda component. Some societies on the Peruvian coast were now as militaristic as those of the sixteenth-century Cauca Valley. Almost certainly they had war leaders, or "nobles by command," whose role it was to subdue their society's enemies. Displays like the temple wall at Cerro Sechín were warnings to potential rivals, letting them know what might happen to them. If this propaganda worked, perhaps some future battles would never need to be fought.

THE RISE OF HIGHLAND RANK SOCIETIES

Some 2,800 to 2,200 years ago, many rank societies of central Peru began to exhibit a pattern seen already in pre-Hispanic Mexico and Colombia: chiefly families began to exchange sumptuary goods over an extremely wide region. Like the Quimbaya goldwork, the Coclé polychrome, and the early Mexican pottery bearing Earth and Sky motifs, the pottery and goldwork of central Peru shared style and symbolic content. Among other things, the shared style emphasized the role of the chief as warrior; it featured not only trophy heads but also dangerous animals such as the jaguar and caiman. These were all creatures of the Amazon lowlands, but the Andes were now so crisscrossed by trade routes that even the irrigators of the coastal desert were familiar with them.

Seven thousand feet up in the Andes, near the headwaters of the Jequetepeque River, lay the site of Kuntur Wasi. Here the builders had turned a natural mountain peak into a stepped pyramidal temple platform, with four terraces and an artificially leveled summit covering more than 30 acres.

A Tokyo University expedition discovered the burials of three highly ranked leaders of Kuntur Wasi society. Each individual's body was found in a tomb, hidden at the base of an eight-foot-deep cylindrical shaft. The first tomb was that of a 50- to 60-year-old man with a gold crown or headband, embossed with images of trophy heads in net bags. He was also accompanied by two pottery bottles and a cup, three large conch shell trumpets, two polished stone ear ornaments, two polished stone pendants, and a variety of stone beads. The individual in the second tomb had a gold crown or headband with an embossed panel of jaguar or puma faces, two gold pectorals with complex feline and snake motifs, two rectangular plaques of embossed gold, and two pottery vessels. The occupant of the third tomb, an elderly woman, had two gold ear ornaments, two stone ear ornaments, a pottery cup, and a distinctive vessel with a stirrup-shaped spout. The floor of her tomb was covered with 7,000 beads that had once adorned a perishable garment.

By this point, in other words, we are dealing with rank societies whose elites wore gold ornaments on the scale of the sixteenth-century chiefs of Panama. Their chiefs liked to be symbolized by dangerous predatory animals and were happy to patronize artisans skilled at chiefly iconography.

Chavín de Huántar

We previously suggested that Peru is a graveyard for theories of environmental determinism. Nowhere is that more true than at the chiefly center of Chavín de Huántar in Peru's Mosna Valley, some 10,395 feet above sea level. One cannot explain the presence of such an impressive center based simply on the agricultural potential of its steep-walled valley. Frost is so common at that altitude that farming is limited to one harvest a year, based mainly on potatoes and other indigenous root crops. To be sure, llamas and alpacas could have been raised in the area, as long as they were moved from one grazing area to another during the year.

To understand Chavín de Huántar, one must realize that it lay along a trade route linking the Pacific coast, the Andes, and the Amazon basin. From Chavín to the Pacific would have been a six-day walk behind a llama pack train. From Chavín to the Amazon jungle would have been a six-day trip going the other direction. Chavín, in other words, was the midpoint for the long-distance movement of goods among three major cultural provinces: coast, highlands, and tropical forest.

Chavín de Huántar was founded 2,900 years ago and declined some 700 years later. Archaeologist Richard Burger estimates that, during its peak, Chavín might have been a community of 2,000 to 3,000 people. At that point its temple center alone exceeded 12 acres, and the associated residences covered more than 100.

Chavín's major temple was unlike any we have examined so far, though it borrowed a few details from earlier societies. Its first version, called the Old Temple, was built from granite slabs some 2,800 to 2,500 years ago. This temple had the form of a U, with truncated pyramids serving as its lateral wings; its builders had also placed a sunken circular court directly in front of it. The most striking feature of the Old Temple, however, was a basal platform honeycombed with secret rooms and windowless passageways known as "galleries." These galleries were connected by stairways, vents, and drains. Given the Andean peoples' long tradition of supplying oxygen to fire pits through subfloor ducts, the architects of Chavín probably had no trouble getting air to the innermost galleries.

Placed deeply into one of the darkest and spookiest interior galleries was a carved stone monolith 15 feet tall—so tall, in fact, that its pointed crown extended through the ceiling of the gallery into a hidden, still higher passageway. The image carved on the monolith was that of a terrifying humanoid with a snarling mouth, its hands terminating in claws and its long hair ending in snakes' heads (Figure 33).

Of the many interpretations of this remarkable monolith, we are most persuaded by the one proposed by archaeologist Craig Morris. Morris, an expert on the later societies of Peru, points out that the fame of many Andean temples derived from their possession of an oracle. Andean oracles (like their Greek counterpart at Delphi) foretold the future and answered travelers' questions in the form of riddles. Morris suspects that the carved monolith hidden in the bowels of Chavín's Old Temple was one of the Andes' earliest oracles. Its answers would have been spoken aloud by an unseen priest, hiding in the upper passageway penetrated by the monolith's crown.

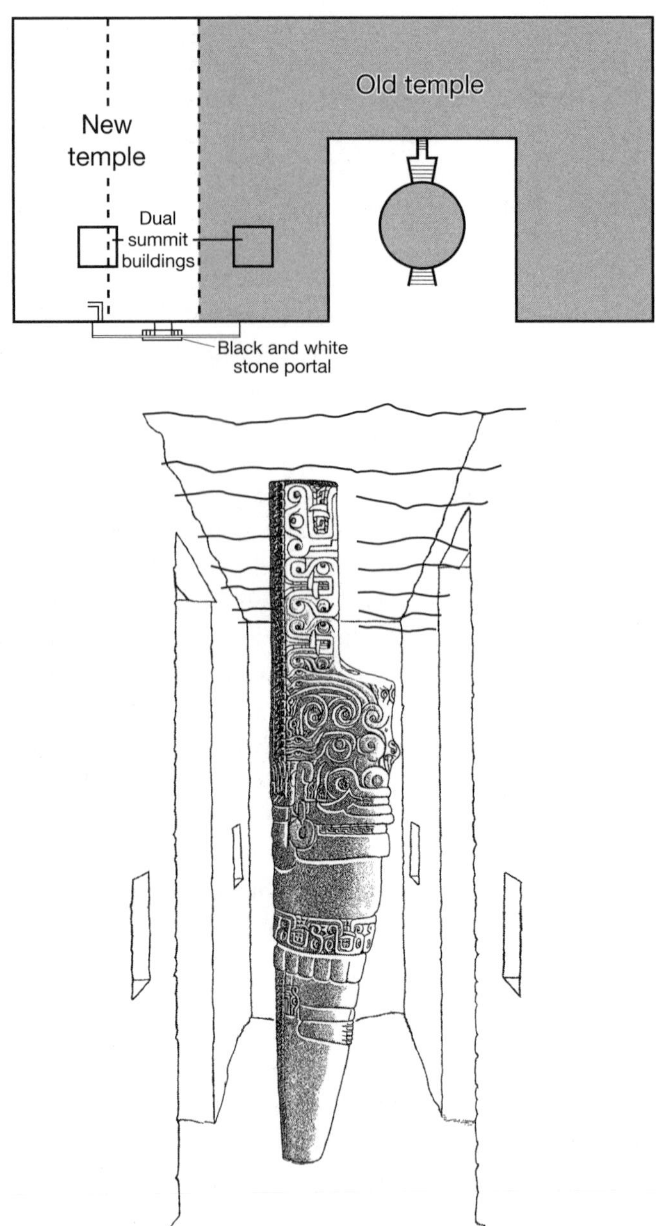

FIGURE 33. The Old Temple of Chavín de Huántar, Peru, was honeycombed with underground galleries. In one of the galleries visitors would have encountered the 15-foot-tall image of a terrifying humanoid that may have been considered an oracle.

Imagine the leader of a llama caravan passing through the canyon of Chavín de Huántar on his way between the Pacific coast and the Amazon jungle. He has questions that need to be answered. In return for payment of some kind (perhaps trade goods, to be deposited in one of the building's other galleries), a ritual specialist lights a torch and leads the traveler down a darkened gallery to the terrifying humanoid image. In the flickering light of the torch, a disembodied voice answers the traveler's question with a riddle. He returns to the surface, pleased not only to have received an answer but also to have returned safely from a place with such dangerously high levels of life force.

Between 2,500 and 2,300 years ago the elite of Chavín de Huántar created a New Temple, one that swallowed up and incorporated the Old Temple. The New Temple was an immense stone masonry building, standing more than 30 feet high; one entered the building through a portal with columns of white granite and black limestone. The original oracle had by now lost its importance, but the New Temple featured two *stelae,* or freestanding stone monuments. One of these stelae, more than six feet tall, featured a grotesque supernatural being with a feline mouth, staffs of authority in its hands, and long hair ending in snakes' heads. The other stela, more than eight feet tall, displayed a pair of caimans flanked by spiny oyster and conch shells, gourds, chile peppers, and manioc plants.

In addition, the entire façade of the New Temple was decorated with tenoned stone heads, frightening human or animal faces projecting out from the building. Many of these heads, displayed as high as 30 feet up on the temple, would have looked down menacingly on all who approached.

The Nature of Hereditary Inequality in Early Peru

Peruvian societies of 2,500 years ago, both on the coast and in the highlands, had invested in all of Irving Goldman's proposed sources of chiefly power. Their elite families claimed descent from supernatural ancestors, possessed dangerous levels of life force, and had the authority to sacrifice and mutilate their enemies. They patronized and rewarded the craftsmen who carved their stone monuments, hammered out sumptuary goods of embossed gold, and produced pottery covered with chiefly symbols. Included among the symbols were stylized references to jaguars or pumas, birds of prey, snakes, caimans, and crocodiles—all animals associated with ferocity and predation.

The religion of this period, centered on the temple, included well-regulated processions, human sacrifice, and the likely use of oracles. Peru had reached the point where a handful of changes in social logic could have led to the creation of a kingdom. We will describe just such a transformation in a later chapter.

THIRTEEN

Aristocracy without Chiefs

We have learned a great deal from societies speaking Tibeto-Burman languages, but there is more to be learned. In this chapter we return to Assam to look at three more societies: the Dafla, the Miri, and the Apa Tani.

In the 1960s there were roughly 40,000 Dafla in the hills of Assam, all tracing descent from a legendary ancestor. The Dafla grew dry rice and millet by slash-and-burn farming and raised pigs, goats, oxen, and mithan cattle.

Like the Etoro of New Guinea, the Dafla lived in longhouses that accommodated up to 12 families. Also like the Etoro, they displayed little in the way of cohesive leadership. They had no hereditary leaders, and their existing headmen and elders could not prevent one longhouse from feuding with another. Some degree of inequality had been introduced into Dafla society by the fact that captives taken in raids were kept as slaves. Such slaves, however, were not prevented from working hard and accumulating sufficient resources to purchase their freedom.

The Miri claimed descent from the same ancestor as the Dafla. Both groups referred to themselves by the ethnic term *Nisü* and shared a wide range of behaviors.

The Apa Tani were related to the Dafla and Miri but had created a strikingly different society by altering traditional social logic. The changes they introduced were supported by a type of agriculture that surpassed that of their Dafla and Miri neighbors.

The valley of the Kele River lies 5,000 feet above sea level in the foothills of the Himalayas. In 1944, according to anthropologist Christoph von Fürer-Haimendorf, there were only seven Apa Tani villages in a valley just two miles wide and six and a half miles long. Expansion was limited by mountains that rose 3,000 feet above the swampy valley floor. In 1961, during a second visit by Fürer-Haimendorf, the population had grown to 10,745 persons, who lived in a total of 2,520 households.

The largest Apa Tani villages had 500 to 700 houses; the smallest had less than 200. Each village was divided into wards, occupied by one or more *halu*, or clans, whose members reckoned descent in the father's line and were required to seek brides from other clans. Each ward maintained a ceremonial building called a *nago*, in which rituals were carried out and trophies (including the severed hands of enemies) were curated. The village also maintained a more secular public structure called a *lapang;* this was a large, open sitting platform like those of the Angami Naga.

Inequality in Apa Tani society was reflected in two types of clans: *mite* and *mura*. The mite were hereditary aristocrats, while the mura were former slaves. Some mura had won their freedom when the British colonial government abolished slavery. Others, like debt slaves in Naga societies, had won their freedom through hard work and loan repayment. However, even when adopted into aristocratic clans as poor relations, the mura were seen as lowly in rank.

Despite the existence of hereditary aristocrats, the Apa Tani did not have chiefs like the great Angs of the Konyak Naga. Their village affairs were managed by a council of aristocrats, all mite citizens, men of character and ability drawn from wealthy lineages. This council, called a *buliang*, included elders, middle-aged men, and a few young men who were regarded as future leaders. In return for their community service, the men of the buliang received gifts of rice beer and meat. These gifts were presented at major feasts, which rotated among village wards so that the costs would be shared by all.

Despite the division of Apa Tani society into aristocrats and former slaves, relations between the two groups were surprisingly accommodating. In fact, many mura clans shared a ceremonial building with an aristocratic clan, on whom they depended for proper instruction in ritual protocol.

In the village of Haja the aristocratic Nada clan shared its sitting platform with two mura clans, the Dusu and Dora. Fürer-Haimendorf often observed aristocrats and former slaves eating together, with minimal regard for rank. Nevertheless, the mura could neither expect to rise to mite levels nor intermarry with mite.

It is possible that, given time, some clans of former slaves might have taken on the status of a caste. For example, as in the case of many Indian castes, certain craft specialties were associated only with mura. Pottery making was monopolized by four former slave clans of the village of Michi-Bamin. Iron-working had become so associated with the mura that even aristocrats who engaged in it found their social status lowered.

Most extreme was the case of the mura women who castrated pigs for each village. This occupational specialty was considered so abhorrent that its practitioners were banned from participation in feasts and religious rituals and could not even enter the house of another family. The way these women were shunned reminds us of the untouchable caste of India, to whom the most unpleasant tasks were relegated. The difference, Fürer-Haimendorf tells us, is that in the cosmology of the Apa Tani, concepts of purity and ritual pollution did not play the powerful role that they did in Indian society.

At the upper levels of Apa Tani society there were subtle differences, like those among the four chiefs of the Tikopia. In theory, all mite clans were equal; in practice, however, some were "first among equals." The differences were attributed to senior versus junior descent. In Hang, a village of 514 households, two mite clans were considered senior to all others. These were the Tenio and Tablin clans, which made up roughly 2 percent of the village. They were considered senior to all other clans as a result of their descent from Ato Tiling, the legendary forefather of all Hang families.

THE APA TANI VERSUS THE DAFLA AND MIRI

Over the centuries the relations between the Apa Tani, Dafla, and Miri oscillated between (1) intense exchanges of goods, requiring peaceful reciprocal visits, and (2) hostile interactions such as raids, kidnappings, and individual homicides. Having no strong central authority, an Apa Tani village often found one of its residential wards trading with the Dafla while another was feuding with them. The sources of friction were many. The Apa Tani, as we explain later, were intensive rice cultivators. To obtain animals for food or sacrifice, they often traded surplus rice to the Dafla. They were also able to use unpaid loans of rice to turn Dafla families into debt slaves. Debts could lead to the confiscation of cattle or the taking of captives for ransom.

Fürer-Haimendorf was told of past raids in which more than ten men were killed and an equal number of women and children captured. Trophies claimed included the enemies' hands, eyes, or tongues, which were ceremonially buried upon the raiding party's return. Raids took the form of sneak attacks at dawn. They were preceded by rituals in which dogs and chickens were sacrificed to ensure success and followed by the negotiation of peace treaties.

Almost certainly there had been a time, long long ago, when the Apa Tani, the Dafla, and the Miri engaged in endless cycles of tribal warfare. At some point along the way, however, the strategy of the Apa Tani began to change. By the time of Fürer-Haimendorf's first visit, they had begun to substitute profit for revenge. After successfully resisting a Dafla attack, they did not kill their prisoners. Instead, they immobilized each Dafla captive by locking one of his ankles inside a hole in an impossibly heavy log (Figure 34). The Apa Tani fed and entertained the prisoner until his family paid his ransom. The ransom was then invested in more rice paddies.

Among the Apa Tani, achieving one's goals by peaceful means rather than violence became a path to prominence. To be sure, the Apa Tani would fight if there were no other option, but they had come to prefer wealth over war.

The Secrets of Apa Tani Success

The cosmos of the Apa Tani was similar to that of other Tibeto-Burman hill tribes. Their world had been created by a celestial couple, Chandun and Didun. Chandun, the husband, made Earth; his wife, Didun, made Sky. Human beings, including the Apa Tani, had been created by a second spirit couple. Lesser spirits were to be found living in rock outcrops and other natural features. Rank survived after death, with aristocrats and former slaves continuing to live in the afterlife as they had on earth.

Chandun had created the Kele Valley as a series of swamps and bogs, crisscrossed by the river and its tributaries. The Apa Tani had turned it into a completely manicured landscape. Twenty square miles of the valley floor were transformed with dams and terraces into a semitropical paradise of wet rice paddies. The Apa Tani stabilized the high ground by planting bamboo, pines, and fruit trees. Still farther upslope lay an untended rain forest of orchids, tree ferns, and rhododendrons. It was, by the 1940s, one of the most carefully managed landscapes on earth.

The Dafla and Miri, limited to dry rice grown on hillsides, suffered chronic shortfalls. Like the other tribal societies so far described, they lacked any concept of private ownership of land. Earth was a living being. One could grow things on its surface but not own pieces of it. A clan could assign its

FIGURE 34. The Apa Tani of Assam were traditionally divided into aristocrats and slaves. On the left we see a young woman wearing the ornaments characteristic of an aristocratic lineage. On the right we see a Dafla prisoner with his foot immobilized by a heavy log. If his relatives did not ransom him, the Apa Tani would convert him into a slave.

members the right to use certain areas, but when they moved on those areas were reassigned. To own individual bits of Earth was as unthinkable as owning individual bits of Sky. A farmer could privatize the basket of rice he had harvested because it was the fruit of his labor. He could not privatize the land on which it grew.

The Apa Tani differed from their neighbors in having three types of land: clan land, village land, and private land. Virtually all the wet rice terraces they created were private, and as such they became a source of private wealth. In the social logic of the Apa Tani, the fact that a family had invested labor in converting a bog to a rice paddy made it their private creation. Apa Tani families also owned granaries, bamboo groves, and garden plots of different kinds. Clan land, on the other hand, was set aside for public buildings, cemeteries, pastures for animals, and forest resources. The latter included pine trees, used for house construction, which were carefully managed.

Wet rice cultivation is labor-intensive. The Apa Tani had six different kinds of rice, which were grown in nurseries, transplanted to irrigated terraces, and fertilized with human and animal manure. Agricultural labor was provided by work gangs called *patangs,* to which every Apa Tani belonged from childhood. Husbands, wives, children, relatives, and former slaves worked together to build dams, canals, and terraces, to transplant, and to garden. It was the responsibility of the paddy's owner to feed the patang, and wealthy families could hire extra labor gangs to farm their large holdings. Such pay allowed the members of a patang to buy more land of their own.

Once having accepted the privatization of land, the Apa Tani began to invest in little else. They discovered that a family with five or six members could produce its yearly supply of rice (about 300 basketloads) on one and a half to two acres. To plant more was to create surplus, and surplus meant even greater wealth. Many Apa Tani families did not have to devote a single acre to grazing land because they could get all the animals they needed from the Dafla in exchange for rice.

The Apa Tani sacrificed cattle, pigs, and dogs, paid for brides with mithan cattle, and hosted feasts of merit with animals purchased from rice-poor neighbors. Even after cotton was introduced to the region, most Apa Tani weavers did not make room for it among their rice paddies. They let the Dafla grow cotton, purchased it with surplus rice, and even returned the cotton seeds to the Dafla after the bolls had been ginned.

While the mura were clans made up of former slaves, new slaves were periodically created by raids and unpaid debts. These slaves, like surplus rice, became a source of wealth. An aristocrat who had invested all his surplus in new paddies could obtain additional mithan for a feast by selling a slave. Even a mura clan, if desperate, might sell one of its members into slavery for cattle.

It was a slave's exchange value that turned the Apa Tani from vengeance to ransom. Any prominent prisoner taken in a raid would be ransomed by his or her family, and captives too lowly to ransom would be sold as slaves. The profit was used to buy more rice land. The Apa Tani did not have bumper stickers, but if they had, the most popular would have read: "Make wealth, not war."

The Logic of Apa Tani Society

Once upon a time the Apa Tani almost certainly shared much of the social logic of the Dafla and Miri. The latter belonged to clans whose members had the right to farm, but not actually own, a portion of the earth's surface. They could privatize their harvests because those were the products of their own labor—the slashing and burning of the wild vegetation and the planting, weeding, and reaping of their crop. They could not, however, privatize Earth. It lay fallow while they cleared another patch of forest, and once its fertility was restored, it could be assigned to another family from the same clan.

To convert an unproductive swamp to productive rice paddies, however, impressive labor is needed. There are check-dams to be built, canals to be dug, terraces to be contoured, and water to be trapped by raising field borders. Having invested this much work in the creation of a paddy, no Apa Tani family was willing to cede its use to another family. They considered it just as much their property as the crop itself, and they therefore maintained its fertility with manure to prevent its going fallow and being reassigned. Over time, while Earth remained the creation of Chandun, the wet-rice paddy came to be seen as a creation of human labor.

Privatization created incentives for intensive agriculture, wealth creation, and a focus on rice production that took land away from other activities. It made many Apa Tani families the equivalent of the Cauca Valley's "nobles by wealth." Privatization, however, undermined long-standing principles of corporate ownership. It relegated clan land to areas of low productivity and converted some slaves from agricultural laborers to a form of capital that could be sold to buy land.

In previous chapters we saw that war was virtually endemic among rank societies. To paraphrase Sean Connery's character in *The Untouchables,* their philosophy was, "They send one of ours to the hospital, we send one of theirs to the morgue."

The Apa Tani, however, show us that endless blood feuds are not inevitable. Instead of applying the principle of social substitution and taking revenge on their prisoners, the Apa Tani turned them into profit. This behavior created a logical contradiction: the desire for wealth now trumped clan loyalty and the principle of social substitution. This contradiction explains the unwillingness of one ward to be drawn into a neighboring ward's feud, especially if it might reduce profits.

The Nature of Apa Tani Inequality

Kachin society had thigh-eating chiefs. Konyak Naga society had great Angs. The Apa Tani had hereditary rank and wealth, but no chiefs at all. They present us with an alternative form of rank society, one that may well have existed in prehistory but would be very hard to detect archaeologically.

The Apa Tani had a hereditary aristocracy that provided all community leaders. That leadership, however, was exercised by a council rather than by a powerful individual. As we saw earlier, the ancient Greeks referred to such a system as oligarchy, or rule by a privileged few.

We have described Apa Tani society as based on rank, but one could argue that it was almost stratified. We raise this possibility because the mite and mura were not allowed to intermarry. Such an impermeable barrier between the elite and nonelite was, as we shall see in later chapters, characteristic of societies that had developed true strata or hereditary classes. The fact that some mite clans included mura families, however, convinces us that the Apa Tani were not fully stratified.

Most achievement-based societies, as we have seen, opposed the accumulation of wealth by individual families and pressured them to distribute it to others. The Apa Tani did not do so, although they did retain the feast of merit which, in other societies, was used to redistribute wealth. Unlike those other societies, the Apa Tani admired wealth and increasingly sought rich men for the village council. They did preserve the concept of clan land but devoted it increasingly to ritual buildings and sitting platforms, while wealthy families bought up the best rice fields. Clan solidarity suffered as wealthy families pursued profit.

How might an archaeologist detect this kind of society? One might rely on sumptuary goods to recognize the burial of a chief, but what about the burial of an aristocratic council member? Trophy heads and hands might be recognized by archaeologists, but what about ransomed captives? Without written texts, would we know that some land was private?

As it happens, these are not just rhetorical questions. We know that one of the world's first civilizations shared some institutions with the Apa Tani. That civilization has been described by one of its leading experts as an oligarchy. It had a council of elders. It had a hereditary aristocracy, commoners with access to land, commoners with no access to land, and slaves. Some of its craft specialties were considered more prestigious than others. This civilization had private land, public land, and land allotted to temples. Slaves might be debtors or war captives and could earn their freedom.

The creators of that civilization were the Sumerians of Southern Mesopotamia, and we know all these things about them because they had writing. The earlier societies out of which Sumerian civilization developed, however, did not have writing. As we will see in the chapter that follows, discovering the institutions of those earlier societies is a task that pushes archaeologists to the limits of their interpretive skills.

FOURTEEN

Temples and Inequality in Early Mesopotamia

Temples, as we have seen, went on to replace men's houses in several parts of the New World. In the cases we have examined, the transition was accompanied by evidence for hereditary inequality. This fact does not surprise us because we have seen that as chiefly elites emerge, they begin to dedicate buildings to the highest celestial spirits in their cosmos.

Now we must search for a comparable transition in the Old World. We have chosen Mesopotamia because its societies were among the first to replace the small ritual house with the temple. Beginning 8,700 years ago with the Terrazzo Building at Çayönü, Turkey, villages of the Tigris-Euphrates drainage built increasingly temple-like structures. For centuries, some early temples coexisted with circular buildings that look like men's houses or clan houses. Finally, between 6,500 and 6,000 years ago, the temples stood alone.

Finding examples of early temples in Mesopotamia is not difficult. The difficulty lies in determining whether the society building the temple also shows signs of hereditary inequality. It will soon become clear why we chose to discuss Tikopia and the Apa Tani before taking on the archaeology of Mesopotamia. Those two societies have prepared us for the fact that some kinds of inequality can be hard to detect archaeologically.

In addition to the overall difficulty of detecting rank, we see differences between Northern and Southern Mesopotamia. A number of ancient villages in Northern Mesopotamia provide us with clues to social inequality, such as elite children buried with sumptuary goods, long-distance exchanges of polychrome pottery among elite families, the clustering of satellite hamlets around chiefly villages, and the burning of elite residences in raids. For Southern Mesopotamia, the evidence for rank is more subtle.

NORTHERN MESOPOTAMIA

The key to Northern Mesopotamian agriculture is the Syrian Saddle, a gap in the Lebanon-Judean mountains that allows winter storms from the Mediterranean to travel east as far as Iraq. As these moisture-laden winds reach the Zagros Mountains, they rise, cool, and rain.

It was this rain that provided one of the key differences between Northern and Southern Mesopotamia. Most areas with rainfall exceeding 12 inches a year can grow cereals without irrigation. Most areas receiving less than 12 inches are required to irrigate. All of Mesopotamia raised wheat, barley, sheep, and goats, but the emphasis was different. Prehistoric villages in the rainier north made greater use of wheat and goats. Prehistoric villages in the hotter, drier south made greater use of barley and sheep. Barley is more tolerant of heat and salinity than wheat. Sheep have an ability to pant, not shared by goats, which allows them to dissipate heat.

Several of the ancient villages considered here lie near the modern city of Mosul, on the Tigris River in northern Iraq. Mosul itself receives 15 inches of rain. The rainiest months are December to March, making winter wheat a favored crop. In days of old it was said that Mosul had three colors: green, bronze, and buff. The green from winter rains lasted until May and then yielded to the bronze of ripening wheat. Summer and fall baked the region to a dusty buff.

The Growth of Extended Households

Roughly 7,500 years ago, villages in the Mosul plain grew wheat, barley, lentils, and peas for food; raised flax to make linen cloth; and herded sheep, goats, cattle, and pigs. These tasks required a division of labor beyond that of the nuclear family. Married sons increasingly remained in the households of their fathers instead of founding their own homes. This created extended families of 15 to 20 persons, capable of dealing with the multitasking of a farming-herding economy. Such families built lots of storage rooms, increasing the privatization of storage we saw at earlier sites such as Çayönü.

The village of Hassuna, 20 miles south of Mosul, exemplifies the transition from single-family to multifamily homes. The earliest mud-brick houses at Hassuna were nuclear family homes of three to five rooms. Somewhat later, the builders began to group these houses around a court or patio. Roughly 7,300 years ago, Hassuna was building irregular complexes of 15 to 20 rooms, flanking two sides of an open court. Often one part of the complex looked more planned than the rest, as if it were the original nucleus to which later rooms were added by accretion. Finally, some 7,000 years ago, there were residential compounds of 15 to 20 rooms whose layout was designed from the outset to accommodate an extended family (Figure 35).

These later residences were composed of three relatively standardized units: courtyards or patios averaging 156 square feet, working or sleeping rooms averaging 108 square feet, and storage rooms averaging 21 square feet.

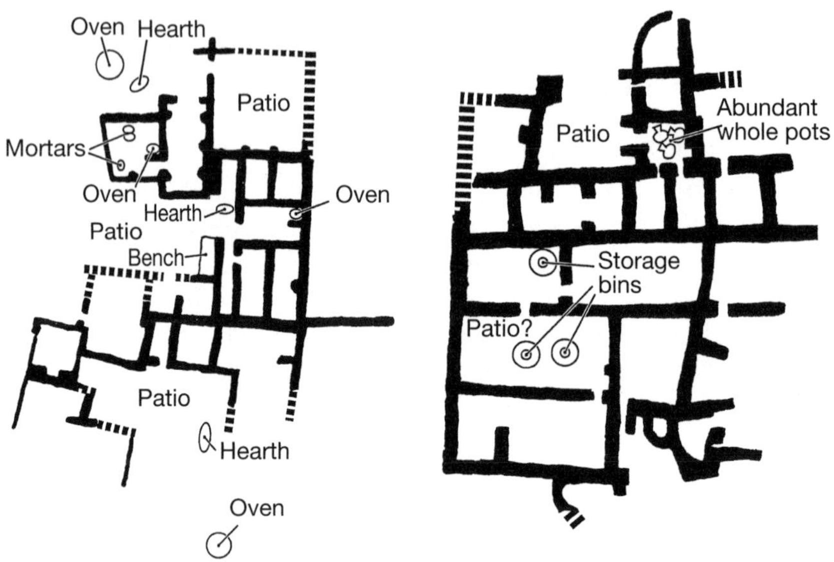

FIGURE 35. Some 7,200 years ago the village of Tell Hassuna in Northern Mesopotamia was building houses for extended families of 12 to 20 people. These families had multiple kitchens and privatized storage bins, holding thousands of pounds of cereal grains. The house on the left is from the village of Level IV; the house on the right is from Level V. Families this large were able to advance socially by building up capital in the form of wheat, barley, sheep, and goats. The patios in their houses averaged 156 square feet, while the working or sleeping rooms averaged 100 square feet.

Houses often had several kitchens. These multiple kitchens provide evidence that more than one married couple occupied the house, with each wife maintaining her own hearth.

According to excavators Seton Lloyd and Fuad Safar, some levels at Hassuna also featured ritual buildings of an unusual type: circular, with a domed roof. Near Eastern archaeologists refer to these buildings with a borrowed Greek term, *tholos* (plural, *tholoi*), despite the fact that they had no connection to the ancient Greek burial chamber of the same name.

There were hints of still larger social units at Choga Mami, a village occupied between 7,300 and 7,000 years ago. Choga Mami lay in the piedmont east of the Tigris River, some 200 miles southeast of Hassuna, and irrigated an alluvial fan created by a small river from the Zagros Mountains. At Choga Mami, excavator Joan Oates found traces of larger, heavily buttressed walls that did not form part of any residence. These thicker walls, she felt, might have allowed a group of related families to separate themselves from the rest of the village. Such walled residential wards imply the presence of larger social segments, such as clans or ancestor-based descent groups.

Evidently Choga Mami also felt the need for defense. At the margins of the village, Oates uncovered what appeared to be a mud-brick watchtower. Any strangers approaching would have been visible from the tower.

Defensive Works and Sumptuary Goods at Tell es-Sawwan

Nowhere during this time period was the need for defense from raiding clearer than at the site of Tell es-Sawwan. Es-Sawwan lay directly on the Tigris River, 100 miles west of Choga Mami and only six miles south of the modern Iraqi city of Samarra. There the Tigris would likely have been 800 feet wide at low water and almost 2,000 feet wide at flood stage. Its course was limited by high conglomerate bluffs. The river was free to meander between these bluffs, leaving areas of floodplain available for growing two irrigated crops a year.

The village of Tell es-Sawwan ran for nearly 750 feet along the east bluff of the Tigris. It also extended back more than 350 feet from the bluff, covering at least six acres. Two dry gullies, spaced roughly 160 feet apart, ran westward through the village on their way to the Tigris. The occupants of Tell es-Sawwan deepened each of these gullies into a defensive ditch and then connected the two with a north-south ditch, cut ten feet into the underlying conglomerate (Figure 36).

The center of the village was thus protected on the west by the river bluff and on the north, east, and south by ditches. Just inside the ditch the villagers built a mud-brick wall, so strongly buttressed that it was still standing three feet high when discovered by Iraqi archaeologists. The total height of the bar-

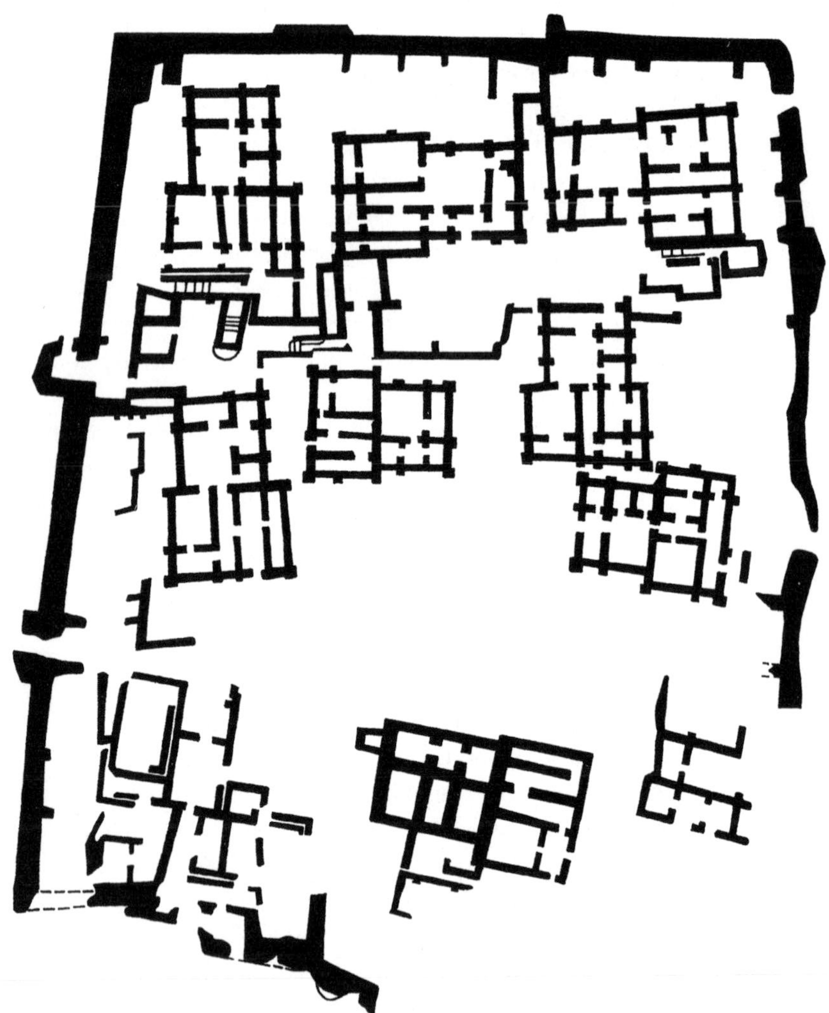

FIGURE 36. The village of Tell es-Sawwan, on the Tigris River near Baghdad, was defended on three sides by ditches and walls and on its remaining side by the bluffs of the river. The distance between its north (left) and south (right) defenses was roughly 160 feet.

rier presented by the wall plus ditch was at least 13 feet, leaving Tell es-Sawwan well fortified. Debris in the ditches left no doubt that there was a need for defense. Included were large numbers of sling missiles, egg-shaped projectiles made of dense clay.

The simple handheld sling, consisting of a leather pad attached to two cords, is probably one of the oldest weapons known. Its origins almost certainly go back to the earliest hunters and herders. Even those of us who have no flocks of sheep to guard are familiar with slings, having heard the story of David and Goliath. It is not generally realized, however, how important an article of warfare the sling once was.

According to a study by archaeologist Manfred Korfmann, "in Mesopotamia, in Persia and in Greece and Rome a slinger was considered a match for an archer." A throw of more than 650 feet was not unusual for slingers of that era, meaning that large numbers of missiles could be launched from a distance. When thrown from only 300 feet away, the missile—coming off the sling at speeds approaching those of a Nolan Ryan fastball—could drop a man in his tracks. The sling missiles of Tell es-Sawwan had been given a streamlined, biconical form that improved their accuracy, velocity, and distance and made them fit more snugly in the leather pad.

Tell es-Sawwan's enemies may have coveted its prime irrigation location. Its main crops were six-rowed and two-rowed barley, cereals that did well under hot-weather irrigation. The flax grown at es-Sawwan had seeds more than 0.16 inches in length, diagnostic evidence for irrigated flax.

The most interesting houses at Tell es-Sawwan were T-shaped, combining long, narrow corridors and wider, more nearly square rooms. These buildings had anywhere from eight to 16 rooms and averaged more than 700 square feet of floor space. Some rooms seemed to yield domestic tools; others held cereals and agricultural implements; and still others may have been household shrines.

It was, however, the burials below the floors that created the greatest excitement at Tell es-Sawwan. Many contained masterpieces of finely painted pottery, stone bowls, marble or alabaster statuettes, turquoise beads, and items of native copper, dentalium, and mother-of-pearl. Significantly, some of the burials richest in sumptuary goods were those of children (Figure 37). Grave 92, for example, contained an infant with three alabaster statuettes (one with eyes of inlaid shell), beads of turquoise and carnelian, and three elegant pottery flasks. Grave 94 belonged to an infant with another alabaster statuette, this one featuring eyes of inlaid shell and its own miniature necklace of stone and asphalt beads. Clearly these children were too young to have earned such sumptuary goods through achievement. Their parents were likely to have been people of rank.

Craft Specialization and Exchanges of Luxury Pottery

Some of the pottery made during this period was elegant enough to compare with Panama's Coclé trade wares. From at least 7,300 to 7,000 years ago this

FIGURE 37. This four-inch alabaster statuette was found with a burial at Tell es-Sawwan. Supplied with asphalt hair, shell-inlaid eyes, and a necklace of turquoise beads, it probably depicts an elite ancestor. The fact that this valuable statuette was buried with a child increases the likelihood that es-Sawwan society featured hereditary rank.

pottery, named "Samarran ware," was distributed over much of Northern Mesopotamia. Samarran potters were the first in Mesopotamia to sign their work with "potters' marks," small painted symbols that identified the maker of the vessel. Many Samarran potters' marks have been found, but we do not know whether they referred to individual potters, lineages of potters, or entire communities.

Despite the obvious expertise of the Samarran potters, between 7,000 and 6,500 years ago their products were gradually replaced by an elegant new bichrome and polychrome painting style. Called "Halaf ware," after an important site in Syria, this style of painting eventually spread over Northern Mesopotamia.

Archaeological sites with Halaf pottery supply us with three different lines of evidence for social inequality. First, they increase our sample of children buried with sumptuary goods. Second, they suggest that some large villages may have had authority over a group of smaller satellite communities. And finally, they provide us with evidence for long-distance exchanges of gifts between elite families, such as those of the Tlingit and Haida.

Halaf burial practices were more complex than those of Samarran societies, implying a greater range of social statuses. Archaeologists have identified "ordinary" graves, special cremations, the reburial of certain people's skulls in finely painted bowls, and a number of shaft-and-chamber tombs that were reserved, we suspect, for individuals of high rank.

Archaeologists Nikolai Merpert and Rauf Munchaev found a number of Halaf burials at Yarim Tepe I and II, a pair of ancient villages in the region west of Hassuna. The variety of burials was as follows:

1. "Ordinary graves" (for example, in Levels 8 and 9 at Yarim Tepe II) contained adults and children laid out full length. A few of these burials had a stone cup or pottery vessel with them, but most were buried with little or nothing.
2. After its abandonment, Yarim Tepe I was used as a cemetery for individuals buried in formal chambers. These burials were often so tightly flexed that one suspects they were wrapped in bundles. They were accompanied by pottery and alabaster vessels, polished pins and axes of iron ore, stone pendants, or beads of seashell or stone. While most of the burials were those of adults, Burial 56 was a four-year-old child buried with a stone macehead—an object which, if found with an adult, would have been considered a status symbol.

Burial 60 was an adult accompanied by the skull of a very large bull, vessels of pottery and alabaster, an iron-ore pin, and about 200 astragali, or tarsal bones, from gazelles. The astragali may have been some kind of kit for divination or casting lots, a prehistoric forerunner of modern dice.

3. Levels 7 through 9 of Yarim Tepe II produced the bulk of the special cremation burials. Perhaps the most revealing was a series of cremated children with sumptuary goods. Burial 40 consisted of the charred remains of a youth 10 to 13 years of age, found in an oval crematorium. After cremation his remains had been placed in a painted Halaf vessel, along with a necklace of 20 polished obsidian beads. Also in the pit were smashed and burned pots, stone vessels, more beads from possible necklaces, and a seal drilled for suspension. The significance of the drilled seal will become clear during our discussion of the site of Arpachiyah.

   Burial 43 consisted of the remains of a cremated ten-year-old child, placed inside a painted Halaf vessel and buried under the floor of a tholos. The crematorium itself, found nearby, included an alabaster goblet with a pedestal base, an alabaster bowl, and three ceramic vessels (Figure 38).

   Once again, it seems unlikely that either of these children had achieved enough in his lifetime to deserve such labor-intensive offerings. Each alabaster vessel had to be cut and polished from a block of stone as hard as marble. The craftsmanship involved makes it likely that Burials 40 and 43 were the children of highly ranked families.

4. Finally, in Level 9 of Yarim Tepe II, there were several burials of isolated human crania. One of these (Burial 56) consisted of three crania, two from adults and one from a youth. We do not know why certain people's skulls were treated this way, but the behavior is consistent with the Near East's long history of curating skulls.

While we do not know the details of Yarim Tepe society, the variety of ways in which these two villages treated their dead, including infants and children, suggests a level of inequality like that of early rank societies elsewhere in the world. In addition, the figurines of Yarim Tepe II reveal that some women were tattooed with the same motifs seen on Halaf pottery. In many rank societies this would be a sign of prestige.

There is circumstantial evidence that the authority of prominent Halaf leaders extended beyond their home villages. In the region east and west of the modern city of Mosul, archaeologist Ismail Hijara found numerous cases

Necklace from Cremation Burial 40

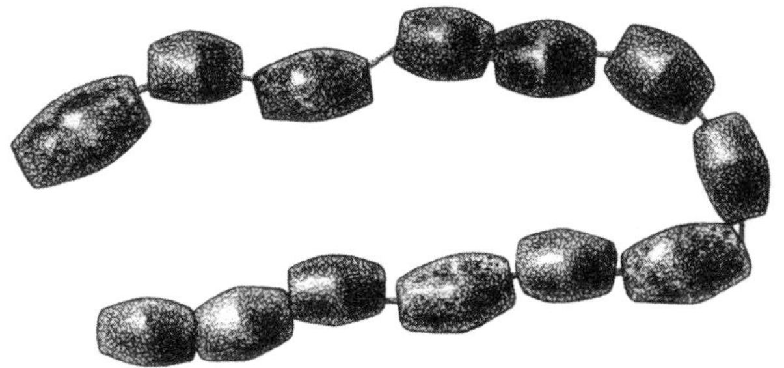

Alabaster vessels from Cremation Burial 43

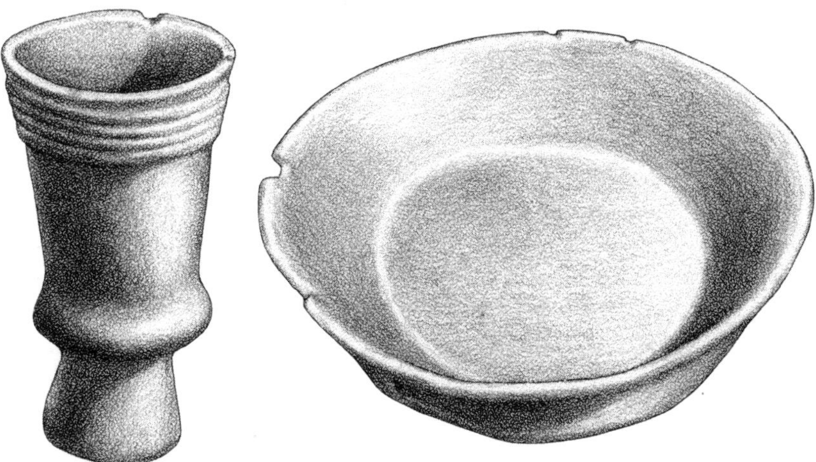

FIGURE 38. Sumptuary goods from the cremation burials of elite children at Yarim Tepe II, Iraq. The necklace of polished obsidian beads was found with a cremated child 10 to 13 years old. The alabaster goblet and bowl were found with a cremated child roughly 10 years old. Such special treatment of children suggests inherited rank.

in which a large (20-acre) Halaf village was surrounded by smaller (two-to-seven acre) communities that may have been subordinate to it.

Archaeologist Patty Jo Watson suspects that this "center-versus-hinterland" pattern may have characterized much of the region using Halaf pottery. At a minimum that region extended for more than 400 miles, from the Euphrates River on the west, past its Balikh and Khabur tributaries in Syria, past the Tigris River in Iraq, and east into the Zagros Mountains. This vast region consisted of farming areas with good alluvial soil, alternating with areas suitable only for grazing sheep and goats.

Long-Distance Elite Exchange

In an effort to learn how all these widespread Halaf societies might have interacted with each other, Watson collaborated with archaeologist Steven LeBlanc. Their study took advantage of the extraordinary repertoire of motifs on Halaf pottery, painted in up to three colors. Watson and LeBlanc suspected that the more closely any two Halaf villages interacted, the more motifs their pottery was likely to share.

LeBlanc selected seven Halaf villages from Syria, Turkey, and Iraq, sites where archaeological excavations had produced thousands of pottery fragments bearing painted motifs. For each pair of villages LeBlanc quantified the degree of sharing of motifs, using a statistical measure of similarity. He then compared this degree of similarity to the distance between each pair of villages.

LeBlanc would not have been surprised to find that the closer any two villages lay to each other, the more motifs they shared. That is not, however, what his results showed. The strongest similarities were found between the largest Halaf villages in LeBlanc's sample, those that were most likely to have been the social and political centers of their regions. For example, the very strongest similarities were (1) between Tell Turlu on the Euphrates and Tell Halaf on the Khabur River and (2) between Tell Halaf on the Khabur and the site of Arpachiyah, just east of the Tigris. The distances between each of these pairs of large villages were on the order of 120 to 170 miles. In contrast, the similarities between Arpachiyah and the smaller village of Banahilk (only 75 miles away) were much weaker.

In other words, exchanges among the largest Halaf villages were stronger than expected, especially given the intervening distances. This situation was similar to that described for chiefly centers in Panama, where young men of high rank established long-distance trading partnerships with Colombian elites before they became chiefs. It is possible that Halaf chiefs exchanged both brides and sumptuary goods, with skilled potters forming part of the chief's retinue. Watson has suggested that the largest Halaf villages were probably "chiefly centers, i.e., places of residence of local strongmen or chiefs."

We can compare the spread of Halaf polychrome pottery to that of Coclé polychrome, Tlingit and Haida crests, Quimbaya goldwork, and the Mexican vessels carved with Earth and Sky motifs. All these products of expertise spread rapidly once they had become appropriate gifts for chiefly families. The Halaf case is especially interesting because it may have involved not only the leaders of sedentary agricultural societies but also the leaders of pastoral societies who occupied the grazing lands between river floodplains.

Halaf Public Buildings

Societies using Halaf pottery built a variety of public buildings. Some of these buildings were for ritual, some probably had secular functions, and some remain enigmatic even to the archaeologists who excavated them.

At Tell Aswad, on a Syrian tributary of the Euphrates, archaeologist Max Mallowan discovered a multiroom building 20 feet by 17 feet in extent; he considered it to be a Halaf temple. At the village of Yarim Tepe II, already described, Merpert and Munchaev discovered a clay-walled Halaf building more than 28 by 14 feet in size. This structure contained none of the domestic debris of a residence and looks like a temple.

An early, 22-room complex in Level 6 of Yarim Tepe II, however, seemed to be a secular public building. This building was laid out in the form of a cross. At its center stood a tholos more than eight feet in diameter, divided into four compartments. The walls of the tholos appear originally to have been lined with alabaster slabs. Merpert and Munchaev suspect that this building may have been a large public storehouse, placed in the very center of the village.

The Halaf tholoi from Yarim Tepe II showed considerable diversity. Tholos 67, the largest, was more than 17 feet in diameter; Merpert and Munchaev considered it a ritual building because it had an offering buried beneath its floor. Some of the smaller tholoi, however, seem to have been used for the storage of household items. Too much use of the term *tholos,* therefore, obscures the fact that not all of these circular structures were built for the same purpose.

Arpachiyah: A Possible Halaf Chiefly Center

One of the Halaf sites chosen for LeBlanc's study was Arpachiyah, a multilayered *tell,* or archaeological mound, not far from Mosul. Arpachiyah has been the scene of repeated excavation, including the work of Max Mallowan in the 1930s and Ismail Hijara in the 1970s.

Arpachiyah was a village of mud-walled houses, tholoi, granaries with traces of wheat and barley, domed ovens, and pottery kilns. It also had long, narrow streets up to four feet wide. These streets were paved with local river cobbles, laid over thick layers of broken pottery in order to improve drainage.

At one point in its occupation Arpachiyah built a special group of tholoi, walled off from the rest of the village. The tholoi in this segregated area were more complex than the circular ones built at Yarim Tepe. They were keyhole-shaped, with an igloo-like entrance that led to the circular chamber (Figure 39). Hijara believed that the entire enclosure, built 7,000 years ago, had been dedicated to ritual.

In a later level at Arpachiyah, occupied during the peak popularity of Halaf polychrome pottery, Mallowan discovered what was almost certainly the residence of a highly ranked family. The dozen or so surviving rooms included a long, narrow court, a number of rectangular living or sleeping rooms, and at least four storage units. This house contained an exquisite collection of bichrome and polychrome pottery; stone vessels; black steatite amulets; figurines; beads of limestone, quartz, and marine shell; pigment-grinding palettes; painters' mixing-and-pouring bowls; and a number of stone seals perforated as pendants. One room of the house was filled with blades and flakes from the chipping of imported Turkish obsidian.

We cannot reconstruct the full extent of this residence because it had, in Mallowan's words, been "sacked and burned by an invader." This elite family—possessing more than their share of polychrome pottery, alabaster vessels, and Turkish volcanic glass—had evidently been the target of a successful raid.

One of the most interesting discoveries at Arpachiyah was that its leaders were closely monitoring exchanges of goods. Their way of doing this was to seal shipments with blobs of clay. While the clay was still wet, someone serving in an official capacity pressed a carved stone seal into it, leaving a distinctive impression. The impressed blob was not supposed to be broken until the shipment had arrived at its destination (Figure 40).

Tholos 137 from Yarim Tepe III

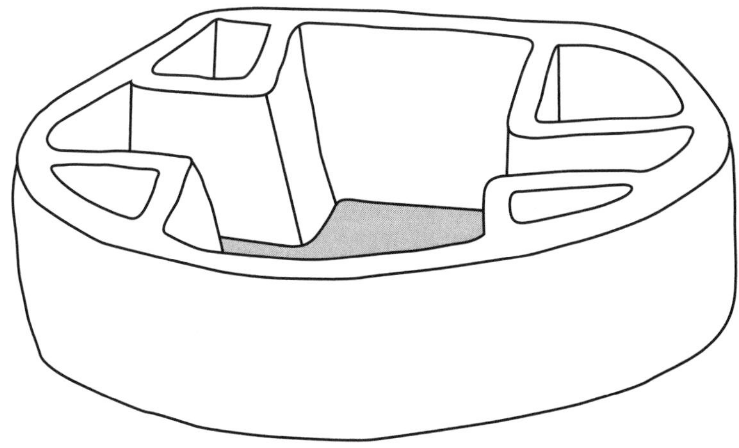

Tholos from Level TT 7-8, Arpachiyah

FIGURE 39. In the villages of the Halaf period not all of the circular or keyhole-shaped buildings called *tholoi* by archaeologists had the same function. Some, like the tholos from Level TT 7–8 at Arpachiyah, appeared to have ritual functions. Others, like Tholos 137 from Yarim Tepe III, had been filled with domestic refuse after abandonment, making it impossible to determine their original function. (The tholos from Arpachiyah was about 60 feet long.)

In the future the sealing of shipments in Mesopotamia would become a complex procedure. In Halaf times it was still relatively simple. The most common seal impressions found at Arpachiyah had been made on oval clay lumps, formed by hand around a knot made in a cord. This knot could not be untied until the clay lump had been broken. Each lump bore one or more impressions made with a seal. Many of the seals found at Arpachiyah bore a loop or had been perforated so that they could be worn around the neck of the person authorized to make the impression.

Now the significance of the seal pendant found with Burial 40 at Yarim Tepe II becomes clear. That cremated youth was too young to have been a village official. The fact that a seal was included with Burial 40 might mean that some youths were being groomed to inherit their fathers' positions within Halaf society.

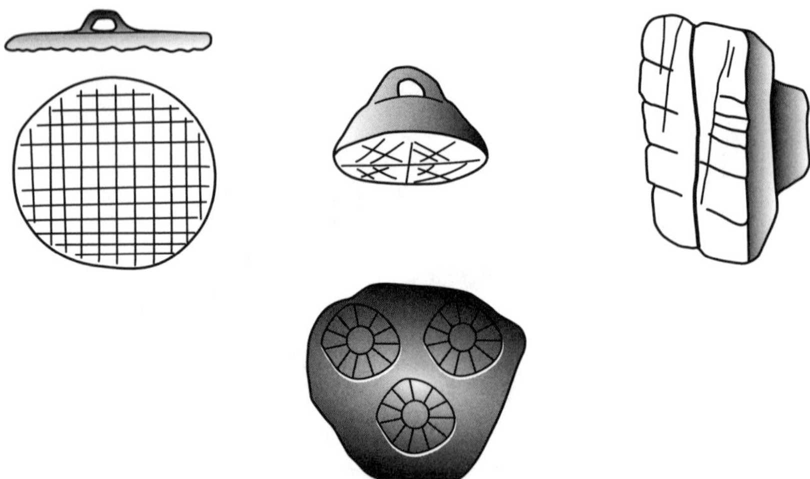

FIGURE 40. At Arpachiyah in northern Iraq, certain shipments of trade goods were closed with blobs of clay; the clay was then stamped with the distinctive seal of the person overseeing the shipment. Many of the seals bore loops, allowing them to be hung on a cord around the owner's neck. Above we see three stone seals. Below we see a blob of gray clay bearing three seal impressions, presumably removed from an incoming shipment. Most seals of this period were no larger than a postage stamp.

The Superimposed Villages of Tepe Gawra

One Northern Mesopotamian community celebrated for its temples was Tepe Gawra. It lay to the north of Arpachiyah, on a rolling plain that would become one of the breadbaskets of the later Assyrian empire.

The village of Tepe Gawra was founded nearly 7,000 years ago, during the heyday of Halaf pottery. It was then occupied and reoccupied for so many centuries and through so many superimposed layers of mud-brick houses that the accumulated remains created a mound rising 70 feet above the surrounding plain.

For eight years E. A. Speiser and Charles Bache dug down through more than 20 building levels, not halting until their excavation was 15 feet below the plain. They exposed a remarkable 100 percent of each village in the upper levels (I–X) and up to a third of each village in the lower levels (XI–XX).

For the purposes of this chapter, only the 11 deepest (and therefore earliest) levels at Gawra are relevant. Those levels were published by Arthur Tobler, restudied by Ann Perkins, and published in more detail by Mitchell Rothman, always with new insights. Tepe Gawra, in other words, is an archaeological gift that keeps on giving.

Level XX, the deepest reached, was occupied during the peak popularity of Halaf painted pottery. Perhaps Speiser's most interesting discovery was a well or cistern that had penetrated the underlying alluvial plain. This well had become the final resting place for some 24 persons, most of whom simply appeared to have been tossed down the shaft. An exception to this hasty disposal of human remains was a single skeleton at the top of the shaft. This individual was buried in a formal position on his or her left side, with knees drawn up and hands to the face, lying on a layer of wooden pole impressions that may represent the remains of a litter or bier. In many societies burial on a litter would be a sign of prestige.

Levels XIX–XV, which date to perhaps 6,500 to 6,000 years ago, represented six consecutive stages in the life of a two-to-three-acre village. By this time the Halaf painting style was in decline. What took its place was a somewhat simpler painted style, named for the site of Tell al-'Ubaid in southern Iraq.

In each of these six levels Speiser excavated roughly a third of the village. Although the details of each level were different, there were at least five types of structures that showed up over and over again. They were as follows:

1. Mud-brick residences of five to 20 rooms, including living or sleeping rooms, kitchens, storage facilities, and courtyards.
2. Pottery kilns, sometimes in the courtyards of residences and sometimes near storage units some distance from the house.
3. Areas of low, parallel walls resembling railroad ties. These walls may have allowed the circulation of air below storage units of perishable material.
4. Tholoi roughly 15 feet in diameter. Some of these were circular, while others were keyhole-shaped. The ground plan of the Level XVII village suggested that there might have been one tholos for each large residential compound (or, to put it differently, one for every 15 to 25 persons). We suggest that whatever public or ritual role these tholoi may have performed, they served only the extended family or lineage and not the entire village.
5. Mud-brick buildings whose ground plan identified them as temples. Each had a long central chamber, which Mesopotamian archaeologists call, by the Latin term, *cella.* This cella was flanked by two rows of small accessory rooms. Each temple was entered through an antechamber that ensured privacy. The cella usually had a podium of hard-packed clay on which burnt offerings could be made.

The earliest of the Gawra temples (in the village of Level XIX) exceeded 30 feet in length. Its central cella alone was 26 feet long. The temple in the Level XVIII village was 35 by 23 feet and had two small storage rooms, probably for ritual paraphernalia. Since there seems never to have been more than one temple in each level, we conclude that these buildings served the entire village.

The ‘Ubaid levels at Gawra, in other words, suggest two kinds of ritual behavior. Each lineage may have built and maintained its own tholos; the entire village probably built and maintained the temple. Tholos ritual may thus have involved each social segment's ancestors, while the temple was dedicated to celestial spirits.

Tepe Gawra underwent a series of dramatic changes from 6,000 to 5,500 years ago. During the period represented by Level XIV, its population fell dramatically. When Gawra was reoccupied in Level XIII, the nature of its occupation had changed. At this point the summit of the mound towered more than 20 feet above the surrounding plain, making it visible for miles on a clear day. Taking advantage of the fact that Gawra was now a local landmark, its leaders built three temples of regional significance.

Although these three temples shared a 60-by-50-foot central patio, careful excavation showed that they were all built at different times, using different sizes of bricks. The corners of the patio, as well as the corners of all three temples, were oriented to the cardinal directions. Tobler reconstructed the temple-building sequence as follows:

1. The easternmost temple was the first built. Its façade exceeded 65 feet, and it had been made with standardized mud-bricks 22 inches long.
2. The northernmost temple was the second built, using standardized mud-bricks 14.2 inches long. This was the smallest but best preserved of the temples, measuring 40 by 28 feet (Figure 41).
3. The third, or central, temple was built in the space between the first two. Its façade was 47 feet long; its other dimensions are unknown, owing to later erosion. This temple had been made with yet a third set of standardized mud-bricks, 19 inches in length. Once this central temple had been completed, it obscured the façade of the northern temple and presumably replaced it.

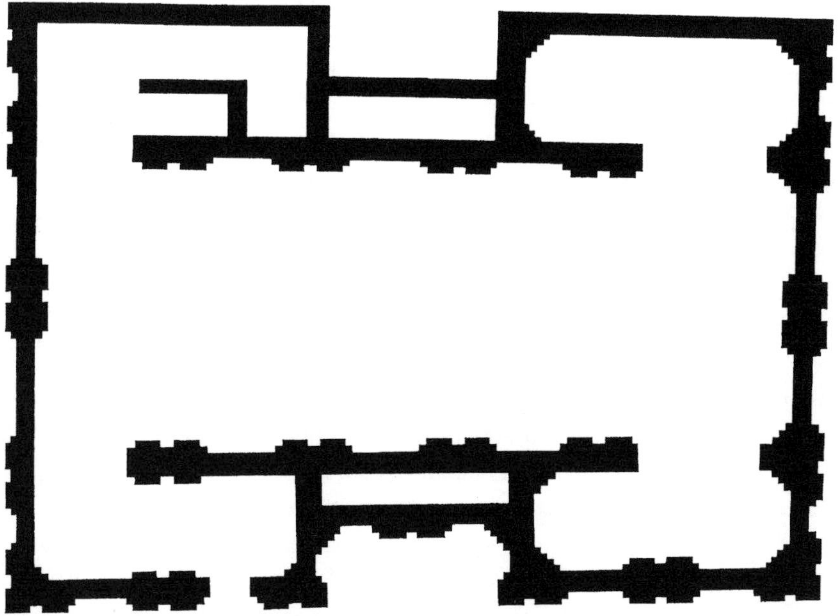

FIGURE 41. The North Temple from Level XIII at Tepe Gawra, northern Iraq, measured 40 by 28 feet. Archaeologists refer to the long, central room of the temple as a *cella*. Note the ornate brickwork used to relieve the monotony of the walls.

When a village builds its temple out of mud-brick, it runs the risk of creating a fairly drab building. The architects of Tepe Gawra avoided this problem by designing a complicated system of decorative brickwork, including recessed piers, pilasters, and wall niches. This design relieved the monotony by creating patterns of light and shadow where bricks protruded or were recessed. The builders also covered the walls with white plaster and, in some cases, bright red to purple paint.

By the time construction began on the central temple, the eastern temple had fallen into disuse and was being used as a place to discard refuse. In the ruins of this building archaeologists made an unexpected discovery: a series of nearly 100 miniature mud-bricks, scaled to one-tenth the size of the bricks used to build the central temple. It was clear to Tobler that these model bricks had been used to work out the most satisfactory methods of building the complicated recessed piers and pilasters found in the temple. In other words we now know how ʿUbaid architects worked in the era before blueprints: they built a scale model one-tenth the size of the final building, giving their bricklayers a template to follow.

A series of fine pottery beakers and an incense burner had been left behind in the cella of the eastern temple. The beakers had been painted with nested geometric figures, reproducing the pattern of niches on the temple façade. The incense burner had a pattern of cutout slots and triangular windows, framed by recesses like those of the temple niches. Almost certainly these were vessels used by the skilled specialists who carried out rituals in the temple—“priests,” for want of a better term.

The Troubled Times of Gawra XII

With the abandonment of Level XIII, Tepe Gawra ceased to function as a temple center. In Level XII-A it returned to being a small village. At this point the mound of Gawra had become a conical, artificial hill with a slightly concave upper surface, rising more than 30 feet above the surrounding plain.

Some 5,700 years ago the occupants of Level XII took advantage of Gawra’s height to turn it into a defensible, densely packed village whose population we estimate at between 130 and 240 persons. All of these people had decided to live together on the one-acre summit of the mound, presumably because its height gave them a measure of safety. One could only enter the Level XII village through one of two narrow, curving streets. Separating these two streets was a substantial watchtower, composed of three mud-brick rooms with a total area of 260 square feet. Behind the tower were six small rooms, arranged in a row that curved back into the village (Figure 42).

Speiser excavated about half of the Level XII village, uncovering at least 103 rooms. Most of these rooms belonged to eight large room complexes with courtyards, living or sleeping rooms, ovens, and storage rooms.

The area south and east of the curving street was the most interesting neighborhood. Its layout suggested a degree of village planning. Originally this neighborhood had been bisected by a street running southeast to northwest. At the northwest end of the street lay the largest and most elaborate residence. From here the street turned west, entering an open plaza more than 25 feet on a side.

Let us look now at the largest residence, the one bordering the open plaza. It had originally been 40 feet on a side and symmetrical, but later additions had compromised its symmetry. Its inner court (called the White Room because of its heavy coating of plaster) covered 576 square feet and had two doors opening onto the plaza. This court was flanked by rooms of different sizes that, according to Tobler, had definitely been used as living quarters. Included were a kitchen with a corner oven, some spacious living or sleeping rooms, and a series of small storage units. We would reconstruct this as the residence of an elite extended family, made up of perhaps 13 to 19 persons.

The second largest residence, only 20 feet farther down the northwest-southeast street, had a similar ground plan. Its interior court (called Room 26) covered more than 330 square feet and was flanked by rooms for living, sleeping, cooking, and storage. Tobler described this building as the dwelling for an "eminent member" of the village.

In spite of its defenses the Level XII village was successfully raided and partially burned, with at least four victims left unburied in the ruins. Significantly, the building hit hardest was the one with the White Room. The whole northern end was burned, its floor covered with ash and charred refuse to a depth of 15 inches. A baby and a child 12 to 14 years old had died in this residence. Two more youths lay unburied near a curving street leading to the watchtower. One of the latter, also a child 12 to 14 years old, is said to have had a stone "perhaps thrown from a sling" still resting between his or her shoulder blades. Following this raid, Tepe Gawra seems to have been abandoned for at least a century.

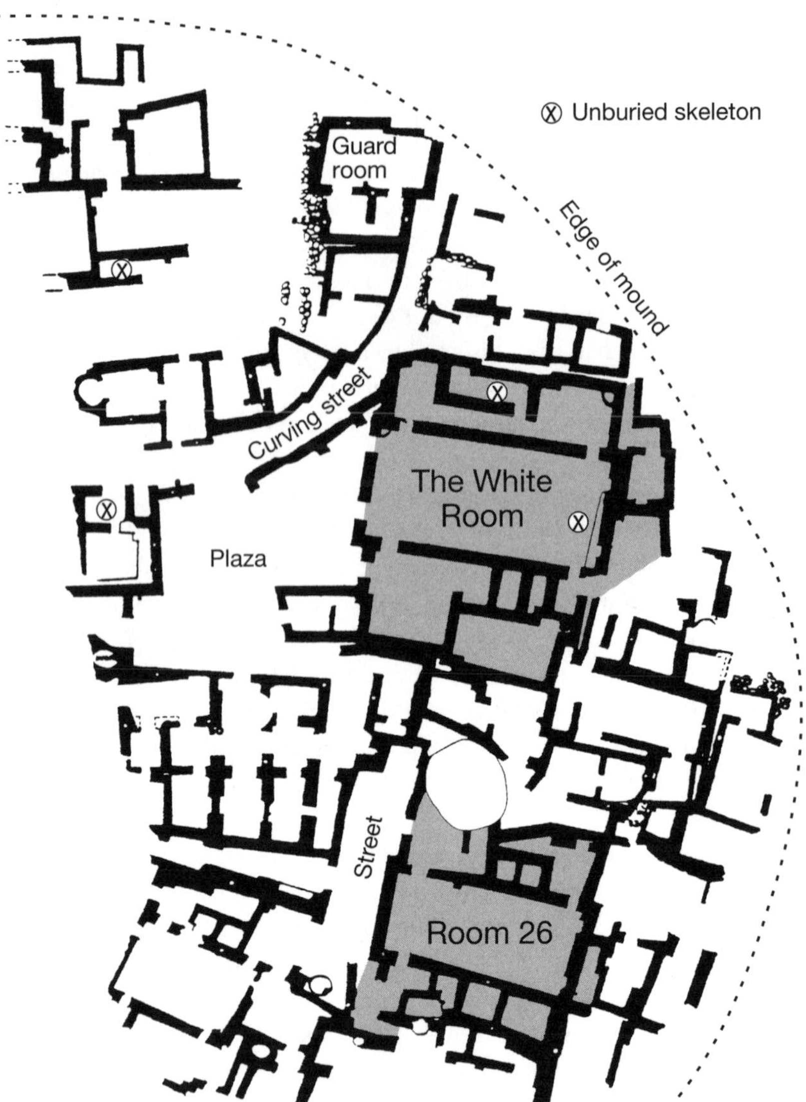

FIGURE 42. In spite of its defensible setting, its restricted access route, and its guard room, the Level XII village at Tepe Gawra was attacked, burned, and abandoned, with unburied corpses left behind. The residences of the two most important families are shown in gray. (The White Room covered 576 square feet.)

As was the case with many of the tribal or chiefly societies we have examined, the enemies of the Gawra XII village seem to have regarded youngsters as fair game. In such societies raids are often sneak attacks, and women and children are frequent casualties. Attacks may be timed to occur when adult men are away from the village, perhaps working in their fields or tending their flocks. Villages whose men are periodically absent may respond by building watchtowers, allowing those left behind a clear view of approaching strangers. Gawra had such a tower.

When rank societies are attacked, the primary target of the raid may be either (1) a prominent temple or (2) the house of a community leader. Truly powerful rank societies may force their victims into a position of subordination. Less powerful rank societies may be content to chase off defenders, do a little killing and burning, and then leave. Given Gawra's apparent abandonment, with some victims left unburied in the ruins, we suspect that the bulk of the Level XII population fled, moved in with their allies, and never returned.

The Nature of Early Rank Society in Northern Mesopotamia

All three of Irving Goldman's sources of power were developed in Northern Mesopotamia. There were buildings infused with sacred life force: household shrines, tholoi for descent groups, and a temple for the whole community. Expertise was evident in the polychrome bowls with potters' marks, polished marble and alabaster vessels, and architectural brickwork designed with scale models. The frequency of defensive ditches, walls, watchtowers, sling missiles, and burned buildings also makes it likely that prowess in raiding and defense was appreciated.

Some degree of hereditary rank is implied by the burial of infants and children with alabaster goblets and statuettes, multiple necklaces of exotic raw materials, and gifts of craft goods between elite families. The inclusion of maceheads and seal pendants with children may mean that some village officials sought to pass on their positions to their offspring.

Finally, there is evidence for secular public buildings. The latter may have included places for public assembly or the corporate storage of grain. The fact that there were persons who had the responsibility of sealing up shipments of trade goods suggests the germ of a bureaucracy. In other words, despite considerable evidence for privileged families who lived in large houses, who patronized craftsmen, and who buried their children with sumptuary goods, there are also hints that many members of society shared in the available power rather than having it concentrated in the hands of a single family like the great Angs of the Konyak Naga.

SOUTHERN MESOPOTAMIA

There is no clear-cut border between Northern and Southern Mesopotamia; the land simply becomes more arid as one travels south of Mosul. Rainfall at Baghdad averages five inches a year. Farther south, on the great alluvial plain between the lower Tigris and Euphrates, it can be drier still.

Despite its bleak appearance, Southern Mesopotamia was home to countless early villages. Their crops had two primary sources of water. Some fields could be irrigated by streams emerging from the Zagros Mountains and flowing west toward the Tigris. Other villages, including some of the largest, could draw water from the lower Euphrates. The lower Tigris, for the most part, was too deeply incised in its floodplain to provide water for the gravity-driven canals of that era.

The Euphrates begins in the snowcapped mountains of Turkey. Its flow is swelled by Syrian rivers such as the Balikh and Khabur. These are its last significant tributaries. Once it leaves Syria and enters Iraq, the Euphrates is carrying virtually all the water it will ever get. Despite this fact, the volume of water in the Euphrates is so great that for most of its southern course its flow is higher than the surrounding plain, held in check only by its natural levees. It begins to create its delta more than 350 miles from the Persian Gulf, near the modern Iraqi city of Hit.

For the prehistoric farmers of Southern Mesopotamia, the challenge was as follows. The great river was at its lowest level in September and October, held steady in November, began to rise in December, and reached flood stage in April or May. The difference between low water and high water at Hit was impressive. The Euphrates' flow was barely 8,830 cubic feet per second in September, but by May the flow had surged to more than 64,000 cubic feet. The problem was that, by May, the barley so crucial to Southern Mesopotamia had already been harvested. It was back in October that the water was needed.

The strategy for Southern Mesopotamian farmers, therefore, was to breach the levees of the Euphrates with flint hoes and fill their canals in October. Early villages in the south have produced hundreds of hoe blades showing the polish caused by alluvial soil. Since the lower Euphrates lies far from most flint outcrops, these hoe blades had to be made from imported Zagros Mountain flint.

The flint blades used to make sickles for harvesting were just as hard to come by. To overcome this problem Southern Mesopotamian villagers came up with a new type of sickle, made of overfired pottery clay. Overfiring vitrified the clay, producing a sickle as sharp as glass and just as fragile; its blade eventually broke in the field.

Since ancient sickle fragments can still be found on the surface of the alluvium, archaeologist Henry Wright set out to calculate how large an area could be cultivated by a village occupied 6,000 years ago. He discovered that pieces of clay sickle could be found up to three miles from the nearest 'Ubaid village.

Archaeologists also find impressive numbers of cattle bones in the refuse of 'Ubaid villages. Their abundance raises the possibility that oxen had now been harnessed to wooden plows, allowing families to cultivate larger tracts of land.

As the Euphrates of that era approached its confluence with the Tigris, it became a braided river, sometimes following a single course and sometimes dividing into multiple channels. Its floodwaters turned natural depressions into marshes filled with reeds, canes, sedges, cattails, and rushes. The main channel of the river, more than 600 feet wide in places, was periodically entered by ocean fish such as mullet, anchovy, sea bream, and even shark. The people of 'Ubaid times not only harvested these fish by boat, but they had also begun to sail into the Persian Gulf to trade with people of the Arabian and Iranian coasts.

Just as many Apa Tani families focused on wet-rice cultivation, some communities of the lower Euphrates appear to have concentrated on irrigated barley, obtaining the other commodities they wanted from their neighbors. Let the people of the Zagros Mountains raise most of the goats and the rainfed, upland wheat. Let the people on the steppe land west of the Euphrates raise most of the sheep. The villagers of the lower Euphrates would use sunlight and irrigation to produce the greatest barley surplus the world had so far seen.

The Temples of Eridu

The Eridu depression is a 20-mile-wide alluvial basin. It lies southwest of the Euphrates River, close to the ruins of Ur of the Chaldees. Long, long ago the depression may have been a marshy basin, closer to the shore of the Persian Gulf than it is today. Towering above the depression is the archaeological mound of Tell Abu Shahrain.

The upper 30 or 40 feet of the mound comprise the ruins of a *ziggurat,* or stepped temple pyramid, belonging to the ancient Mesopotamian city of Eridu. This ziggurat was either built or restored between 2112 and 2094 B.C. by Ur-Nammu, a king of the Third Royal Dynasty of Ur.

When archaeologists Fuad Safar and Seton Lloyd arrived at Tell Abu Shahrain in 1946, they knew they were standing on the ruins of ancient Eridu. Nothing, however, prepared them for the extraordinary sequence of prehistoric communities that lay beneath the southern corner of the ziggurat. Safar proceeded to dig down more than 40 feet, through at least 19 superimposed villages, until he had reached the sand dune on which the first arrivals had settled. In the course of his amazing descent through the mound he discovered traces of no fewer than 17 temples, built virtually one upon another for nearly 2,000 years.

Safar numbered these temples from the top down, in the order in which they were found. The oldest, Temple 17, was a one-room structure of mudbrick, roughly nine feet on a side. It may have been built more than 7,000 years ago, at a time when Samarran pottery was still popular in Northern Mesopotamia. The Samarran painting style, however, was not used at Eridu. Instead, Southern Mesopotamia had its own painting style and perhaps its own ethnic identity. Archaeologist Joan Oates has referred to the pottery of this period as 'Ubaid 1 because, in her view, it was an early precursor of the 'Ubaid style seen later in the region.

The oldest temple at Eridu for which Safar recovered a complete ground plan was Temple 16. It consisted of a rectangular one-room building, seven by ten feet in extent, with an additional alcove more than three feet on a side (Figure 43, top). The main room had a clay podium for the placement of offerings, and the alcove had a clay altar. The podium, when first found, was still covered with ashes from burnt offerings.

Next in the sequence was Temple 15, larger than Temple 16 but not nearly as well preserved. It may once have measured 24 by 27 feet.

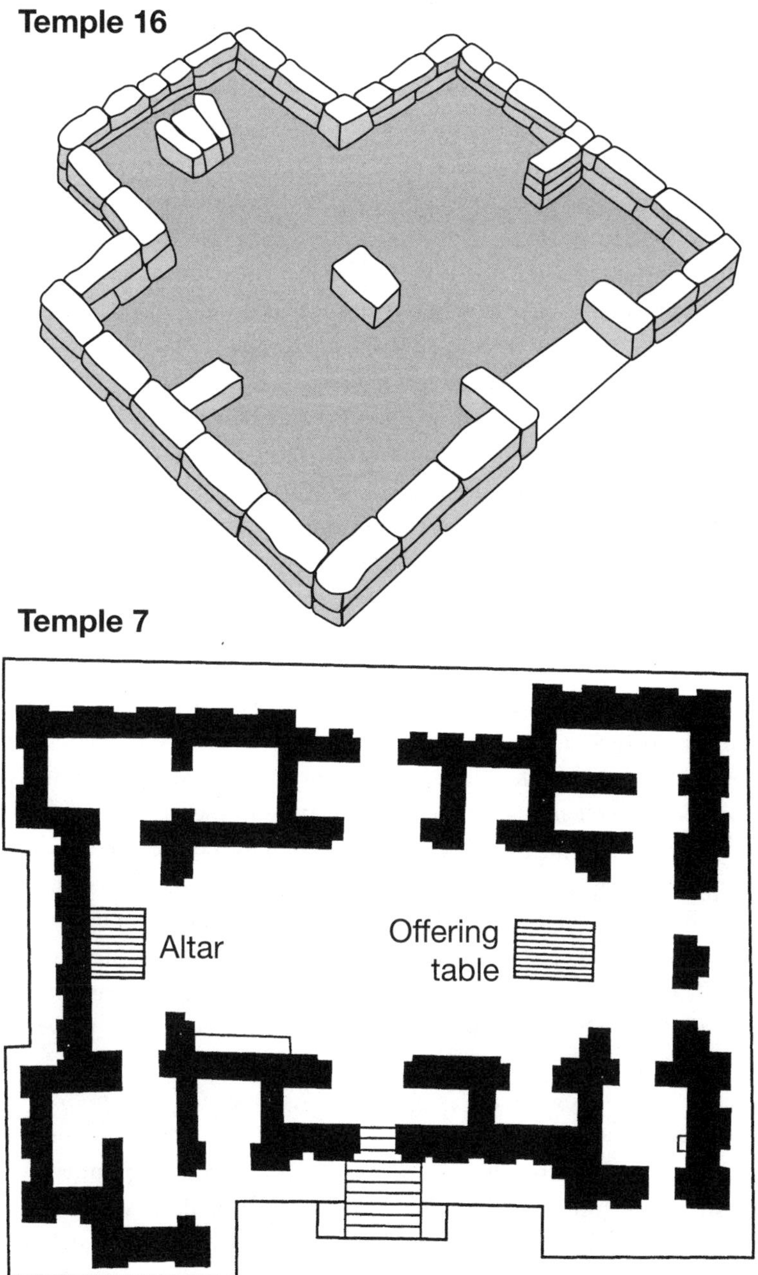

FIGURE 43. The site of Eridu in southern Iraq has produced the longest sequence of prehistoric temples known from Mesopotamia. Above we see Temple 16, one of the earliest, which measured seven by ten feet. Below we see Temple 7, contemporary with the North Temple from Level XIII at Tepe Gawra, which measured 65 by 49 feet.

For the next few centuries, at a time when Halaf pottery was reaching its peak popularity at northern villages such as Arpachiyah, the architects of Eridu continued to build temples. Significantly, however, they do not seem to have built tholoi like their Northern Mesopotamian counterparts. And while they could not have been ignorant of what Halaf potters were producing, they continued to pursue their own Southern Mesopotamian painting style, referred to by Oates as ʻUbaid 2.

By the time Temple 11 of Eridu was built, its architects had adopted the decorative pilasters, buttresses, and recesses that would become standard on Mesopotamian temples. Temple 11 had a central chamber more than 40 feet long. Similar to several of the earlier temples, it had ash from burnt offerings on its podium. Its pottery, defined by Oates as ʻUbaid 3, dated to perhaps 6,000 years ago.

Sometime between the building of Temple 9 and the building of Temple 8, the architects of Eridu began to create temples like those of Level XIII at Tepe Gawra. Temple 8 had walls more than two feet thick and displayed a long narrow cella flanked by smaller rooms.

One of the most nearly complete temples in the sequence at Eridu was Temple 7 (Figure 43, bottom). Measuring 65 by 49 feet, it had been built on a mud-brick platform nearly five feet high. Its four corners were oriented to the cardinal directions. Seven steps provided access to the main door, which was on the southeast side. Near the southern corner of the building was a second door, which may have been a private entrance for the priests who served the temple. This private door led directly to the altar end of the cella. Worshippers using the main door, on the other hand, would have had to wait in an antechamber before entering the cella. Scattered around the clay podium of Temple 7 were fish bones that may have been the remains of burnt offerings. All in all, the ground plan of Temple 7 at Eridu was remarkably close to the plan of the northern temple in Level XIII at Tepe Gawra (compare Figure 41).

Temple 7 and its successor, Temple 6, were associated with pottery whose style of painting was called ʻUbaid 4 by Oates. This was the final stage in the development of the ʻUbaid style, and there were now strong similarities between the pottery vessels of Eridu and Gawra. Particularly similar were the incense burners, a number of which were found in Temple 6 at Eridu. In other words the evidence suggests that Northern and Southern Mesopotamia were largely on the same page when it came to ritual architecture and pottery.

A Possible Fishermen's Ward

In an effort to learn more about the residential architecture of Tell Abu Shahrain, Safar carried out a series of excavations at some distance from the ritual buildings. Most residences seemed to have been built of mud-bricks similar to those used in the temples. A residential ward less than 90 yards southeast of the ziggurat, however, was different. The 'Ubaid 3 residences in this ward, built around the time of Temples 11 through 9, had reed walls plastered on both sides with a thick layer of clay. They were, in other words, the Mesopotamian equivalent of the reed-and-clay houses of Mexico and Peru.

Safar found it significant that these reed-and-clay houses contained none of the vitrified clay sickles described earlier in this chapter. The reed houses did, however, contain two dozen fishing-net weights, as well as significant accumulations of fish bone. Safar and Lloyd therefore suspected that they had uncovered a residential ward of families that specialized in taking fish and waterfowl from the marshes and river channels of the lower Euphrates. It was not clear whether this type of house was associated with a separate ethnic group or only an occupational specialty.

The 'Ubaid Cemetery at Eridu

In the course of their excavations outside the temple area, Safar and Lloyd also uncovered a cemetery of the 'Ubaid 4 period, roughly contemporaneous with Temple 6. The area chosen for the cemetery lay near the northwest outskirts of the village. The archaeological crew had time to excavate only 193 burials. Safar estimated that the cemetery might have included four or five times that many.

By far the greatest number of graves had been excavated down to the clean sand below the village. A kind of rectangular box was then created around the corpse by building four mud-brick walls. The deceased lay full length on the clean sand layer, accompanied by his or her grave offerings. The box was then filled with earth and sealed with a mud-brick lid.

Most graves contained only one individual. There were, however, a significant number of family burials, cases in which a mud-brick box had been reopened so that the husband or wife of the original occupant could be added. Occasionally the body of a child accompanied one or both parents. More often, however, children were given their own small brick boxes and offerings of miniature pottery.

Exceptions to the normal burial ritual included a number of individuals laid to rest without brick boxes. Safar believed that these burials might represent people "of more humble social status." Burial 97 was also unusual. It consisted of the complete skeleton of an adult, accompanied by a dozen skulls from other people.

In the Eridu cemetery it was not uncommon to find burials accompanied by a substantial "table service" of dishes, cups, chalices, beakers, or spouted flasks. The beakers were among the finest products of the 'Ubaid 4 potter. Almost as thin as an eggshell, they had a gracefully flaring rim and horizontal bands of painted motifs.

Many burials wore bead necklaces or bracelets, and one (believed to be a woman) had an ornamental belt and the beaded fringe from a long-disintegrated skirt. One adult man was accompanied by the clay model of a sailboat, with a socket for the mast and holes for the stays. Burial 185 tugged at Seton Lloyd's heart because it contained a youth 15 to 16 years old, accompanied by his faithful dog; the dog had even been buried with a bone near its mouth.

In the 1980s archaeologists Henry Wright and Susan Pollock reanalyzed the 'Ubaid burials at Eridu. They found no conclusive evidence for differences in rank. Wright and Pollock caution, however, that only 20–25 percent of the cemetery has ever been excavated. Their caution is wise, given what we saw earlier at Yarim Tepe in Northern Mesopotamia.

Recall that at Yarim Tepe, the highly ranked and "ordinary" individuals were not even buried in the same cemetery. The shaft-and-chamber tombs with the finest sumptuary goods were found at Yarim Tepe I. Most ordinary graves (including an Eridu-like family grave with two adults and a child) were found at Yarim Tepe II. There is thus no guarantee that the graves Safar excavated reflect Eridu's full range of social categories.

Now let us consider another reason for caution, based on what we saw among the Apa Tani. The Apa Tani had an aristocracy without chiefs; they were led by a council of wealthy men drawn from aristocratic clans. A cemetery of Apa Tani aristocrats would include many wealthy clansmen; we doubt that an archaeologist could determine which ones had been council members.

Making the archaeologist's task more difficult is the fact that some aristocratic Apa Tani clans had adopted former slaves as poor relations. Such poor relations might have been treated like Safar's "people of more humble social status" at Eridu.

We would be skeptical of any claim that Southern 'Ubaid society had no differences in rank. We would not, however, be surprised to learn that Southern 'Ubaid society had forms of rank that were hard to detect archaeologically.

A Secular Public Building at Tell Uqair

Fifty miles south of Baghdad, and roughly equidistant from the Tigris and Euphrates, lay the 'Ubaid 4 village of Tell Uqair. Its ruins consisted of two mounds, A and B, separated by a linear depression through which a canal might once have run.

Mound A of Tell Uqair, the older of the two, covered about 12 acres. Apparently founded on marshy ground, its deepest levels had thick layers of reed or bulrush matting. Some 5,600 years ago it had grown to be a substantial 'Ubaid village with streets and mud-brick houses. Here Seton Lloyd and Fuad Safar recovered many of the chipped flint hoes and vitrified clay sickles that one might expect in an agricultural community. They also, however, found stone weights for fishing nets and impressive deposits of freshwater mussel shells.

The walls of the 'Ubaid houses at Uqair were usually only one mud-brick wide. Across the main street from the ordinary residences, however, Lloyd and Safar discovered a more impressive building with walls almost three feet thick. Its mud-bricks were laid not in simple horizontal courses but interdigitated in order to strengthen the walls. This building may have had more than ten rooms; some were long and narrow, but there was nothing to suggest the ground plan, podium, or altar of a temple (Figure 44).

Archaeologists believe that this massive structure, located on what may have been the main street of Tell Uqair, was a secular public building. To be sure, we do not know what kinds of activities took place inside. The importance of the building lies in its hint that 'Ubaid society had both secular and religious hierarchies. As we have seen in previous chapters, even the partial separation of these two paths to power could be a source of dynamic rivalry, an engine that drove political ambition.

Elite Houses at Tell Abada

The Diyala River, a major tributary of the Tigris, begins in the high mountains of Iran. Descending the long parallel ranges of the Zagros, it emerges from the piedmont and runs west to the Tigris. The Diyala region lies near the transition between Northern and Southern Mesopotamia. Restricted to ten inches of rain a year, the farmers of the Diyala relied on the irrigation of wheat and barley. The nearby Zagros piedmont provided pastures for sheep and goats.

Irrigation of the Diyala basin was under way in Samarran times, and by the late ʻUbaid period a few villages exceeded 14 acres in size. Tell Abada, 20 feet deep and covering about seven acres, would be considered larger than average. During the late 1970s archaeologist Sabah Abboud Jasim was able to excavate an extraordinary 80 percent of the site.

From roughly 7,000 to 6,000 years ago, Tell Abada had grown emmer wheat, bread wheat, and barley and had raised sheep, goats, cattle, and pigs. With ready access to the abundant flint sources of the Zagros Mountains, its farmers had harvested their cereals with flint-bladed sickles rather than the overfired clay versions used on the lower Euphrates.

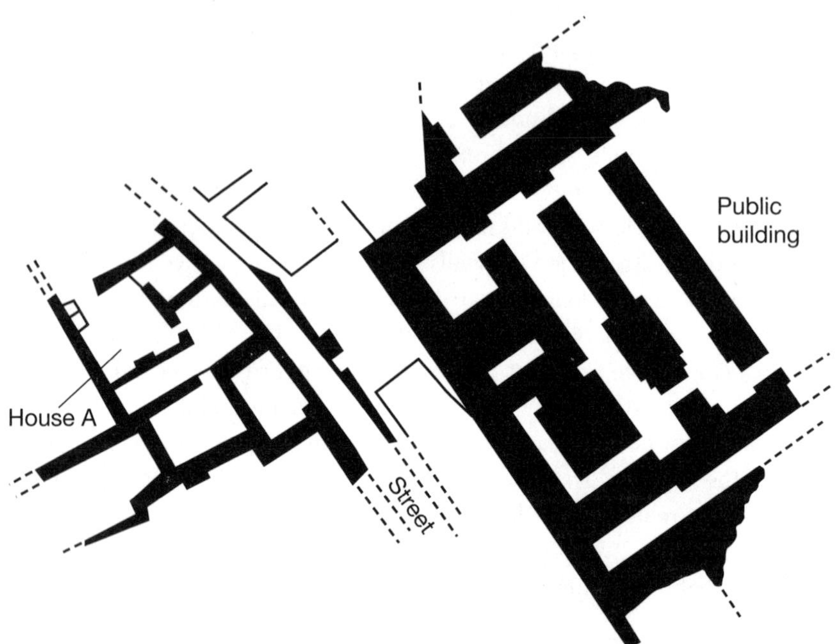

FIGURE 44. Not all public buildings of the ʻUbaid period in Southern Mesopotamia were temples. One public building at Tell Uqair, found across the street from House A, lacks the ground plan of a temple; it may have had a secular function. This building's full dimensions are unknown, but its walls were almost three feet thick.

The village of Level II, dating to 6,500 years ago, was especially well preserved. Here Jasim found 11 independent residential units separated by streets and narrow alleys. These were clearly houses for large extended families and showed repetitive use of a module consisting of small rooms surrounding T-shaped patios (Figure 45). While only one story had been preserved, there were stairways leading suggestively upward, implying that there had once been a second story.

One of the largest houses had a buttressed outer wall and three T-shaped patios or courtyards, flanked by 20 to 25 rooms. It also had a large backyard for outdoor activities, walled off in such a way that it could only be entered from the house. An antechamber gave indirect access to the residence, providing additional privacy.

The artifacts from Tell Abada reinforce the idea that we are dealing with families of privilege: included were six elegant scepters, polished from fine marble, which must have been symbols of office or rank. There were also stone cosmetic palettes, reflecting the kind of personal grooming one expects from elite families.

It is not surprising that the painted pottery of Level II was of craft-specialist quality. What is most interesting is that the pottery reflected two schools of painting. According to Joan Oates, it is clear that during this period in the Diyala region "there were potters working in both the Halaf and 'Ubaid traditions, perhaps even side by side in the same villages." This is neither our first nor last hint that some large prehistoric societies were conversant with the distinctive art styles of more than one ethnic group.

The Tripartite Building at Tell el-'Oueili

Thirty miles north of Eridu, and on the opposite side of the Euphrates, lay the 'Ubaid village of Tell el-'Oueili. The region of el-'Oueili is one that would have been irrigated by canals running east from the great river. The main cereal recovered by excavator Jean-Louis Huot was six-rowed barley. More than half of the identified animal bones belonged to cattle, a beast possibly used for plowing as well as meat. The villagers of el-'Oueili also took advantage of the carp, catfish, mullets, and occasional sharks to be found in the local watercourses.

In the uppermost level of the site, Huot discovered an interesting two-story building. Its walls were not only heavily buttressed but also supported by caissons and a terrace wall. The building's upper story was divided into three

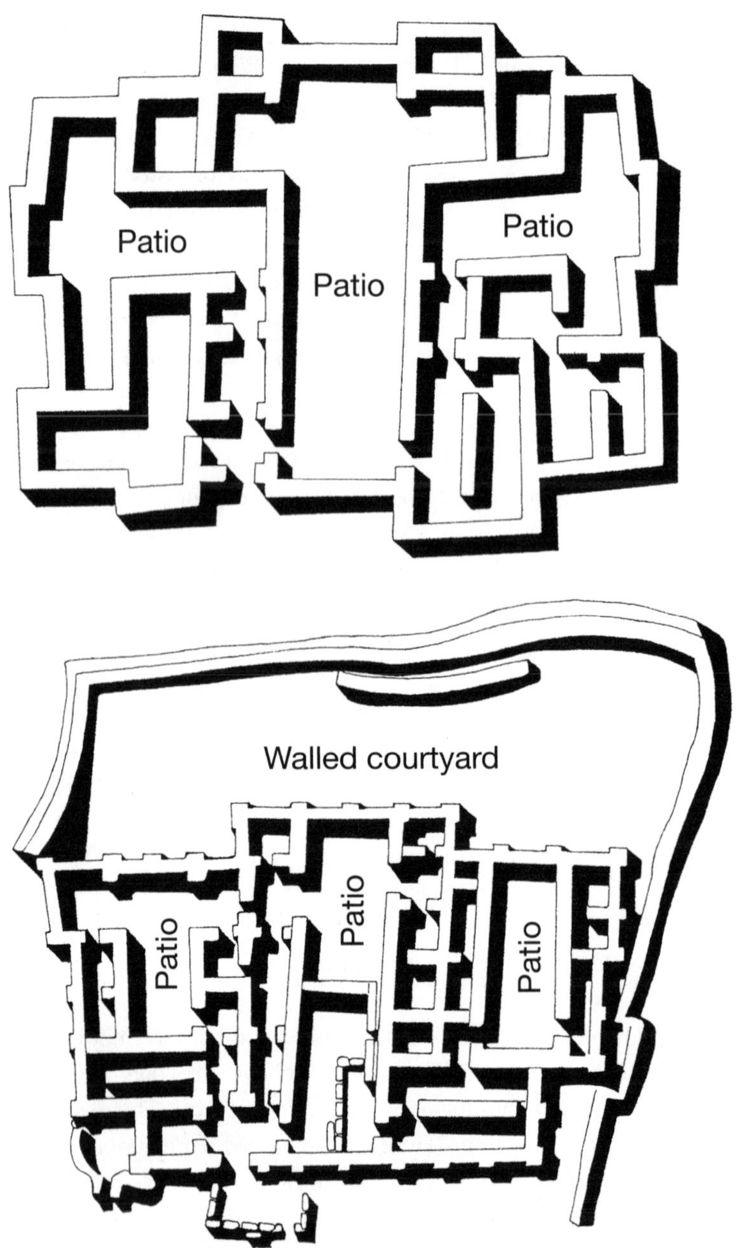

FIGURE 45. The site of Tell Abada, in the Diyala River region of Iraq, featured a series of residences for elite families of the ‘Ubaid 3 period. All of the houses had multiple patios or courtyards, and some may have had a second story above the one shown. (The house at the top was 48 feet wide.)

symmetrical rooms, each measuring 31 by 9 feet. The lower story was divided into dozens of what were most likely grain-storage rooms.

Archaeologists differ in their interpretation of this building, built between 6,000 and 5,600 years ago. Most, however, see it as a public building, with its lower story devoted to tons of stored grain. If this interpretation is correct, it indicates that 'Ubaid 4 society had large public institutions, perhaps supported by contributions of grain from every family.

Long-Distance Exchange in 'Ubaid 4 Times

It will come as no surprise that 'Ubaid 4 communities engaged in exchange. That exchange, however, apparently went beyond visits to trading partners in other regions. 'Ubaid 4 societies had actually begun to place trade enclaves in the upper Euphrates region of Syria and Turkey.

The people of Southern Mesopotamia knew that Turkey's Taurus Mountains had exposures of copper, silver, lead, and gold. The route to those mineral resources followed the Euphrates to its upper tributaries. It was simply a matter of convincing your trade partners to let you build houses in their village.

Değirman Tepe, a village on the upper Euphrates in Turkey, had a residential ward with painted pottery that Joan Oates has declared to be "pure 'Ubaid." Tell Abr, on the Great Bend of the Euphrates in northern Syria, also had evidence of 'Ubaid 4 residents. These villages were not situated at the actual mineral outcrops, however; they occupied places where trade goods could be placed on boats and sailed down the Euphrates.

Villages will not permit an enclave of visitors in their midst unless there is something in it for them. We cannot simply say, therefore, that the 'Ubaid 4 elite wanted gold, silver, lead, copper, timber, turquoise, and lapis lazuli. We also have to ask what the societies of the Euphrates headwaters wanted in return. Fortunately, the behavior of many of the societies described in earlier chapters of this book suggests a two-part answer.

The first part of the answer is based on the Tlingit and their Athapaskan trading partners. How did the Tlingit get furs from the Athapaskans? They married Athapaskan women and betrothed their own sisters and daughters to Athapaskan headmen. How did the Athapaskans respond? Many became "Inland Tlingit," claiming hereditary rank through their in-laws and negotiating the right to display Tlingit heraldic crests.

The second part of the answer involves the Shan and the Kachin. When the Shan elite wanted jade from the mines of the Kachin hills, what did the Kachin leaders want in return? Prestigious brides and irrigated Shan rice. And how did Kachin leaders react when they received the brides and rice? They began dressing and acting like their elite in-laws, converting to Buddhism and adopting Shan symbolism.

If I am sitting on a Turkish silver mine, I will let someone put an enclave in my village if he brings me a bride who raises my prestige. Occasional gifts of Mesopotamian barley to increase my beer supply would not hurt either.

The Nature of Society in Southern Mesopotamia

The forensic evidence for social inequality in Southern Mesopotamia is less compelling than the evidence we saw in Northern Mesopotamia—no infants buried with alabaster statues, and no youths interred with alabaster goblets, obsidian necklaces, stone maceheads, or official-looking seals. The marble scepters and cosmetic palettes from Tell Abada provide our best evidence for sumptuary goods.

We are skeptical that Southern Mesopotamian society had only achievement-based differences in prestige. Unfortunately, we are forced to rely on circumstantial evidence for inequality.

Consider the differences in residence. Some families at Tell Abada lived in two-story mud-brick houses with 20 to 25 rooms on the lower floor alone. Their desire for privacy was such that their houses were entered indirectly through antechambers, and some rooms could be reached only by passing through eight doors and several patios.

Standing in contrast to these large brick houses were the reed huts in the fishermen's ward at Eridu. To be sure, the occupants of these reed houses may have had an occupational specialty. It is nevertheless hard to see them as equal in status to the occupants of the large houses at Tell Abada.

Then there is the cemetery at Eridu, where most people were laid to rest in brick boxes, but a minority were simply buried in the sand. It was a subtle difference, but not one to be dismissed out of hand.

Finally, there is Temple 7 at Eridu. This well-made temple had two entrances. One, at the head of the stairs, was obviously for general worshippers. The other, leading to the altar end of the inner sanctum, appears to have been for the priestly staff of the temple. Most preindustrial societies whose temples were managed by actual priests had hereditary inequality. Head priests, in fact, tended to be drawn from the elite and underwent training unavailable to commoners.

In terms of Irving Goldman's three sources of power, we believe that ritual and religious authority was most heavily stressed in the Southern Mesopotamian 'Ubaid. The shift from men's house to temple relegates clan ancestors and lesser spirits to the background and brings deities or celestial spirits to the foreground. It is in the interest of the aristocracy that temples be built for the highest deities, to whom they owe their right to lead society.

To continue with Goldman's sources of power, expertise was probably second only to religious authority. Seals and seal impressions in clay suggest the emergence of officials whose expertise lay in controlling the movement of commodities. Fishermen, leather workers, alabaster carvers, Halaf potters, and 'Ubaid potters all suggest expertise at crafts. The creation of secular public buildings implies councils or assemblies with the expertise to share the burden of decision making.

Without the defensive walls, ditches, watchtowers, and burned elite residences that we saw in the north, we cannot be sure how important Goldman's third source of power—military prowess—was to Southern Mesopotamia. In later chapters we will learn that it would become very important in Mesopotamia's future.

THE NATURE OF LEADERSHIP IN EARLY MESOPOTAMIA

We now need to place the differences between Northern and Southern Mesopotamia into a broader context, one that permits comparisons with other parts of the world.

Archaeologist Colin Renfrew has called attention to some interesting differences among the prehistoric European rank societies of 5,000 to 3,500 years ago. Some of these societies, he notes, produced impressive public monuments but left almost no evidence for the personal aggrandizement of their leaders. Other European societies filled the graves of their leaders with objects of wealth and rank but left fewer impressive public monuments. Renfrew wisely decided not to regard these as alternative types of societies; he treated them as two extremes of a continuum and called them "group-oriented" and "individualizing."

In the years since Renfrew's original suggestion, many of his fellow archaeologists have proposed similar schemes, sometimes using contrasting terms such as "corporate strategy," "individual negotiation," or "networking." Unfortunately, many of these later scholars have committed the very mistake Renfrew avoided. They act as if they have discovered mutually exclusive types of societies when they are actually looking at the extremes of a continuum in which all available strategies were used.

We can provide examples of Renfrew's continuum without even stepping outside the Tibeto-Burman language family. In their thendu, or rank, mode the Konyak Naga were led by an individualizing chief called a great Ang. At the opposite end of the continuum were the Apa Tani, for whom all leadership decisions were group-oriented. Both the Konyak Naga and the Apa Tani had hereditary inequality, and both kept slaves. But while the names of individual great Angs lived on in legend, the Apa Tani councillors were the group-oriented members of an oligarchy.

The wealth of a great Ang came from tribute and sumptuary goods. The wealth of an Apa Tani aristocrat came from his privately owned rice paddies. In both societies these disparities in wealth were tolerated because the rich were of aristocratic birth. Renfrew never suggested that his group-oriented societies were egalitarian. He understood that both extremes of his continuum had hereditary inequality, albeit expressed in different ways.

In Mesopotamia our impression is that while the north and south shared many institutions, the societies in the north were more individualizing and the societies in the south more group-oriented. With the wisdom of hindsight, we know that the future of Southern Mesopotamia was to become an oligarchy with a ruler whose decisions were informed by a council of elders. This later Mesopotamian society would have a multilevel hierarchy of administrators, some of whom would be trusted commoners.

To be sure, a few Mesopotamian archaeologists regard Southern 'Ubaid society to be so group-oriented as to have been egalitarian. We consider this unlikely on several grounds, one of which was the obvious replacement of men's houses with temples. As we have seen in a number of living societies, this replacement reflects a change in social logic, similar to that of the gumsa or ranked Kachin. While families of lower rank continued to involve clan ancestors and lesser spirits in their rituals, the Kachin elite were permitted to pray directly to higher deities or celestial spirits. Higher deities, as mentioned earlier, do not visit men's houses, tholoi, or any other kind of clan house. They visit temples or shrines where they are given offerings, some of them burnt on altars or podia.

Finally, we come to the problem of wealth. Were the families who lived in the grandest 'Ubaid houses a hereditary aristocracy, or did they simply own lots of irrigated land? The answer, we suspect, is "both." Egalitarian societies, as we have repeatedly seen, have a low tolerance for disparities in wealth. Hereditary rank, on the other hand, provides justification for such disparities. After all, a society that has no concept of nobility cannot have "nobles by wealth."

FIFTEEN

The Chiefly Societies in Our Backyard

From Memphis to New Orleans, the Mississippi takes a winding route past wetlands and antebellum mansions and the oxbows of its former course. With his windows shut and his air conditioner on, the traveler passes signs for barbecues and po' boys, Delta blues, and the well-groomed battlefields of the War between the States. A few miles south of Natchez, Highway 61 crosses a tributary called St. Catherine Creek. If the traveler picks this moment to text message, he misses a chiefly center of historic importance.

The Fatherland site, as the former chiefly center is known today, was partly defended by the natural bluffs of St. Catherine Creek. It was one of a number of settlements under the authority of a Natchez chief called the Great Sun. Founded roughly 800 years ago, the Fatherland site was still occupied when French explorers arrived in 1682. The French built a garrison called Fort Rosalie. They traded with the Natchez and wrote useful eyewitness accounts. By 1729, however, the French had become so annoying that the Indians decided to massacre them and leave.

In 1698 the population of the Natchez was estimated at 3,500, a thousand of whom may have been warriors. The French calculated that there might have been somewhere between nine and a dozen Natchez settlements, all modest in size except for one. Archaeologists believe that the largest community, known as the Grand Village of the Natchez, was located where the Fatherland site sits today. Fatherland fits the French description of a chiefly center with an elite residence, an old temple, and a new temple, all occupying platform mounds.

In 1718 a French engineer named Antoine le Page du Pratz visited the Grand Village. Both the Great Sun and his brother, a War Chief named Tattooed Serpent, lived there at that time. Du Pratz made friends with Tattooed Serpent, who lived in a cane-and-clay house 30 feet long and 20 feet high, overlooking the village from the crest of an earthen mound. Next to this mound was a ritual plaza, described by one French writer as 300 paces long and 250 paces wide. At the south end of the plaza was another earthen mound, this one supporting a two-room, cane-and-clay temple that measured 65 by 40 feet. On the opposite side of St. Catherine Creek were the homes of an estimated 30 to 40 extended families, totaling more than 400 persons.

Du Pratz lived among the Natchez for four years and was present at the funeral of Tattooed Serpent in 1725. His account is so detailed that several archaeologists have attempted to confirm it by finding the remains of Tattooed Serpent. The problem is that Natchez nobles were only buried long enough for their flesh to decay, after which their bones were cleaned and kept in a hamper in the temple. It would be difficult to determine which hamper of curated bones belonged to Tattooed Serpent.

Archaeologists believe that the southernmost mound at the Fatherland site is the place to look. That mound has, in fact, produced the remains of several superimposed temples. Below the floors of those temples were more than 20 burials, ranging from complete skeletons to reburied limb bones and isolated skulls. None can be definitively identified as Tattooed Serpent.

Burial 15 of this mound has been described by archaeologist Robert S. ("Stu") Neitzel as "easily the most important individual buried in the mound." Given his importance, this adult male may have been a Great Sun rather than a War Chief. Since his bones had never been gathered up and placed in a hamper, he may have died just before the village was abandoned in 1730.

The sumptuary goods that accompanied Burial 15 remind us that the Natchez had been trading actively with the French. For one thing, this man's ear ornaments were coiled wire springs of European manufacture. For another, his weapons included a flintlock pistol, three penknives, and an iron axe. His burial offerings included a brass oven, an iron pot, a tinned brass pan, an iron hoe, and strings of glass beads. Some of his arrow points were of native flint, while others were of copper. Perhaps his most indigenous possessions were two lumps of galena, a lead ore prized by earlier Native American societies.

The Natchez were an impressive people during the colonial period and are believed to have been even more powerful prior to 1682. We should stress, however, that the Natchez were hardly unique to the southeastern United States. Beginning with the evolution of Eastern Flint corn 1,200 years ago, the Southeast became an incubator for chiefly societies. More than 100 flamboyant, expansionist, individualizing rank societies took over the lower Mississippi, dozens of its tributaries, and scores of rivers flowing to the Atlantic from Virginia to Florida.

All these chiefly societies arose in our backyard. And, sadly, each year we see more of their remains bulldozed away, covered with tract homes and shopping malls, or submerged by the waters of hydroelectric dams.

THE HISTORIC NATCHEZ

The Natchez, like most rank societies of the southeastern United States, reckoned descent in the mother's line. This meant that, just as with the Crocodile clan of the Bemba, the chief's son could not succeed his father. When a chief or Great Sun died, his title passed to the son of his most important sister, who was called White Woman. This was not a reference to fair skin but to the color that symbolized peace.

In the cosmology of the Natchez the first Suns were a man and woman from the Upper World. The male member of this primordial pair was the actual younger brother of the sun. He ordered the Natchez to build a temple. Once it had been built, he brought down fire from the sun and asked that it burn forever in the temple. According to French eyewitnesses, an attendant was charged with making sure that the temple fire would never go out.

After explaining to the Natchez how his successor should be chosen, the younger brother of the sun turned himself into stone rather than having to endure earthly death and putrefaction. Similar stories of self-petrification may have been widespread in the cosmology of southeastern Indian societies. At the ancient chiefly center of Etowah, discussed later in this chapter, archaeologist Lewis Larson recovered a pair of stone statues that may represent a petrified couple from the Upper World.

By analyzing colonial accounts, historian Charles Hudson has reconstructed the Natchez system of rank. The Natchez cosmology just described was used to justify the Suns' right to rule. A Sun was required to marry a woman from another lineage in order to avoid incest; since he was already a member of the highest lineage, that meant marrying down. The children of female Suns were Suns, but the children of a male Sun who married down were only regarded as Nobles. It was for this reason that the next Great Sun had to be born to White Woman. White Woman, like women of the Bemba Crocodile clan, was allowed to be promiscuous and to marry and divorce husbands at will. After all, she outranked them all.

Just as the children of male Suns were only Nobles, the children of male Nobles were only Honored People. While they could not become chiefs, both Nobles and Honored People could rise in prestige through their exploits in war, like the "nobles by command" of Colombia's Cauca Valley. Some could even work their way up to the office of War Chief, who was second in command to the Great Sun.

At the bottom of the ranking system were commoners known as Stinkards. Honored People were allowed to marry Stinkards, which provided some flow of genes and privileges between ranks, preventing the Stinkards from becoming a separate social stratum. Additional exchanges of genetic material resulted from the fact that unmarried Natchez girls were encouraged to be generous with their sexual favors. If her favors led to a child out of wedlock, a girl was allowed to perform infanticide. The child's father, however, was not allowed to participate in its death. He would have been killing a member of another clan, which could precipitate a feud.

The Great Sun possessed impressive quantities of life force; according to Hudson, however, he "reigned more than he governed." He shared power with a council of advisers, and much of his actual administration was carried out by lower-ranked overseers.

The Great Sun wore a special headdress of white feathers set in a red diadem. No commoner could eat with him or touch the vessels from which he had eaten. Anyone approaching him had to show deference, shouting "hou" three times to announce his or her arrival. Upon leaving the chief's presence, one had to walk backward and continue to shout "hou." These acts make the chief seem almost as sacred as the temple, to which one also had to shout "hou" as he or she passed.

Despite his life force, even the Great Sun could not enter the temple of the Grand Village without performing a preliminary ritual. First he stopped in the plaza before the temple and bent down low in a position of obeisance. He then turned slowly to all four of the great World Directions and, while facing each, humbled himself by throwing handfuls of dirt on his head. This was the Natchez equivalent of a Tikopian chief's offer to eat his deity's excrement. It tells us that even for a man with the authority of a Great Sun, the invisible celestial spirits still were the alphas in his dominance hierarchy.

While he may have humbled himself to his deity, the Great Sun was virtually above the law when it came to his fellow humans. Neither women, nor children, nor Stinkards could enter his house. His people were expected to supply him with large quantities of food, and he in turn was expected to be generous to them when they were in need. To finance his largesse, his leading warriors cultivated a special field of Flint corn for his chiefly storehouse.

Food for the Natchez came from the floodplain of the Mississippi and its tributaries, where the Indians intercropped corn, beans, squash, gourds, and sunflowers. They hunted deer in large groups, surrounding the animals in a U-shaped formation that gradually closed to a circle. The deer were presented to the Great Sun as tribute; he in turn showed generosity by distributing the meat to the organizers of the hunt. The Natchez also ate wild turkeys, fished in the rivers and bayous, and gathered hickory nuts, persimmons, and other wild fruits. They smoked a specially grown variety of strong tobacco and drank a caffeine-filled ritual tea brewed from holly. Visitors were treated to the Natchez' favorite comfort food, hominy mixed with chunks of venison.

War Chief was an important office for the Natchez, because chiefly rivalries and intergroup revenge triggered endless cycles of raiding and peacemaking. Envoys made tentative offers of payment for casualties. If an offer was judged insufficient, the War Chief tied a flag to a pole painted red—the color of war—and pointed it in the enemy's direction. Armed with arrows and war clubs, the raiding parties returned with scalps or entire heads. The Natchez kept some captives as slaves, while others were tortured to death.

Burial ritual reflected the differences in rank. The corpses of Stinkards were exposed on wooden platforms until only the bones remained. The corpse of the Great Sun, on the other hand, was carried around on a litter, much as the chief had been carried in life. Dozens of people might be sacrificed to accompany him in death.

In 1725 du Pratz witnessed the funeral of his old friend, the War Chief Tattooed Serpent. Such was the grief of the Great Sun that he threatened to kill himself at his brother's funeral. All fires in Natchez territory were extinguished in anticipation; they were rekindled once the Great Sun had been persuaded to go on living.

Dressed in his feather headdress, his face painted red and his feet placed in the moccasins he needed for his journey to the other world, Tattooed Serpent lay in state for three days. His guns, war clubs, and bows and arrows were tied to his bed. Surrounding him were his ceremonial tobacco pipes and a chain of 46 cane hoops symbolizing the enemies he had killed.

Finally, a priest in elaborate costume began the burial ritual. Tattooed Serpent's corpse was placed on a litter and carried by six guardians of the temple. They followed a looping course toward the temple, with each successive loop bringing them closer. This circuitous route provided time for the sacrifice of numerous people, all of whom had volunteered or been chosen to accompany Tattooed Serpent in the afterlife.

Dressed for sacrifice were two of Tattooed Serpent's multiple wives; one of his sisters; his most prized warrior; his leading servant and that servant's wife; the craftsman who made Tattooed Serpent's war clubs; and two ritual healers, described by du Pratz as the War Chief's "doctor" and "nurse." Some people of lesser rank volunteered for glory by giving their own lives. Other people, reluctant to die, offered their children as substitutes.

Each sacrificial victim was given six balls of tobacco to swallow. This dosage stupefied them, after which they were garroted by a pair of executioners. Tattooed Serpent and his two wives were buried in the temple, while other dignitaries were buried nearby. Any Stinkards who had been sacrificed were placed on scaffolds at a greater distance. Eventually, the bones of Tattooed Serpent, his wives, and most noble associates were exhumed, cleaned, and stored in the temple near the remains of previous Suns.

While the Natchez were unrelated by language or history to the chiefly societies of Panama, the burials of their most highly ranked citizens show striking convergence. In both cases hereditary chiefs were so infused with life force that many of their closest supporters volunteered to accompany them in death. A special treat, in fact, awaited noble Natchez women who were sacrificed: in the afterlife, there would be no prohibitions that kept them from sharing meals with the Great Sun.

MOUNDVILLE: PROVIDING AN ANTHROPOLOGICAL SCENARIO FOR AN ANCIENT CHIEFLY CENTER

In 1904 anthropologist Frank Speck traveled to Oklahoma Territory to visit a Native American group called the Chickasaw. The Chickasaw had lived in Mississippi before the U.S. government forced them to move to Oklahoma. They had a rank society with a hereditary chief called a *minko.* Their matrilineal clans were ranked relative to one another, and within each clan the various lineages or subclans were ranked as well.

Like many southeastern Indian societies, the Chickasaw spent much of the year dispersed in farmsteads and small villages. From time to time, however, all subclans convened at a common campground for the purpose of holding a strategic council. At these periodic encampments, the leaders of each subclan took up positions that reflected their relative ranking.

A Chickasaw elder named Ca'bi'tci drew Speck a diagram showing the layout of a traditional council camp. The drawing began with a rectangular plaza, at the center of which was a sacred council fire. A north-south line divided the camp into two opposing divisions, called the Intcukwalipa and the Imosaktcan. The six house groups of the Intcukwalipa lay to the west of the line; the seven house groups of the Imosaktcan lay to the east. The most highly ranked subclan of each division occupied the most northerly position, then came the second most highly ranked subclan, then the third, and so on. The most lowly ranked subclan occupied the most southerly position. The layout of the council camp is shown on the left in Figure 46.

Upon reading Speck's report, archaeologist Vernon James Knight Jr. realized that the diagram of the Chickasaw council camp would be useful for interpreting the arrangement of earthen mounds at the prehistoric chiefly center of Moundville.

Moundville is a 185-acre archaeological site on the Black Warrior River near Tuscaloosa, Alabama. Founded 1,100 years ago and still occupied when Spanish explorers arrived, Moundville was naturally defended on its north side by the bluff of the river. Its southern border was fortified with a palisade of wooden posts. Bastions with watchtowers were placed 100 to 130 feet apart along the palisade.

Like so many chiefly centers, Moundville went through cycles of expansion and contraction. During its first two centuries, from 1,100 to 900 years ago, it was still relatively modest in size. After that it began to grow rapidly, reaching its heyday between 800 and 700 years ago. For the next 150 years its leaders struggled to retain power in the face of competition from rival groups. Moundville society collapsed 550 years ago but had managed to reorganize itself by the time Hernando de Soto reached Alabama in 1540.

During its peak, perhaps 800 years ago, Moundville may have controlled more than 30 miles of the Black Warrior floodplain. Its hinterland included seven or eight villages important enough to build mounds of their own, plus a greater number of farmsteads without public architecture. Archaeologists consider this pattern diagnostic of a society with three administrative levels, like that of the Bemba—a chief's village, smaller villages run by subordinate nobles, and still smaller settlements occupied by people of low rank.

Excavations indicate that Moundville's palisade was built some 800 years ago. The area inside the palisade incorporates at least 29 artificial mounds.

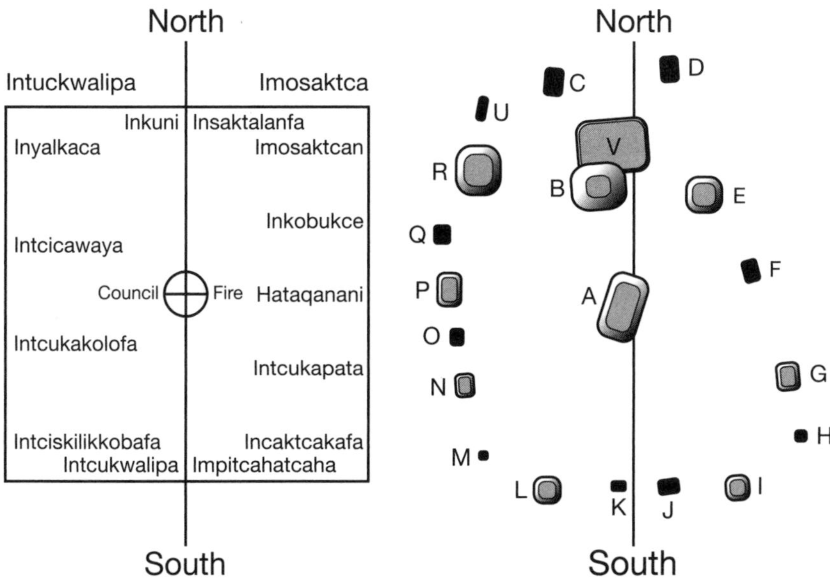

FIGURE 46. Many Native American societies of the U.S. Southeast had two major divisions, in which the hereditary rank of clans or subclans reflected their genealogical distance from the chiefly line. Sometimes the rank order could be detected in the layout of a settlement.

On the left we see the layout of a "council square" occupied by the Chickasaw a century ago. A north-south line separated the two major divisions, called Intuckwalipa and Imosaktca. Six subclans on the west and seven on the east were ranked in descending order, from north to south.

On the right we see the layout of Moundville, a prehistoric chiefly center, as it probably looked 750 years ago. A north-south line, passing through the earthen mounds that supported the chief's residence (B) and the central temple (A), creates two major divisions of residential and mortuary mounds. Mounds R, P, N, and L (on the west) and E, G, and I (on the east) decrease in cubic volume of earth from north to south, perhaps reflecting the relative rank of the families whose houses they supported. The mounds shown in black are believed to have supported mortuary temples. (Moundville's plaza measures 1,600 feet from north to south.)

Many of these mounds outline a plaza running more than 1,600 feet from north to south.

On the right, in Figure 46, we see Moundville as it would have looked between 750 and 700 years ago. Looking at the arrangement of its earthen mounds, Knight saw some interesting similarities with the layout of the Chickasaw council camp.

On the northern edge of Moundville's plaza lies Mound B, which probably supported a paramount chief's residence. Mound A, which supported Moundville's main temple, occupies a central position similar to that of the Chickasaw council fire. A north-south line through Mounds A and B divides the plaza in half.

Running down both sides of the plaza are residential mounds that decrease in cubic volume of earth as one moves from north to south. Mounds R, P, N, and L line the west side of the plaza; Mounds E, G, and I line the east side. These mounds may have supported the houses of families whose rank decreased from north to south. Alternating with these houses were mounds containing high-status burials, decapitated skeletons, skulls, and infant burials. These mounds may have belonged to mortuary temples associated in some way with the residences.

While Knight is cautious in his interpretation, Moundville's layout does resemble that of the Chickasaw council camp. To be sure, the Chickasaw camp had no major temple and no chiefly residence equivalent to Mound B, which stands 56 feet high and required three million cubic feet of earthen fill. That difference, however, may reflect the much higher population and centuries-long occupation at Moundville. Knight's use of Ca'bi'tci's diagram is an example of the way social anthropology can be used to reconstruct the living society that created an archaeological site.

Social inequality at Moundville was also reflected in burial ritual. Mounds C and D (to the north of the plaza) contained numerous burials of highly ranked people. Mounds M1 (to the south) and U (to the north) contained dense concentrations of commoner burials.

The most impressive sumptuary goods occurred with people buried between 700 and 550 years ago. One adult male in Mound C wore bracelets and anklets of copper-covered beads, three gorgets of sheet copper, a pearl necklace, an amethyst pendant, a copper ornament fixed to his hair with a pin made of bison horn, and a copper-bladed axe, which seems to have been a favored possession of elite men. Other highly ranked people were buried with copper ear ornaments, stone cosmetic palettes, lead ore crystals, and seashell beads. Such prestige goods make it clear that, in Colin Renfrew's terms, Moundville was an individualizing rank society.

CHIEFLY CYCLING IN NORTHERN GEORGIA

The ancient chiefly societies of northern Georgia underwent cycles like those of the Kachin and Konyak Naga. According to archaeologist David Hally, few north Georgia rank societies remained at peak strength for more than a century or two. Among the reasons for their periodic collapse were factional disputes, military defeats, revolts against overly demanding chiefs, and episodes of weak leadership. Such problems may occasionally have been worsened by drought.

Chiefly centers tended to endure longer than their satellite communities, but even they went through periods of decline and reorganization. A frequent strategy for retaining power was to form confederacies with neighboring rank societies through chiefly intermarriage or military alliance.

The greater a north Georgia society was, the larger the sparsely occupied buffer zone left around it. At the slightest sign of weakness or decline in a preexisting rank society, ambitious rivals tried to establish themselves in one of these buffer zones. Such a pattern was not unique to Georgia. In studying analogous African societies, anthropologist Igor Kopytoff found that ambitious leaders often attempted to establish new territories in the unoccupied frontiers between preexisting societies.

Hally describes most chiefly territories as only 10 to 12 miles wide. Exceptional chiefly centers like Etowah (described later) might have had territories 18 to 20 miles in extent. The sparsely occupied zones between rival societies could be five to 20 miles wide. In addition to serving as military buffers, these zones were often filled with the second-growth forest preferred as a habitat by white-tailed deer.

Three of the best-known rank societies in north Georgia were the one centered at ancient Etowah (which peaked between 750 and 675 years ago) and the historic Coosa and Ocute (which were visited by European explorers between A.D. 1500 and 1580). In this chapter we look at Etowah and the Coosa.

The Growth of Etowah

The remains of the prehistoric chiefly center of Etowah lie on the river of the same name near Cartersville, Georgia. The site's semicircular defensive ditch (both ends of which once reached the Etowah River) enclosed more than 50 acres (Figure 47, top). In its heyday Etowah was the paramount center for a chiefly territory 20 to 30 miles across, separated from its rivals by buffer zones. There were at least three smaller villages across the river from Etowah,

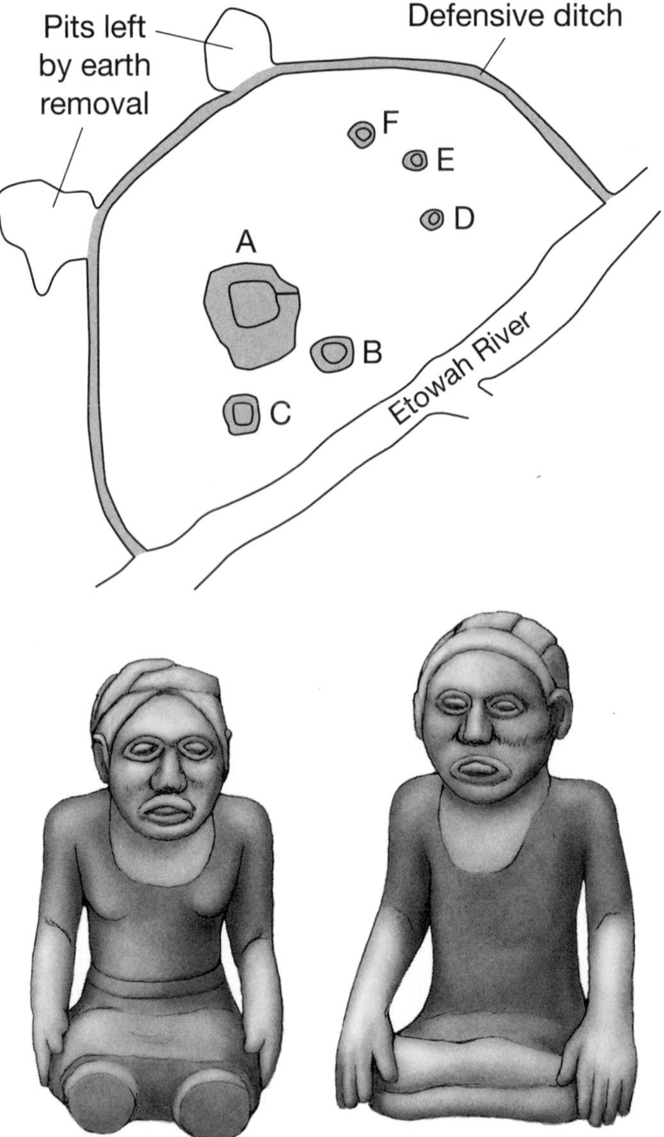

FIGURE 47. The archaeological site of Etowah was the chiefly center for an important prehistoric rank society of northern Georgia. Above we see the overall layout of Etowah, with its defensive ditch and Mounds A-F; the area enclosed by the ditch was 50 acres.

Below we see a pair of marble statues, found hidden in a log tomb at the base of Mound C. These statues are believed to represent the mythological couple who founded Etowah's chiefly lineage.

three more upstream, and three more downstream. Its hinterland also included shifting farmsteads.

Archaeologists have studied Etowah for more than 100 years. Work began with Cyrus Thomas in the 1890s and Warren G. Moorehead in the 1920s, continued with Lewis Larson in the 1960s, and then the torch was passed to Adam King in the early twenty-first century. We do not know the name of the ethnic group that founded Etowah, but it lay in Coosa territory at the time Spanish explorers arrived.

The largest earthen construction at Etowah was Mound A, which once stood 66 feet high and contained an estimated four million cubic feet of fill. An earthen ramp led from the summit of Mound A to a ceremonial plaza. To the south of Mound A were Mounds B and C, both large. On the opposite side of the plaza were smaller mounds, called D, E, and F. Early excavators concluded that Mounds B, D, and E had supported residences. Mound C attracted the most attention because it had once supported a temple; in the mound below the temple there were many burials with fabulous sumptuary goods.

Etowah was founded some 1,000 years ago, and, according to King, its beginnings were modest. None of the large mounds had yet been built at that time. The original occupants were just beginning to enjoy the benefits of Eastern Flint corn, which they combined with the hunting of deer and the taking of turtles, catfish, drum, and gar from the river.

In levels that antedated the building of Mound C, Larson uncovered half a dozen houses and a series of ritual buildings, built one atop another. One of the ritual buildings was more than 100 feet long and may have been a charnel house. Another building was 40 feet long, and its floor had been coated with red ocher pigment. Mounds A and B were built between 900 and 800 years ago, after which Etowah went into decline.

New leaders appear to have rejuvenated Etowah between 750 and 675 years ago, and the site entered an impressive phase of public building. Mounds A and B were enlarged, and Mound C was raised in multiple stages to an estimated height of 18 feet. At this time the defensive ditch was dug, and just inside this newly created perimeter the occupants erected a wooden palisade with bastions.

Mound C was filled with burials of the hereditary elite. Larson believed that he could detect both "major nobles" (buried in log tombs) and "minor nobles" (buried in simpler graves). Many Etowah burials were accompanied by standardized elements of costume and paraphernalia that probably reflected their rank.

Let us look now at a few notable burials from Mound C. Burial 57 was a robust adult male, placed in a log tomb with a pole roof and a floor of walnut planks. He wore copper-covered wooden spools in his earlobes, multiple shell necklaces, a shell gorget, a headdress with copper ornaments, traces of a disintegrated feather robe, and masses of pearls that had once been attached to a garment. He had been buried with eight conch shell cups, five or six embossed copper plates, and two copper axes.

Burials 25 and 64 provided evidence for inherited rank. Burial 25, an adult, had its head covered with a copper sheet. Burial 64, a five-year-old child, had its head covered with a similar, but miniature, copper sheet. Since Burial 64 had surely not lived long enough to achieve renown, he or she must have been entitled by noble birth to be buried with a copper sheet.

Larson reveals that offerings of engraved shell, symbolic "swords" chipped from Tennessee flint, stone cosmetic palettes, lead ore, shark teeth, beaded arm and leg bands, stone and copper axes, cutouts of mica and sea-turtle shell, and embossed copper plates were standard burial goods for the Etowah elite.

Finally we come to Burial 15, a log tomb at the base of the ramp leading to the summit of Mound C. Inside were the dismembered skeletons of four individuals, accompanied by copper-covered earlobe spools, copper hair ornaments, antler points for weapons, shell beads, and tobacco pipes. Added to this tomb was the remarkable pair of statues shown at the bottom of Figure 47. One of these marble statues represents a man seated cross-legged; the other is a kneeling woman.

King's interpretation of these marble statues takes us back to the Natchez, whose creation myth held that the first Suns were a man and woman who came to earth from the Upper World. When his work on earth was done, the primordial man turned himself into stone.

King suggests that the marble statues in Burial 15 depict an analogous "founder couple" and may originally have been kept in a mortuary temple. The disarticulated occupants of the log tomb seem to have been hurriedly buried, at roughly the time that Etowah's palisade was burned; they might therefore represent the elite victims of a raid. Under such circumstances, their tomb may have been seen as a convenient place to hide the sacred statues.

Following the burning of its palisade and the hiding of its founder couple's effigies, Etowah was abandoned for most of a century. Although it eventually rose again from the ashes, its moment of greatness was over.

The Coosa Confederacy

Some rank societies of northern Georgia lasted long enough to be seen and described by sixteenth-century Spanish explorers. Such was the case with the Coosa, who were visited by de Soto in 1540 and Juan Pardo between 1566 and 1568. Coosa history has been the subject of collaborative research by Charles Hudson, David Hally, and their colleagues.

The Coosa, like several other rank societies in the Southeast, had made themselves stronger through confederacy. To do this they used military alliances and noble intermarriage. The resulting confederacy was able to demand tribute from a much larger area. Leaders sent each other gifts, smoked tobacco together, and exchanged wives. Unfortunately, since they included multiple factions with conflicting agendas, most southeastern confederacies were doomed to collapse.

According to Hudson the territory of the sixteenth-century Coosa extended from the village of Chiaha (near Knoxville, Tennessee) to the village of Talisi (near Birmingham, Alabama). This is a distance of roughly 250 miles, too large by far to represent the territory controlled by a single chief. The Coosa had surely formed a confederacy of some kind, and Hally has identified approximately 11 clusters of Coosa sites that may have been participants.

The main town of the Coosa lay near Carters, Georgia, at the confluence of the Coosawattee River and Talking Rock Creek. Known today as the Little Egypt site, this community covered ten acres at its peak. The Spaniards described the paramount chief of the Coosa confederacy as traveling in a litter, accompanied by hundreds of warriors. He is said to have owned three houses atop pyramidal mounds, but only two mounds survive at Little Egypt today. Archaeologists believe that these mounds occupied the north and east sides of a plaza measuring almost 350 by 200 feet.

Many Coosa villages were fortified. The King site, farther down the Coosa River in Floyd County, Georgia, may represent the ancient village of Piachi. Etowah, only a one- or two-day trip south of Little Egypt, was greatly reduced in size during the sixteenth century; it may at that time have been a Coosa satellite village named Itaba. One of the most interesting villages in the confederacy was ancient Chiaha, which marked the northern limit of Coosa territory. Chiaha lay on an island in Tennessee's French Broad River and was defended with a palisade. To the north of the river was a buffer zone 30 miles wide, separating the Coosa from their enemies the Chisca.

The rank societies discussed in this chapter were all impressive. British eyewitnesses, however, suggest that in 1607, the most powerful southeastern rank society may have been the one led by a chief named Powhatan. The village from which he ruled, known as Werowocómoco, lay beside Purtan Bay in Gloucester County, Virginia. Powhatan, his subchiefs, and his allies controlled vast areas west of the Chesapeake Bay, yet the archaeological remains of Werowocómoco are superficially unimpressive. Lacking the platform mounds of sites such as Moundville, Etowah, or Little Egypt, Werowocómoco shows us that archaeologists cannot always rely on a site's monumentality to reflect its political importance. It is only because of British eyewitnesses that we know how powerful Powhatan was. And the British even threw in the romantic story of Powhatan's daughter, a maiden named Pocahontas.

THE EXTENT OF INEQUALITY IN THE SOUTHEASTERN UNITED STATES

At their peak, many of the chiefly societies in our backyard were on a par with those of Panama and all but the greatest of Colombia's Cauca Valley. Southeastern chiefs were carried on litters and expected a retinue of subordinates to accompany them to the afterlife. Below them were major and minor nobles, some of whom increased their prestige through achievement. Prowess in war provided upward mobility, and hereditary nobles patronized and rewarded skilled artisans. Like the Bemba of Zambia, the rank societies of the Southeast had male chiefs, although nobility descended in the female line.

Southeastern society shows us two widespread strategies of ambitious nobles. One was the founding of new rank societies in the buffer zones between preexisting chiefdoms. The other was the formation of confederacies to create more powerful societies. Despite their size, it appears that none of these confederacies went on to become kingdoms like those of ancient Mexico and Peru. One obvious reason is that they were decimated by Old World diseases once European colonists began to arrive. Had they been left alone, some southeastern societies might eventually have become kingdoms through the processes described later in this book.

SIXTEEN

How to Turn Rank into Stratification: Tales of the South Pacific

In most of the world's chiefly societies rank formed a continuum from the chief to the lowliest free citizen. Under the right conditions, however, rank societies sometimes made the transition to stratification. This amounted to drawing an invisible line across the continuum, thereby establishing a sharper break between the rulers and the ruled.

Social strata were usually kept separate by a behavior anthropologists call *class endogamy*. That simply meant that members of each stratum were only supposed to marry their peers. If a man of noble birth took a commoner wife, any children she bore him were less than noble. Frequent consequences of stratification were that (1) nobles began to keep lengthy genealogies; (2) rulers competed to marry the most elite spouses; and (3) society might be willing to ignore the usual incest prohibitions if one of the ruler's siblings would make the noblest marriage partner.

To be sure, there were gradations of prestige within the noble stratum, just as there are gradations from prince to duke, earl, baron, and marquis within traditional European nobility. Many of these gradations were survivals of the continuum of rank that had preceded stratification.

Let us make clear that the terms *stratum* or *class* used in this book are quite different from the phrases "upper class," "middle class," and "lower class" used to describe today's American society. American society has no stratum of hereditary nobles. We use terms such as upper, middle, and lower class to refer to arbitrary divisions of a continuum of wealth. Genealogy is not the criterion; all it takes to move from one economic class to another is an increase or a decrease in wealth.

THE CATEGORIES OF POLYNESIAN SOCIETY

As we have seen, Irving Goldman divided Pacific Island societies into three categories. In his "traditional," or least powerful, category the authority of the chief was based mainly on his greater quantities of mana, or life force. "Open" Polynesian societies combined mana with military force. The most powerful, or "stratified," societies used every source of power in a chief's arsenal; many drew the aforementioned line across the continuum of rank. Any move to stratification, of course, had to be validated by modifying society's logic.

Tahitians traditionally chanted that the *ari'i nui,* or sacred great chiefs, were direct descendants of the deities Ti'i and Hina. The *manahune,* or commoners, on the other hand, had merely been conjured into being by the gods, so that there would be someone to do the manual labor. This did not result in full stratification, because there remained an intermediate category: the *ra'atira,* who were well-to-do commoners or landed gentry. Tahitian logic explained the ra'atira as the result of intermarriage between nobles and commoners. There were also Tahitians called *ari'i ri'i,* "small chiefs," whose relationship to the ari'i nui was like that of small Angs to great Angs among the Konyak Naga. Tahiti's logic explained the ari'i ri'i as the offspring of marriages between great chiefs and ra'atira.

Tahitian society could have been converted from a ranked to a stratified one by eliminating the intermediate category of landed gentry. That, as we will see later in this chapter, is exactly how the later Hawai'ian chiefs created social strata.

HOW WESTERN SAMOA TOOK THE FIRST STEP

Archaeologists believe that the islands of Samoa, Tonga, and Fiji were colonized before the rest of Polynesia. A kind of ancestral Polynesian society then arose on these islands, before canoeloads of colonists sailed farther into the Pacific. The ancestral society is thought to have resembled Goldman's traditional type, where sacred life force was the main basis for chiefly authority.

Samoa is made up of multiple islands, none covering more than 700 square miles. The larger islands fall into two groups, called Western Samoa (Upolu, Savaii, and Tutuila), and Manu'a, or Eastern Samoa (Ofu, Olosenga, and Tau). Early Samoa was considered a land of plenty, producing up to three crops of taro each year. The Samoans also raised yams, sweet potatoes, plantains, pigs, and chickens and were skilled at fishing.

Prior to A.D. 1200, all of the archipelago seems to have been occupied by traditional rank societies, led by *ali'i* (the Samoan equivalent of Tahiti's ari'i and Tikopia's ariki). Samoa's chiefly lineages allegedly descended from the sky god Tangaroa, which gave them more mana than anyone else.

Samoans believed that their system of rank had begun on the islands of Eastern Samoa. Its chiefs were considered the first to descend from the Sky God, with all other ali'i branching off as junior lineages. It followed logically, therefore, that the most illustrious of all the chiefs should be Eastern Samoa's Tui Manu'a—literally, "the lord of Manu'a." His home village was a kind of capital for Eastern Samoa. Beyond this capital lay a series of semiautonomous villages, each with its own lesser chief, power-sharing council *(fono),* and spokesman/administrator *(tulafale).* Within a village there might live 300 to 500 people, of whom only 10 to 20 were the heads of elite families.

In traditional Samoan society most inter-ali'i competition was over *a'o,* or titles. This competition reminds us of the battles for sacred names in Avatip. It was a process that probably began in the competitive atmosphere of achievement-based society and then found new outlets in rank society.

Let us look for a moment at the Tui Manu'a, the most highly ranked of 15 to 20 sacred chiefs in the Samoan archipelago. His subjects prostrated themselves before him. His glance could wither fruit on the tree. His body, his house, his personal possessions, and even the vessels from which he ate were so charged with mana as to be dangerous. In addition to his councillors and spokesmen, his retinue included stewards, cupbearers, trumpeters, messengers, barbers, and jesters. He made sacrifices to the sky gods, whose temples were tended by *taula,* or priests.

About 800 years ago, according to oral history, something happened that would change Samoan society profoundly. Canoeloads of warriors, having sailed more than 500 miles from the Tongan archipelago, invaded the Western Samoan islands of Upolu, Savaii, and Tutuila. The Tongans managed to dominate those islands for four centuries. Eventually the Samoans were able to expel them by force.

In the course of developing the military prowess needed to overthrow the Tongan invaders, Western Samoa became an open society in Goldman's terms. War leaders became more important, and a'o titles became the spoils of war. The island of Upolu created a new war conquest title, Tafa'ifa. The man bearing this title became the Western Samoan equivalent of Eastern Samoa's Tui Manu'a.

According to Goldman the Manu'a island group, never having been invaded by Tongans, developed along more peaceful lines. While chiefs continued to compete, it was along the more traditional routes of sacred life force, persuasion by eloquence, and accumulation of titles.

The Western Samoan case shows us one possible way that a rank society based on sacred authority can become more militaristic. Toa, or prowess in battle, allowed lesser chiefs to rise by conquest, weakening the mana of sacred chiefs. Because of the overall richness of the Samoan environment, the battles rarely focused on resources; they were all about accumulating prestigious titles.

Eventually some chiefs sought to conquer all the islands of Samoa. An eighteenth-century noble named Tamafainga claimed to have done that, but he was soon assassinated. A later chief named Malietoa Vaiinupo did succeed in subduing the entire archipelago. True to tradition, however, his goal was to seize all of Samoa's four major titles rather than its resources. Later in this chapter, we will see that his Hawai'ian counterparts were far more interested in garden land.

Converted to a more militaristic society by the Tongan invasion, the Samoans became legendary warriors. No one who follows American professional football will be surprised by this fact. Picture a war canoe paddled by Manu Tuiasosopo (6′3″ and 255 pounds), Tiaina "Junior" Seau (6′3″ and 250 pounds), Edwin Mulitalo (6′3″ and 340 pounds), Chris Fuamatu-Ma'afala (5′11″ and 252 pounds), and Joe Salave'a (6′3″ and 290 pounds). At the rear are the smaller, but no less menacing, Mosi Tatupu and Troy Polamalu. Now run for your life.

STONE MONUMENTS, BURIAL MOUNDS, AND DESPOTIC POWER: CHIEFLY CYCLING ON TONGA

We have now learned that both Tikopia and Samoa were invaded by Tongan warriors. In fact, at one time or another, the islands of Tikopia, Samoa, Futuna, Rotuma, and 'Uvea all paid tribute (we might call it "protection money") to Tongan chiefs. It is estimated that a quarter to a third of the population of Tonga's six main islands were warriors. Who were these Tongans, and how had they become so aggressively expansionist?

Tonga, an archipelago of more than 100 islands, was part of the Samoa-Tonga-Fiji homeland. Tongan society undoubtedly began as Goldman's traditional type, but over time it moved closer and closer to his stratified type. By the time they were first seen by Europeans, Tongan chiefs were as politically powerful as those of Colombia's Guaca and Popayán. Tonga stood, in other words, at the threshold of becoming a kingdom.

The three largest islands of the archipelago were Tongatapu, Vava'u, and 'Eua. At its peak, the population of these three islands may have approached 25,000. Tongatapu, the largest, is 25 miles long and has a superb lagoon. It was on this lagoon that Lapaha, the greatest of the Tongan chiefly centers, was established.

Tongatapu supported plantations of yams, taro, and sweet potatoes; there were also groves of coconut trees, plantains, bananas, and breadfruit. The islanders raised pigs and chickens and hunted native pigeons. They built fishing dams in channel outlets.

Happily, from our perspective, Tongatapu is another of those places where social anthropology and archaeology have worked hand in hand. In our brief look at Tongan society we rely on studies such as those of social anthropologist Edward Gifford and archaeologist W. C. McKern, as well as overviews by social anthropologist Irving Goldman and archaeologist Patrick Kirch.

According to Kirch, colonists reached Tongatapu more than 3,000 years ago. For many centuries their settlements hugged the shores of the lagoon, moving inland only in response to gradual population growth. Not until A.D. 900 or 1000 did Tongan society show signs of the monumental earthworks for which Tongatapu is famous.

Archaeologists believe that some of these earthworks are early examples of *langis,* or burial mounds for high chiefs and their relatives. Langis came in several shapes and consisted of earthen fill held in place by walls of coral limestone. There were also *fa'itokas,* or communal burial mounds for members of lower-ranking social segments. (This is the type of mound that the Tikopians began to build, following the arrival of Tongan visitors.)

Oral history in Tongatapu takes us back to A.D. 950, about the same time that langis began to show up in the archaeological record. Prior to the tenth century, the legends claim, Tongatapu was led by men called "worm rulers." Finally, the Sky God began mating with mortal women to produce a semidivine elite. The Sky God's favored son, Ahoeitu, was named the first Tui Tonga, or "lord of Tonga." His half brothers, angry at having been relegated to the status of lesser chiefs, assassinated Ahoeitu while he was still in the sky. He was resurrected before reaching Earth, where he replaced the last worm ruler.

From 950 to 1865 the office of Tui Tonga passed through 39 men, most of them firstborn sons of firstborn sons. The first Tui Tongas exercised both religious and secular authority. In many cases the oral history specifies which Tui Tonga is buried in a particular langi. Having such information is every archaeologist's dream.

The dynamic and competitive nature of Tongan leadership is exemplified in the lives of the Tui Tongas. Little is known of the semidivine Ahoeitu. The tenth Tui Tonga, Momo, established his paramount village at a place called Heketa.

Momo's successor, Tuitatui, built an exceptional monument at Heketa. Called the Trilithon, this monument was like a gateway arch composed of three immense stones. The two upright stones represented the chief's sons, Lafa and Talaihaapepe; each stone was 15 to 17 feet tall and weighed 30 to 40 tons. The lintel spanning the two uprights symbolized the two sons' inseparable bond; it was 19 feet long and 4.5 feet wide. These stones had been cut with stone axes, transported on sleds like those used by Naga stone-pullers, and set upright using earthen ramps. As long as the arch stood, it was said, Tuitatui's sons dared not quarrel.

The 12th Tui Tonga moved his paramount village to Lapaha on the shore of the lagoon, where it remained for six centuries. The 15th Tui Tonga is suspected of being the one who sent canoeloads of warriors to Western Samoa between A.D. 1200 and 1250. Both Havea I (the 19th) and Havea II (the 22nd) were assassinated. These murders were probably plotted by rival Tongan chiefs; reportedly, however, they had to be carried out by Fijian "hit men" because it was taboo for a Tongan to touch his own paramount chief. By 1450 this taboo seems to have weakened, because the 23rd Tui Tonga was assassinated by his own countrymen. His son, Kauulufonua I, avenged his father's death by running the killers to ground.

Around 1470 Kauulufonua I became the 24th Tui Tonga and announced that his firstborn son, Vakafuhu, would be his successor. He then did something that would alter Tongan history: he created the title of Tui Haa Takalaua for his second son, Moungamotua. By so doing, Kauulufonua I divided rulership between a sacred chief (the Tui Tonga) and a secular chief (the Tui Haa Takalaua).

Having dual chiefs seemed like a great way to frustrate future assassins, since it would be difficult to murder both men. With the passage of time, however, the secular chiefly line would become increasingly active and powerful; meanwhile, the sacred chiefs, in Goldman's view, deteriorated into indolent, self-indulgent womanizers. For example, Uluakimata I, the 29th Tui Tonga, assembled a harem of 200 women and built himself Lapaha's most spectacular burial mound.

Here we see a widespread process in rank societies: a junior lineage splits off from a senior lineage and, after years of effort, overtakes the latter in power and influence.

This was not the last bifurcation of chiefly Tongan lineages. Sometime before 1610, the 7th Tui Haa Takalaua gave his younger brother the title Tui Kanokupolu, assigning to him the day-to-day administration of chiefly affairs. Unfortunately, the Tui Haa Takalaua lineage became extinct after only 13 generations. This left only the sacred Tui Tonga and secular Tui Kanokupolu lineages. In 1865 the 39th and final Tui Tonga died. That left the archipelago in the hands of the 19th Tui Kanokupolu, who promptly renamed himself King George I.

The Tongan Social Hierarchy

Let us look now at Tongan society. Goldman assigns it to his stratified category, a reasonable decision given how great the inequality was between high chiefs and commoners. Chiefs were considered so different from commoners that different terms were used for the parts of their body. Commoners kneeled or assumed postures of obeisance in the chief's presence. Chiefs sometimes sat cross-legged on a stack of woven mats, letting commoners touch the soles of their feet as a sign of subservience. Many chiefs wore feather headdresses and were tattooed with special symbols. Like Africa's Bemba chiefs, they had the power to mutilate commoners who offended them.

The mana possessed by the Tui Tonga was so dangerous that the Tongans created a separate earthen mound for the burial of his hair clippings, blood, and bodily wastes. Chiefs had their own special bathing holes, fans, and fly whisks. They received the first fruits of every harvest and the first fish of every catch. Only the back, head, chest, and rump of a pig were suitable for a chiefly meal; any part that routinely touched the earth was rejected.

The children of high chiefs maintained separate houses and servants. Chiefly daughters were kept out of the sun, scented with flowers, rubbed with candlenut oil, and prevented from overeating. Because these young women were crucial to the creation of chiefly marriage alliances, their legs were tied together at night to prevent them from taking lovers.

Gifford supplies us with an anecdote that illustrates the Tongan reverence for hereditary authority. We have already mentioned that the 19th Tui Tonga was assassinated, possibly by Fijian contract killers. In fact, he was cut in half while bathing, and only his upper half was recovered. This left his corpse incomplete for burial. A lesser chief named Lufe offered to be killed and bisected so that his lower half could be joined to the dead paramount's upper half. His relatives took him at his word.

When a high chief died, his *matapules,* or titled ceremonial attendants, took charge of the funeral. One served as undertaker while another supervised the quarrying and hauling of stones for the chief's burial mound. The chief's body, anointed with oils and fanned continuously with a fly whisk, lay in state for days. His brain and intestines were removed, much as the ancient Egyptian embalmers did with their pharaohs. Undertakers were normally paid for their work with gifts of palm mats and *tapa,* or bark cloth; when the tomb of the eighth Tui Haa Takalaua was reopened, however, it was discovered that his undertaker had also been honored by being buried with his chief.

A chief's funeral required an initial spectacular feast, followed by 10 to 20 days of lesser feasting. The favored meat was pork, while the ritual beverage of choice was *kava,* a drink brewed from the root of an aromatic pepper plant. Commoners singed off their hair in mourning; elite mourners sang songs of grief; torches were lit all night, and the chief's burial mound was decorated with colored stones.

However stratified these descriptions make the Tongans sound, the fact is that their society was still not completely divided into two class-endogamous strata. Instead, Tongan society has been described as a great tree with limbs, branches, and twigs extending outward from its noble trunk. In Gifford's words each patrilineage consisted of "a nucleus of related chiefs about whom are grouped inferior relatives, the lowest and most remote of whom are commoners."

Tongan chiefs, however, referred to themselves as if they belonged to a separate stratum. They called themselves *tui* (major lords) or *eiki* (lesser chiefs), while commoners were called *tua.* There were also *popula,* or slaves, mostly prisoners of war, and *hopoate,* or "strangers," mostly shipwreck survivors. There were, in addition, terms for what seem to have been bureaucratic offices: matapule (titled attendants who were allowed to wear tapa cloth); *takanga* (untitled attendants with fewer privileges); local governors and stewards (who kept the chief apprised of problems in their districts); and *eikisi'i* (renowned warriors who, like ancient Colombia's "nobles by command," were granted the privileges of a petty chief).

Tongan Power Sharing

We saw in earlier chapters that even the most powerful chiefs shared power with councillors or advisers of some kind. The Tui Tonga was advised by a quartet of ministers known as the *falefa* ("four houses"), who built their residences near his. They attended the Tui Tonga, influenced his decisions, supervised work on his personal garden plot, drank kava with him, and sometimes could be bribed to orchestrate his assassination.

Further power sharing resulted from the premise that, within any group of siblings, the females were thought to possess more mana than the males. Thus while the office of Tui Tonga went to a man, he was outranked by his sister, who was known as the Tui Tonga Fefine. She was treated like a queen throughout Tonga, and her firstborn daughter, the Tamaha, also outranked her uncle. Gifford reports that even the most powerful Tui Tonga, who was carried from place to place on a litter and had the power to mutilate his subjects, allowed the Tamaha to place her foot on his head.

This inequality between noble brothers and sisters provided logical contradictions when selecting a spouse. That is, the firstborn son of the Tui Tonga could be outranked by the firstborn sons of the Tui Tonga Fefine and the Tamaha. To avoid such problems, Patrick Kirch reveals, the Tui Tonga Fefine might be married to a Samoan or Fijian chief who was outside the Tongan system. Another strategy was to marry the Tamaha to a secular chief like the Tui Haa Takalau, whose lineage was separate from that of the sacred Tui Tonga.

There were several outcomes to this jockeying for rank. One is that, according to Kirch, noble houses on Tonga, Fiji, and Samoa established long-term patterns of intermarriage. The resulting familiarity with all three archipelagoes probably facilitated some of the Tongan invasions mentioned earlier. Another outcome was the relaxation of incest taboos; first cousins could marry, even though they were called "brothers" and "sisters" in the Tongan language.

Land and Power in Tonga

At the heart of the Tongan invasions of other islands was an important difference between the aspirations of Samoan and Tongan chiefs. Samoan ali'i, as we have seen, wanted to accumulate noble titles. Tongan chiefs wanted garden land.

In Samoa and Fiji, land was the corporate property of clans or villages. In Tonga, all agricultural land was controlled by the Tui Tonga but could be delegated to lesser chiefs. Commoners might be given permission to create gardens on the high chief's land, but in the end the chief could take anything he wanted. On Tongatapu alone, seven tracts were set aside for the Tui Tonga.

Most chiefs also controlled the best fishing areas. Only commoners who lived on the coast could fish; those who lived inland were limited to trading yams, taro, and fruits for fish. The first fish from each catch, like the thighs of sacrificed animals delivered to Kachin chiefs, went to the Tui Tonga.

The wealth generated by gardens and fishing stations gave Tongan chiefs added incentive to invade resource-rich areas such as Western Samoa. Chiefly monopoly of resources also denied lesser nobles a way to support their followers, undermining the traditional continuum of rank based on sacred life force.

The Role of Tongan Warfare

As in many societies we have considered, toa gave nonelite warriors a route to prominence. Anyone who brought back the heads of ten or more enemies might be raised to the rank of titled attendant or petty chief. Such a man was allowed to drink kava with the high chief.

Tongan warfare was more formal than that of Samoa. Up to 200 men could be used at one time, organized into *matanga* (companies) and *kongakau* (regiments). The battle commander was usually a chief of medium rank, accompanied by assistants (lesser chiefs or titled attendants) who transmitted his orders. Chiefs watched the battle from their litters, but they rarely took part.

To the beat of drums the regiments followed scouting parties into enemy territory. Battles were preceded by lots of kava drinking and trash-talking. Attackers carried bows and arrows and spears and clubs, and defenders built camouflaged traps filled with sharpened stakes. The greatest military insult was to drink coconut milk while sitting on the chest of a captive.

The Logic of Inequality in Tonga

We are now in a position to list some of the premises of Tongan social logic. Many of these premises were identified by Goldman years ago; our contribution has merely been to add a few more.

1. Our sacred chiefs are descended from the Sky God and a mortal woman.
2. Their semidivine origin entitles them to the products of all garden land and prime fishing areas.
3. Senior lineages outrank the junior lineages that split off from them.
4. A firstborn offspring outranks later siblings of the same sex.
5. Among siblings, sisters and their daughters outrank brothers and their sons.
6. The head of each family, however, should be a male.
7. Political offices, as well, should be filled by males.
8. No two persons in the same family hold the same rank.
9. In order to ensure the high rank of one's children, cousins can marry each other, even though they are classified as siblings in the Tongan language.
10. In order to ensure the supremacy of a chief's children, his more highly ranked sister should marry into a different chiefdom.
11. A chief's principal wife should be the most highly ranked woman available.
12. To satisfy premise 11, a chief may have to import his principal wife from a different chiefdom or marry a close relative (premise 9).
13. No Tongan dares assassinate his own chief, owing to the latter's high levels of mana.
14. Assassins from other islands, however, can be hired to kill a Tongan chief.
15. Dividing authority, by creating a line of secular chiefs that will coexist with sacred chiefs, makes political assassination more difficult.
16. Secular chiefs, however, pose the threat of usurpation.
17. To reduce the risk of usurpation, the sacred chief should limit the land (and other resources) allocated to the secular chief.

Tongan society, as we suggested earlier, stood at the threshold of true stratification. All that remained was to draw a line across the continuum of rank, separating the elite trunk of the tree from the commoner branches and twigs.

We have one more task to perform before moving on to the creation of true social strata: we need to look in detail at the Tongan chiefly center of Lapaha. So much is known of Lapaha's buildings and pyramidal mounds that the site can be used as a framework for understanding spectacular chiefly centers elsewhere in the ancient world.

Archaeology and History at Lapaha

Lapaha, the civic and religious center founded by the 12th Tui Tonga, stretched 1,600 yards along the Tongatapu lagoon. In the beginning it had consisted only of an oval enclosure 600 yards long, flanked on one side by the shore and protected elsewhere by a defensive ditch 10 feet deep and 20 feet wide. The earth from the ditch was used to create a high embankment with a palisade of wooden posts, increasing the height of the barrier. This enclosure, known as "Old Lapaha," belonged to the Tui Tonga lineage.

Three changes led to an enlargement of Lapaha. One change, presumably natural, was the migration of the shoreline to the west, making new land available for the Tui Haa Takalaua lineage. The second change was the work of human agents: following the creation of the chiefly Tui Kanokupolu lineage, additional land to the south was turned into New Lapaha. A third change involved the creation of more langis, or elite burial mounds.

Figure 48 is a modified version of W. C. McKern's 1929 map of Lapaha. Note that from the air the site would have appeared as a linear arrangement of plazas, temples, pyramidal mounds, stone monuments, and elite residences. What makes Lapaha special is the fact that we have eyewitness accounts of its buildings and occupants, beginning with the visits of Captain James Cook in 1773 and 1777.

Let us look first at Old Lapaha. Its centerpiece was a large, grassy plaza 100 yards long. To the south of this plaza were two important residences. One was the house of the Tui Tonga, which was 50 feet long and flanked by the smaller houses of his servants. This chiefly residence sat within a private enclosure, shielded from public view by a reed fence taller than a man's head. The gateway to this enclosure was accompanied by a stone monument: a stela of volcanic rock three feet tall and a foot thick, a gift from the chief of Houma on the island of 'Eua.

Two other landmarks in the vicinity of the plaza are worth mentioning. Some 35 yards west of the Tui Tonga's private enclosure lay the house of the priest assigned to Taufaitahi, the patron deity of the sacred chief's family. A short distance north of the chiefly enclosure was the earthen mound where the Tui Tonga's hair clippings, blood, and bodily wastes were ritually buried.

To the north of the grassy plaza were several additional constructions. Four of them, called Langis J1-J4 on McKern's map, were the rectangular burial mounds of past Tui Tongas. Immediately to the west of the mounds lay the Tui

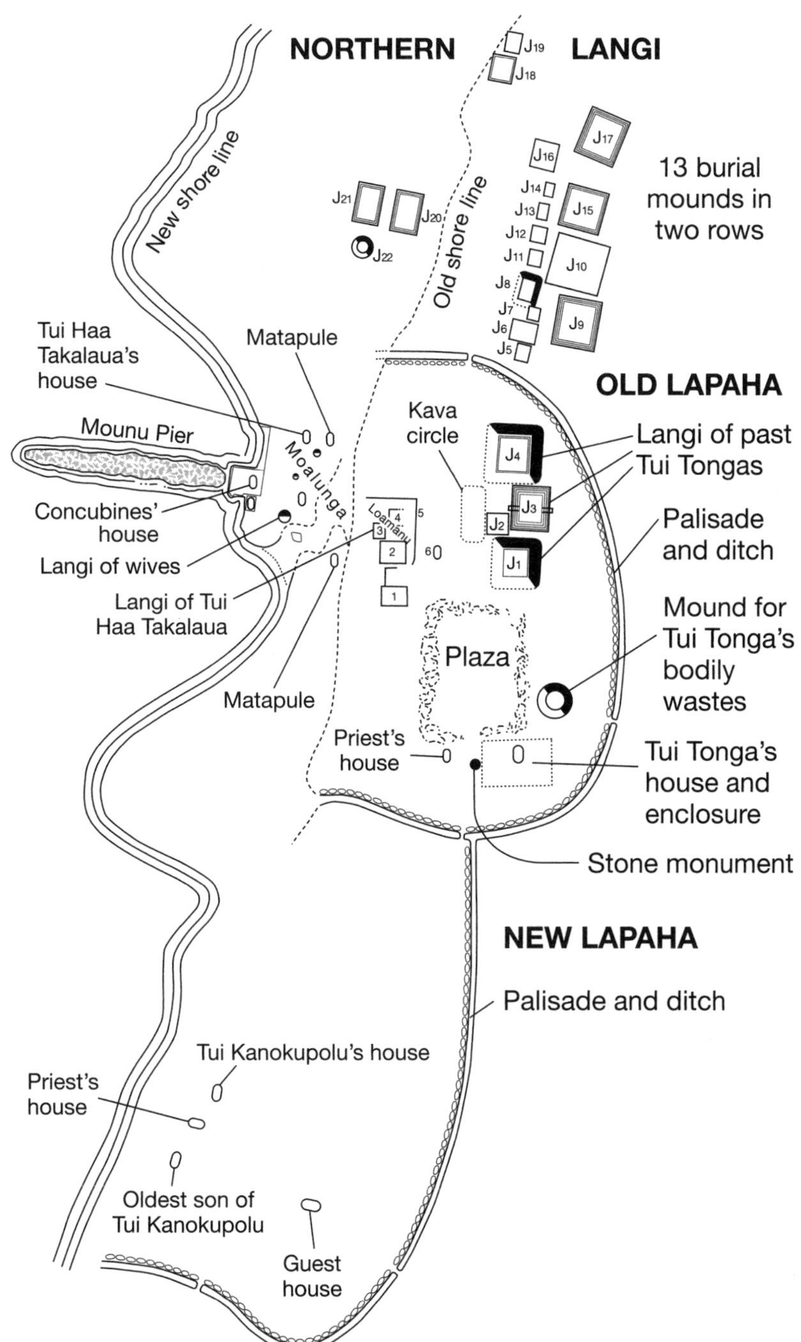

FIGURE 48. The civic-ceremonial center of Lapaha, Tonga, stretched for 1,600 yards along a lagoon.

Tonga's kava circle, a plaza where he could have a drink with his advisers and subordinate chiefs.

Oral histories give the occupants of several of Old Lapaha's burial mounds. Langi J4 is said to hold the remains of Tui Pulotu I, the 33rd Tui Tonga. He was reportedly buried facedown, with his brother Tokemoana at his back. Tui Pulotu I's sister (who outranked him) was given her own langi. Tui Pulotu II, the 35th Tui Tonga, was reportedly buried in Langi J1.

One last landmark of Old Lapaha is a walled enclosure called Loamanu, near the former shoreline of the lagoon. This was the burial place of past Tui Haa Takalauas, whose langis were smaller than those of the Tui Tongas. Evidently, although the Tui Haa Takalaua chiefs lived to the west of the former shoreline, they wanted to be buried in Old Lapaha.

Let us look next at Moalunga, the area created for the Tui Haa Takalaua lineage. It appears on McKern's map as a cluster of houses and mounds just west of Old Lapaha, built on land exposed by the migration of the shoreline. One of its major features was an artificial pier called Mounu, which extended 500 to 600 feet into the lagoon. This pier is said to have been created from stone slabs native to the island of 'Uvea, boatloads of which had been brought 500 miles to Lapaha for langi construction.

The Tui Haa Takalaua's house lay 50 yards inland; one of his matapules, or titled attendants, lived nearby. At the approach to the pier was a large house occupied by the secular chief's numerous concubines. Immediately to the south was a kind of beach house, to which the chief could take any of his concubines for a tryst. This "cohabitation house" was flanked by a stone monument ten feet in diameter, the gift of a neighboring chief. Farther south lay a burial mound for the Tui Haa Takalaua's legitimate noble wives. The southernmost landmark of Moalunga was the house of a second matapule.

The area called New Lapaha also had its landmarks, though they were not as densely clustered as those in Old Lapaha. New Lapaha contained the residence of the Tui Kanokupolu and, perhaps 35 yards to the south, the house of the priest assigned to his family's patron deity. There were also houses for the Tui Kanokupolu's firstborn son and daughter. The largest structure in New Lapaha, however, was a guest house.

Finally, there were at least 18 langis to the north of Old Lapaha. Fifteen of these burial mounds, including a group of 13 arranged in parallel rows, were built east of the lagoon's old shoreline.

Langi J9, one of the four largest of these northern burial mounds, is believed to have been the first one created at Lapaha. It is said to have been built by Talatama, the 12th Tui Tonga, who became the first sacred chief to establish his paramount village at Lapaha. This suggests that the first langis at Lapaha were built outside the defensive ditch.

The Sacred Landscape of the Tui Tonga

Throughout this book we have argued that archaeology and social anthropology contribute more when they work together. We have also seen that principles of behavior stand out more clearly when we can show them operating (1) in different parts of the world and (2) in the past as well as in the present. To illustrate this we will look first at what Gifford's social anthropology and McKern's archaeology combine to tell us about Lapaha. Next we will look at an analogous chiefly center from another part of the ancient world.

To begin with, Lapaha was huge; from north to south, it measured almost a mile. At the same time, it was in no sense a "city." There is no evidence that it included large concentrations of commoners or urban wards of craftsmen. Lapaha seems to have been what archaeologists call a "civic-ceremonial center."

Like many civic-ceremonial centers, Lapaha grew by accretion. Old Lapaha was dedicated to the Tui Tonga lineage. Hundreds of yards of burial mounds were gradually added to the north. The westward migration of the shoreline added land for the newly created Tui Haa Takalau lineage. The area later assigned to the Tui Kanokupolu lineage required a 600-yard extension to the south. This growth tells us that each of the three major chiefly lineages insisted on its own space.

Tonga fits Goldman's most powerful category of Polynesian society, the one whose chiefs combined militarism and expertise with mana. McKern's plan of Lapaha, however, gives the impression of a sacred landscape. We do not see military barracks or concentrations of artisans, as we will see later when we discuss kingdoms. We see long alignments of ritual plazas and burial mounds, the largest of which were for sacred, not secular, chiefs. Only the Tui Tonga had a private enclosure surrounding his residence. The use of open space to create a tranquil, parklike atmosphere is unmistakable.

Among the people and objects the Tongan chiefs wanted close at hand were their wives; their concubines; their servants; their matapules, or titled attendants; the stone monuments they had received as gifts; and the priests in charge of their patron deities. Not shown on McKern's map, but described by

Gifford, were the houses of the Tui Tonga's falefa advisers and the temples to the chiefs' patron deities. Apparently not included in Lapaha were the residences and fa'itokas (communal burial mounds) of lower-ranking lineages.

Anthropological and historical accounts stress the secular and military power of Tongan chiefs. What the archaeology of Lapaha shows us, on the other hand, is a "built environment" dedicated to the privacy and self-indulgence of a privileged few; places for ritual processions and kava drinking; and memorials to semidivine rulers in the form of elegant pyramidal mounds.

LAPAHA AND LA VENTA: COMPARISONS AND CONTRASTS

Some 3,000 to 2,400 years ago the tropical coast of the Gulf of Mexico was home to several spectacular rank societies. Some of the most flamboyant occupied the drainages of the Papaloapan, Coatzacoalcos, and Tonalá Rivers in the Mexican states of Veracruz and Tabasco.

The coast of western Tabasco is prograding, or expanding, toward the Gulf, even as it subsides under the alluvium of the Tonalá. A dozen miles inland, the archaeological site of La Venta occupies a residual hill almost buried by the recent alluvium of the coastal plain. La Venta was once surrounded by swamps and sloughs draining into the Tonalá, leaving it naturally defended. Inland from La Venta lay a tropical forest receiving more than 100 inches of rain each year.

Archaeologist Robert Heizer once calculated that 18,000 people lived close enough to La Venta to consider it their region's main civic-ceremonial center. Those people likely grew two crops of corn each year, a main crop sown just before the rains began in May and a minor crop planted just before the February dry season. Small farming villages would have moved periodically as the tropical forest was cleared, burned, planted, and left fallow for years. In this landscape of shifting settlements, La Venta would have been the one fixed point.

Like Lapaha, La Venta took the form of a long, linear complex of pyramidal mounds, flat-topped platforms, stone monuments, and chiefly burials. Also like Lapaha, La Venta grew by increments. Complex A (the counterpart to Old Lapaha) was only about 300 yards long and oriented eight degrees west of true north. This complex, which covered about five acres, was built in stages between 3,000 and 2,600 years ago. Immediately to the south is a large earthen pyramid called Complex C. From this point southward the mounds and plazas of La Venta have been less fully investigated, and some date to periods too recent to be discussed here. Suffice it to say that La Venta's linear sequence of public constructions eventually reached 1,600 yards in length, similar to Lapaha at its peak.

At least four generations of archaeologists have worked at La Venta. Included were Frans Blom and Oliver LaFarge in the 1920s; Matthew Stirling in the 1940s; Philip Drucker, Robert Heizer, and Robert Squier in the 1950s; and Rebecca González Lauck in the twenty-first century.

Figure 49 shows the layout of Complex A. Let us begin in the south, where we find an elongated plaza between two long, low earthen mounds called A4 and A5. The northern half of this plaza was occupied by Mound A3. Immediately to the south of A3, the people of La Venta buried a large mosaic pavement formed with blocks of imported serpentine, a hard metamorphic rock.

To the north lay a second plaza, this one outlined by a stone fence. The posts of the fence were of columnar basalt, a volcanic rock that occurs naturally as a series of multifaceted columns. Within this second plaza stood five earthen platforms, numbered A1-c through A1-g.

Buried below Platforms A1-d and A1-e, on the south edge of the plaza, archaeologists found two nearly identical mosaics composed of serpentine blocks. Each resembled a giant, anthropomorphized mask of Earth, with additional elements representing the four great World Directions. Buried near the north edge of the same plaza was another massive offering of serpentine blocks.

Buried below Platform A1-f, Drucker, Heizer, and Squier found a group of 16 stone figurines (15 of them jadeite) and six tongue depressor-like jadeite celts, all arranged in a ritual scene. All of the figurines appear to represent males with deliberate cranial deformation, perhaps a sign of rank. The lone sandstone figurine stood with his back against one of the jadeite celts, which may represent the basalt fence posts of the very plaza in which they were found. The scene seems to show four jadeite men filing past the sandstone figure, while the remaining 11 jadeite men watch. This scene may commemorate an actual event, but opinions differ on its meaning. Were the four men, walking single file, lucky initiates or captives destined for sacrifice?

Immediately to the north of the plaza just described rose Mound A2, an earthen pyramid resembling some of the langis at Lapaha. This mound did, in fact, contain a chiefly burial. Tomb A had walls and a roof of basalt columns,

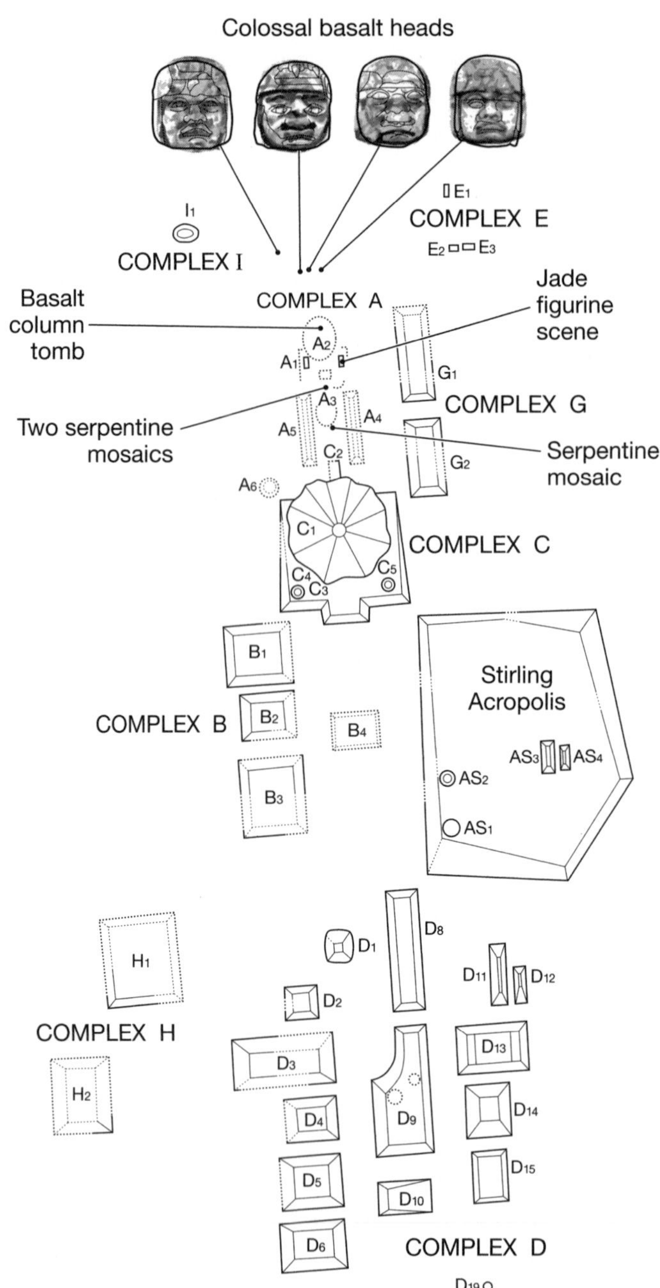

FIGURE 49. The civic-ceremonial center of La Venta, Mexico, grew to exceed 1,600 yards; Complex A alone measured 300 yards.

looking for all the world like a box made from giant Lincoln Logs. The occupant of the tomb had evidently been laid to rest in a carved sandstone sarcophagus. To the disappointment of the archaeologists, at some point Tomb A had been reopened and the occupant's remains taken elsewhere. Still lying on the floor were the remains of two young people; they had been buried previously, later exhumed and bundled up, and then added to the tomb. Found with these bundled remains were four jadeite figurines, one sitting cross-legged and wearing an iron-ore mirror; a jadeite bloodletting tool, carved to represent a stingray spine; several genuine stingray spines; a jadeite pendant in the form of a clam shell; an iron-ore mirror; and countless smaller baubles of jadeite. Left behind in the sarcophagus were additional sumptuary goods, including two jadeite ear ornaments, two jadeite pendants in the form of jaguar teeth, a jadeite awl or bloodletter, and a jadeite figurine. Whoever the original occupant of the tomb was, he must have been as full of sacred life force as a Tui Tonga.

Like the Tongans, the chiefs of La Venta liked to erect stone monuments. Most of the stones they used were foreign to the coastal plain of Tabasco. Notable among these were basalt columns and blocks from the Tuxtla Mountains, 50 miles to the west. Archaeologists believe that the basalt monuments, some of which weighed 20 to 25 tons, had been hauled to the coast and transported by raft to the Tonalá River. The distance involved seems impressive, until one considers the 500-mile ocean voyages needed to bring stones for some of Lapaha's langis.

The most talked about of La Venta's basalt monuments were the four human heads found 50 yards north of Mound A2. Standing up to eight feet tall, these three-dimensional monuments depicted broad-nosed, thick-lipped men in helmets. The chiefs of La Venta have understandably been credited with commissioning these giant heads. We should bear in mind, however, that some of the most impressive stone monuments at Lapaha were gifts to the Tui Tonga from neighboring or subordinate chiefs.

Among the other stones brought to La Venta were flagstones of marly limestone, available 35 miles to the west, and serpentine from the mountains of Puebla and Oaxaca, hundreds of miles farther west. While we have no doubt that La Venta's chiefs inherited sacred authority and claimed semidivine ancestry, a chief's ability to negotiate for massive quantities of exotic stone was undoubtedly a way to achieve still further renown.

González Lauck believes that Complex E of La Venta, a cluster of buildings some 200 yards northeast of Mound A2, was residential in nature. This spacing suggests that chiefly families at La Venta, like those at Lapaha, lived just far enough from the ritual buildings to ensure privacy.

We would like to know whether La Venta maintained separate residences for sacred and secular chiefly lineages, as was done in Tonga. Additional stone monuments in the Tonalá and Coatzacoalcos drainages hint at such a division. Some seem to depict chiefs seated cross-legged, accompanied by symbols of sacred authority; others show chiefs with what appear to be military headgear, body armor, or war clubs; and still others show serene, priestly individuals presenting infants to an unseen audience. While these monuments might represent the multiple roles of one paramount lineage, it is just as likely that they reflect the division of sacred and secular authority seen in the most powerful rank societies.

Like Lapaha, La Venta was an artificially created, sacred landscape. Despite its size, it was no more a "city" than was Lapaha. It was a civic-ceremonial center that grew by accretion as new chiefly lineages were added to the old. It differs from Lapaha in one respect, namely, that the architects of La Venta were committed to maintaining one consistent astronomical alignment for their buildings throughout. This was less important to the Tongans.

Both chiefly societies created linear arrangements of plazas, mounds, stone monuments, and elite burials. Both imported tons of stone; Lapaha wanted it for the facing of its huge langis, La Venta for its massive mosaic pavements and stone fences. La Venta produced more three-dimensional stone monuments, though none were as big as the Trilithon built by the 11th Tui Tonga. Let us close with this lament: what a shame that we lack anthropological and historical data on La Venta society, comparable to Gifford's data on Tonga.

FROM RANK TO STRATA IN HAWAI'I

Hawai'i is one of the largest, yet most remote, of the Polynesian archipelagoes. Archaeologists believe that it may not have been colonized until A.D. 300 or 500. The earliest colonists showed up on O'ahu, Kaua'i, Moloka'i, and the "Big Island" of Hawai'i. There are reasons to suspect that these new arrivals had traveled 2,300 miles from the Marquesas Islands.

According to archaeologist Patrick Kirch, the earliest Hawai'ians came from a society that already had differences in rank. On O'ahu, for example, the colonists buried a nine-year-old girl with likely sumptuary goods. The girl's burial also featured a red stain that appeared to come from dyed tapa cloth, which could be worn only by people of noble birth or their titled attendants.

Hawai'i's earliest occupants had reached a fertile tropical paradise. One could grow yams, sweet potatoes, taro, breadfruit, coconuts, and pigs. To maximize access to resources, the Hawai'ians created resource territories called *ahupua'a,* each of which was shaped like a slice of pizza. The crust, or widest part, began at the ocean; the slice narrowed as one moved inland over the coastal plain, across the piedmont, and into the mountains. Each ahupua'a therefore included all the locally available land types and was set aside for use by a corporate, kin-based landholding group. The manager of each triangular slice was a hereditary chief called an *ali'i* (the Hawai'ian equivalent of Tikopia's ariki and Tahiti's ari'i), who was essentially the highest-ranking man of his lineage.

All five of the largest islands (Hawai'i, Maui, Moloka'i, O'ahu, and Kaua'i) had multiple rank societies. Smaller islands such as Kaho'olawe, Lana'i, and Ni'ihau rarely supported more than one chief and sometimes came under control of a larger island. For its part, the Big Island of Hawai'i was divided into five to seven districts (North and South Kona, North and South Kohala, Hilo, Ka'u, and Puna), each with its own chief. The most powerful were the chiefs of the Kona and Kohala districts.

Between 1100 and 1400, chiefly cycling in Hawai'i became as dynamic as that of the Kachin, Konyak Naga, Samoans, and Tongans. Most volatile were the relationships of junior and senior chiefly lineages. Despite having less sacred power, the heads of junior lineages kept accumulating supporters until they could overthrow senior chiefs by force. Some usurpers even killed or sacrificed their own half brothers. This is exactly the kind of fratricide that the builder of Tonga's Trilithon had hoped to prevent.

Two key transformations took place in the late prehistoric period. First, Hawai'ian chiefs began marrying their own sisters and half sisters, whose levels of mana ensured the high rank of their offspring. They revised Hawai'ian cosmology to indicate that such sibling marriage was legitimate, since the gods from whom the chiefs descended had married their siblings as well.

The second transformation created true social strata. Like Tonga, early Hawai'i had a continuum of rank from high chiefs, to lesser chiefs, to landed gentry, to landless commoners, and so on. What the Hawai'ian chiefs did was eliminate the landed gentry by declaring all garden land the property of major chiefs. The result was two de facto strata, the *ali'i* (hereditary nobility) and the *maka'ainana* (commoners), with a gap between them that had formerly been bridged by the landed gentry. From that point on, Hawai'ian chiefs would use garden land to reward their allies and deny their rivals.

Hawai'i's creation of a landless commoner class reminds us that the worst inequality results not from the granting of new privileges to the people on top but from the removal of existing privileges from the people on the bottom.

From 1450 onward, the details of Hawai'ian society become richer. For one thing, we have the oral histories of the Hawai'ians themselves, compiled by authors such as Samuel Kamakau, Kathleen Mellen, and Herbert Gowen. For another, we have research by archaeologists such as Patrick Kirch. We have, in addition, theoretical frameworks for Hawai'ian society provided by social anthropologists such as Marshall Sahlins, Valerio Valeri, and Irving Goldman. We draw on all those authors in this chapter and the next.

In late prehistoric Hawai'i, stratification was justified on the grounds that commoners had merely descended from human lineage founders, while the genealogies of the ali'i went back to the Sky God and Earth Goddess. Goldman stresses, however, that the ali'i were not considered a monolithic class; their vocabulary reflected almost a dozen internal gradations like those of European nobility. The offspring of sibling marriages between the highest ali'i were called *niaupio* and were considered semidivine. Subordinates prostrated themselves not only before these high chiefs but even before displays of their sumptuary goods. Under penalty of death, no one was allowed to handle a high chief's garments or cast a shadow on his possessions. We consider a curtsy before the Queen of England an act of obeisance, but it pales in comparison to the acts demanded by niaupio.

The offspring of a niaupio was merely a *pio*. The offspring of a niaupio or pio man and a woman from a junior chiefly line was called *wohi*. The offspring of a niaupio or pio woman and a man from a junior chiefly line was called *papa*. From that point on, titles continued to decline in prestige until they reached persons of mixed ali'i/maka'ainana descent (for example, the child of a chief and his commoner concubine). Even though the latter were not noble, they were treated with more respect than the average commoner.

To paraphrase Kirch, Hawai'ian society now consisted of a multilevel pyramid of ali'i of different titles, superimposed on a permanent underclass of commoners that worked the land and paid tribute to its hereditary lords. The tribute paid to paramount chiefs included pigs, dogs, chickens, tapa cloth, pearls, ivory, and the feathers of tropical birds. Payment of tribute was timed to coincide with a harvest ceremony, making it more palatable by placing it in the context of religious ritual.

Below the Hawai'ian paramount chiefs, or *ali'i-ai-moku,* were *ali'i-ai-ahupua'a,* subchiefs who ruled the pizza-slice strips running from the mountains to the sea. Below them were stewards called *konohiki,* minor nobles who directly managed the gardens and fish ponds. These officials gave Hawai'i an administrative hierarchy of at least three levels, considered typical of the most powerful chiefly societies.

Religion and Politics in Hawai'i

The escalation of inequality in protohistoric Hawai'i affected religious roles as well. Early Hawai'ian society, like that of Tikopia, had a transitional mix of men's houses, shrines, and temples. Just as in Tonga, priests had traditionally come from families specializing in ritual expertise. As competition and usurpation increased, however, so did the need for priestly legitimization of a ruler's genealogical credentials. Oral history records that Liloa, a chief of the Big Island in 1420, bribed a cooperative priest to legitimize him with an offer of chiefly land. Liloa is also believed to be the first paramount chief to split his inheritance, granting his firstborn son control of the land while his second son was given religious authority.

By the fifteenth and sixteenth centuries the Hawai'ians were building temples called *heiaus,* which stood on stone masonry platforms. Most early heiaus were dedicated to agriculture, but Kirch describes at least one such structure that was rebuilt as a *luakini,* or war temple, and dedicated to human sacrifice. The priests serving in these temples were drawn from both social classes. Minor priests were still commoners, but the high priests or *kahuna nui* ("big kahunas") were drawn from the ranks of nobles.

Admit it. All your life you have heard the phrase "the big kahuna" and never knew what it meant until now.

War and Politics in Hawai'i

In Hawai'i, as on Tonga, armed conflict was used for both political usurpation and territorial conquest. So great was the role of toa that it created "nobles by command" like those of Colombia's Cauca Valley. Each paramount chief was a titular commander in chief. He referred to himself metaphorically as "the head"; his subchiefs were his "shoulders and chest"; the priests were his "right hand"; his spokesman/minister was his "left hand"; his warriors were his "right foot"; the farmers who worked the gardens of his chiefdom were his "left foot." Renowned warriors received grants of land and were allowed to marry women who outranked them.

Honaunau: Kona's Lapaha

In 1475 the paramount chief of the Kona district on the Big Island created his own civic-ceremonial enclosure, 715 feet long and 404 feet wide. Naturally defended by the ocean on its north and west sides, the Honaunau enclosure was protected on the east and south by a wall 12 feet high and 330 yards long. Like Lahapa on Tonga, Honaunau grew by accretion and included several temples. One of these temples, Hale-o-Keawe, played a role similar to that of a Tongan burial mound: it preserved the remains of past Kona chiefs. A larger temple, called Ale'ale'a, underwent six renovations after its original construction.

One major difference between Honaunau and Lapaha was that Kona's chiefs did not live permanently at the civic-ceremonial center. During the year, Kirch reveals, Hawai'ian chiefs moved from district to district within their chiefly territories, so that the burden of tribute and corvée labor would be evenly distributed among their subjects. This behavior tells us that chiefs were expensive to support and helps explain why they needed to keep adding followers through conquest.

The Nature of Hawai'ian Inequality

By the eighteenth century, as Goldman points out, society on the Big Island of Hawai'i was well prepared for the transition from chiefdom to kingdom. It had converted a continuum of rank into two de facto classes. By declaring all garden land the property of the chiefly lineage, it had eliminated the landed gentry. This allowed the ali'i-ai-moku to use land as a reward for political and military loyalty instead of a genealogical entitlement. Like Bemba chiefs, Hawai'ian paramounts often appointed loyal commoners to important governmental posts, bypassing ali'i who might be potential usurpers. This act created a corps of bureaucrats whose constituents were not their kinsmen. Such officials were less easily pressured or bribed.

Hawai'ian chiefs had begun to transcend the district of their birth, moving periodically so that the burden of supporting the ali'i-ai-moku would be diffused. When a chief learned from local officials that his subjects were grumbling about his tribute and labor demands, it gave him incentives to take land and labor away from other ali'i.

Such ambitious chiefs were seeking only to increase the size of their chiefdom. However, as we will see in the next chapter, a series of particularly aggressive leaders eventually converted Hawai'i to a monarchy.

IV

Inequality in Kingdoms and Empires

SEVENTEEN

How to Create a Kingdom

We come now to a multigenerational process that changed Hawai'i forever. Beginning at least 800 or 700 years ago, certain Big Island chiefs began trying to expand their territories to include other islands. The earliest attempts rarely succeeded, but later chiefs kept trying.

Around A.D. 1270, for example, a Big Island chief named Kalanuihua is said to have conquered Maui and Moloka'i and invaded O'ahu. He overextended himself by attacking Kaua'i, where he was taken prisoner.

To be sure, before expanding to other parts of the archipelago it was necessary for a chief to solidify control of his home island. This was especially tough in the case of the Big Island, which had so many districts. One of the most detailed stories of Hawai'ian unification is that of a man named 'Umi, who was born in the fifteenth century.

'Umi was the second son of Liloa, an *ali'i-ai-moku,* or paramount chief, of the Big Island, already mentioned in our discussion of stratification. Liloa split his inheritance. He left his first son, Hakau, all his garden land, while 'Umi was given religious authority.

According to oral historian Samuel Kamakau, Hakau was Liloa's son by his legitimate noble wife. 'Umi, on the other hand, was a love child. One day, or so the legend of 'Umi begins, Liloa saw a commoner woman named Akahi bathing. Captivated by her beauty, he seduced her. Before their affair ended, Liloa gave Akahi his chiefly loincloth, ivory pendant, and feather cape, telling her to give them to any son he had caused her to bear.

Akahi gave birth to 'Umi but kept his paternity secret for years. Eventually, when he was old enough, she sent him to Liloa bearing his pendant, cape, and loincloth. Liloa recognized the sumptuary goods he had given Akahi and accepted ʻUmi as his son.

As Liloa's two sons grew, ʻUmi became bigger than Hakau and superior at virtually every task. On his deathbed, however, Liloa named Hakau his chiefly heir and made ʻUmi the guardian of the god Kuka'ilimoku. ʻUmi knew that Hakau resented him, so after Liloa's death he sought asylum with a chief of the Hilo district.

Over time ʻUmi earned great respect, while Hakau came to be hated as a despot. Eventually a group of big kahunas and lesser chiefs conspired to replace Hakau with ʻUmi. For pragmatic reasons, they were willing to overlook the fact that ʻUmi's mother was a commoner. With the priests' help, ʻUmi and his supporters smuggled weapons into a ceremony at which Hakau was assassinated.

Once the nobles of Kona, Kohala, Hilo, Ka'u, and Puna heard of Hakau's death, they declared their districts autonomous. If ʻUmi wanted to control the whole island, he would now have to conquer it on his own.

ʻUmi's strategy was as follows. First he laid the corpses of Hakau and his bodyguards on the altar of a temple, alleging that their death was a sacrifice. He then married his half sister Kapukini, who was of noble birth. ʻUmi also married the daughter of the chief of Hilo, the district that had once given him refuge. After the chief of Hilo had an argument with him, however, ʻUmi used their dispute as an excuse to overthrow his father-in-law. He next subdued the district of Puna; then came a long campaign against Ka'u. ʻUmi took over Kona and Kohala more easily. Sometime around the mid-fifteenth century, ʻUmi was in control of the whole Big Island.

To be sure, the legend of ʻUmi (like most oral histories) is romantic and idealized. Despite these shortcomings, it shows us many premises that the Hawai'ians themselves considered important. For example:

1. Chiefly half brothers were destined to become rivals.
2. If the senior heir was a despot, his overthrow by a junior heir was justified.
3. Even a junior son with a commoner mother, assuming that he was popular, could usurp the senior son's position.
4. However, the usurper should then marry the most highly ranked woman available and demonstrate high levels of achievement.
5. Conquering a neighboring chief's territory, and then incorporating it into one's own chiefdom, was considered an achievement.

6. The greater the territory he conquered and incorporated, the greater the renown of a chief.
7. Both usurpation and conquest required the support of lesser chiefs, priests, warriors, and loyal commoners. It was acceptable to reward their loyalty with land grants, even when they did not have the genealogical credentials to deserve garden land.

Might these premises have relevance beyond Hawai'i? Yes indeed; and in the pages that follow, we will see that many of the first kingdoms created in other parts of the world were also the handiwork of usurpers.

Neither Kalanuihua nor 'Umi succeeded in unifying all of the archipelago. Hawai'ian leaders, however, did not stop trying. During the late eighteenth century, one usurper finally gained a strategic advantage that allowed him to do what earlier chiefs had been trying to do for 500 years. His story is just as romantic and idealized as 'Umi's, but with one important difference: much of it is independently confirmed by Euro-American eyewitnesses.

THE SUCCESSFUL UNIFICATION OF HAWAI'I

Early in the eighteenth century, according to various oral histories, a noble named Alapai rose to be chief of the Big Island's Kohala district. Like so many before him, he set out to unify all districts of the island; by the mid-1700s, he had done so.

Alapai's military success did not, however, make him the most respected chief in the entire archipelago. That distinction belonged to Kahekili, the paramount chief of Maui, whose genealogical credentials were superior. Like the sacred high chiefs of Tonga, Kahekili possessed such mana that his subjects prostrated themselves in the presence of his spear, his feather cloak, and even the spittoon containing his noble saliva.

Alapai had a nephew named Keoua who lived in the Kona district of the Big Island. Keoua fell in love with Alapai's beautiful niece Keku'iapoiwa; with Alapai's blessing, he married her.

Now the soap opera began. Kahekili of Maui (he of the mana-filled spittoon) heard of Keku'iapoiwa's beauty. He invited her to visit his court. Kahekili was married, but his main wife, Namahana, was frequently away visiting relatives. During a brief window of opportunity, or so the legend goes, Kahekili got Keku'iapoiwa in a family way.

Keku'iapoiwa returned to Kona as if nothing had happened. Her pregnancy was officially attributed to her husband, Keoua. Rumors of her affair, however, had reached Alapai.

All pregnant women are alleged to have unusual food cravings. Keku'iapoiwa's cravings were beyond unusual. During her sixth month of pregnancy, she allegedly asked to be fed the eyeball of a chief; she had to settle for the eyeball of a man-eating shark. Upon learning of this episode, Alapai asked the kahunas to interpret Keku'iapoiwa's craving. He was alarmed when they concluded that her child was destined to become "a slayer of chiefs."

Just as 'Umi had once been hidden from his half brother Hakau, Keku'iapoiwa and her baby would now have to be hidden from Alapai. Sometime around 1758, Keku'iapoiwa gave birth to a son. He was given to a foster mother and hidden for years in the sacred, landlocked Waipio Valley of North Kohala.

Finally, as he grew old and feeble, Alapai's fear of the child diminished. He allowed Keku'iapoiwa's son—officially his grandnephew—to be brought to his paramount village and given a chiefly title. Alapai named him Kamehameha, "The Lonely One."

Then Keku'iapoiwa's husband fell ill. He asked his brother Kalaniopu'u, the chief of the Ka'u district, to raise Kamehameha as if he were his own. Kalaniopu'u assigned his greatest warrior the task of training the boy, and the soap opera continued. Kamehameha (1) mastered every chiefly skill; (2) had an affair with one of Kalaniopu'u's younger wives; and (3) met a young girl named Ka'ahumanu, who would one day become his favorite wife.

When Alapai, paramount ruler of the Big Island, died, he was succeeded by his son Keaweaopala, who turned out to be an unstable despot. The subchiefs of his own Kohala district decided to overthrow him, and the rebellion spread to Ka'u. The leader of the Ka'u insurgence was Kalaniopu'u, and he went to battle with his protégé Kamehameha at his side. Together they pursued the unpopular Keaweaopala to the coast of Kona and then killed him. This left Kalaniopu'u in charge of both Kona and Ka'u.

Word of this conflict reached the sacred chief Kahekili on Maui. Kahekili was convinced that he was Kamehameha's biological father. He therefore sent his twin half brothers to Kona to make sure that no harm befell Kamehameha.

Kalaniopu'u was already making plans to invade Maui and claim Hana, its beautiful eastern district, for himself. It took him a year to build enough canoes and assemble the warriors needed. During that year, 1778, Captain

James Cook arrived in Hawai'i. From that point on, we have British and American texts to complement Hawai'ian oral history.

Kalaniopu'u invaded Maui and established a foothold on Hana. By 1781, however, he was nearing death, and so decided to split his inheritance three ways. Kiwalao, his son by his most highly ranked wife, would inherit his title. Keoua-of-the-Flaming-Cloak, his son by a less highly ranked wife, would inherit land. Kamehameha, his protégé/nephew, was named custodian of the war god Kukailimoku and also of the sacred Waipio Valley, where he had hidden as a child.

Keoua-of-the-Flaming-Cloak envied Kamehameha's share of the inheritance and thus attacked him. When Kamehameha counterattacked, general war broke out. Kiwalao was killed, and Keoua-of-the-Flaming-Cloak fled. By 1783 Kamehameha was the most powerful noble on the Big Island, and supporters began to flock to him.

Seeking advice from the big kahunas, Kamehameha was told to build a new temple and lay on its altar the body of a chief. He commissioned a big temple and invited Keoua-of-the-Flaming-Cloak to hold peace talks there. No sooner had Keoua stepped from his canoe than he was killed by one of Kamehameha's warriors. His corpse, along with those of his noble followers, was then laid on the altar of the newly dedicated temple.

Next for Kamehameha came marriage. For this he turned to Ka'ahumanu, whom he had met when she was a little girl at Kalaniopu'u's village years ago. Of course, Ka'ahumanu was all grown up now. According to oral historian Kathleen Mellen, Kamehameha's bride was six feet tall and weighed close to 300 pounds.

Kalaniopu'u died in 1782 with his dream of unifying the islands unfulfilled; he had lost even his beachhead on Maui. Now it was up to Kamehameha, and he knew that he would need a military advantage that no previous Hawai'ian leader had enjoyed. That advantage came in the form of muskets and cannons from the Western ships now docking at Hawai'i's ports. Some of these ships had originally been British but became American property as a result of the Revolutionary War.

British sailor John Young had been boatswain of the good ship *Eleanor,* but he stayed ashore when the ship sailed. Angry because the crew of the *Eleanor* had killed some unarmed islanders, the Hawai'ians retaliated by confiscating a second ship, the *Fair American.* They stripped that ship of its guns and ammunition and took prisoner one of its mates, Isaac Davis. Kamehameha realized that Young and Davis knew Euro-American weaponry and military tactics. He not only spared their lives but made them two of his most trusted advisers. They fought by his side during his campaign to unify Hawai'i, and he later awarded them governmental positions, wives, and grants of land.

In 1790 Kamehameha invaded Maui. Paramount chief Kahekili had by then retired to O'ahu, leaving Maui to his son. After a taste of Kamehameha's artillery, Kahekili's son decided to join his father in O'ahu, leaving Maui to the invaders.

Kamehameha encountered even less resistance on Moloka'i. Then, following the death of Kahekili in 1794, he invaded O'ahu with 1,000 war canoes, 12,000 warriors, 16 foreign advisers, and abundant logistic support. He defeated Kahekili's son and offered the latter's heart on the altar of a temple.

Kamehameha now was master of all but Kaua'i and the smaller islands. While Ka'ahumanu would always remain his favorite wife, he knew that to establish a dynasty he would have to marry the most noble woman available. That turned out to be Keopuolani, the eight-year-old granddaughter of a Maui noblewoman. Keopuolani's grandmother betrothed her to Kamehameha in return for economic support and political protection.

Kamehameha then began building war canoes and assembling provisions for an assault on Kaua'i. He chose for this task the Anahulu Valley on the northwest coast of O'ahu, an ideal jumping-off place. A collaborative study by social anthropologist Marshall Sahlins and archaeologist Patrick Kirch reveals that the entire valley was artificially terraced from top to bottom, providing intensive agricultural support for Kamehameha's labor force. Realizing that resistance was futile, the chief of Kaua'i capitulated in 1810. This left Kamehameha in command of the entire archipelago.

Kamehameha's Kingdom

Kamehameha now found himself in charge of a territory that was too large to be administered like a chiefdom. Paramount chiefs on the Big Island, as we have seen, had traditionally moved from district to district during the year. These moves distributed the burden of support among all the chief's subjects while making him intimately familiar with each district.

Frequent rebellions against past chiefs, however, show us that it was not always easy to control the 4,028-square-mile Big Island. Now Kamehameha needed to control an archipelago 1,500 miles long, with 6,423 square miles of dry land and big stretches of ocean. He would have to appoint a governor for each island, someone loyal to him rather than to the natives of the island. He sent his favorite wife's father, a native of Ka'u on the Big Island, to be governor of Maui. And when Kamehameha went to war, he often left John Young in charge of his home district.

It is significant that the archipelago now had a political hierarchy of four administrative levels. In olden days each island had been ruled by a paramount chief (Level 1), below whom there were subchiefs (Level 2), who in turn supervised minor nobles (Level 3). Now Kamehameha was all alone in Level 1; the governors of each island occupied Level 2; the subchiefs occupied Level 3; and minor nobles occupied Level 4.

Ali'i-ai-moku was no longer an adequate title for Kamehameha. If his unification of Hawai'i had occurred in the absence of Euro-American visitors, he might have created a new Hawai'ian term for his office. Owing to his extensive contact with English speakers, however, he decided to call himself King Kamehameha I.

We have described the unification of Hawai'i at length for a reason: even though Polynesian in its details, it is an example of a widespread process by which monarchies were created from smaller-scale societies. We use the term "process" because it often took a succession of leaders to complete the transition. Rarely were the efforts of one ruler sufficient. In the case of Hawai'i important roles were played by 'Umi, Alapai, Kalaniopu'u, and Kamehameha. Opinions differ on the exact moment in this sequence when a kingdom existed. As we shall see later, similar disagreements apply to the rise of monarchy on the coast of Peru.

We also will see, in this and later chapters, that among the Zulu of southern Africa, the Asante of west Africa, the Merina of Madagascar, and the Hunza of the Pakistani-Kashmir borderlands, indigenous kingdoms arose in the context of elite rivalry. For a substantial period of time—centuries, in some cases—a series of rival rank societies competed with one other. Despite moments of political unification, the long-term outcome was a stalemate. Eventually the aggressive leader of one rank society (often a highly motivated usurper) gained an unforeseen advantage over his neighbors. He pressed his advantage relentlessly until he had subdued all his rivals. He turned their chiefdoms into the provinces of a society larger than any previously seen in the region. To consolidate power, he broke down the old loyalties of each province and replaced them with an ideology stressing loyalty to him. He rewarded priests who were willing to verify his genealogical credentials and revise his group's cosmology, ensuring his divine right to rule.

While his rise to power may have been brutal, the new king then cultivated an image of beneficence. Kamehameha, for example, decided that his rule would emphasize peace and prosperity. He encouraged his people to intensify agriculture and tried to serve as a role model by working publicly in his own gardens.

Kamehameha died in 1819, leaving his kingdom to Liholiho, the son of his most highly ranked wife, Keopuolani of Maui. Kamehameha's remains were hidden in an undisclosed location, reportedly a cave visible only from the sea. From such a burial site, his mana would continue to draw schools of fish to the coast.

Because of Liholiho's youth, his mother was made regent to ensure an orderly succession. One of the most dramatic changes in Hawai'ian social logic is attributed to Liholiho but may well have been his mother's idea.

In many hierarchical societies the ruler symbolized order in a world plagued by disorder. Hawai'i was no exception. The inauguration of each new ruler was preceded by a period of deliberate chaos, during which his subjects violated all ritual taboos. It was, among other things, taboo for men and women to dine together. After sufficient time had elapsed, the new ruler would appear and restore order, reinstating all taboos.

At the appropriate moment, Prince Liholiho appeared. At a feast sponsored by his mother, however, he defied the taboo and dined with the women. By this and other acts, according to scholars such as Ralph Kuykendall and William Davenport, Liholiho separated rank from religious protocol and made Hawai'i a more secular kingdom.

THE UNIFICATION OF THE ZULU

In our discussion of the Bemba we mentioned Africa's Bantu migration, the dispersal of ironworking farmers and herders from their homeland north of the Congo. Some 1,700 years ago these people had spread south to the Limpopo River, on the border between Zimbabwe and South Africa. By A.D. 800 they had crossed the Limpopo and entered the acacia grasslands of the south. There they found an environment suitable for cattle herding, one where the tsetse fly was less of a problem.

We believe that these migrating Iron Age societies already had a degree of hereditary inequality. Among the Bantu speakers crossing the Limpopo were the ancestors of the Zulu people. They spread into the grassland of eastern

South Africa, a province called Natal by Portuguese explorer Vasco da Gama. Because the early Zulu were a clan-based society with chiefs and warriors, they had little difficulty displacing the hunters and gatherers whose ancestors had occupied Natal for millennia.

By the late Iron Age (A.D. 800–1200), according to archaeologist Tim Maggs, the societies of Natal had become greater in scale and complexity, with many settlements moving to defensible hilltop localities. There is evidence, in other words, for ongoing competition among chiefly societies, not unlike the competition we saw in protohistoric Hawai'i. By the late eighteenth century, a period for which European colonists left us written texts, there may have been as many as 50 different rank societies in Natal. Perhaps the most powerful, according to historians John Wright and Carolyn Hamilton, were the chiefly societies known as the Mabhudu, the Ndwandwe, and Mthethwa. Their neighbors, the Zulu, were a smaller and less imposing rank society.

The average Zulu settlement was a farmstead built by a senior man with multiple wives. Each wife and her children occupied a beehive-shaped hut; all huts were set in a protective circle around the corral where the cattle spent the night. The entire farmstead was further enclosed by a stockade. While the men herded cattle, the women raised millet, sorghum, and melons. We know these farmsteads not by their Zulu name but by the Afrikaans word *kraal*.

The rise of the Zulu has been reconstructed from oral histories, European eyewitness accounts, and anthropological research. According to anthropologist Max Gluckman, the families of many kraals were united into clans that reckoned descent in the father's line. Multiple clans were united under a chief, who was the hereditary leader of the most senior descent group. Chiefs ruled by dividing their territory into sections. Each section was commanded by one of the chief's brothers or half brothers, who served as a subchief.

Quarreling among sections was common and usually ended with one brother declaring his independence and moving his subjects to a new location. The alternatives were fratricide and usurpation. To guard against the assassination of an heir, polygamous chiefs instructed their wives to live in different sections of their territory, surrounded by loyal followers. With the death of the chief, each wife lobbied for her son to become his successor.

Cattle were the main source of wealth in protohistoric Natal, and no chief, lineage, or clan ever had enough. Chiefs distributed cattle to subchiefs and other officials and used them to reward outstanding warriors. Such was the demand that stealing cattle became the main reason for raiding one's neighbors.

The intensification of warfare gradually modified social behavior in Natal. According to Wright and Hamilton, a chief periodically rounded up all young men of appropriate age who lived in his territory. These youths were organized into a group called an *ibutho* (plural *amabutho*), and all went through initiation together. Although the young men came from different kraals, each ibutho was given its own name and insignia, creating corporate solidarity for a lifetime. Chiefs gradually came to rely on the amabutho as regiments of warriors, enforcers, and tribute collectors. In times of peace the amabutho could be sent out to hunt elephants, increasing a chief's supply of ivory. What had begun as a ritual association became a means to expand a chief's wealth and territory.

In 1787 Senzangakhona, chief of the Zulu, had an illegitimate son by a woman named Nandi. The boy was given the sarcastic name "Shaka," a reference to an intestinal parasite that simulates pregnancy by causing a woman to miss her menstrual period. Senzangakhona made Nandi his third wife, but she was mistreated by the Zulu and banished to her home village when Shaka was six. The mistreatment continued until Nandi sought refuge with the neighboring Mthethwa in 1802.

The chief of the Mthethwa was Dingiswayo, a man with his own violent past. The son of a previous chief, Dingiswayo had once fled the Mthethwa under charges of plotting to kill his father. When he finally returned, he found his father dead and his brother installed as chief. Dingiswayo killed his brother, made himself chief, and set about expanding Mthethwa territory. His rationalization for subduing his neighbors was that their constant fighting was "against the will of the Creator," and he intended to "make them live in peace."

Shaka, by now a strapping, athletic teenager, became one of Dingiswayo's most trusted warriors. His rise to power has been described not only by Gluckman but also by scholars such as E. A. Ritter and John Selby.

Dingiswayo defeated the Zulu and many of his other neighbors. As his victories mounted, his doting mother began to keep the skulls of his decapitated rivals in her hut. True to his goal of establishing peace, however, Dingiswayo did not always press his advantage. In 1813, for example, Dingiswayo defeated chief Zwide of the Ndwandwe, even though his Mthethwa forces were outnumbered 2,500 to 1,800. Dingiswayo declined to execute Zwide, a decision that would cost him dearly.

Dingiswayo's sparing of Zwide was probably influenced by the traditional chivalry of warfare in Natal. The list that follows gives some of the principles of warfare at that time.

1. Warriors in Natal wore sandals, advanced until they stood at a reasonable distance from the enemy, and then proceeded to hurl iron-tipped spears.
2. Military formations were simple, consisting mainly of lines of warriors from a series of age-based ritual societies.
3. An enemy who threw down his spear to concede defeat was spared.
4. Women and children came out to watch the battles and were left unharmed no matter which side won.

One of the reasons Dingiswayo was able to defeat the Ndwande was because the men under Shaka's command fought so fiercely. Dingiswayo was appreciative; as a result, when Senzangakhona died in 1816, Dingiswayo made Shaka the new chief of the Zulu. Now the 29-year-old Shaka controlled 100 square miles and an army of 500 men.

One of Shaka's first acts was to punish the Zulu who had mistreated his mother when he was a boy. The lucky ones had their skulls bashed in. The unlucky ones were taken to a hill frequented by hyenas. There they were impaled on stakes and left to die slowly, while the hyenas closed in for lunch.

As Shaka's military experience grew, so did his dissatisfaction with traditional warfare. Soon he was at work on the following strategies that would give the Mthethwa and Zulu an edge:

1. Tired of throwing and retrieving javelins, Shaka had his blacksmiths create a new short-hafted, broad-bladed stabbing spear. He called his new weapon *ixwa,* after the sucking sound it made when pulled from an enemy's chest.
2. He had his warriors shed their sandals so that they could run faster. In the future they would surprise their enemies by sprinting toward them to fight at close range.
3. He trained his men for close combat with the ixwa and a tough cowhide shield.
4. He created a new battle formation, composed of four groups of warriors standing shield to shield. In the center was a block of seasoned veterans known as "the head," who did the bulk of the fighting. Behind them was a block of reserves called "the chest," who waited for a signal to join the fray. Extending out from "the head" were two curving columns called "the horns," designed to encircle the enemy.
5. He pioneered military formations that minimized the effect of being outnumbered. One was an unbroken circle, used when his warriors

were surrounded. The other, called "the millipede," was a linear formation of men with interlocking shields, used to cross territories where one was likely to be ambushed.

6. Shaka also modified the amabutho. For years they had been age-based ritual societies, made up of youths from different kraals within the same chiefdom. Shaka, who by now had subjugated a number of formerly autonomous societies, expanded the system by creating regiments into which all warriors of the same age were placed, regardless of the chiefly territories from which they had come. Under Shaka's direct control, these feared "age regiments" broke down old territorial loyalties and produced warriors beholden only to him. Shaka trained them puritanically, forbidding the youths to marry without his permission.
7. Traditionally, an enemy who threw down his spear had been spared; this is probably why Zwide had not been killed by Dingiswayo. Shaka's new policy, called *impi ebomvu,* or "red war," abolished such chivalry. His regiments finished off the enemy, wounded, pursued, and killed all the retreating warriors they could catch, and slaughtered the enemy women and children who had come to watch.

Fighting alone or beside Dingiswayo, Shaka used these innovations to defeat all comers. In 1816 he crushed the E-Langeni; in 1817 he conquered the Butelezi. By now his sphere of influence was 400 square miles.

In 1817 Dingiswayo assembled an estimated 4,500 warriors for an assault on the Ngwane; he was joined by Shaka and 1,000 of his men. Dingiswayo was victorious this time, but in 1818 he was captured and killed by his old nemesis, Chief Zwide of the Ndwandwe, whose life he had spared in the past.

Shaka replaced Dingiswayo as leader of the combined Mthethwa-Zulu forces. Zwide, emboldened by his defeat of Dingiswayo, now made plans to eliminate Shaka as well. An estimated 8,000 Ndwandwe warriors forded the White Umfolozi River. Before them stood Gqokli Hill, which Shaka and his 4,000 warriors had turned into a fortress.

Outnumbered two to one, Shaka won this crucial battle by superior military strategy. In a completely independent move, according to Selby, Shaka hit upon "the famous British square tactics used by Wellington at Waterloo." Drawing his veteran forces into a tight circle, Shaka instructed his reserves to hide in the brush. He then sent off a decoy force of 700 men who pretended to flee with a cattle herd, kept at the fortress for food. The Ndwandwe were taken in by the ruse, sending a third of their force to chase the herders. The remaining two-thirds charged the hill, unaware that there were twice as many defenders there as they could see.

The Ndwandwe threw their javelins; Shaka's troops avoided most of them and charged and killed 1,000 enemies at close range. Shaka's highly disciplined warriors held firm through repeated attacks, and at the appropriate moment his reserves came out of hiding and joined the fray. The Ndwandwe were defeated.

Through this and other victories, Shaka turned 30 chiefly societies into the provinces of a unified kingdom covering 7,000 square miles. In only 12 years (a period so short as to be virtually invisible to an archaeologist) he had gone from the illegitimate son of a minor chief to the king of the Zulu.

Shaka established his capital in 1820 at a place called New Bulawayo. There he built a royal kraal a mile in diameter, defended by a stockade and containing some 1,500 residences. He stationed a portion of his 50,000 warriors at New Bulawayo, which had 130 acres set aside for cattle.

Shaka's revisionist ideology portrayed the Zulu as ruling by right of genealogical seniority rather than conquest. To his inner circle of advisers, however, he confided his belief that "you can only rule the Zulus by killing them . . . only the fear of death will hold them together."

Shaka, whose genealogical credentials were shaky, worried constantly about being usurped. He often said that a king "should not eat with his brothers, lest they poison him." Shaka kept many concubines but was so afraid of being overthrown by a son that he executed any lover who became pregnant. He would leave no heir.

In 1827 Shaka's beloved mother, Nandi, with whom he had endured so much abuse in childhood, passed away. In order to mourn her properly, Shaka ruled that for one year no crops would be planted, no cows would be milked, and no married couples would engage in sex. He executed 7,000 of his own subjects who did not appear to be grieving sufficiently.

This "year of hell" caused such grumbling in Natal that two of Shaka's half brothers, Dingane and Mhlangane, were persuaded to assassinate him. At a meeting with them in 1828, an unsuspecting Shaka was fatally run through with a spear. Dingane was then named king of the Zulu.

By then, of course, European colonization of Natal had become an irreversible process. In 1838 the Boers drove the Zulu north of the Tugela River. In 1880, after a very bloody war, the British conquered the Zulu; thirty years later, in 1910, Zululand was ceded to the Union of South Africa.

In 1994 the province of KwaZulu-Natal was created as a homeland for the Zulu. Its parliament is based in Pietermaritzburg, and the Zulu king receives a government stipend. Each year the king is allowed to take an additional wife, but he usually declines. Instead, the current Zulu king uses the annual ceremony to promote abstinence and the prevention of HIV/AIDS.

Shaka's Kingdom

From time to time we hear one of our colleagues say, "Wasn't so-and-so the worst president in the history of this country?" After listening politely, we add, "but at least he didn't execute 7,000 of his own citizens because they refused to mourn the death of his mother."

Let us look briefly at some differences between Shaka's kingdom and Kamehameha's. While some of Kamehameha's predecessors may have created political hierarchies of four administrative levels, the society in which Shaka was raised had only two levels above the headmen of each kraal. Leadership in Natal relied heavily on military force; not only did it lack the sacred aspects of Polynesian leadership, it also lacked a continuum of rank based on the differential possession of sacred life force.

The citizens of the Zulu kingdom did, however, display four descending levels of prestige, based largely on the length of time each group had been loyal to Shaka. On the most prestigious level were the king, the Zulu ruling lineage, and the ruling lineages of allied groups who (like the Mthethwa) had embraced Shaka from the beginning. The second level consisted of the more important chiefs and notables of the societies subjugated during the middle stages of Shaka's career. As a result of aligning themselves with the Zulu king, they were left in charge of their old territories.

The third level of Zulu citizenry consisted of the lower-ranking members of Shaka's most favored subjugated societies. Although these people had once belonged to different ethnic groups, they were now encouraged to think of themselves as sharing a common origin. Those who distinguished themselves would be appointed to bureaucratic posts.

The fourth, or lowest, level was composed of people who had either sought refuge with the Zulu or were subjugated late in Shaka's career. According to Wright and Hamilton, such people were often referred to as "destitute," "menials," "people with strange hairstyles," or the like. They never became full

Zulu citizens and were regarded as ethnically inferior even if they had once been led by hereditary chiefs.

THE UNIFICATION OF THE HUNZA

Let us now turn to the roof of the world. The Hunza River, an upper tributary of the Indus, rises in the Karakoram Mountains of the Pakistani-controlled region called the Northern Areas. From some of the snowcapped 25,000-foot peaks, one might be able to see Afghanistan and the Hindu Kush in the distance. To the south lies Kashmir and to the east lies China's Xinjiang province.

Three hundred years ago, according to anthropologist Homayun Sidky, the territory of the Hunza consisted of three fortified villages. Each of these villages—Baltit, Altit, and Ganesh—had its own small irrigation system and grew barley, wheat, buckwheat, millet, apricots, and vegetables. Their fields lay at 8,000 feet above sea level, and the Hunza pastured goats, sheep, and cattle at still higher elevations. They used manure from both humans and animals as fertilizer on their crops and grew alfalfa for their herds.

Prior to A.D. 1790, Hunza society had male lineage heads, clan elders, and a village headman called a *trangfa.* All three villages were nominally under a chief called a *thum,* though he shared power with the elders and headmen. The thum, like the chiefs in Tikopia society, derived his authority from sacred life force. He was alleged to have a special relationship with the *pari,* or supernatural mountain spirits. This relationship gave a thum the power to melt glaciers and bring rain, both essential to agriculture.

The thum's supernatural powers were validated by religious practitioners called *bitan.* The bitan were not formal priests and maintained no temples. They served more as oracles, soothsayers, and earthly spokesmen for the mountain spirits. Travelers to Hunza territory report that the bitan entered ecstatic trances, drinking goats' blood and inhaling the smoke of burning juniper.

While the thum tended to come from one elite lineage, there were no firm rules of succession. When a chief died, his sons and brothers often began a struggle that ended only when one of them had murdered his rivals or forced them into exile. Even the winners in this power struggle could not relax. A thum who proved unable to bring rain or melt glaciers might be assassinated or overthrown.

Periodic attempts to consolidate power among the Hunza began in the 1500s and took the form of eliminating rival factions. During the late 1600s, a Hunza chief named Mayori massacred the Diramheray faction with the aid of the Hamachating and Osenkutz factions; the three allied factions divided up the victims' land and animals. Mayori's son, Ayasho II, then massacred his father's former allies the Hamachating with the aid of the Osenkutz; again, the victors divided up the victims' land and animals. The cycle of bloodshed was extended when Ayasho II went on to massacre his former allies the Osenkutz, seizing all their land and livestock for himself.

At this point the Hunza had been unified under one faction but were still no more than a rank society. Whatever plans the thum may have had for further aggrandizement were put on hold in the mid-eighteenth century. In 1759 the Chinese emperor Kien-lung brought the Karakoram Range and neighboring Turkestan under his control. From 1760 onward the Hunza were forced to send gold to China as tribute. In return the Chinese gave Hunza leaders tea, silk, and horses.

Around 1790 a man named Silim Khan usurped the position of thum from his brother Ghuti Mirza. Unfortunately for Ghuti Mirza, his term as chief had been plagued by drought, and his subjects had ceased to believe that he could control the mountain spirits. According to oral history, his brother Silim Khan then caused snow to fall "to the depth of an arrow shaft" in mid-summer. This feat was enough to swing public support to Silim Khan.

Silim Khan took on the title of *mir,* the Pamir or Persian equivalent of thum. With most rival factions already eliminated by Mayori and Ayasho II, he moved quickly to bring Baltit, Altit, and Ganesh under his control. Applying a "top-down" strategy, he named a series of loyal subordinates to be the headmen of each village. He then installed a *wazir,* or vizier, to oversee for him all three villages.

Mir Silim Khan fully intended to expand against his neighbors. He knew that to do so he would need a series of advantages not shared by his predecessors. He built hilltop forts, watchtowers, and fortified granaries. He then set out to create the greatest system of irrigation canals ever seen in the Karakoram Mountains. Some of these canals would bring water from glaciers on the region's high mountains, delivering it to previously uncultivated stretches of the Hunza River valley.

The first canal took seven years to complete. Silim Khan demanded that each household commit one male member to his labor force, using apricot-wood shovels and crude picks made from the horns of mountain goats. The mir worked his diggers from dawn to dusk and required highly ranked families to provide them with food. When finished, the canal brought water from a stream above Baltit and irrigated a former wasteland down valley from Ganesh. There Silim Khan founded a new village, called Haidarabad.

The second and longest canal, called the Samarqand waterway, brought water from Ultar glacier, high above the river. This canal was designed to irrigate a wasteland even farther downstream. Here Silim Khan founded the new village of Aliabad. The mir allowed the people of Ganesh to divert some of the Samarqand water to arid lands closer to their homes. Soon a new village, called Hasanabad, had split off from Ganesh.

Finally, Mir Silim Khan directed work on a third canal, bringing water from another mountain glacier. This canal was used to irrigate a wasteland upstream, near Altit. The increased water allowed the people of Altit to found a new village, called Ahmadabad.

These canals ushered in a period of unprecedented prosperity, but they also altered traditional Hunza society. Because the irrigation system was the creation of Mir Silim Khan, it became the property of an embryonic Hunza monarchy. None of the new villages, established on formerly useless land, had a previous history. They were occupied by new followers of Silim Khan, ethnically and genealogically heterogeneous, united by place of residence rather than common ancestry. These new people were the clients of an emerging Hunza kingdom, and their loyalties were only to the mir.

Aware of his growing political importance, Silim Khan turned his back on the soothsayers who had formerly validated his sacred vital force. In a move reminiscent of the Kachin chiefs who adopted Buddhism, the mir converted to Islam. From now on, water would be provided by hydraulic expertise instead of supernatural power. Silim Khan's monarchy would promote Islam, and the ecstatic trances of the soothsayers would be reduced to folk religion.

The irrigation system brought about enormous immigration to Silim Khan's realm. Soon he expanded east to the neighboring Shimshal Valley and north to the Pamir Range. Most of all, he longed to expand southeast into the territory of the Nagar. Years before the building of his canal system, the mir had been "dissed" by a Nagar chief who asserted that his virile member was larger than Silim Khan's entire chiefdom.

By expanding into the Pamirs, Silim Khan compelled the Kirghiz nomads to pay tribute to him rather than to China. Expanding still farther to the north, he attacked the small kingdom of Sarikol and turned many of its villagers into slaves.

Now Silim Khan had a vantage point from which to raid caravans along the Silk Route. Instead of paying tribute to China as his predecessors had, he was soon receiving protection money from the Chinese. The luxury goods stolen from caravans further enriched Hunza society. The loot produced by raiding created another route to prominence, analogous to that of Colombia's "nobles by wealth."

In 1824 Silim Khan was succeeded by Mir Ghazanfar Khan. The latter enlarged the Hunza canal system and extended the kingdom's territorial control, finally subduing the Nagar society so hated by his predecessor.

The Hunza Kingdom

The entire process of creating a Hunza monarchy may have taken 150 years, beginning with Mayori's slaughter of rival factions and climaxing with the territorial expansion of Ghazanfar Khan (1824–1865). This is further evidence that the transition from chief to king is a process rather than an event, with a long succession of rulers contributing to the final outcome.

Like the Hawai'ian and Zulu cases seen earlier, the unification of the Hunza required one aggressive lineage to achieve an advantage over its rivals. The ultimate advantage in this case was an irrigation system, turning barren tracts into fields controlled by the mir. Three significant consequences were (1) a reduction in the authority of clan elders and lineage heads, (2) the replacement of supernatural legitimacy by true political power, and (3) the triumph of centralized control over ethnic loyalties. The result was a monarchy with a royal lineage; wealthy nobles; a bureaucracy including a vizier, heads of districts, heads of villages, tax collectors, and multilingual diplomats; and a commoner workforce consisting of peasant farmers, herders, and slaves.

THE UNIFICATION OF MADAGASCAR

Madagascar, the world's fourth largest island, lies in the Indian Ocean some 400 miles east of Africa. Over the years it has become a laboratory for studying the creation of kingdoms out of rank societies, one where collaboration between social anthropologists and archaeologists has been exemplary.

Like Hawai'i, Madagascar has a rich oral history, much of which has been compiled in a manuscript called *Tantàran 'ny Andrìana*, "The History of the

Kings." According to historian Mervyn Brown, this manuscript traces Merina rulers back to the fourteenth century, "where tradition becomes legend, with the first king said to be the son of God."

During the 1960s social anthropologist Conrad Kottak began working in Madagascar and immediately saw the potential for collaborating there with archaeologist Henry Wright. Wright teamed up with archaeologists Robert Dewar, Susan Kus, and Zoe Crossland, as well as Malagasy scholar Jean-Aimé Rakotoarisoa. We draw on the results of their work in the pages that follow.

The rise of the Merina began during a period of chiefly cycling like that of other rank societies we have described. Between the fourteenth and sixteenth centuries, European visitors reported finding powerful rank societies in the highlands of Madagascar. Their existence has been verified by Dewar and Wright's archaeological surveys, which reveal that chiefly centers were each surrounded by five to ten subordinate villages.

Oral histories claim that conflicts broke out between two ethnic groups called the Vazimba and the Hova. These conflicts escalated once the Hova acquired iron axes and crossbows, giving their warriors what Brown calls "a decisive superiority" over the neighboring Vazimba.

Sometime around the end of the sixteenth century, an ambitious Hova chief named Ralambo attempted to unify the Malagasy highlands. Western historians attribute his victories over the Vazimba to the acquisition of European firearms, an advantage similar to Kamehameha's in Hawai'i. In the logic of Ralambo's society, however, his success was credited to his possession of a *sampy,* a powerful talisman that could, like a Mandan sacred bundle, increase a person's life force. Here we see that, as among the Hunza and Polynesians, Hova leaders were seen as possessing or acquiring superior amounts of sacred power.

Ralambo was the first chief to refer to his territory as Imerina, and his successor was the first to move his paramount village to Tananarive, the current capital of Madagascar. At this point it was said that there were four ranks of Imerina nobles, based on their genealogical relationship to the paramount chief.

During the seventeenth century a young noble named Andriamasinavalona managed to usurp the position of paramount chief from his older brother by promising his supporters to share more power. In a move reminiscent of Silim Khan's creation of irrigation canals, Andriamasinavalona rounded up corvée labor to convert a huge marsh near Tananarive into rice paddies.

The intense level of raiding at this time is reflected in the archaeology of the region. Surveys by Dewar and Wright reveal a landscape dominated by large polygonal fortresses, often with multiple defensive ditches, and surrounded in turn by smaller fortified communities. There are also signs that many small valleys in the region had been converted to rice paddies. Like Kamehameha's terracing of O'ahu's Anahulu Valley, the Merina chiefs' intensification of rice production underwrote their territorial expansion.

Problems with chiefly succession, as we have seen in earlier chapters, contribute to the periodic collapse of rank societies. Oral history records that Andriamasinavalona created such a situation. Rather than leaving his territory to his oldest son (primogeniture) or his youngest son (ultimogeniture), he divided it among all four of his heirs. Soon they were all in competition, each seeking to take over his brothers' provinces.

This breakdown of centralized control provided opportunities for usurpation. Around 1745, in the northernmost of the four disputed provinces, a young man named Ramboasalama was born. He was the nephew of Andrianjafy, the current chief of the province, and soothsayers predicted great things for him. These predictions worried his uncle so much that Ramboasalama, like the young Kamehameha, went into hiding for a time.

Andrianjafy turned out to be a hated despot. In 1787 12 Merina chiefs rallied to Ramboasalama, giving him the support he needed to usurp Andrianjafy's position. Ramboasalama then changed his name to Andrianampoinimerina, "Prince Desired by Imerina."

Because usurpers achieve their titles by strategy or force rather than genealogical entitlement, they must often work hard to establish their legitimacy. Andrianampoinimerina's new name implied that the Merina people wanted him. He also argued that it was his destiny to rule. To support this notion he relied on the Merina concept of *vintana,* a process by which events are predestined through the ordering of time and space. He added special talismans to increase his *hasina,* or vital force.

Andrianampoinimerina began his rule modestly. His first "royal residence" was a wooden hut 20 by 12 feet in extent; its largest piece of furniture was said to have been the bed for his 12 wives. Over time, however, he succeeded in reunifying the provinces that Andriamasinavalona had divided among his four sons. During the next decade Andrianampoinimerina extended his political control to the entire central Malagasy plateau. "The sea," he is said to have boasted, "shall be the limit of my rice fields." His son Radama I, who took over in 1810, made good his father's boast.

As with other cases we have seen, the Merina creation of monarchy was a gradual process involving a series of aggressive rulers and some occasional setbacks. Along the way there were appropriate changes in social logic. Recall that the Hova attributed Ralambo's earlier victories to a powerful sampy, or talisman. Later Merina rulers appropriated all local sampy and created a class of "royal talismans" for their use alone. These royal talismans were called *sampy masina,* which implied a significant change in the distribution of hasina or vital force. According to Kottak, Merina nobles now claimed to have been born with much more hasina than commoners, providing the justification for social stratification.

The upper stratum of Merina society was called the *andriana,* or hereditary nobility. Commoners were divided into "true *hova*"—the descendants of central Imerina's ten original villages—and *mainty,* the royal servants and former slaves. These ranks within the commoner class remind us of the Zulu distinction between respected commoners and menials. Denied Merina citizenship was one group known as *andevo,* or slaves, mostly enemies captured in battle.

Dewar and Wright found that the structure of the Merina kingdom was visible in the archaeological record. Between 1760 and 1810 its heartland was reorganized and featured a settlement hierarchy of four levels. At the top was a fortified, 86-acre capital city. Below this city was the second level, a series of fortified 25-acre towns whose entrances could be closed with multiton stones. The third level of the hierarchy consisted of villages, while the fourth level was made up of smaller hamlets. There were also specialized military settlements at the frontiers between the Merina and other ethnic groups. Hereditary nobles were buried in tombs within each town's defensive walls, while the graves of the commoner class were left outside the walls.

Finally, just as in other monarchies we have examined, loyal commoners were sometimes appointed to bureaucratic positions. These commoner appointments were made because rulers did not always trust other members of the nobility, and because commoners realized from the outset that they did not have the genealogical credentials to usurp a higher position.

THE NATURE OF INEQUALITY IN KINGDOMS

Having looked at early monarchies in four different regions of the world, let us now consider this question: Was inequality any greater under a despotic king than under a despotic paramount chief?

Slavery, after all, was practiced even by foragers like the Tlingit. Certain villagers in New Guinea were treated as rubbish men. Hawai'ian chiefs rendered thousands of their own subjects landless. Bemba chiefs mutilated people who annoyed them.

Kingdoms continued many of those forms of inequality. In addition, partly as a result of the processes by which they formed, kingdoms created new types of inequality and enhanced others.

In the four cases we examined, not one kingdom was the offspring of a rank society that simply got bigger. There is apparently no social steroid that can trigger that kind of growth. Instead, all four kingdoms arose through the forced unification of a group of competing rank societies. It would seem that competition among chiefs, like the confrontations that produce an alpha chimp, was one of the engines driving the process.

In many parts of the ancient world, including Alabama and Panama and Colombia, such chiefly competition continued indefinitely. In Hawai'i, Natal, Madagascar, and the Hunza Valley one of the competing societies eventually gained an advantage. That advantage could be new weaponry, new military strategy, a new irrigation system, or thousands of new rice paddies. In addition, the ruler pressing the advantage seems to have been very aggressive, often a usurper with a chip on his shoulder, a man of elite ancestry but not in line to be heir, someone prepared to kill his half brother and marry his half sister if necessary.

This man and his heirs succeeded in subduing their neighbors, turning rival rank societies into the provinces of a larger territory. Neighboring chiefs who capitulated might be allowed to stay on as governors of their own provinces. Those who resisted were killed or exiled and then replaced with one of the victor's trusted allies.

Many conquered provinces still preserved the three levels of administrative offices left over from their days as chiefdoms. The man who unified the provinces now got tribute from them all. He had created an overarching administrative level and needed a higher title—"king"—because "chief" was now only a provincial title.

Some newly created kings turned their residences into palaces; moved or enlarged their capitals; turned their chiefly retinue into courtiers; had monuments erected to themselves; and ordered that their tombs be greater than anyone else's. All such acts help archaeologists identify kingdoms.

Kings also designed strategies to break down their subjects' former loyalties to their respective territories, replacing them with loyalty to the royal family. In the case of the Zulu this process began with Shaka's expansion of the amabutho to include youths from all the societies he had conquered. Once Shaka had become a king the process expanded further, and he endeavored to turn all commoners into citizens of a Zulu state.

It was here that a new form of inequality—ethnic discrimination—came to the fore. We have seen that ethnocentrism is universal; even villagers in egalitarian societies consider their behavior superior to that of their neighbors. Kings like Shaka and Andrianampoinimerina, however, had incorporated many neighbors into their realms. Certain commoners would be treated as full citizens, "true Zulu" or "true Hova." Other commoners, however, would be considered "destitute," "menials," "people with strange hairstyles," and so on. The eagerness of kings to incorporate foreigners into their labor force was greater than society's ability to tolerate their ethnic differences. Second-class citizenship was the result.

Some of the kings discussed in this chapter also increased inequality by weakening power-sharing. Recall that among the Bemba, members of the council inherited their positions, while the chief was chosen by his fellow aristocrats. Since the councillors were not under his thumb, the Bemba chief had to take their advice seriously.

Some early kings, however, handpicked their major advisers. Twelve Merina chiefs had helped Andrianampoinimerina usurp his uncle's position; he made them his inner council of advisers. Once he had become king, he also added a council of 70 aristocrats called "husbands of the earth." Every year he gave a public speech called the *kabary*, allegedly to "share decision making with the people." Such public displays of power-sharing, however, were largely cosmetic.

Kamehameha turned five chiefs of his native Kona district into councillors and initially sought their approval on important decisions. According to Ralph Kuykendall, however, once the five original councillors retired, their successors had much less influence on Kamehameha. Mir Silim Khan delegated power to his vizier and was supposed to share power with a Hunza council called the *marika*. However, as Homayun Sidky points out, the mir himself presided over the marika, and "nobody dared speak out of turn."

All first-generation kings, no matter how despotic, needed political support. They often obtained this support by making at least a pretense of power-sharing. Kings were, however, more likely to handpick or bypass their advisers than were the chiefs who preceded them.

THE PROBLEMS OF BEING THE FIRST

The creation of all four of the kingdoms discussed in this chapter required the consolidation of formerly independent societies by force. But was force always required?

We have seen that several of the largest chiefly societies in the southeastern United States, such as the Coosa, were voluntary confederacies. Might not some early kingdoms have formed voluntarily as well?

Without closing the door to that possibility, we doubt it. We note that the Coosa—and the famous League of the Iroquois, for that matter—were not the first chiefly societies in their regions. By the time those confederacies arose, there had probably been rank societies in North America for more than 1,000 years. We suspect that once a particular type of society has existed for a while and the template for its organization is understood, more alternative routes to its creation are possible—including voluntary routes.

Anthropologist Robert Carneiro makes the point that most societies do not surrender their autonomy willingly. We have seen that every clan who tried to seize ritual or secular control of its society met with resistance from other clans. The various Catío villages of Colombia's Cauca Valley did not submit to the authority of a single chief unless they were threatened by hostile neighbors. And many archaeologists suspect that the Coosa and the Iroquois formed their confederacies only after Spanish, French, or English colonists came to be seen as a threat.

The first rank society in each region had no template to follow; neither did the first kingdom. One of the problems of being the first kingdom in your region is that your neighbors see nothing to be gained by giving up freedom to become one of your provinces. The question of how the first kingdoms were created is, therefore, quite separate from the creation of second-, third-, and fourth-generation kingdoms.

Archaeologist Charles Spencer has presented mathematical support for the idea that the first state to form in a region will likely require the kind of territorial expansion we saw among the Hawai'ians, Zulu, Hunza, and Merina. Since a kingdom is one kind of state (almost certainly the earliest kind, with military dictatorships and parliamentary democracies arising later), Spencer's work is worth discussing here. We will, however, leave out the mathematical details.

Borrowing an equation from the zoological literature on predator-prey relations, Spencer demonstrates that as a chief reaches the limit of the resources he can extract from his followers, and the growth curve of his society goes from steeply rising to horizontal, one of three things must happen. Such chiefs must either:

1. Step up demand for resources from their own subjects, which may lead to revolt.
2. Intensify production through technological improvement, which will likely increase wealth but not necessarily sociopolitical complexity.
3. Expand the territory from which they get their resources, which will probably require the subjugation of neighbors.

When alternative 3 is chosen, and the expanded territory grows beyond the limits that a chief can administer through the usual methods, he is compelled to make changes in administration and political ideology, and a state begins to form. That change is less likely with alternatives 1 and 2.

The reason military force so often seems to be involved in the creation of the first kingdoms is because rival chiefs are unwilling to surrender their territory and independence voluntarily. In the four cases we saw in this chapter, state formation involved thousands of deaths, and thousands of other people were converted to slaves.

Sorry, but no one said that creating the first kingdoms would be pretty.

The Unanswered Question

Monarchies, as we have seen, can be created out of rank societies. And, as we learned from Irving Goldman, rank societies come in several forms. The chiefs of Tikopia relied on sacred authority. The wealth of the Quimbaya flowed from the expertise of their goldsmiths. The Zulu were secular warriors who took what they wanted by force. And the Tongan chief could wither a commoner with his glance, but his own sister could place her foot on his head.

It turns out that monarchies, too, come in several forms. Anthropologist Herbert Lewis once took a close look at the traditional monarchs of Ethiopia. The king of the Kaffa, he reveals, was considered a divine monarch. The king of Abyssinia was not considered divine, but he was surrounded by taboos and an element of the sacred. The king of the Galla (or Oromo) was considered only a powerful mortal.

These different forms of kingship, as we learn in later chapters, were widespread. The Egyptian pharaoh was considered divine. The Maya king had sacred qualities and was sometimes portrayed as the reincarnation of a mythological ancestor. The early Mesopotamian king, on the other hand, was only a powerful mortal, sometimes referring to himself as the "tenant farmer" of a patron deity. Only the later Mesopotamian kings sought to be considered divine.

Here, then, is an important unanswered question: Was there a logical connection between a particular type of monarchy and the chiefly societies out of which it was created? Did divine monarchs result from the unification of rank societies in which religious authority was paramount? Did secular kings result from the unification of largely militaristic rank societies, where religious specialists were little more than witch doctors? Or could any type of monarchy be created by uniting rank societies of any type?

We have no answer to this question, because social anthropologists and archaeologists are not working on it. But they should be.

EIGHTEEN

Three of the New World's First-Generation Kingdoms

We have just described the birth of four kingdoms. Each was created by a series of ambitious leaders, who kept trying until they had unified a group of formerly independent rank societies. The leaders involved had no model to follow and no template to show them what a monarchy should look like. All four cases, therefore, qualify as first-generation kingdoms for their regions.

Admittedly, the reason we know so much about the Hawai'ian, Zulu, Hunza, and Merina cases is because there is written documentation. All four cases took place late in world history, while Western observers were watching. Wouldn't it be nice to have some earlier cases, unaffected by Western contact?

In this chapter we go back more than 2,000 years in search of pristine cases of kingdom formation. The good news is that one can find such cases in the archaeological records of Mexico, Guatemala, and Peru. The bad news is that, precisely because those cases are so ancient, we lack eyewitness descriptions. We must therefore rely heavily on archaeological inference.

While we believe that most first-generation Mexican and Peruvian states were kingdoms, we will sometimes use the generic term "state" when we are not sure that actual monarchs were involved. Some early Mexican states—for example, the one centered on Teotihuacan near modern Mexico City—did not portray monarchs in their art. In contrast, we know that early Maya rulers were monarchs, because their hieroglyphic inscriptions use the Maya word for "king."

There is a reason we have chosen to begin our search in the New World. Many archaeologists working in the Mexican highlands, the Peruvian coast, and the Maya lowlands have specifically designed their research to determine how and when the earliest kingdoms arose. In other words, they have prepared the way for us.

THE CREATION OF THE ZAPOTEC STATE

Roughly 2,500 years ago Mexico's Oaxaca Valley was occupied by at least three rank societies, seemingly in conflict with each other. To the north lay the chiefly center of San José Mogote, in command of a society with at least 2,000 inhabitants. To the south lay San Martín Tilcajete, the chiefly center for a society of 700 to 1,000 persons. To the east lay Yegüih, the chiefly center for another society of perhaps 700 to 1,000 persons. Despite the differences in population, none of these rank societies were able to subjugate their rivals; they therefore left a sparsely occupied buffer zone between their territories. They periodically burned each other's temples, or commissioned stone monuments to portray the sacrifice of enemy leaders. These chiefly rivalries remind us of those on the Big Island of Hawai'i before it was unified by men such as 'Umi, Alapai, and Kamehameha.

Eventually the leaders of San José Mogote did something that gave them an advantage. Gathering up 2,000 people from their paramount center and its satellite villages, they moved to the summit of a mountain in the buffer zone just mentioned. This mountain, rising 1,300 feet above the valley floor, was protected by steep slopes on the south and east. On the more easily climbed slopes to the north and west, the new arrivals began building two miles of defensive walls. They turned the mountain into a stronghold from which they could subdue their rivals.

As so many other rank societies did, the first occupants of the stronghold had created a new chiefly center in an unoccupied buffer zone. Some 200 years after its founding the mountaintop community had grown to an estimated 5,000 people, probably through a combination of internal population growth and the attraction of additional immigrants. It became Oaxaca's first city, whose ruins we know today as the archaeological site of Monte Albán.

It is probably no coincidence that the pottery of this newly created city shows us the first use of the *comal*, a griddle for mass-producing corn tortillas. We suspect that the workers building and maintaining the city received rations of tortillas from their leaders.

Thanks to archaeologists Charles Spencer and Elsa Redmond, we know a lot about Monte Albán's ongoing conflicts with San Martín Tilcajete, the rival chiefly center one day's travel to the south. Both communities laid out rectangular civic-ceremonial plazas, reminiscent of the ones at Lapaha and La Venta. Monte Albán's plaza, however, was laid out true north-south, while Tilcajete's was laid out 25 degrees east of true north. The chiefly lineages of these two communities evidently disagreed on which astronomical alignment was appropriate, suggesting that leadership was based at least partly on celestial authority.

Tilcajete's response to the growth of Monte Albán was to double its own size, from 60 to 130 acres. This increase may have included the deliberate drawing in of defenders from its satellite villages. About 2,280 years ago, however, Monte Albán raided Tilcajete and torched the buildings on its plaza.

Tilcajete refused to capitulate. Between 2,250 and 2,000 years ago it increased its size to 177 acres and moved its plaza uphill to a more easily defended ridge; the new plaza defiantly maintained the same orientation as its predecessor. Tilcajete also added defensive walls to its most easily climbed slopes.

Monte Albán, however, was prepared for a long campaign. Its leaders concentrated thousands of farmers, craftsmen, and warriors into a ring of 155 satellite villages, most of them within a half-day's walk of Monte Albán. Many of these villages occupied piedmont settings, where canal irrigation could be used to intensify the production of corn. This strategy was reminiscent of Kamehameha's terracing of O'ahu's Anahulu Valley.

Eventually Monte Albán attacked Tilcajete again, burning its ruler's palace and the community's major temple. Charcoal from the destruction of both buildings dates to about 2,000 years ago. Tilcajete did not recover from this second attack. It was abandoned, and on a mountaintop nearby its conquerors built a Level 2 administrative center linked to Monte Albán. It would seem that by this point Monte Albán had subdued the entire Oaxaca Valley and turned its former rivals into the districts of a first-generation kingdom.

The palace at Tilcajete measured 52 feet on a side and consisted of eight large rooms around a central patio (Figure 50). The walls were of adobe brick, set on a stone masonry platform several courses high. This elite residence appeared to have been built by corvée labor. At least three different work gangs must have been involved, because Spencer and Redmond could distinguish bricks of three different sizes, colors, and clay sources.

During this same period, Monte Albán carved a series of stone monuments that reflected its military unification of the valley. One of the most important structures on the west side of Monte Albán's central plaza was Building L. Its façade was covered with a display of slain enemies analogous to those at Cerro Sechín, Peru, though not as numerous or gruesome. The largest of these carved stones depicted sprawling corpses, some with evidence of heart removal or genital mutilation. A few of the smaller stones showed severed heads. Several monuments appear to include the hieroglyphic names of important victims (Figure 51).

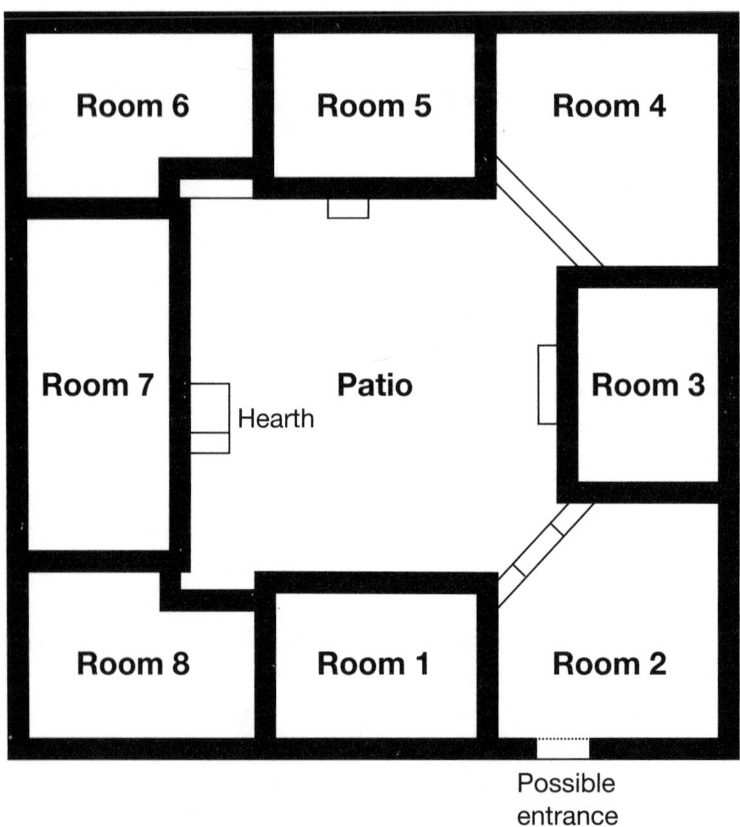

FIGURE 50. Competition between Monte Albán and Tilcajete ended with the conquest of Tilcajete and the burning of its palace. The palace, measuring 52 feet on a side, consisted of eight rooms around a central patio. It had been built by multiple work gangs, using adobe bricks from different clay sources.

Set in the southeast corner of Building L were two more carved stones that provide an inscription eight hieroglyphs long. Included are glyphs for dates in the two known Zapotec Indian calendars, a 365-day secular calendar and a 260-day ritual almanac. This text seems to refer to a ruler who, in addition to commissioning Building L, claims credit for the slain enemies depicted. Such credit-taking by one man strengthens the likelihood that the early Zapotec state was a monarchy rather than an oligarchy.

The unification of the 1,290-square-mile Oaxaca Valley was a significant accomplishment, though it pales in comparison to the 4,028-square-mile Big Island of Hawai'i. The rulers of Monte Albán, however, had barely begun to expand. Indeed, the available radiocarbon dates suggest that they may have begun subduing some of their weaker neighbors even before their conquest of Tilcajete.

Some 1,800 years ago the city of Monte Albán covered 1,028 acres, and archaeologist Richard Blanton estimates its population at 15,000. Its period of aggressive military expansion may have gone on for several centuries, until Zapotec rulers were receiving tribute from an estimated 8,000 square miles.

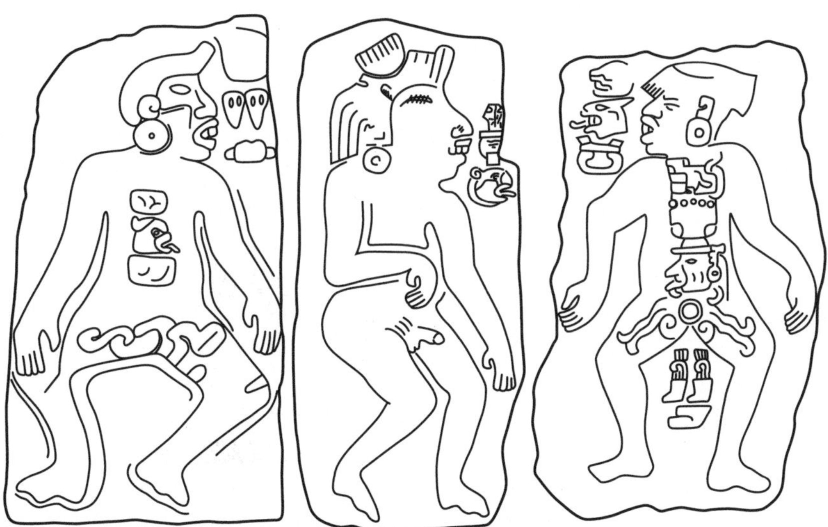

FIGURE 51. The façade of Building L at Monte Albán was covered with carved stones depicting slain or sacrificed enemies. Most were shown naked, and a number had blood scrolls indicating genital mutilation. These three examples also bear hieroglyphic inscriptions that may include the victims' names.

A significant problem in the study of first-generation states is that 1,000-acre cities and 8,000-square-mile tribute territories are too large for individual archaeologists to investigate in detail. What the archaeologists working in Oaxaca did was form a consortium of researchers. Members of this consortium then collaborated on a full-coverage survey of the Oaxaca Valley, including the city of Monte Albán and the hundreds of towns, villages, and hamlets in the political hierarchy below it. Some veterans of the original survey have now extended their efforts beyond the limits of the valley, into neighboring regions with names such as Cuicatlán, Ejutla, Miahuatlán, Sola de Vega, Peñoles, Tilantongo, and Huamelulpan. A bare-bones list of collaborators would include Richard Blanton, Stephen Kowalewski, Gary Feinman, Linda Nicholas, Laura Finsten, Andrew Balkansky, Charles Spencer, and Elsa Redmond.

Strategies for Territorial Expansion

The archaeological record makes it clear that the Zapotec had three different strategies for territorial expansion. The region of Sola de Vega, 40 miles southwest of Monte Albán, was so sparsely occupied that it could be annexed simply by sending in colonists. The Ejutla region, 30 miles south of Monte Albán, appears to have been taken over peacefully, possibly as the result of strategic marriage alliances among noble families. Ejutla thrived as the result of its incorporation by Monte Albán; its craftsmen imported seashells from the Pacific coast and converted them to ornaments for the Zapotec capital.

Yet another strategy was required by the region of Cuicatlán, 50 miles to the north of Monte Albán. In contrast to the temperate Oaxaca Valley, whose alluvial floor averages 4,800 to 5,500 feet above sea level, Cuicatlán lies in an arid tropical valley whose elevation averages 1,600 to 2,200 feet. With irrigation, Cuicatlán could produce both cotton and tropical fruits unavailable in Oaxaca Valley.

The leaders of Cuicatlán's villages chose not to surrender their autonomy to the Zapotec. Unfortunately for those leaders, their population was organized only as a rank society. Monte Albán's more experienced warriors made short work of Cuicatlán and left behind a wooden rack displaying the skulls of 61 local victims. Such a rack, known in Zapotec as a *yàgabetoo,* was designed to intimidate anyone who resisted incorporation.

The Zapotec reorganized the Cuicatlán landscape, moving its surviving population from the river floodplain to the piedmont. This left the floodplain free to be irrigated with newly built canals and aqueducts. The Zapotec intensified irrigation in Cuicatlán much as Mir Silim Khan had intensified it in Hunza, and the outcome was presumably similar: he who built the canals got to decide how and by whom they were used.

Recall that Mir Silim Khan also built forts in his territory. The Zapotec did something similar. At the northern extreme of the Cuicatlán district, near a mountain pass leading to the valley of Tehuacán, they built a hilltop redoubt called the Fortress of Quiotepec. The pottery and tombs of this fortress were in typical Monte Albán style. To the north of the fort, however, Spencer and Redmond discovered a no-man's-land some four miles wide. Beyond this buffer zone, dating from roughly 2,200 to 1,800 years ago, the pottery no longer resembled that of Monte Albán.

Whom was the Fortress of Quiotepec designed to discourage? We believe that it marked the frontier between Monte Albán's expansion and that of Teotihuacan, an even larger first-generation state that we will discuss later in this book.

The Chain Reactions That Create More Kingdoms

Monte Albán, to be sure, was not able to annex every region it desired. The Zapotec expanded most successfully against their weaker neighbors, relying on diplomacy to bring the others in line.

In the mountains to the northwest, however, lay a series of well-populated valleys whose inhabitants had no intention of being incorporated into the Zapotec state. Included were the valleys of Tilantongo, Nochixtlán, and Huamelulpan. At the time of the Spanish conquest, all these valleys were occupied by speakers of the Mixtec language.

Full-coverage surveys by Andrew Balkansky, Stephen Kowalewski, and Verónica Pérez Rodríguez suggest that the movement of the Zapotec to the fortified summit of Monte Albán set off a chain reaction among its neighbors to the northwest. Soon the leaders of these valleys were concentrating their supporters on defensible summits as well.

One of the first hilltop communities to appear was La Providencia in the Tilantongo Valley, which seems to have been founded at roughly the same time as Monte Albán. A few centuries later, while Monte Albán was working to subdue Tilcajete, La Providencia lost population to Monte Negro, a larger community on an even higher mountain. Monte Negro's occupants built elite residences for their leaders and erected a number of temples. We believe that

Monte Negro was in the process of creating its own kingdom when it was abruptly abandoned. Contemporaneous with Monte Negro was Cerro Jazmín, a hilltop community in the Nochixtlán Valley.

The largest of all these defensible mountaintop settlements, however, was Huamelulpan, located in the valley of the same name. Balkansky suggests that Huamelulpan took over from an earlier chiefly center called Santa Cruz Tayata, much the way that Monte Albán took over from San José Mogote. Neighboring Mixtec societies, in other words, nucleated and fortified themselves to keep Monte Albán at bay; the resulting political consolidation allowed them to create embryonic kingdoms of their own.

Inequality and Administrative Hierarchy in the Zapotec State

Let us turn now to the internal workings of the Zapotec state. One of the benefits of the full-coverage survey just mentioned is that it reveals multiple levels in the administrative hierarchy.

Eighteen hundred years ago—two centuries after Monte Albán's defeat of Tilcajete—there were 518 communities in the Oaxaca Valley. The largest was Monte Albán itself (Level 1), with an estimated 15,000 inhabitants. Level 2 of the hierarchy consisted of six towns with estimated populations of between 900 and 2,000. There were palaces and elegant tombs at Monte Albán and its Level 2 towns, indicating that people of noble birth were in charge there. All six towns were less than a day's walk from Monte Albán, making intercommunication easy. All these larger settlements had multiple temples.

Level 3 had at least 30 villages with populations estimated at 200 to 700. There were no palaces at these smaller communities, but several had at least one temple. Finally, the fourth, or lowest, level of the hierarchy consisted of 400 small villages, with no evidence of temples or palaces. These settlements brought the estimated population of the valley to more than 40,000.

The stone monuments, ceramic sculptures, and tomb murals of Monte Albán all confirm that the Zapotec state was a monarchy. Rulers are shown sitting on thrones, sometimes costumed as jaguars or wearing the feathers of quetzal birds from the distant cloud forest (Figure 52). Not until the Spaniards conquered Oaxaca in A.D. 1521, however, did the rest of the world get eyewitness descriptions of Zapotec society.

FIGURE 52. This funerary urn from Tomb 103 at Monte Albán portrays a royal ancestor in his role as mighty warrior, carrying the severed head of an enemy by its hair. In addition to jadeite ornaments and a headdress of quetzal tail feathers, he wears a mask made from the dried skin of a flayed enemy's face. The urn is 20 inches high.

From the Spanish accounts we conclude that Zapotec society was divided into at least two major strata, hereditary nobles and commoners. At the apex of the noble stratum was a king *(coqui)* and his principal wife *(xonaxi)*. A major ruler, who might be referred to as a *coquitào*, or "great lord," lived in a *quihuitào*, or "beautiful royal palace."

Archaeologist Alfonso Caso excavated a number of palaces and royal tombs at Monte Albán. A typical palace consisted of eight to 12 rooms around a central patio. Under the patio floor was the royal tomb, reached by a stairway that allowed the king's descendants to make additional offerings on the anniversaries of his death.

Tombs 104 and 105 of Monte Albán, laid out roughly 2,400 years ago, were two of the most magnificent. Tomb 105's walls bore polychrome murals that show royal men and women (perhaps relatives or ancestors of the deceased lord) accompanied by their hieroglyphic names. The door of Tomb 104 was closed with a large stone, carved with the hieroglyphic names of what are almost certainly royal ancestors. Such carving or painting of royal genealogies helped the descendants of the deceased ruler confirm their right to rule.

The ruling class was made up of *tija coqui*, the royal lineage; *tija joana*, lineages of major nobles; and *tija joanahuini*, lineages of minor nobles. The Spaniards compared these lineages to the various ranks of European nobility.

The stratum of Zapotec commoners also had its gradations in prestige. There were landed commoners, landless serfs, and slaves. Free commoners belonged to *tija peniqueche*, "lineages of townspeople," and held corporate rights to parcels of land dispersed through the valley floor, piedmont, and mountains. These varied parcels were the Zapotec equivalent of Hawai'i's pizza-slice transects from coast to mountain.

Social inequality was expressed in terms of address, clothing, diet, and other forms of behavior. Nobles were addressed with terms equivalent to "your grace." Even they, however, had to bow and remove their sandals in the presence of the king. Nobles wore bright cotton mantles, feather headdresses, and jade ornaments in their earlobes and lips. Some male nobles had 15 to 20 wives. They dined on venison and enjoyed drinks flavored with chocolate, a plant imported from the lowlands.

Commoners, on the other hand, wore agave fiber mantles and were allowed much less ornamentation. Instead of venison, they ate the flesh of dogs, turkeys, rabbits, and local small game. Only the wealthiest of commoner men could afford a second wife.

Zapotec rulers sometimes appointed trusted commoners to bureaucratic posts. In addition, there were some social institutions in which both nobles and commoners regularly collaborated. Within the military the officers were of noble birth, while the foot soldiers were conscripted commoners. Officers wore body armor of quilted cotton, and their valor was rewarded with costumes depicting them as pumas, jaguars, hawks, or eagles. Foot soldiers went to battle in loincloths.

Within the religious establishment, the highest priests were of aristocratic birth. Often they were the younger sons of nobles, outranked by their older brothers and therefore unlikely to inherit their father's title. Like all members of the noble stratum, they were given a religious education that was denied to commoners. This difference in education helped maintain inequality.

The high priest's assistants were commoners who underwent special training. These minor priests, according to the Spaniards, virtually "lived in the inner room of the temple."

The temple itself was called *yohopèe,* "the house of the vital force," reminding us that the Zapotec had a concept of sacred life force like mana among the Polynesians or hasina among the Merina. Anything that moved, including lightning bolts, flowing blood, and the foam on a cup of hot chocolate, possessed pèe. Any plant that induced visions of the spirit world, such as jimson weed *(Datura),* morning glory, strong tobacco, or hallucinogenic mushrooms, was considered sacred. Zapotec priests were trained in bloodletting, human and animal sacrifice, and ritual drug use. Nobles had their own special rituals, because after death they would metamorphose into semidivine ancestors, living among the clouds and serving Cociyo—Lightning—the most powerful being in the Zapotec cosmos.

While Zapotec rulers were extremely powerful, some of this power was shared with a governmental council. This council, presumably composed of nobles, met in a special building called a *yohohuexija,* which ought to be archaeologically identifiable.

PERU'S MOCHE STATE

When we last looked at Peru, we were struck by how precocious its rank societies were. They were also militaristic, and armed conflict remained a factor in the rise of Peru's largest first-generation kingdom. That conflict took two forms:

neighboring coastal chiefs fought with each other, and highland chiefs sought to take over the irrigation systems of coastal valleys.

Some 2,500 years ago, rank societies operated in all the major river valleys of the northern and central Peruvian coast. Many important chiefly centers lay upstream in what is called "the middle valley," the point where rivers emerge from the canyons of the Andes.

The middle valley was narrower than the lower valley, requiring less labor and expertise to irrigate. There were places where water could easily be diverted from the river and brought downhill to potato and manioc fields. Still farther upstream, at elevations of 2,000 to 4,000 feet, conditions were often satisfactory for the irrigation of two crops: cotton for textiles and coca leaves for ritual.

Archaeologists have subjected many of Peru's coastal valleys to full-coverage surveys, making possible a tentative reconstruction of what happened next. In no case, however, do we have the step-by-step chronology of events that we would like to have.

Roughly 2,400 years ago, according to archaeologist Brian Billman, a wave of violence swept the coast as highland societies began to raid coastal valleys. It would appear that the highland chiefs controlled enough warriors to take over some of the cotton and coca lands just mentioned.

Coastal chiefs were occasionally able to unite a number of rank societies under a single leader in response to this threat from aggressive highland societies. Surveys by David Wilson provide evidence that a large society, with a political hierarchy of four administrative levels, formed in the Casma Valley between 2,200 and 2,000 years ago. This society eventually broke down, however, suggesting that its enemies were too strong to expel.

Between 2,000 and 1,800 years ago, according to Billman, highland invaders had penetrated the middle portions of the Moche, Virú, Nepeña, Casma, Chillón, and Lurín Valleys. These invaders caused the abandonment of a number of coastal population centers. One of those centers was Cerro Arena in the Moche Valley, a settlement on a defensible mountain ridge. At its peak, according to archaeologist Curtiss Brennan, Cerro Arena had 2,000 visible structures crammed into three-quarters of a square mile.

With the abandonment of centers such as Cerro Arena, the indigenous population of the Moche Valley moved closer to the ocean and consolidated into a single, highly nucleated community called Cerro Oreja. Apparently now united under one leader, this coastal society became one of the first to fight back successfully against highland invaders.

After driving out their highland enemies and reclaiming their upstream irrigation lands, the leaders of Cerro Oreja moved their capital to a new locality, known to archaeologists as Las Huacas de Moche. There they established a 250-acre urban center that served as the capstone of a four-level political hierarchy. Below it were five towns in the 35- to 125-acre size range; a half dozen large villages in the 12- to 35-acre range; and scores of smaller villages and hamlets.

The creation of Peru's Moche kingdom seems to have been a centuries-long process, involving both highland invaders and neighboring coastal societies. Archaeologists working on Peru's north coast are still not in agreement about whether Cerro Arena (200 B.C.–A.D. 1), Cerro Oreja (A.D. 1–200), or Las Huacas de Moche (A.D. 200–400) was the capital of the valley's first kingdom. We see this situation as analogous to the one described for the Oaxaca Valley. Archaeologists working there sometimes disagree about whether the first Zapotec kingdom emerged when Monte Albán was founded (500 B.C.), when it began to attack its rivals (300 B.C.), or when it finally defeated Tilcajete (30 B.C.). Similar questions have been asked about the sequence of Hawai'ian rulers comprising 'Umi, Alapai, Kalaniopu'u, and Kamehameha. What seems to happen in such sequences is that each ruler pushes the system closer to monarchy until the evidence becomes incontrovertible.

The analogies between the Moche and Zapotec did not end there. The leaders of Las Huacas de Moche, like those of Monte Albán, were not content with local victories. They now had the most effective military apparatus on the coast. Instead of continuing to pursue their highland enemies, they began to use that apparatus against their coastal neighbors. Between A.D. 200 and 600, the Moche state expanded until it came to dominate 15 coastal valleys. The result was a narrow strip of empire stretching 360 miles, from Piura in the north to Huarmey in the south. The Moche succeeded in part because they knew it would be easier to subdue other coastal valleys than to confront belligerent highland armies on the latter's home turf.

The Southern Moche

Between the Jequetepeque and Chicama Rivers lies an expanse of desert called the Pampa de Paiján. This waterless barrier divided the Moche empire into northern and southern halves. The Moche Valley itself was the heartland of the southern half. Its capital, Las Huacas de Moche, featured two immense pyramids separated by densely occupied residential areas. One of the most extraordinary features of the Moche empire was its lavish use of labor. Billman estimates that in the Moche Valley alone, labor gangs dug more than two million cubic feet of new irrigation canals and built more than 44 million cubic feet of monumental public architecture.

The *huacas,* or sacred pyramids, of the Moche capital represented a huge undertaking. Archaeologist Michael Moseley estimates that the smaller of the two huacas was originally 312 feet long, 279 feet wide, and 66 feet high, requiring more than 50 million adobe bricks. It was wholly artificial, its builders eschewing the earlier practice of dressing up natural hills to resemble pyramids.

The larger of the two Moche pyramids was the biggest adobe structure ever built in the New World. Before its partial destruction by the Spaniards it was 1,122 feet long, 522 feet wide, and 131 feet high, requiring more than 143 million adobe bricks. Each of the dozens of work gangs assigned to the pyramid impressed the upper surfaces of their bricks with a distinctive "maker's mark," presumably to prove that their quota of bricks had been met.

The Northern Moche

To the north of the Pampa de Paiján, another five or six valleys made up the northern Moche region. Each of these valleys supported a district capital, and because later rulers occasionally moved their capitals, some valleys display the ruins of several large administrative centers.

Any question about which type of state the Moche created has been answered by the archaeological site of Sipán in the Lambayeque Valley. There the excavation of a series of spectacular tombs by Walter Alva and Christopher Donnan reveals that the Moche state was not only a monarchy but an individualizing monarchy, in Colin Renfrew's terms.

The royal tombs of Sipán were hidden deep within a platform of adobe bricks. Their principal occupants may represent three generations of Moche rulers. The graves include spectacular sumptuary goods, sacrificed animals, prisoners, and individuals who might be servants and/or relatives of lower rank.

Tomb 3 was the earliest. Its central figure was a male ruler wrapped inside several layers of cloth and a woven mat. He was accompanied by objects of gold, silver, and copper, including two scepters and the standing figure of a warrior. One of his necklaces consisted of ten large gold beads, each depicting a spider with a human head caught in its web. Another set of beads portrayed owls' heads, and there was a half-human, half-crab figure in gilded copper. The symbolism of the king as warrior was unmistakable.

The principal occupant of Tomb 2 was a 35- to 45-year-old man; his body lay in a plank coffin rather than the mat used in the earlier tomb (Figure 53). Wearing nose and ear ornaments of gold, silver, and turquoise, the Tomb 2 ruler was accompanied by hundreds of copper discs, thousands of shell beads, copper bells, copper slippers, necklaces of miniature copper trophy heads, and a headdress with a gilded copper owl.

To the far left of the ruler was a teenage boy in a cane coffin, accompanied by two large copper discs. A smaller cane coffin that contained a child eight to ten years of age lay at the ruler's feet; sharing the coffin were a dog and a

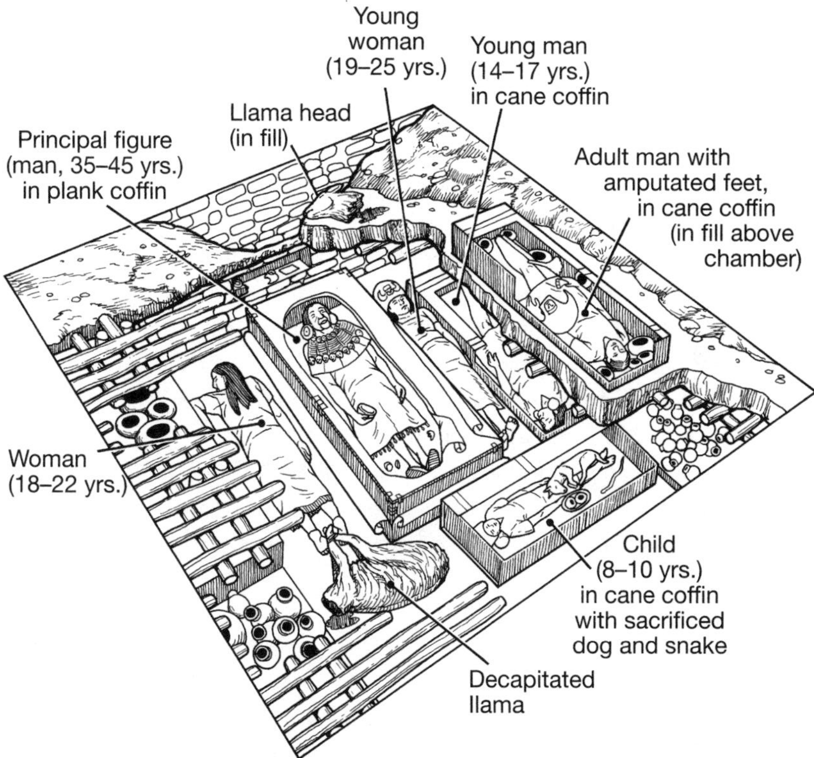

FIGURE 53. Tomb 2 of Sipán, Peru, contained a Moche ruler in his plank coffin, accompanied by numerous sacrificed people and animals.

snake. To the ruler's immediate left lay a woman 19 to 25 years of age, wearing a copper headdress and a textile to which discs of gilded copper had been sewn.

A number of humans and animals had been sacrificed to accompany the ruler. To his right lay a woman 18 to 22 years of age, perhaps a servant or slave, sprawled facedown. Near her feet was a decapitated llama. Buried above the roof of the tomb was another cane coffin, this one bearing a man whose feet had been cut off at the ankles. He was accompanied by a copper crown and a feather headdress with a large copper shaft, indicating that he might have been a mutilated captive of relatively high rank.

Tomb 1 was the final one in the sequence. The burial chamber had adobe benches along its walls. The ruler's plank coffin was securely fastened with copper straps so that it could be lowered into the tomb with ropes, a ceremony sometimes depicted on painted Moche vessels. Two sacrificed llamas lay on the tomb floor to either side of the coffin.

The ruler in the plank coffin was so covered with gold, silver, copper, and turquoise that it took the excavators a long time to reach his actual corpse. He wore a copper-plate headdress, and his necklaces were of gold and silver beads, designed to look like unshelled peanuts. His scepter was also of gold and silver. Piled on his chest were scores of miniature crabs and trophy heads made from precious metals.

Half a dozen other people accompanied this ruler in death. Directly atop one of the sacrificed llamas was the cane coffin of a powerfully built man, possibly a warrior. His coffin included a large war club, a circular shield, and a crescent-shaped head ornament. Atop the other sacrificed llama was the cane coffin of a 35–45-year-old man accompanied by a dog. This man wore a beaded chest covering and had several copper offerings.

Three other cane coffins held women between 15 and 20 years of age. Their skeletons were partially disarticulated, suggesting that the women had died earlier and were exhumed in order to be reburied with the ruler.

Inequality and Administrative Hierarchy in the Moche Empire

It seems likely that the Moche expanded until they had incorporated a number of formerly autonomous kingdoms, occupied by different ethnic groups and speaking non-Moche languages. Such a takeover of other kingdoms is one criterion of an empire.

Just as we saw in the Oaxaca region, some smaller kingdoms probably allied themselves with the Moche peacefully, while others had to be conquered. As a result, the capitals of some provinces retained their own distinctive architecture and pottery style, while others appear to be brand-new administrative centers imposed by the Moche.

All empires have an expiration date, and the Moche began to lose power to a series of second-generation states after A.D. 600. The Moche had no writing system, and they disappeared long before the arrival of European eyewitnesses. All we know about them is what we can infer from their archaeological remains.

One source of information consists of detailed scenes of Moche activities, executed in fine-line painting on luxury pottery (Figure 54). In some scenes we see the Moche ruler carried on his litter or seated prominently at the top of a ramp or stairway. We see warriors seizing captives by the hair, crushing their skulls with war clubs, or dismembering them. We see male priests slitting captives' throats and female priests catching the blood in copper goblets. We know that some of these scenes have real-world validity, because Christopher Donnan and Luis Jaime Castillo found two of the priestly women,

FIGURE 54. In this detail from a finely painted Moche vessel, nude prisoners of war are presented to a victorious ruler who sits atop a pyramid. The painted scenes on other Moche vessels show captives being sacrificed, with their blood being used to fill goblets like the one held by the ruler in this scene. Reproduced by permission of Christopher B. Donnan and the estate of Donna McClelland.

dressed in full regalia and accompanied by their copper goblets, buried at San José de Moro in the Jequetepeque Valley.

The chief of the Bemba mutilated subjects who insulted him, but compared to the Moche kings, he barely qualified for a learner's permit.

Moche rulers had fabulous treasure in their plank coffins, and they clearly wanted to be portrayed as warriors. Nobles from lower-ranked lineages were buried in cane coffins, with fewer sumptuary goods. Judging by their graves, the highest priests, whether men or women, were of noble birth. Exceptional warriors seem to have been honored by being entombed with their rulers. The body parts of enemies were scattered on royal tombs like croutons sprinkled on a Caesar salad. It is likely that some ethnic minorities within the Moche state were treated as second-class citizens, but we cannot be sure without eyewitness accounts.

CALAKMUL: AN EARLY MAYA KINGDOM

Our next first-generation kingdom was created in the Mirador basin, a forested depression in the Maya lowlands. This basin straddles the border between northern Guatemala and the Mexican state of Campeche. The tropical setting contrasts with the temperate highlands of Oaxaca and the desert coast of Peru, but the social and political dynamics involved were similar.

Some 2,800 years ago there were large villages in the Maya lowlands, some covering more than 100 acres. Over the next millennium the evidence for rank societies grew. Sumptuary goods of jade and mother-of-pearl circulated widely. Leaders used corvée labor to build temples atop stone pyramids. Military competition was evidently intense, because archaeologists have found mass burials of young men with injuries typical of warriors.

One raid at the site of Cuello in Belize, carried out 2,400 years ago, left the façades of some public buildings destroyed and their perishable superstructures burned. Nearby was a mass grave of 26 men who showed signs of having been butchered. Some of these men showed healed fractures of the wrist and forearm, injuries likely sustained in previous battles.

At about this time, Nakbe emerged as a chiefly center in the Mirador basin. Laid out roughly east-west, Nakbe had two complexes of public buildings connected by a causeway. Archaeologist Richard Hansen found that these complexes included platforms as much as 24 feet high and pyramids rising 150 feet. Nakbe's earliest stone monument depicts two men in chiefly regalia, one of them pointing toward the head of an ancestor.

One key to Nakbe's rise was its intensive agriculture. Hansen discovered special garden plots, framed by low stone walls and filled with organic soil carried in from nearby swampy depressions. Rather than supporting themselves entirely with slash-and-burn agriculture, which requires constant clearing of the forest and the fallowing of old fields, the chiefly lineage at Nakbe had opted to create a man-made landscape of continuous productivity. By the time it began to decline in importance, 2,200 years ago, Nakbe had established a pattern for the next generation of Maya chiefly centers: huge masonry pyramids, plaza groups linked by causeways, the court for a ritual ball game, and a carved stone monument.

The paramount center that took over from Nakbe was El Mirador, only eight miles to the northwest. Some 1,850 years ago El Mirador had grown to be the largest community in the region, covering an estimated 107 acres. Archaeologists William Folan and Ian Graham and remote sensing expert W. Frank Miller discovered a series of roads radiating out from El Mirador and leading toward its satellite communities. One of the roads leads southeast toward Nakbe, while another leads north toward a place called Calakmul. This situation likely reflects chiefly cycling: El Mirador, once a satellite of Nakbe, had turned the tables and made Nakbe its satellite.

El Mirador copied the east-west layout seen earlier at Nakbe. Its leaders took advantage of a natural hill at the western limit of this ceremonial axis, crowning it with one of the largest stone temple complexes ever seen in the Maya region. The centerpiece of this complex was a pyramid 180 feet high. On its flat summit sat three smaller pyramids with temples. The largest of these temples had a stairway flanked with eight grotesque stucco masks, all featuring jaguar claws; these claws suggested the nickname El Tigre for the building. A second complex called Danta (Tapir) stood more than 200 feet high.

Elsewhere at El Mirador was an acropolis that may include the earliest palace known from the Maya area. This possible palace, combined with the evidence for high levels of corvée labor and a raised road system linking El Mirador to its subject communities, suggests that a consolidation of power was under way in the region.

We have seen that it often took several generations of aggressive rulers to create a kingdom. In this case a succession of rulers at Nakbe and El Mirador had brought Maya society to the threshold of monarchy. Roughly 1,750 years ago, however, El Mirador suffered the same fate as Nakbe. One of its own satellite communities, Calakmul, rose to prominence and seized control of the basin from its former overlords.

As so often happens, nearby satellite communities had learned key lessons of statecraft by observing El Mirador. Calakmul carried the process one step further by creating a kingdom that endured for seven centuries. And because its leaders erected hieroglyphic monuments to themselves, we can be sure that the Calakmul state was an individualizing monarchy.

The Size of the Calakmul Kingdom

Calakmul was founded on a hill rising more than 100 feet above the surrounding lowlands. We do not know how large Calakmul was when Nakbe and El Mirador came to power. All we know is that once El Mirador had begun to fade, Calakmul grew aggressively until it was one of the largest cities in the Maya lowlands.

Calakmul's golden age was the period A.D. 400 to 700. The city came to include more than 6,250 buildings spread over 11 square miles, with a population estimated at 50,000. Calakmul erected 117 stelae (free-standing stone monuments), the most of any Maya city. The hieroglyphic texts on many of these stelae mention kings and their accomplishments, using a calendar more accurate than the one employed by the sixteenth-century Spaniards.

Calakmul was the capital of a territory with an administrative hierarchy of four levels. At its peak, the city was surrounded by six Level 2 towns, including settlements with names such as Naachtun, Oxpemul, Balakbal, and Uxul. These Level 2 towns all appear to be spaced one day's walk (about 20 miles) from each other and from the capital. Calakmul was linked to these towns by raised roads made of *sascab*, or crushed limestone. Each Level 2 town was in turn surrounded by Level 3 settlements, and so on down the hierarchy (Figure 55).

The heartland of the Calakmul kingdom was the 1,500-square-mile territory controlled by its Level 2 towns. Calakmul's political influence, however, could be felt over an area greater than 10,000 square miles. We know this because subordinate Maya towns often referred to their overlords when they carved hieroglyphic texts. Calakmul's subordinate towns did this by using an expression that can be translated "under the auspices of the Sacred Lord of Calakmul." On occasion, places as much as 150 miles away used similar phrases. Calakmul's attempt to annex such distant places, however, put it in direct conflict with other first-generation Maya kingdoms.

Three huge buildings dominated Calakmul's main plaza. Structures 1 and 2 were pyramids. Structure 2 was similar to El Tigre at El Mirador; it measured 395 feet on a side at its base and rose 150 feet above the plaza.

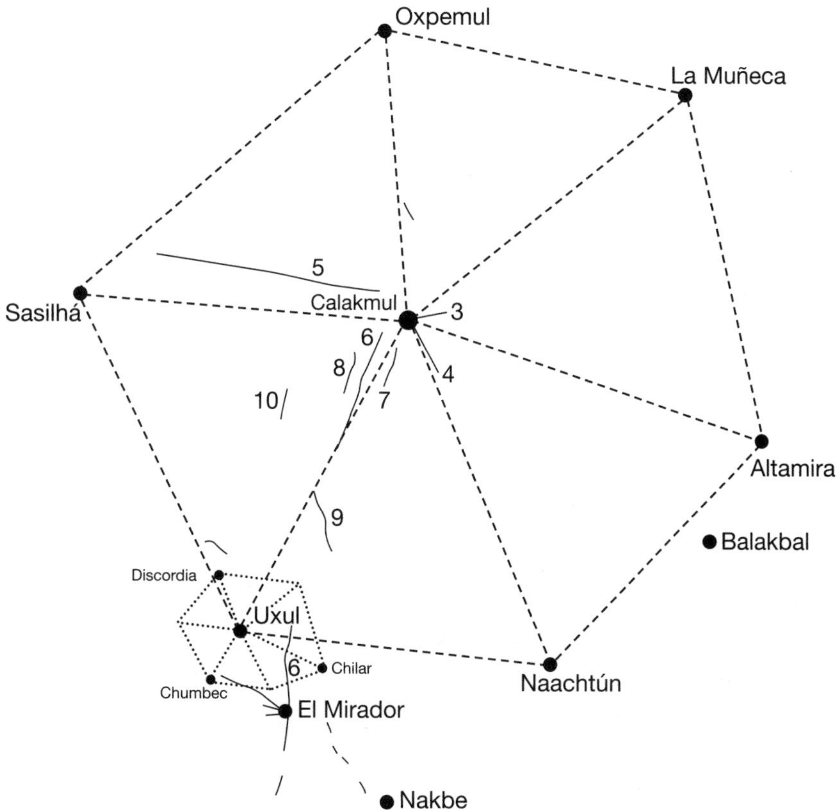

FIGURE 55. During its heyday the Maya city of Calakmul was encircled by six smaller cities that constituted the second level in its political hierarchy. In this drawing the dashed lines indicate straight-line distances among cities; the solid lines, accompanied by numbers, represent sections of actual pre-Hispanic roads that linked the capital to its subordinate centers. Also shown are Nakbe and El Mirador, two earlier paramount centers. (The distance from Calakmul to Level 2 towns such as Naachtún was about 20 miles.)

Structure 3 was an unmistakable palace atop a 16-foot-high platform (Figure 56). This royal residence measured 85 by 55 feet and was divided into a dozen rooms. Its ceilings were vaulted, and small windowlike ventilators promoted the circulation of air while preserving privacy. At least eight of the rooms might have been residential, while others were halls of some kind. In the rear was a likely throne room, accessible only after traversing three stairways and four doorways.

Under the floor of an inner room, William Folan and his colleagues found the tomb of an unnamed king interred around A.D. 400. The 30-year-old ruler wore jade ear spools and had been buried with three jade mosaic masks. Three jade plaques on his chest were incised with hieroglyphs. Other offerings included elegant pottery, a pearl, two spiny oyster shells, and a stingray spine for ritual bloodletting. The tomb was also equipped with a psychoduct, an opening in the wall through which the soul of the ruler could leave and return. We will see that the ancient Egyptians made similar arrangements for the pharaoh's soul.

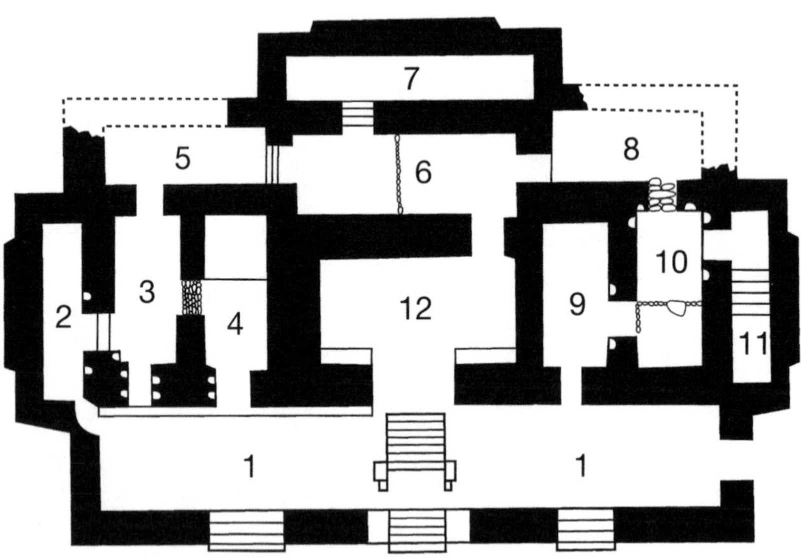

FIGURE 56. Structure 3 at Calakmul was a 12-room Maya palace, measuring 85 by 55 feet. Rooms 2–5 and 8–11 were probably residential. Room 12 was an entrance hall, and Room 7 may have been a throne room. A Maya ruler was buried in a tomb below Room 6.

Calakmul's Relations with Other First-Generation States

A chain reaction, similar to the one described earlier for the Zapotec and Mixtec, rippled through the Maya lowlands. Once first-generation kingdoms had begun to form, neighboring societies hastened to nucleate in order to avoid being taken over. The largest early kingdoms in the Maya lowlands were headed by Calakmul and Tikal, two cities located 80 miles apart. Their stormy relationship is recorded in hieroglyphic texts, with precise dates given in the Maya calendar.

The alleged founder of Tikal's royal lineage was a king named Yax Ehb Xook, who ruled just before A.D. 100. Burial 85, beneath Tikal's North Acropolis, is believed to be his skeleton. We do not know the name of this ruler's counterpart at Calakmul because so many early monuments are badly eroded.

During the sixth century Calakmul engaged in military alliances to gain the upper hand. During the period 562–572 a Calakmul ruler named Sky Witness called upon the distant city of Caracol to help him defeat Tikal. As a result of two attacks, one in 579 and another in 611, Calakmul's next ruler claimed victory over Palenque, a city in the Mexican state of Chiapas. In 619 his successor, Yuknoom Ti' Chan, reinforced Calakmul's earlier military alliance with Caracol by attending an event with the ruler of that city.

In A.D. 650 a Calakmul ruler named Yuknoom Ch'een II defeated the town of Dos Pilas, some 70 miles southwest of Tikal. At that time the lord of Dos Pilas was the son of the Tikal ruler, so this defeat was an indirect slap to Tikal. The young prince of Dos Pilas fled the scene, taking refuge in the neighboring city of Aguateca. According to the hieroglyphic texts, by the time he returned to Dos Pilas, this prince had become a vassal of Calakmul.

Emboldened by success, the Calakmul ruler attacked Tikal in 657, forcing its ruler, Nuun Ujol Chaak, into exile. This was the apogee of Calakmul's political power. In 677 the vassal ruler of Dos Pilas joined the Calakmul ruler in another attack on Tikal. This time, their hieroglyphic texts claim that they captured the Tikal ruler's second in command.

Unfortunately for Calakmul, its golden age came to an end in August 695. Hieroglyphic texts allege that the Tikal ruler Jasaw Chan K'awiil I defeated a Calakmul ruler whose name has been translated as "Claw of Fire." From that point on, Tikal held the upper hand over Calakmul.

In 849 one of Calakmul's final rulers, a man named Chan Pet, attended a summit meeting with the kings of Tikal, Seibal, and Motul de San José. By then his kingdom's political power was greatly diminished. Although Calakmul managed to put up one last monument in 909, that city faded from the scene not long afterward.

Inequality in Maya Society

Maya society was divided into hereditary nobles and commoners. There were differences in rank within each class. All lineages reckoned descent in the father's line and were ranked relative to one another. Most rulers were men, but there were exceptions. A woman from the highest lineage of a Level 1 city usually outranked a man from the highest lineage of a Level 2 town. Royal women could serve as regents when their sons were too young to assume power.

The Spaniards wrote down a hierarchy of titles for sixteenth-century Maya nobles. Because some of these titles appear in earlier hieroglyphic texts, we know that they had been in use for more than 1,000 years. Examples include the terms *ahaw* or *ajaw* ("lord") and *k'uhul ahaw* ("divine lord").

One royal woman from Calakmul, who married the ruler of Yaxchilán in the Mexican state of Chiapas, was known by three different titles: *ix Kaan ahaw* ("Lady of Calakmul"), *ixik k'uhul* ("Holy Woman"), and *lak'in kaloomte'* ("East Ruler"). She bore the Yaxchilán ruler a son. After her husband's reign ended in 741, the Calakmul woman held onto the throne of Yaxchilán for almost a decade, ensuring that her son would become ruler in 752. During those years her son was gaining military experience, so that when the time came he could demonstrate prowess in war. That prowess would allow him to outmaneuver his competitors—for example, any half brothers born to his father's other wives.

In the territory around Yaxchilán the lords of subordinate towns were called *sajalob'* ("provincial sublords"); the most highly ranked among them was designated *b'aah sajal*, "head sublord." Among the duties of a sublord were the collection of local tribute and its delivery to the capital and the delivery of war captives to the palace of the ahaw.

Sublords often fought in battle beside their overlords. For example, one battle scene carved on a lintel at Yaxchilán shows the divine king Bird Jaguar the Great and his "head sublord" Yellow Flint capturing two noble enemies. After the battle each of the victors was given the title "captor of [prisoner's name]." Such titles were awarded only in the case of elite captives and amounted to entries on a noble's resumé.

FIGURE 57. This carved stone wall panel uses the height of a throne and the steps of a staircase to communicate the Maya social hierarchy. The divine king of Yaxchilan, Itzamnaaj B'ahlam, sits cross-legged on his throne. His sublord, Aj Chak Maax, kneels on the top step of the staircase, presenting his ruler with three elite prisoners. At the bottom of the scene we see the prisoners, with their upper arms bound and their jadeite ear ornaments replaced with strips of cloth.

Some monuments depict obeisance and humiliation among the Maya. A carved wall panel from Yaxchilán shows a sublord presenting war captives to his divine king in 783 (Figure 57). The Yaxchilán king sits on a throne; his sublord kneels on the top step of a stairway; the prisoners kneel on a lower step, with their arms bound and their jade ear ornaments replaced with strips of cloth. The ruler's face expresses Zenlike serenity, while the captives grimace. The hieroglyphic text informed the elite minority who could read. The majority of Maya, who had not been taught to read, either had to rely on the scene to get the gist of the message or attend performances at which a minor priest chanted the text.

Literacy was not the only skill denied to Maya commoners. Even the most renowned war captain was prevented from usurping a sublord's position by his ignorance of sacred lore. The Maya lords who vetted him would rapidly discover that ignorance. The Spaniards report that in sixteenth-century Yucatán, any usurper who could not answer a set of official riddles would quickly be exposed.

Power-sharing among the Maya was exemplified by the noble councillors who met in the *popol naah,* or council house. Skilled commoners could fill important posts such as master craftsman, head sculptor, monument carver, scribe, or painter. Within the religious hierarchy they could also hold lower-order posts such as "keeper of the fire" or "keeper of the sacred book" (a reference to hieroglyphic texts on bark paper or deer hide).

The lineages of Maya commoners, often known by surnames inherited from specific male ancestors, held corporate rights to agricultural land. The Maya also had landless commoners, serfs, and slaves; the latter were usually war captives.

SOME THOUGHTS ON THE NEW WORLD'S FIRST-GENERATION KINGDOMS

The Zapotec, the Moche, and the Maya all created monarchies out of rank societies. They did so in the absence of European visitors and without any template for what a monarchy should look like. It is therefore significant that these ancient societies did it much the way that the Hawai'ians, Zulu, Hunza, and Merina did it: by forcibly uniting a group of rival societies.

All these cases began with societies that already possessed a degree of hereditary inequality. The engine that drove kingdom formation was competitive interaction among multiple elite actors. The balance was tipped when one of the actors achieved a competitive advantage. Whether set in the temperate highlands, the coastal desert, or the tropical forest, the process was similar. It was independent of environment or ethnic group.

Our three New World monarchies had something else in common. Once having created the apparatus of a kingdom, they expanded against neighboring groups. Expansion was facilitated by the fact that many groups could not defend themselves against the centralized control and military strategy of a newly formed monarchy. As a result, many pristine New World kingdoms reached their maximum size early in their history. The Zapotec, for example, reached their territorial limits about A.D. 200. After that, their outer provinces began to break away and form their own kingdoms. The same thing happened to the Moche around A.D. 600.

Among both the Maya and the Zapotec, as we have seen, the process of political consolidation set off a chain reaction. Monte Albán, Monte Negro, and Huamelulpan kept each other at bay. Calakmul and Tikal attempted to pick off each other's Level 2 and Level 3 communities; only rarely did they go head-to-head. By A.D. 900, the repetitive cycles of military conflict had weakened both kingdoms and contributed to their eventual collapse.

We now need to see whether the earliest kingdoms in the Old World were created in the same way. We will begin in the land of the pharaohs.

NINETEEN

The Land of the Scorpion King

The world's longest river has two main branches. The White Nile begins at Lake Victoria in Uganda. The Blue Nile begins near Lake Tana in Ethiopia. Both branches are fed by the summer rain of the tropics. White and Blue join at Khartoum in the Sudan and then pick up more water from the River Atbara. It is the last water the Nile will receive on its journey to the Mediterranean.

The Nile is almost 3,900 miles long, surpassing both the Amazon and the Mississippi. So full is its Blue branch at flood stage that its water dams up that of the White. Only after the Blue Nile has crested does the White deliver its burden, prolonging the high-water season. Once, before the era of hydroelectric dams, the flooding Nile carried almost six trillion cubic feet of water.

The bad news, of course, is that the Nile carries more water at Khartoum than it does at Cairo. As one travels north from Uganda, precipitation declines dramatically. Khartoum receives five to seven inches of rain each year; Cairo gets only one or two. For 1,300 miles, from Khartoum to the Mediterranean, the Nile crosses a desert whose evaporation exceeds its rainfall.

The result is an environmental contrast like that of Peru's coastal valleys: a long strip of green alluvium flanked by desert. Geographer George Cressey once estimated that all but 4 percent of Egypt was desert. Before the adoption of gasoline or electric pumps, the Nile could irrigate only seven million acres in a land of 386,000 square miles.

The ancient Egyptians called the Nile alluvium *kemet,* "the black land." They called the "red land" to either side *deseret,* from which we get our word "desert." The foragers of Wadi Kubbaniya and Jebel Sahaba gathered plants on the black land and hunted game on the red.

THE RISE OF FARMING AND HERDING

Eleven thousand years ago some Nile Valley hunter-gatherers had begun to live in camps like those of the Natufians. One of these camps lay in the Wadi Or, a dry canyon running toward the Nile just south of Wadi Halfa, Egypt. Its occupants lived in circular huts ten feet in diameter, their conical roofs supported by wooden posts. Their tools suggest that they were gathering edible seeds and bulbs and hunting game with bows and arrows. In addition to wild members of the horse and cattle families the foragers of Wadi Or ate hippopotamus and turtle from the Nile.

When domestic plants and animals were added to the economy of the Nile Valley, they came from two sources: local and foreign. The local species included native African sorghums and millets, cattle and pigs, and date palms. The foreign species included wheat, barley, flax, sheep, and goats, all of which were native to the Near East. These foreign species probably reached Egypt from the region occupied today by Jordanians, Israelis, and Palestinians. We will refer to this region by the generic term Southern Levant.

The Fayum Oasis

Egypt's Fayum is a large, wind-eroded depression, 60 miles south of Cairo and 15 miles west of the Nile. Its source of water is a side branch of the Nile that breaks off from the main river, runs parallel to the Nile for 120 miles, and forms a lake in the Fayum. The lake attracted both wild game and early farmers.

Several generations of archaeologists have investigated the early farmers. Gertrude Caton-Thompson and geologist Elinor Gardner pioneered Fayum archaeology in the 1920s. Willeke Wendrich and René Cappers have since discovered that agriculture was under way in the Fayum more than 7,000 years ago.

The Fayum farmers lived in circular huts with clay floors. Taking advantage of humid soils at the lake margin, they grew wheat, flax, and two races of barley. They also herded sheep and goats, fished the lake, and hunted wild game. Owing to the aridity of the surrounding desert, even their baskets and wooden implements are sometimes preserved. The Fayum farmers prepared the ground with stone hoes, perforated it with digging sticks, planted their cereals, harvested them with flint-bladed sickles, threshed them with wooden pitchforks, and ground the grain on stones. The meal was then made into porridge by mixing it with water and heating it in earthenware pots.

One of Caton-Thompson's most interesting discoveries was a series of 165 underground granaries, excavated in a dry cliff above the lake. At least 56 of these storage pits had been lined with basketry; some were still sealed with mud-plastered lids. Caton-Thompson estimated that each granary could have held the wheat or barley produced by a two-to-three-acre field.

Bir Kiseiba and Nabta Playa

While cereal agriculture spread over the Fayum and the Nile delta, a different lifeway was taking shape to the south. In the desert west of Wadi Halfa, near the border between Egypt and the Sudan, some of the world's first cattle pastoralists were learning how to use an inhospitable environment.

The desert near Bir Kiseiba featured a series of huge, shallow depressions that temporarily filled with rain runoff in the summer. Geologists refer to such depressions as *playas*, the Spanish word for "beach." More than 7,000 years ago, according to archaeologists Fred Wendorf, Romuald Schild, and Angela Close, groups of herders were pasturing their cattle at Bir Kiseiba.

When the depressions became playa lakes in the summer, grasses and herbs provided forage for the cattle. When the lakes dried up, herders moved to the center of the depression and dug shallow wells to tap the subsurface water. These herders traveled between playas and oases, using ostrich eggshells as water bottles and hunting gazelles, desert hares, and the wild North African donkey.

At one wind-deflated desert basin, called Nabta Playa, Wendorf and his group found evidence for clearly defined huts and storage pits. The stored foods included seasonally available wild sorghum, grass seeds, tubers, legumes, and fruits. The prehistoric wells dug in the center of Nabta Playa were larger and deeper than those at Bir Kiseiba. Some were walk-in wells, with steps cut in the sides.

The huts and wells did not surprise Wendorf's team. What surprised the group was that Nabta Playa had evidently served as a ritual center for the region. One ritual feature consisted of a north-south alignment of upright sandstone blocks. To the north of this alignment was a circle of smaller upright stones, carefully embedded in the playa sediments. There were, in addition, seven artificial mounds at Nabta Playa. One contained the burial of an adult cow, while the others held the remains of several cattle.

Let us consider the cultural legacies of the Fayum, Nabta Playa, and Bir Kiseiba. In the Fayum we see an early stage of the cereal economy that supported the later pharaohs. At Nabta Playa and Bir Kiseiba we see plausible ancestors for the cattle-keeping societies of eastern and southern Africa. The mitochondrial DNA of modern African cattle tells us that they are a domestic form of wild African cattle. Later societies such as the Nuer, Turkana, Fulani, Somali, and Zulu owe a debt to the early cattle herders of the playas.

THE APPEARANCE OF RANK SOCIETIES IN EGYPT

For two millennia, from roughly 7,000 to 5,000 years ago, the population of the Nile Valley grew at a spectacular rate. Each year the flooding Nile carried tons of organic mud north from tropical Africa, overflowing its banks and providing a new layer of fertile alluvium. These floods also flushed out the accumulated salts from the previous year, preventing the problems of salinization that accompany canal irrigation.

From January through April the Nile was low. By mid-May villagers near Khartoum could tell that the water was rising. Downstream near Aswan, in southern Egypt, it took until June for the flooding to show. By early July high water had reached the location of modern Cairo. Here the Nile broke up into dozens of channels called distributaries, fanning out over the river's delta. By September the flood was so high that all farmland was inundated, and low-lying depressions had become marshes. Finally, by the end of November, the water had receded, leaving a new layer of mud on which to plant.

Today we know that the water is coming from Uganda. The ancient Egyptians did not. In their cosmology all water was connected to *Nun,* a vast reservoir beneath the earth. The Nile flooded every year because water from Nun bubbled up from caverns near Aswan. The Egyptians predicted the start of the flood by observing a star named Sothis, which disappeared for 70 days each year and reappeared in the predawn sky around June 23.

In the first century A.D., long after the events of this chapter, the Roman naturalist Pliny the Elder visited Egypt. He reported that the Egyptians of that period measured the height of the Nile flood with instruments called *nilometers.* The units of rising water were called *ells,* and they were considered so accurate a predictor of harvest size that they determined the taxes paid by farmers along the Nile. In the words of Pliny a flood of only 12 ells meant hunger; 13 ells, sufficiency; 14 ells, joy; 15 ells, security; and 16 ells, abundance.

At various times between 7,000 and 5,000 years ago, societies displaying wealth and rank appeared along the Nile. Like the rank societies of Northern and Southern Mesopotamia, these Egyptian societies were sufficiently varied in architecture, burial ritual, and craft activity to be assigned different names. To put them into context, we must establish a few geographic landmarks.

The arid region along the border between Egypt and the Sudan is referred to as Nubia. Here the Nile cuts through some very hard stretches of granite bedrock, leaving a series of rapids called the Five Cataracts of the Nile. The Second Cataract, near Wadi Halfa, lies to the east of Bir Kiseiba and Nabta Playa. The First (or northernmost) Cataract, near Aswan, lies at the point where the waters of Nun allegedly bubbled to the surface. From this cataract, it is 750 miles to the Mediterranean.

North of Aswan the Nile runs straight for a time and then turns east, north, and west again in a loop called the Great Bend of the Nile. The Great Bend is the scene of famous landmarks such as the Valley of the Kings and the ancient cities of Karnak, Luxor, and Abydos. North of the Great Bend the Nile passes Tell el-Amarna, the capital of the ruler Akhnaten.

This stretch between the First Cataract and Amarna is Upper (as in upstream) Egypt, and it played a major role in the creation of Egypt's first-generation kingdom. Lower (as in downstream) Egypt lies to the north of Amarna and includes such landmarks as ancient Memphis, the pyramids of Giza, the Nile delta, and the modern city of Cairo.

The early villages of Upper and Lower Egypt had a number of features in common. Included were circular houses; the growing of wheat, barley, and flax; the raising of sheep, goats, and cattle; coppersmithing; active trade with the Southern Levant; and the ritual burial of animals.

Now let us consider the differences. Most villages of Upper Egypt buried their dead facing west, toward the setting sun. Their burials showed precocious use of silver, lapis lazuli (a semiprecious blue stone from the Near East), gold from Nubia, coral from the Red Sea, and ivory from East Africa. Their elite used abundant cosmetics, which they ground on effigy palettes of slate or siltstone.

Many early villages of Lower Egypt buried their dead facing east, toward the rising sun. Their use of sumptuary goods was modest compared to Upper Egypt. Some villages were large and appear multiethnic, with several distinct house types. Their copper came from southern Jordan and the Sinai Peninsula. Perhaps because of their well-watered delta setting, they were able to raise many more pigs than the Upper Egyptians.

Between Upper and Lower Egypt was a stretch of river with fewer villages per square mile, like a sparsely occupied frontier between competing rank societies.

LOWER EGYPT

The 60-acre village of Merimde Beni-Salame once occupied a Nile distributary in the western delta. Its occupants grew wheat, barley, and lentils, herded sheep, goats, and cattle, and raised hundreds of pigs, abandoning Merimde only after the distributary changed its course 6,000 years ago. The Nile supplied Merimde with hippopotamus, crocodile, and waterfowl; the desert to the west supplied it with antelope, gazelle, and ostrich.

Originally discovered by archaeologist Hermann Junker in the 1920s, Merimde has been extensively excavated over the years. There are hints that extended families occupied clusters of circular huts. The wickerwork roof of each hut was supported by a large central post and a series of small outer posts.

In some cases huts were grouped into a kind of residential compound, set apart from neighboring compounds by a reed-bundle fence. Included within these compounds were the following features: basketry-lined granaries like those of the Fayum; large water jars set in the ground with their mouths at the surface; and threshing floors where cereal grains could be separated from chaff.

The compounds of circular huts at Merimde remind us of those built by many clan-based societies of central and East Africa today. Such compounds contrast strongly with the residences of the Samarran and Halaf societies in Northern Mesopotamia, with which they were contemporary. The Northern Mesopotamian societies had more signs of rank and built large, rectangular, extended-family houses.

One of the largest villages of Lower Egypt was Ma'adi, on the east side of the Nile near Cairo. Because of encroachment by modern Cairo, the site's full extent will never be known. Occupied between 6,000 and 5,000 years ago, Ma'adi overlapped in time with the 'Ubaid societies of Mesopotamia.

At least three types of houses were built at Ma'adi, suggesting differences either in rank or ethnicity. The simplest houses, amounting to round huts or oval shelters, had posts of tamarisk wood. Floors were often below ground; to make it easier to descend into the hut, the occupants used hippopotamus bones as steps.

Interspersed with these oval huts were rectangular buildings made from logs and mud; the largest of these was roughly 17 by 10 feet. We do not know whether these were ritual houses or the residences of privileged families.

Finally, in one area of Ma'adi, there was a group of truly unusual houses. These were subterranean residences, excavated six to ten feet below ground. Their entrances were slanting passageways with steps cut into them. Around the margins of the residential chamber, the builders had driven posts into the floor to support a roof of woven mats. In the center of each floor was a sunken hearth.

These subterranean houses at Ma'adi were remarkably similar to those of Shiqmim, a village near Beersheba in the Negev district of the Southern Levant. This similarity was no coincidence; the Ma'adi houses also contained pottery similar to that found at Shiqmim. It is likely, therefore, that a small enclave of traders from the Negev lived at Ma'adi.

Any trade, of course, was a two-way one. A prehistoric village in the Gaza district of the Southern Levant, known simply as "Site H," seems to have been a trade enclave filled with goods from Lower Egypt.

One of the resources Ma'adi wanted most was copper, for which there was a foreign source. Analyses of the trace elements in Ma'adi's copper artifacts, according to Andreas Hauptmann, indicate that most of the copper came from the mines of Feynan in southern Jordan. Some of the copper ore was made into ingots for transport. These ingots would later be melted and cast into adzes, axes, fishhooks, pins, wire, or copper sheets.

Copper is heavy, but the people of Ma'adi had a new form of transportation: the donkey. Wild donkeys are native to North Africa and had been hunted for thousands of years. Roughly 5,500 years ago the Egyptians domesticated them. Pack trains of donkeys escalated trade, the way pack trains of llamas did in the Andes. In the twinkling of an eye, archaeologically speaking, donkeys spread to the Levant and were on their way to Mesopotamia.

In addition to its varied house types there are other signs that Lower Egypt may have been a multiethnic society. There were at least three cemeteries at Ma'adi, and although their occupants had been buried at somewhat different moments in time, the skeletons show a range of anatomical features. Burials in the south cemetery, for example, seemed to represent people who were taller and more heavily built and had biological ties to central or eastern Africa. Cemeteries elsewhere at Ma'adi contained skeletons whose biological ties were to the populations of the Mediterranean basin. In later periods the

Egyptian state would make a point of depicting both African and Mediterranean people in its art.

Many burials in the south cemetery at Ma'adi had little in the way of sumptuary goods. Other people in the same cemetery, however, were buried with ivory combs in their hair, polished stone vases, and elegant cosmetic palettes. It is likely that Ma'adi society not only had differences in rank but neighborhoods of people with different ethnic identities and beliefs about the afterlife. Such diversity probably resulted from the fact that the highly productive Nile delta was a magnet for settlers in an otherwise desolate region. One way that such diverse groups could be integrated was by occupational specialization: farmers, herders, potters, traders, and coppersmiths all needed each other's products.

The discovery of a likely defensive palisade and ditch at Ma'adi suggests that not everyone attracted by the delta was peaceful. Like the cotton and coca lands of Peru's middle river valleys, the Nile delta was coveted by ambitious neighbors. With the wisdom of hindsight, we know that some of those ambitious neighbors lived in Upper Egypt. Let us now look at them.

UPPER EGYPT

In Upper Egypt the Nile alluvium was a ribbon of green between dusty cliffs. Among the preferred sites for villages were rocky spurs descending from the cliffs, close to alluvium but too high to be flooded.

Typical of the smaller villages was Hemamieh, to the east of the Nile between Amarna and the Great Bend. Here the villagers grew cereals, legumes, and flax, raised sheep, goats, cattle, and pigs, fished the river and harvested palm fruits, and collected the same sedge bulbs seen earlier at Wadi Kubbaniya. They lived in circular huts three to eight feet in diameter, with domed, mud-plastered roofs. A section of mud wall 30 feet long extended out from Hut 242, perhaps marking the limits of a residential compound. Another hut was found to be filled with dried sheep or goat dung, which is used as fuel in the region today.

Archaeologists Fekri Hassan and T. R. Hays conducted a survey of sites within the Great Bend of the Nile. They found that 5,800 to 5,500 years ago most villages had populations of 50 to 250 people. Some 5,500 to 5,200 years ago there were fewer but larger villages and more evidence of sumptuary goods. Evidently an emerging Upper Egyptian elite was concentrating its followers in larger and more defensible settlements.

One of the largest communities in the Great Bend was Naqada, some 20 miles north of Luxor. The people of Naqada had discovered that gold could be mined in the mountains of Nubia, between the Nile and the Red Sea coast. Some 5,600 to 5,400 years ago Naqada's elite families were occupying rectangular houses with courtyards and mud-brick walls, while commoner families continued to live in circular huts. The people of Naqada were weaving both flax and wool, using the potter's wheel, and brewing beer from barley and wheat.

During the 1890s pioneering archaeologist Sir Flinders Petrie began work at Naqada. He detected an ancient division of the community into a "north town" and "south town" and found three different prehistoric cemeteries. Petrie excavated a staggering 2,200 burials from these cemeteries and left no doubt that Naqada society had hereditary rank. Its elite members were buried in mud-brick tombs with gold and silver baubles; combs and bangles of ivory; lapis lazuli; obsidian from Turkey or the Aegean; amazonite, a blue-green stone from the Sahara; flint daggers with silver or ivory handles; cosmetic palettes in the shape of fish and other creatures; and vases carved from attractive stone.

Some nobles were buried with lower-ranking individuals who may have been servants or slaves. Commoners at Naqada were buried in simple graves, with pottery vessels but little or no evidence of sumptuary goods.

Tomb 1863 at Naqada was noteworthy. It held the remains of a young girl buried with a stone vase, two cosmetic palettes, ivory bracelets, a bone comb, a pottery dish imported from the Sudan, and a seal imported from Mesopotamia. It is unlikely that this girl had achieved enough in her short life to deserve such offerings; she must have been born to a family of high rank.

In Naqada's south town Petrie found what he believed to be a thick, mud-brick fortification wall. This wall tells us to expect archaeological evidence for chiefly competition, dense concentrations of farmers, craftsmen, and warriors, and the intensified agriculture necessary to support them all.

CHIEFLY CYCLING AND UNIFICATION

Upper Egypt presents us with clear examples of chiefly cycling. According to Egyptologist Barry Kemp, at least three competing rank societies arose between 5,500 and 5,200 years ago (Figure 58). One lay in the Great Bend and had Naqada as its paramount center. The second lay downstream from the

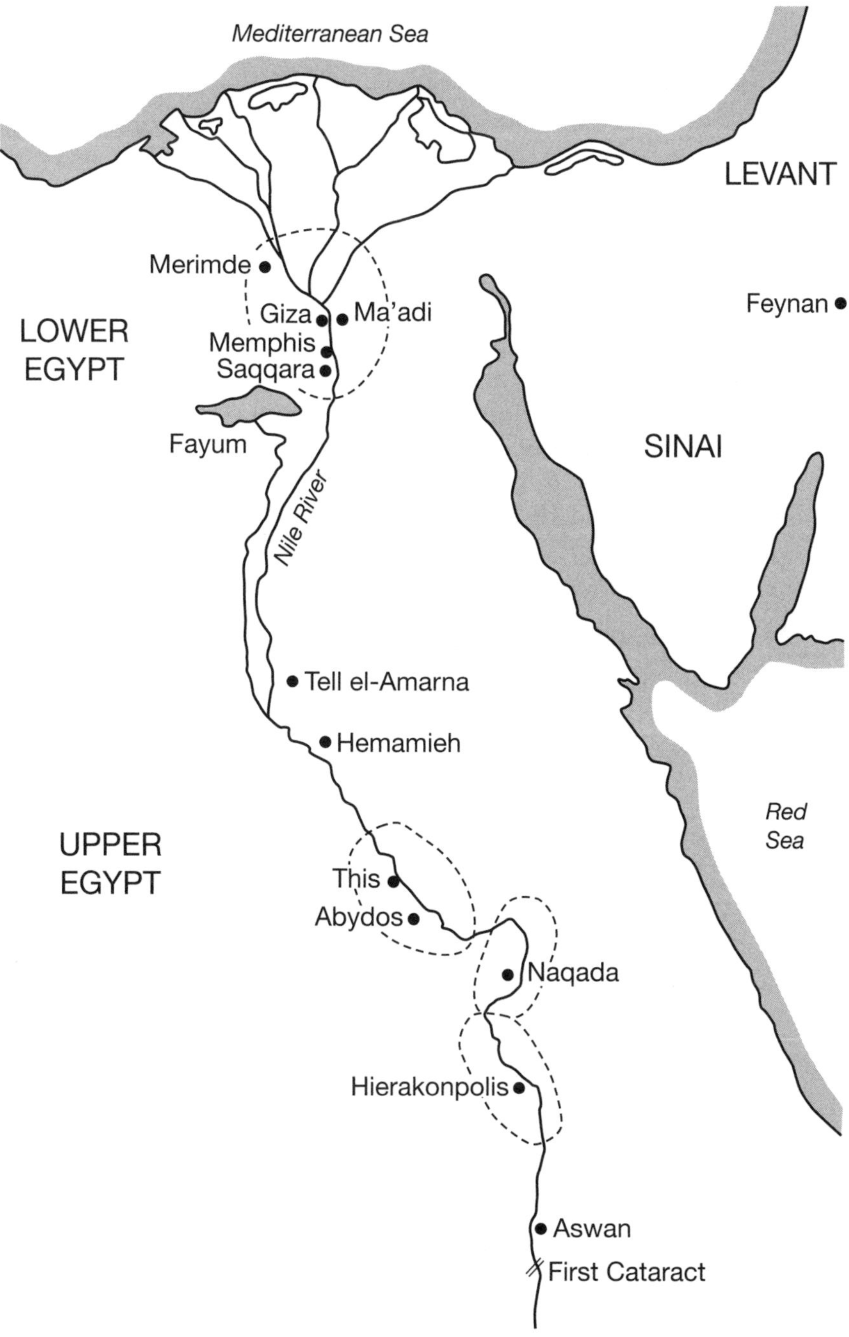

FIGURE 58 Egypt's first kingdom was created by the gradual unification of formerly independent societies. First came the unification of the territories of Hierakonpolis, Naqada, and This in Upper Egypt. Next came the unification of Upper and Lower Egypt, leading to Dynasty 0. On this map, dashed lines surround some of the best-known territories involved in this process. (The distance from Ma'adi to Aswan is 430 miles.)

bend, in the Abydos region, and had a town called This as its paramount center. The third society lay upstream, between the Great Bend and the First Cataract. The paramount center of this society was Nekhen, a town better known by its Greek name, Hierakonpolis. These societies seem to have been very powerful, so powerful that Irving Goldman would probably have assigned them to his stratified category.

Naqada appears to have been the first to reach its peak—5,400 to 5,200 years ago it was likely the most powerful of the three societies. At the same time the defensive wall in Naqada's south town may reflect concern over aggressive neighbors like Hierakonpolis. Those fears may have been well founded, for Hierakonpolis does seem to have outstripped Naqada approximately 5,200 to 5,000 years ago.

Archaeologists Michael Hoffman, Hany Hamroush, and Ralph Allen studied the region of Hierakonpolis and presented a step-by-step scenario for its rise to prominence. The scenario covers an area of 56 square miles surrounding Kom el-Ahmar, the sprawling archaeological site that once was Hierakonpolis.

After centuries of slow colonization by farmers, the area studied by Hoffman's team underwent a period of rapid growth. Between 5,700 and 5,400 years ago an estimated 5,000 to 10,000 people occupied nine settlements in the area.

The bulk of the population lived at the town of Hierakonpolis itself, which overlooked an alluvial floodplain six miles long and up to two miles wide. This town had the largest and most richly furnished tombs in the area. Under the patronage of their chiefs, the craftspeople of Hierakonpolis produced maceheads for elite warriors, plum-colored pottery, stone vases, flint daggers, cosmetic palettes, linen textiles, and lots of beer. Although thousands of domestic animals were raised at Hierakonpolis, a microscopic analysis of its coprolites, or desiccated human feces, reveals that commoners' meals consisted mainly of cereals. The evidence, in other words, suggests a society in which elite families had greater access to meat.

Between 5,500 and 5,200 years ago the leaders of Hierakonpolis were concentrating their followers into fewer—but larger and more defensible—settlements. This change probably reflects increased competition with neighbors such as Naqada.

Early in this period Hierakonpolis built a temple 42 feet on a side, equipped with a large, oval courtyard. The presence of wine jars from the Southern Levant indicates that the temple staff was importing wine, which the Egyptians appreciated but did not have the proper climate to produce. Among the creatures sacrificed by the priests of Hierakonpolis were cattle, sheep, goats, Nile crocodiles, turtles, and fish. Unfortunately, we know little about the ritual structures that preceded the first Egyptian temples.

Late in this period the leaders of Hierakonpolis ordered the building of a mud-brick defensive wall. Like the early rulers of Monte Albán, they were now ready to expand against their neighbors. It would seem that their first move was to take over the territories controlled by Naqada and This, unifying Upper Egypt into a first-generation kingdom. That kingdom was so powerful that its next step was to move against Lower Egypt, creating an even larger state.

We suspect that the unification of Egypt was a process that took centuries and proceeded step-by-step. Unfortunately, because of the large area involved, we cannot yet see the details of the process. We must infer what happened by combining bits of archaeological data from several different places.

Hoffman's team believes that Hierakonpolis achieved supremacy between 5,200 and 5,100 years ago. During that period it continued to grow, building large palaces and temples at Hierakonpolis itself and establishing an isolated royal cemetery in the desert. Such a royal *necropolis,* or "city of the dead," suggests that the hereditary elite of Hierakonpolis belonged not simply to an elite lineage but to a separate social stratum. The precedent had been established, in other words, for later royal cemeteries like those in the Valley of the Kings.

In 1898, long before Hoffman's survey, archaeologists James Quibell and Frederick Green discovered a broken limestone macehead at Hierakonpolis. Carved on the macehead was a scene showing a ruler digging an irrigation canal. Since it is doubtful that rulers performed such manual labor, the scene is presumed to be symbolic. Most interesting is the fact that the ruler's hieroglyphic name is given as "Rosette Scorpion."

Might this "Scorpion King" be buried at Hierakonpolis? The site has produced a number of royal tombs, but none can be linked to Rosette Scorpion. Perhaps the most elegant was Tomb 11, which measured 16 by 8 feet. This tomb included a wooden bed whose legs ended in bull's feet; ornaments of gold, silver, copper, carnelian, garnet, turquoise, and lapis lazuli; carvings of ivory; and pottery effigies of humans and animals.

Complicating our search for the elusive Scorpion King is another tomb at Abydos, in the downstream territory once headed by This. Known as Tomb U-j, it was made of mud brick and measured 29 by 23 feet; its wooden roof beams have been dated to 5,150 years ago. Tomb U-j had a large chamber for the coffin and 11 smaller rooms for the burial offerings. Excavator Günter Dreyer believes that this tomb may be a smaller version of the palatial residence that the ruler had occupied during his life. He had been supplied with an ivory shepherd's crook, a symbol both of political authority and of the pastoral legacy of some of his subjects.

We know what some of this ruler's priorities were, because his offerings included numerous jars of locally made beer and 700 jars of wine from the Southern Levant. One remarkable discovery in Tomb U-j was a series of ivory tags, each with a hole drilled in it so that it could be attached to a funerary offering. The hieroglyphs on these tags indicate the names of the places from which the offerings had come. They also provide the name "Scorpion."

Who was the Scorpion buried at Abydos? Was he the same Scorpion mentioned on the macehead at Hierakonpolis? Was he the ruler who unified Upper Egypt? Does he belong to history, legend, or both?

The Palette of Narmer

In addition to the Scorpion macehead, Quibell and Green found a unique siltstone cosmetic palette at Hierakonpolis. The palette was two feet long and shaped like a warrior's shield (Figure 59). On both sides of the palette we see the hieroglyphic name of the ruler, placed within the rectangular symbol for "palace." The stone carvers used rebus writing for his name, combining the glyphs for "fish" *(nr)* and "chisel" *(mr).* Since Egyptian writing lacked vowels, readers of the text would have supplied them; when we do so, his name becomes "Narmer."

To understand the scenes on the Narmer palette, we must consider the later symbolism of Egypt. The Upper Egyptian ruler wore a white crown and was referred to as *nswt,* "the sedge ruler." Upper Egypt was symbolized by the lotus; its patron deity was Nekbet, the vulture goddess; its paramount center was Hierakonpolis. The Lower Egyptian ruler wore a red crown and was referred to as *bity,* "the bee ruler." Lower Egypt was symbolized by the papyrus, a sedge growing in the delta marshes; its patron deity was Wadjet, the cobra goddess; its paramount center was Buto.

On one side of the palette from Hierakonpolis we see Narmer grasping a captive by the hair, preparing to crack his skull with a mace. A servant stands behind him, carrying the ruler's sandals and water jar. Below the ruler are two sprawling enemies. Near his face is a falcon, standing on the papyrus symbol for Lower Egypt and holding a captive by a cord through his nose. Narmer wears the white crown of Upper Egypt.

On the opposite side of the palette Narmer wears the red crown of Lower Egypt. Still accompanied by his sandal bearer, he inspects the scene of a battle. That scene includes ten decapitated enemies, laid out with their heads between their ankles. Elsewhere on this side of the palette a bull is depicted battering down the walls of a fortified town. The town is identified by the hieroglyph *sh,* thought to be a reference to the Southern Levant.

We cannot date either the Narmer palette or the Scorpion macehead because they were found in a cache of objects that were heirlooms from an earlier era. The Narmer palette suggests that the Egyptians themselves considered the unification of Egypt to have involved bloodshed. First, a ruler from Upper Egypt conquered the delta; once in charge of the delta, he expanded into the Southern Levant.

FIGURE 59. Two sides of a carved cosmetic palette, found more than a century ago at Hierakonpolis. Both sides show a ruler whose hieroglyphic name would have been pronounced "Narmer." On the left we see Narmer, wearing the white crown of Upper Egypt, seizing a captive by the hair. On the right, wearing the red crown of Lower Egypt, he inspects a battlefield where ten bound and decapitated enemies lie with their heads between their feet. These scenes are thought to symbolize the unification of Upper and Lower Egypt.

There is independent confirmation of Egypt's expansion into the Southern Levant. According to Thomas Levy and David Alon, Egyptian pottery bearing the hieroglyphic name Narmer has been found at archaeological sites along the border between Judea and the Negev. The Egyptian presence in the Southern Levant may have lasted only 50 to 100 years, however; after that, the settlements in the Levant fortified themselves and defended their autonomy.

Dynasty 0

The first-generation Egyptian state was clearly a monarchy—and an individualizing monarchy at that. Unfortunately, the earliest monarchs were shadowy figures, part history and part legend. Egyptologists consider 3100 B.C. to be an educated guess for the date by which Egypt was unified.

That date would presumably mark the beginning of Egypt's first royal dynasty, which moved its capital to Memphis near the head of the Nile delta. Now that the Egyptian state stretched from the First Cataract to the Southern Levant, Hierakonpolis lay too far to the south to serve as its capital. Once the capital had moved to Memphis, Hierakonpolis declined in importance but continued to serve as a provincial center for Upper Egypt.

For years it was assumed that King Narmer was the founder of Dynasty 1. Increasingly, however, there are hints that there may have been an even earlier, semilegendary set of rulers, including the mysterious Scorpion King. This possibility has forced Egyptologists to propose a Dynasty 0. The Egyptians of Dynasty 0 show us many behaviors typical of later dynasties. Their rituals included burying wild animals such as hippopotamuses, crocodiles, and baboons. Like the people of Nabta Playa, they also buried domestic cattle.

As early as 3100 B.C., some of the scenes painted on Egyptian pottery show boats rowed by more than a dozen oarsmen. Some of these boats feature an individual seated under a sunshade in a kind of cabin. The implication is that early rulers made journeys up and down the Nile, visiting other nobles or checking up on the officials who ran the provinces of their kingdom.

Dynasties 1–30

Little by little the rulers of Egypt emerged from legend and into the archives of history. Perhaps the first historian to divide the Egyptian kings into dynasties was Manetho, a priest of the third century B.C. He coined the term *dynasty* to refer to a multigenerational sequence of kings related by common descent. Each break between dynasties provided an opportunity to move the capital or add new territory. Sometimes, however, the new ruler proved weaker than the old. The result was a series of cycles similar to those seen in rank societies.

The Egyptian state, as we have seen, was created by unifying a number of formerly autonomous territories. These territories became the *hesps,* or administrative districts, of the Egyptian state. Upper Egypt was divided into 22 hesps, Lower Egypt into 20. Each hesp (or *nome,* as a hesp was called by Greek historians) had a governor who was supposed to be loyal to the king. Under strong kings, governors were more subservient; under weak kings, they exercised more autonomy. Such cycles of strength and weakness probably characterized most long-lived states. They directly affected the levels of inequality between kings and their governors.

Manetho divided the Egyptian kings into 30 dynasties by drawing on a number of earlier king lists in the temple archives. Interestingly enough, none of these early texts mention Narmer or Scorpion. They list the founder of Dynasty 1 as Men (in Egyptian) or Menes (in Greek). Such ambiguity is not unexpected at the boundary between history and legend.

Egyptologists group Dynasties 4–6 (4,700 to 4,300 years ago) into a cycle of powerful rulers called the Old Kingdom. They call the next cycle of powerful rulers, Dynasties 11–14, the Middle Kingdom (4,000 to 3,600 years ago). They group Dynasties 18–20 (3,500 to 3,000 years ago) into another cycle of powerful rulers called the New Kingdom. Between these three cycles of centralized power were "Intermediate Periods," when the governors of provinces had greater autonomy. We use this framework in the pages that follow.

COSMOLOGY AND DIVINE KINGSHIP

Ancient Egypt lay at one extreme of Herbert Lewis's continuum of monarchies: the king, or *nesw,* was considered a deity. In at least one version of Egyptian cosmology, Re, the Sun, was one of the divine creators of the world and also the first ruler; all subsequent kings were on a par with him. The king was also strongly identified with the palace and by Dynasty 18 had come to be called "pharaoh," a word derived from *per-aa,* "palace."

Whenever the king appeared at public events, he was described by the same verb—*khay,* "to shine forth"—used to describe the Sun at sunrise. Any statement made by a ruler exuded *ma'at,* "truth," "order," and "justice."

In another cosmology, Geb, the male Earth deity, and Nut, the female Sky deity, produced four divine children: Osiris, Isis, Seth, and Nephthys. Male Osiris represented the underworld and female Isis the throne of Egypt. They mated to produce Horus, the falcon we already saw associated with the ruler on the Narmer palette. The mating of siblings Osiris and Isis provided cosmological justification for the king's marriage to his sister or half sister, in the event that she were the most highly ranked bride available. We have seen similar justification for chiefly sibling marriages in powerful Polynesian societies.

The ancient Egyptians had no word for "state." No such word was needed, because all aspects of the state were concentrated in the ruler. His welfare was so important that even his corpse continued to receive deliveries of food. This fact leads us to one of Egypt's most talked-about practices: the entombment of the ruler's mummified body.

There were three parts to the ruler's soul: the *ka,* the *ba,* and the *akh.* The ka was a vital force that would stay alive as long as it was fed. As a result, rulers frequently paid in advance to have food brought to their tombs.

The ba, often pictured as a human-headed bird, had the power to leave the ruler's body by day and return at night. The ba could inspect the ruler's kingdom by flying over it but would have no place to sleep if his body decayed. The tombs of some Egyptian rulers had psychoducts (like the tomb we saw in the Maya city of Calakmul) to facilitate the coming and going of the ba.

The akh was that part of the ruler's soul that rose to dwell eternally among the stars. Because a king could "shine forth" like the Sun, his akh would twinkle from the night sky forever.

Like the ancient Panamanians, who preserved their chiefs' corpses by smoking them, the Egyptians sought to preserve the corpse to which the ruler's ba returned at night. Unfortunately, placing a body in a tomb like King Scorpion's at Abydos was incompatible with preservation. Even worse was burying the ruler beneath a *mastaba,* a monumental stone or mud-brick platform, as was done to many kings of Dynasties 1 and 3. The sealed mastaba retained too much moisture, and the ruler's body decomposed.

After centuries of trial and error, Egyptian morticians hit upon several methods of preserving corpses. Which one they used depended upon the rank and wealth of the deceased. Rulers had their internal organs removed and their bodies desiccated in *natron,* or Glauber's salt, a form of hydrated sodium bicarbonate. The word natron comes from the Wadi el-Natrun, a desert canyon where this salt was available by the ton. After dehydration the body was wrapped tightly in natron-soaked strips of linen.

Not every attempt at dehydration produced a lifelike corpse. Some bodies became rigid, and others turned so black that they appeared to have been dipped in pitch, or natural asphalt. Early archaeologists overheard their workmen describe these blackened corpses with the Arabic word for pitch, *mumiya*. This gave us the word "mummy."

While the morticians were experimenting with mummification, the kings' architects experimented with grander and grander mastabas. Finally, one team of Dynasty 3 architects stacked a series of increasingly smaller mastabas one upon another, producing a stepped pyramid.

Some 4,800 years ago, Zoser, a ruler of Dynasty 3, had a stepped pyramid built to cover his future burial site at Saqqara. His architect, a man named Imhotep, at first designed a pyramid of four steps, which was to rise directly over the mastaba. This building was later incorporated into a larger pyramid of six steps, 204 feet tall and measuring 411 by 358 feet at the base.

The Age of Pyramids had begun. Egyptian kings would eventually commission more than 90 pyramids at places such as Saqqara, Dahshur, Meidum, Giza, and Abu Sir, all west of the Nile in Lower Egypt. Some 4,600 years ago, architects learned how to build pyramids with sloping sides instead of steps. Their efforts culminated in the building of three huge Dynasty 4 pyramids at Giza. These monuments reminded Greek visitors of the peaked loaves of bread they called *pyramidia*, giving us the term we use today.

Egypt's Giza pyramids and Peru's Moche pyramids had two things in common. Both were built by multiple work gangs that took credit for their contributions. Moche laborers incised their bricks with makers' marks; Egyptian laborers painted names such as "The North Gang," "The Victorious Gang," or "The Drunken Gang" on the stones they contributed. The last name reminds us how often, throughout world history, labor has been rewarded with rations of beer.

Another similarity is that in both regions the largest pyramids were built by first-generation kingdoms. No later Peruvian state produced pyramids equal to the Huacas de Moche, and no later Egyptian dynasty matched the Great Pyramid of Khufu, 478 feet high and 756 feet on a side. Lavish use of corvée labor on monuments was typical of first-generation states; later states usually had other priorities.

THE CHANGING LOGIC OF POLITICAL ADMINISTRATION

The Egyptians were not aware that the kingdom in which they lived had developed from earlier rank societies. They believed that the institution of kingship was as old as Earth. This view fit with a cosmology in which the universe was unchanging, a view to which we will return in our discussion of Egyptian religion.

For most of the societies we have seen so far, the alphas in the dominance hierarchy were supernatural spirits or deities; the betas were ancestors; and the most highly ranked living humans were gammas. The Egyptian monarchy is the first we know of in which the ruler was, in effect, one of the supernatural alphas.

Divine kingship created an enormous gap between the ruler and the next most important authority figure. The governors of the 42 hesps might be hereditary nobles, but they were not the Sun incarnate.

There was a lengthy administrative hierarchy below the ruler. To be sure, it changed over the course of 30 dynasties as offices were raised, demoted, or combined. No summary, including the one we are about to give, could do it justice.

As in the later Hunza state, the king's second in command was his vizier *(wazir)*. During the Old Kingdom, kings tended to name their uncles, brothers, or sons to this post. By the time of the New Kingdom, many rulers had learned that it was better to appoint a loyal commoner than an ambitious noble. The latter might become a usurper.

In addition to the vizier, many Old Kingdom officials were relatives of the king. Close male relatives became state treasurers and high priests; more distant relatives became district officials. The governors of hesps inherited their positions until roughly Dynasty 12, at which point expertise became a more important criterion than noble birth.

The rise of the skilled commoner is exemplified by Uni, an official of Dynasty 6. Uni began his career as "undercustodian of royal domains," worked his way up to "superior custodian," and later became a district judge in the hesp of Hierakonpolis. When the king discovered a conspiracy in his own harem, he bypassed the vizier and made Uni his confidential investigator. Uni was next placed in charge of an army and led five campaigns against the Sinai and Southern Levant. Finally, the king named Uni the Governor of the South, an extraordinary honor for a commoner.

Another important office in the Egyptian state was that of the scribe. Thanks to numerous hieroglyphic texts, we know that the chain of command was (1) the vizier, then (2) the scribe, then (3) the overseer, and finally (4) the common laborer. Each hesp had a similar chain of command, allowing orders for corvée labor to travel down the hierarchy while tribute traveled up. So important was tribute that from Dynasties 15–17 the royal treasurer temporarily became more powerful than the vizier.

Because the king was a deity, he was not seen as needing to share power with anyone. Egypt did have a council called a *kenbet,* but its job was to advise the vizier.

A huge staff of servants attended the ruler, including a sandal bearer like the one depicted on the Narmer palette. There were, in addition, keepers of the king's robes and crown; the king's barbers; the king's physicians, cooks, and messengers; and entertainers of various kinds.

At the bottom of the social ladder were slaves, mostly captives taken in war. While slaves were assigned a variety of tasks, Cecil B. DeMille's notion that they built the pyramids was pure Hollywood. Archaeologist Mark Lehner's work has shown us that the pyramids at Giza were built by teams of loyal Egyptian commoners, conscripted for the task and housed in special barracks at the state's expense. Similar teams were drafted to quarry stone, hunt elephants for their ivory, act as porters on royal trading expeditions, and serve as foot soldiers in war. All such workers received standard rations from the state. By the Middle Kingdom, soldiers were being issued wooden tokens shaped like loaves of bread; these tokens could be redeemed for actual bread.

Finally, we should say a word about the impact of divine kingship on the economy. Like protohistoric Hawai'ian chiefs, the Egyptian kings controlled all land, all important resources, and all foreign trade. The economy of Egypt depended on the distribution of raw materials and goods through the king and his agents. This applied not only to gold from Nubia, cedar from Lebanon, wine from the Levant, and spices from Eritrea but also to locally produced commodities such as wheat, barley, cattle, and linen. The word for merchant *(swy.ty)* was unknown before Dynasty 18, and even then it was applied mainly to temple officials who had been granted special permission to engage in foreign trade.

To be sure, there were local markets in which surplus crops, birds, fish, and wild game could be bartered. Such free enterprise, however, remained marginal to the top-down, command economy of the ruler. Working through the governor of each hesp, the king demanded his cut of every cereal harvest, every domestic herd, and every fisherman's catch. The vast resources brought to his storehouse were used to support the huge staff below him.

The Tension between Palace and Temple

The most significant aspects of the universe for the Egyptians were those that were timeless and unchanging. While Western societies tend to celebrate unique events and individuals, the Egyptians celebrated the static and eternal. This is one reason animals were considered so meaningful. Humans were seen as having unique individual attributes; animals, on the other hand, seemed unaltered generation after generation. The Egyptians respectfully mummified thousands of animals, from wild species such as the ibis, hippo, and crocodile to their own house cats, an animal first domesticated in North Africa.

One of the most sacred creatures was the scarab or dung beetle. Zoologists today know that scarabs create balls of animal dung and lay their eggs in them. The Egyptians were unaware of the eggs. When new scarabs hatched and ate their way out of the ball of dung, the Egyptians thought they were witnessing spontaneous generation. Nothing says immortality like a self-generating creature, hence Egypt's countless images and amulets of scarabs.

Egypt, as we have seen, was created by unifying the formerly independent districts that became its hesps. Each of these districts had once worshipped its own patron deities and sacred animals. All of these deities and animals were accepted by the Egyptian state. The combined inventory gives the impression of a pantheon of more than 80 gods, but no one person or hesp would have worshipped all of them. Bastet, the cat, was honored at Bubastis in the delta; Sobek, the crocodile, was honored in the Fayum; other districts honored ibises, bulls, vultures, baboons, and so on. In addition to the major deities who were worshipped in temples, there were lesser gods whose images were kept in commoner households. Like Egyptian society itself, the deities had their hierarchy.

The Egyptian temple was known as *hwt ntr,* "the mansion of the god." As in the Zapotec state, the high priests came from noble families, while their assistants were trained commoners. Many minor priests served on a rotating basis, working three months for the temple and nine months at a secular profession. Among a priest's duties were directing rituals, sacrificing animals, interpreting dreams, and making astronomical calculations.

In a land where rulers were divine, religion was extremely important. The state funneled grain, oil, beer, wine, and precious metals from the tax collectors to the temple. Temples owned productive land but were not themselves subject to taxation. Many rulers, in fact, bequeathed land to the temple in return for a promise that priests would bring food to their tombs. As the dynasties rolled by, temples became wealthier and high priests more politically powerful.

Some 3,500 years ago the priests at state, district, and local temples began to unite into an integrated network. One high priest of the god Amun ("The Hidden One"), stationed at Luxor in the Great Bend of the Nile, became a grand vizier to the Dynasty 18 ruler Amenhotep III. Another high priest became Amenhotep's royal treasurer. As if this were not enough power, the priests of Amun acquired control of the gold mines of Nubia. Priests now held political power and wealth second only to the king, creating a level of power-sharing that no previous ruler had been forced to endure.

Sometime around 1380 B.C., Amenhotep III was succeeded by his son Amenhotep IV. What happened next has been analyzed by many scholars and interpreted in different ways. The interpretation we follow here is that of anthropologist Leslie White, whose views lend themselves to an explanation based on changing social logic.

Re, the Sun, had long ago been a hesp-level deity. By Dynasty 5 he had become a state-level deity, and by Dynasty 12 his supremacy was unquestioned. Many hesps wanted to share in Re's power, so they added his name to that of their district's patron deity. For example, the crocodile god Sobek became Sobek-Re. The god Amun, supreme deity at Amenhotep's capital city of Luxor, became Amun-Re.

Amenhotep IV and his supporters hit upon a plan to reduce the growing power of the priests. He changed his name from Amenhotep to Akhnaten ("Beloved in Life Is Aten") and officially embraced the worship of Aten, the Sun Disk, an updated version of Re. He moved his court from Luxor and created a new capital downstream at Akhetaten, known today as the archaeological site of Tell el-Amarna.

Under Akhnaten, Egypt became briefly monotheistic. Only the worship of the Sun Disk was appropriate. Akhnaten disenfranchised the priests of Amun, closed their temples and those of other deities, confiscated their temple lands and gold mines, and directed their resources to himself as the head of the cult of Aten.

It was once common to hear scholars praise Akhnaten as a "visionary" who "created monotheism." In White's analysis he was just a shrewd politician who, by tweaking the premises of Egyptian society, prevented the priestly establishment from continuing to encroach upon the divine ruler's power. Akhnaten's strategy anticipated the actions of later kings such as Henry VIII, who defied the Vatican, bypassed powerful priests and bishops, and made himself head of a new Church of England.

The analogy of Henry VIII is particularly apt, because Henry's actions created an angry backlash. His daughter Mary I reinstituted Catholicism with ferocity, burning and beheading Protestants until she had earned the nickname "Bloody Mary." Something analogous but less bloody happened in Egypt.

When Akhnaten died, his son Tutankhaten was only nine years old. The boy king was unable to stand up to the angry priests of Amun. Forced to change his name to Tutankhamun ("Beloved in Life Is Amun"), he allowed the cult of Amun to be restored, along with some of the priests' former wealth.

Unfortunately for Tutankhamun, he lived for only 18 years. A later ruler named Horemheb returned the capital to Luxor, ordered that his own name replace Akhnaten's on the latter's monuments, and rewrote history so that he would appear to be the direct successor to Amenhotep III.

Ethnic Stereotyping in King Tut's Tomb

Because Tutankhamun died at 18, he was denied the opportunity to do more than capitulate to the priests of Amun. He is remembered mainly as the occupant of "King Tut's tomb" in the Valley of the Kings. The treasures of his largely unlooted tomb are a matter of record. Less often mentioned is the fact that his burial offerings provide evidence for ethnic stereotyping during Egypt's Dynasty 18.

We have already seen that in the Zulu kingdom created by Shaka, citizens of non-Zulu appearance were referred to as "menials" or "people with strange hairstyles." New Kingdom Egyptians seem to have had similar stereotypes about people from Nubia and the Levant.

When Tutankhamun's tomb was opened in 1922, archaeologist Howard Carter at first thought he had stumbled upon a royal storehouse. The antechamber was stacked high with beds, disassembled chariots, a golden throne, alabaster vases, mummified ducks and sides of beef, and wooden chests filled with clothes and jewelry. Beyond the antechamber was the burial chamber, where Tutankhamun's sarcophagus was guarded by two life-size statues. Wall paintings in this chamber depicted Tut's funeral procession, the transport of his sarcophagus by sledge, and Tut's ritual reanimation by his successor.

Beyond the burial chamber were two more rooms, the treasury and the annex. In the treasury were 113 statuettes of the servants who would work for Tutankhamun in the afterlife. Also stored in the treasury were the internal organs removed from the boy king prior to his mummification, as well as a carefully preserved lock of hair. The hair turned out to belong to Tut's grandmother, the wife of Amenhotep III, whose mummy was fortunately available for DNA comparison.

Figure 60 shows how Egyptian artists depicted Tutankhamun, the people of Nubia, and the people of the Levant. The image of Tutankhamun, taken from the backrest of his gold-plated throne, gives him the handsome and serene profile considered appropriate for a young pharaoh. The figure of a Nubian, carved on a ceremonial baton, is done in ebony to make his skin black; he is given an iron earring and a gold armband. The figure of a man from the Levant, done in ivory, is given pale skin and a jet-black beard. These same stereotypes were used to depict prisoners of war on the sides of Tutankhamun's chariot.

Gender Inequality in the Egyptian State

Scholars have identified thousands of words in Egypt's hieroglyphic texts. "Queen" is not among them. Egyptian kingship, like its patron deities Re and Horus, was male. The phrases "king's wife" and "god's wife" are all we can find. Perhaps four out of an estimated 300 Egyptian rulers, however, are believed to have been women.

In 1492 B.C. Thutmose I, a king of Dynasty 18, gave up the ghost. Thutmose II, his heir, married and had a son named Thutmose III. He later married his royal half sister, a woman named Hatshepsut. When Thutmose II died, his son, Thutmose III, was still too young to rule.

Thutmose III was both a stepson and a nephew to Hatshepsut. The usual practice would have been for her to act as regent until the youth was older. Hatshepsut, however, usurped the throne, made Thutmose III her junior regent, and ruled Egypt for 20 to 22 years. She backdated her reign to the death of her half brother, Thutmose II.

FIGURE 60. Eighteenth-Dynasty Egyptians engaged in ethnic stereotyping. Above we see Tutankhamun and his wife, Ankhesenamun, shown as serene and handsome examples of Egyptian aristocracy. On one carved baton, lower left, we see a man from the Levant carved in ivory, with white skin and a black beard. On a second baton, lower right, is a Nubian carved in ebony, with black skin and an iron earring.

Usurpers often go to great lengths to legitimize their reigns. Hatshepsut was no exception, commissioning 200 statues in her image. On many of her monuments, as noted by Gay Robins and Lana Troy, she was shown in a male ruler's attire, including a *nemes* head cloth, a *chenjyt*, or kilt, and a false beard. When shown naked to the waist, she had herself depicted as flat-chested. Her hieroglyphic texts refer to her as "he."

Hatshepsut rewrote history to claim that her father, Thutmose I, had crowned her king before his death. In preparing her future funerary temple at a place called Deir el-Bahri, she had herself portrayed either as the offspring of Amun or the cow goddess Hathor.

Hatshepsut retained her father's trusted steward Senenmut, solidifying his loyalty by promoting him to governor of her palace. She sent Senenmut to the granite quarries of the First Cataract to procure stone for two huge obelisks. These obelisks were set up at the Temple of Karnak in the Great Bend of the Nile; one was inscribed "The King himself [*sic*] erected two large obelisks for his [*sic*] father Amun-Re." On another monument, Hatshepsut had herself portrayed as a sphinx, trampling the bodies of Nubian enemies.

Hatshepsut bore a daughter named Neferure. She prepared her daughter to succeed her by having her portrayed on her monuments as a boy, right down to the single braid worn by male children. She did this even while referring to Neferure as *hmt ntr*, "the god's wife."

Unfortunately, things did not work out for Neferure, who died young. Hatshepsut was succeeded instead by her stepson/nephew, Thutmose III. The new king ordered that Hatshepsut's name be obliterated from her monuments and replaced with his. Egyptologists such as Donald Redford do not believe that this defacement was done out of resentment toward his stepmother; rather, Thutmose III was trying to legitimize his own reign by linking himself to earlier male rulers.

We know a lot about Hatshepsut because her funerary temple has survived. However, an official list of Egyptian kings, prepared more than 150 years later by the ruler Seti I, did not even mention her. Hatshepsut was, after all, a woman.

INEQUALITY IN THE EGYPTIAN STATE

The fact that its rulers were considered gods gave Egypt one of the highest levels of inequality of any first-generation state. Pharaohs were immortal and required food even after burial. By contracting with priests to have those meals delivered, Egyptian kings so added to the wealth of the priestly establishment that the latter came to be seen as political rivals. Akhnaten was strong enough to curtail priestly power. The priests were strong enough to make Tutankhamun restore it.

Egyptian rulership was lopsidedly male. Strong women occasionally served as regents in both the Egyptian and Maya states, but with this notable difference: royal women continued to be glorified on Maya monuments long after their death, while Hatshepsut's name was obliterated from her monuments and ignored by later king lists.

Like so many kingdoms, Egypt eventually discovered that a skilled commoner makes a better official than a corrupt or an incompetent noble. During periods when the central government was strong, called Kingdoms, the new king was usually one of the old king's sons. During times when the central government was weak, called Intermediate Periods, it was more likely that a noble usurper could maneuver his way to the throne.

An Unanswered Question

We consider Egyptologists to be among the luckiest of archaeologists. The body of data available to them is so enormous that they have the potential to answer almost any question. We would like, therefore, to ask them a question that may not have occurred to them.

In our discussion of the way kingdoms form, we asked whether there might be continuity in the sources of power between a first-generation monarchy and the chiefly societies out of which it was created. Given its history of divine kingship, Egypt should have data relevant to this question.

Were the chiefs of Naqada and Hierakonpolis already seen as divine, or did they simply possess Irving Goldman's combination of sacred life force, expertise, and military prowess? If the latter, then exactly when did divine kingship first appear? Was it created anew to justify the title—nesw, or king—given to the man who now ruled a group of formerly independent regions? Or did the Egyptians create divine monarchy in two stages—first (1) claiming that kings metamorphosed into gods after death, and later (2) claiming that they had been born divine to begin with?

It would not surprise us to learn that divine kingship represented a deliberate change in social logic, designed to justify one man's rule over what had once been a group of independent rank societies. But no one is sure, and Egypt would be the perfect test case. We know that Egyptologists have a lot on their plates already, but it would be great if they could put this question on their "to do" list.

TWENTY

Black Ox Hides and Golden Stools

The first kingdom on the African continent was that of Egypt. More would follow. Some, like Aksum on the upper Nile, borrowed strategies from their Egyptian neighbors. Others, like the Zulu kingdom, were created by the iron-working, cattle-herding descendants of the Bantu migration. Still others arose among the matrilineal, horticultural societies of central and western Africa. Some African kingdoms, prevented by tsetse flies from relying on cattle, found that their wealth could be based on ivory, gold, or slaves.

Once the first kingdom has appeared in a region, it provides a model for later generations of kingdoms. Archaeologists can date those later-generation kingdoms but often have only mythical accounts of their origins. In Africa, however, a great many new kingdoms formed during the eighteenth and nineteenth centuries, a time when European eyewitnesses could write down what happened. In such cases we have both the indigenous and Western versions of events.

In this chapter we look at two African kingdoms for which both native and European accounts are available. One was part of a chain reaction that included the Zulu. The other was created by an ambitious noble whose people had become tired of paying tribute to more powerful neighbors.

THE RISE OF THE SWAZI

By the late Iron Age thousands of Bantu speakers had crossed the Limpopo River into southeast Africa. Eighteenth-century European travelers reported more than 50 different rank societies in Natal alone. Among the most powerful were the Ndwandwe and Mthethwa.

North of the Ndwandwe lived a group called the Dlamini, named for the founder of their chiefly clan. From roughly A.D. 1500–1700 the Dlamini had lived in the region known today as Mozambique. In order to preserve the high rank of their heirs, Dlamini chiefs were allowed to marry women from within their own patrilineal clan, while commoners still had to marry women from other clans.

Eventually the Dlamini, whose economy relied heavily on cattle herding, formed an alliance with a group called the Tembe, whose members relied more strongly on farming and crafts. This alliance created a more diversified economy. It was seen as a logical partnership because Tembe chiefs also married their "sisters" (that is, women of their own clan) and therefore "were one with the Dlamini."

In Mozambique the Dlamini lived in kraals like the Zulu, herding cattle and working the land. A chief named Dlamini III eventually moved his residence inland to the Pongola River. In the late 1700s his people moved again, establishing their capital at a place called Lobamba.

This move placed the Dlamini in closer contact with the expansionist Ndwandwe and Mthethwa, and under a chief named Ngwane II they began their own cycle of conquests. Soon they had begun to refer to themselves as "the People of Ngwane" and to their expanding territory as "Eshiselweni," or "the Place of Burning." Ngwane II, who died in 1780, was buried in a cave near Lobamba. This act established a traditional burial place for Eshiselweni rulers, and Ngwane II came to be revered in legend.

The People of Ngwane were now part of a chain reaction involving the Mthethwa, the Ndwandwe, and the Zulu. Out of the resulting competition, conquest, chiefly intermarriage, and emigration of refugees, at least two kingdoms would arise: the Zulu and the Swazi.

Ngwane II was succeeded by his son, Ndungunye, who was considered a tyrant. Ndungunye was a contemporary of the Mthethwa chief Dingiswayo (under whom Shaka learned his military skills) and a Ndwandwe chief named Zidze (presumably the "Zwide" who was eventually defeated by Shaka). Many people fleeing Shaka's conquests sought refuge with the People of Ngwane, increasing their numbers.

Around 1815, Ndungunye died and his son, Sobhuza I, rose to power. While Sobhuza continued his father's despotism, his cruelty is said to have been buffered by his influential mother. Sobhuza sought marriage alliances with other powerful leaders. He took Zidze's daughter as his principal wife and sent two of his own daughters to marry Shaka. Unfortunately, both of the latter became pregnant, and, given Shaka's paranoia about being usurped by a son, he had both of them put to death.

This interplay of Dlamini, Mthethwa, Ndwandwe, and Zulu leaders fed the chain reaction. All of the ruling families of southeastern Africa intermarried, and each knew the political and military strategies of the others. If they chose to copy a rival's strategy, they could; if they chose to do the opposite, they could.

When he was threatened by the Ndwandwe, Shaka confronted and defeated them with superior military strategy. When the Ndwandwe began to feud with him over the rich garden land of the Pongola River, however, Sobhuza I used a different strategy. Aware that his warriors would be outnumbered, he chose to flee to the north with his wives, followers, and cattle. Sobhuza entered the land of the Nguni, Sotho, and Tonga, people he regarded as politically and militarily weak. He then demanded tribute from his new neighbors. Groups that submitted retained their chiefs, land, and autonomy. Groups that resisted had their men slaughtered and their women assimilated. Small ethnic groups that fled were pursued and punished.

The conquests of Sobhuza I made him one of the most powerful rulers in the region, and his northward emigration had created some space between his people and the aggressive Zulu. Like Shaka, he began to refer to himself as a king. He moved his royal kraal back to Old Lobamba, the capital of his ancestor Ngwane. He assigned districts of his realm to his kinsmen and incorporated willing foreigners into his army. It was widely believed that Sobhuza possessed extraordinary magic. He had the power to wage war, decide issues of life and death, reward allies, and punish enemies.

When Sobhuza I died, in 1839, he left his kingdom to his second son, Mswati. Mswati went on to become the greatest king and warrior of his people, causing them to change their name yet again. Today we know his people as the Swazi (a Westernized version of "Mswati") and their country as Swaziland.

The Regime of Mswati

Mswati took advantage of his unprecedented power and influence to make a number of changes in the principles of Swazi society. Among his key changes were the following:

1. Armies had formerly been organized on a local basis, with each chief calling up men from his own district's kinship groups. Mswati conscripted men from all districts and reorganized them into Zulu-style age regiments. He was clearly emulating Shaka's strategy and was encouraged to do so by his Ndwandwe mother.
2. Special royal villages were established as mobilizing centers for each district. All young men were now considered citizens of Mswati's state and had to abandon their district or ethnic loyalties.
3. Military outposts were established to facilitate raids on other societies.
4. Neighboring groups were raided for cattle and captives. Elite captives were then ransomed for cattle or other valuables. Commoner captives were used in prisoner exchanges. The Swazi were not as interested in acquiring new land as they were in increasing their wealth and military renown.
5. All plunder was brought to Mswati, who redistributed it to his *emaqawe*, or "heroes." This made the Swazi one more society in which commoners could rise through military prowess.
6. Mswati was increasingly called upon to resolve factional disputes among less powerful societies.
7. The legitimacy of their neighbors' hereditary leaders was recognized by the Swazi. Whenever he conquered a neighboring group, Mswati made a point of raising one of the defeated chief's sons as if the boy were his own. Mswati also allowed any legitimate chiefly heir to rebuild his shattered society.
8. Refugees fleeing the Zulu (or any other African society) were welcomed and protected by the Swazi, on the condition of loyalty. It was widely said that refugees "fled to the safety of Mswati's armpit."
9. The Swazi king was established as the central figure in both secular government and religious ritual. Mswati was at once commander of the army, supreme legal authority, highest ritual leader, appointer of all officials, and official redistributor of wealth.
10. The Swazi king could seize any unmarried woman he wanted for his harem. His wealthy kinsmen were also allowed many wives. As a result, the ruling Dlamini clan soon became the largest in the kingdom.

As the Dlamini clan grew, so did the desire to emulate it. Its mode of dress, earlobe decoration, spoken dialect, and clan rituals were so widely imitated that, over time, they became Swazi national characteristics.

Of the 70 clans that now made up Swazi society, about a fifth were considered "true Swazi." Roughly a seventh were respected as "prior inhabitants" of the region. The remainder were considered immigrants. Each clan had its history, but all clans had to concede the superior position of the Dlamini.

Mswati died around 1870. Mbandzeni, his successor, convinced both the British and Boer colonists to sign treaties recognizing Swazi autonomy. The flood of European immigrants could not be slowed, but the Swazi had the satisfaction of a nation that bore their name.

Swazi Society

During the 1930s anthropologist Hilda Kuper came to live among the Swazi. Kuper was years ahead of her time; long before it became fashionable to do so, she had drafts of her reports from the field critiqued by members of Swazi society. She even gave an early draft of her manuscript to King Sobhuza II, a monarch who subscribed to anthropological journals. Sobhuza gave Kuper feedback on her research and opened many doors for her. "Anthropology," he ventured, "makes possible comparison and selection of lines of further development. European culture is not all good; ours is often better. We must be able to choose how to live, and for that we must see how others live. I do not want my people to be imitation Europeans, but to be respected for their own laws and customs."

The Swaziland of Kuper's study was made up of grassy plains, mountains rising to 6,000 feet, and alluvial river valleys. Eighteen inches of annual rainfall made the grasslands productive for cattle herding. Men and boys did most of the herding, and, as among the Zulu, cattle were a source of wealth. Rain rose to more than 50 inches a year in the mountains, producing denser vegetation where the Swazi could graze goats and hunt game.

Taking advantage of the riverine alluvium, Swazi women cultivated native sorghum and millet and introduced New World crops such as peanuts, pumpkins, and maize, or Indian corn. Among the important crafts was ironworking, which produced both weapons and agricultural tools.

During Kuper's stay at least 25 clans considered themselves "true Swazi," with the Dlamini clan serving as a ruling stratum. At least another eight clans were made up of people who had already been living in the area when the Swazi arrived. Another 35 clans consisted of Sotho, Nguni, and Tonga people who had been incorporated into Swazi society.

Crosscutting the clans were age classes for young men and women. Just as among the Zulu, the age classes for young men led to military regiments in which sons were separated from fathers and older brothers from younger brothers. The commander of each age regiment was a commoner who had risen through the military; he was appointed by the king but lived at the queen mother's homestead. New regiments were inaugurated when old regiments had served for five to seven years and were ready to marry. Normally this took place when men had reached ages 25 to 35.

Women's age classes were keyed more to their physiological stages of development than to their absolute ages. Young women in each class worked together in teams to weed crops, thresh harvests, winnow grain, brew beer, and plait ropes. Service ended when a woman was chosen as a bride. Until then she was discouraged from getting pregnant, and, were a girl to do so, her whole family was punished by having one of its animals confiscated and eaten by the other young women in her age class.

The Swazi king inherited his eligibility to rule from his father. Which of the eligible princes was selected as the royal heir, however, was often determined by his mother's rank as the principal wife of the harem. The other princes were placed in charge of various provinces of the Swazi kingdom.

Once a prince was chosen as king, his mother became the queen mother. This was such an important position that droughts and floods were blamed on quarrels between the king and his mother. The Swazi king was referred to as "the Lion" or "the Child of His People." The queen mother was referred to as "the Lady Elephant" or "the Mother of Her People."

The king presided over the highest Swazi court and could pronounce death sentences. One king, in fact, is said to have had his own mother executed for plotting to overthrow him. The queen mother presided over the second highest court and could provide sanctuary for men whom the king had sentenced to death. The king was supposed to control the entire Swazi army, but his commander in chief lived at the queen mother's homestead because that was considered the Swazi capital. Both the king's and the queen mother's residences were guarded by age regiments of warriors.

The queen mother served as one check on the king's power, and the Swazi had other institutions of power-sharing. There were two councils. One, the Inner Council, was composed entirely of aristocrats from the Dlamini clan; when the king traveled, these councillors became part of his entourage. There was also a larger General Council composed of chiefs from Level 3 of the administrative hierarchy, prominent headmen from Level 4, and any other adult male who chose to attend.

In addition to all his councillors, the Swazi king had two special aides called *tinsila* (singular, *insila*). The tinsila were chosen from respected lineages of the Mdluli and Motsa clans, and the king was bound to them as a kind of "blood brother."

Insila is another of those abstract concepts from which other premises flow. Kuper defines it as "an essential part of the self which, even when it has been removed by washing or scraping, remains intimately linked with the person." Anyone who gained possession of someone's insila could influence its owner.

Before the future king reached adulthood, two boys his age were chosen from the "true Swazi" clans mentioned earlier. At a secret ceremony, a ritual specialist made cuts on the boys' bodies and matching cuts on the future king. These cuts were made on the right side of the Mdluli youth (who was to be the king's "right-hand insila") and the left side of the Motsa youth (who was to be his "left-hand insila"). The ritual specialist then rubbed blood from the future king's right side into the cuts on the Mdluli youth, and vice versa; the process was repeated for the Motsa youth, using blood from the left side.

Once the blood transfer was complete, the king possessed two loyal assistants. It was also believed that any danger threatening the king would strike his tinsila instead. The tinsila mediated between political factions on behalf of the king, helping to keep the peace. In Swazi logic the fact that they had exchanged blood with the king allowed them to speak on his behalf.

The tinsila ceremony was not the only occasion on which the king exchanged blood; a similar ritual accompanied his first marriage. The king's first two wives were carefully chosen from two "true Swazi" clans, the Matsebula and Motsa. These women were known as his "right-hand queen" and his "left-hand queen." In the ritual hut of a newly built harem enclosure, special medicine men sliced into the right side of the king and his Matsebula bride, and their blood was mingled. The king married his Motsa queen a few weeks later, accompanied by ritual but without the transfer of blood. Once the king had married, his tinsila were required to wed as well.

While the king's first two queens were expected to dominate all others in his harem, he was encouraged to form political alliances by marrying other noble women from neighboring groups. For their part, the princesses born to the Swazi king's wives were given in marriage to allies, including both foreign rulers and the heads of Swazi clans other than the Dlamini.

The Swazi had no official state religion. They built no great temples to deities but did conduct rituals honoring their ancestors. Like the Zulu, the Swazi believed that their king had magical powers beyond those of ordinary men. Members of the royal family continually reinforced this belief by carrying out rituals that confirmed their magic. Swazi elites also worked hard to appear generous, confirming the fact that generosity remains a first principle even in monarchies.

Some Swazi commoners became wealthy by accumulating cattle, entering into politically advantageous marriages, or winning choice bureaucratic appointments. Such prominent commoners could mobilize their own large work parties, rewarding them with feasts and lavish amounts of beer.

Finally there were the *tifunjwa,* captives of war. The concept of slavery was foreign to the Swazi, but war captives presented to the king were often assigned as servants to his wives or his most renowned warriors. There were also children in Swazi society known as *tigcili,* whose fathers had been executed as evildoers. These children might be given to foster parents.

The Swazi Kingdom

Kuper was mainly concerned with the social organization of living Swazi, but she also collected information of use to archaeologists. For example, she diagrammed the administration of the Swazi kingdom and revealed that it had a hierarchy of four levels (Figure 61). It is worth noting that Kuper outlined the four-level organization of the Swazi state more than 30 years before archaeologists began to identify similar four-level state hierarchies in the ancient Near East.

At the apex of the Swazi hierarchy were the king and his mother. The king maintained two royal homesteads; he placed his mother in one and resided mainly in the other, thereby avoiding friction with her. The settlement where the queen mother lived was considered the capital of Swaziland. The settlement where the king spent most of his time was called "the king's village." These two settlements occupied Level 1 of the hierarchy.

Level 2 consisted of a series of "royal villages." These settlements were occupied by the princes who oversaw the provinces of Swaziland. Each prince had power over only one province and was kept at a distance from the king to lessen the chance of usurpation. Level 3 and Level 4 settlements were found in every province, and a prince could reorganize his subjects as he saw fit.

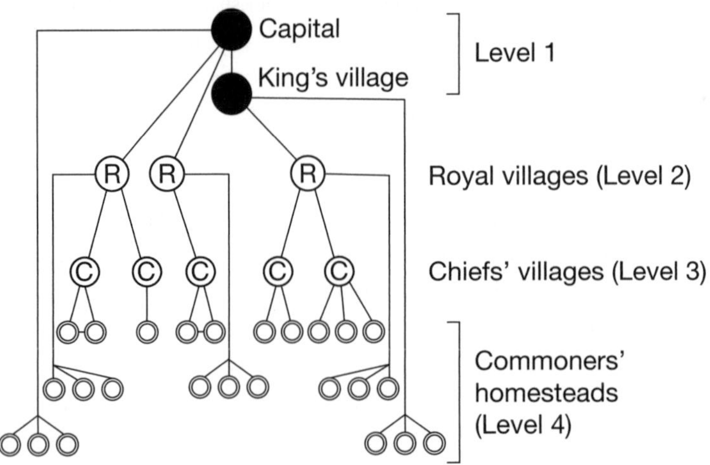

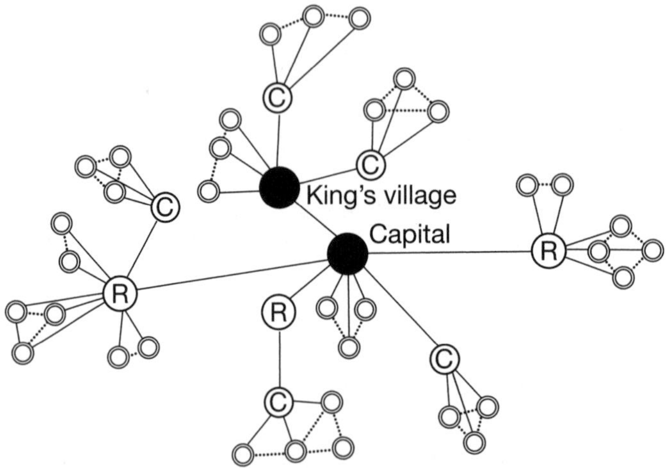

FIGURE 61. The queen mother's village was considered the capital of the Swazi kingdom. Her village, and the king's, comprised Level 1 of the political hierarchy. Below them were royal villages (R), chiefs' villages (C), and commoners' homesteads. This drawing shows both the hierarchical arrangement of levels and the way settlements were spread out on the landscape.

Level 3 consisted of "chiefs' villages," of which each province had several. Each local chief reported to the prince of his region. In turn, he oversaw a series of *umuti,* or commoner homesteads.

Level 4 consisted of hundreds of these umuti, which were the basic building blocks of Swazi society. Each consisted of the residential compound of an extended family. The household head took multiple wives if he could afford to. His homestead usually consisted of a *sibaya,* or cattle corral; his *indlunkulu,* or "great hut"; the huts of his wives; and a *lilawu,* or bachelors' quarters. There might also be a small hut for ritual activities. Each homestead's grain supply was carefully hidden from outsiders.

Commoner homesteads averaged 7.2 occupants, while those of chiefs and princes averaged 22 to 23. Polygamy accounted for much of the difference. Some 60 percent of all married men could not afford a second wife, while King Sobhuza had 19 wives and 30 children in his compound.

Kuper's sketch of the queen mother's compound at Lobamba provided the template for our Figure 62. In Kuper's day Lobamba's population was only 265, providing a striking contrast to the huge cities seen in some ancient kingdoms.

The first structure created in any homestead was the cattle corral. The corral at Lobamba was the largest in Sobhuza's kingdom, measuring 180 feet in diameter and facing east toward the rising sun. To either side of the corral were barracks for the senior age regiments of warriors who guarded the capital. A third barracks, filled with more junior warriors, was placed behind the experienced troops as a second line of defense. Protecting the rear of the homestead was a semicircle of more than 30 households, some of which belonged to men of high rank. This semicircle created a barrier that enemy raiders would have had to penetrate.

In the center of the homestead was an open yard called the *sibuya,* and here the indlunkulu and the sigodlo were built. The indlunkulu in this case was more than a "great hut"; it was actually a great walled enclosure consisting of 11 huts under the supervision of the queen mother. This enclosure included a ritual hut 15 feet in diameter, storage huts for meat and grain, and a platform for the drying of pumpkins.

The sigodlo was the harem for Sobhuza's wives. This harem was deliberately located in the queen mother's homestead, because one principle of Swazi behavior was that new wives were supposed to spend time serving their mothers-in-law. This was a departure from the old African hunter-gatherer principle, which required the groom to serve his mother-in-law.

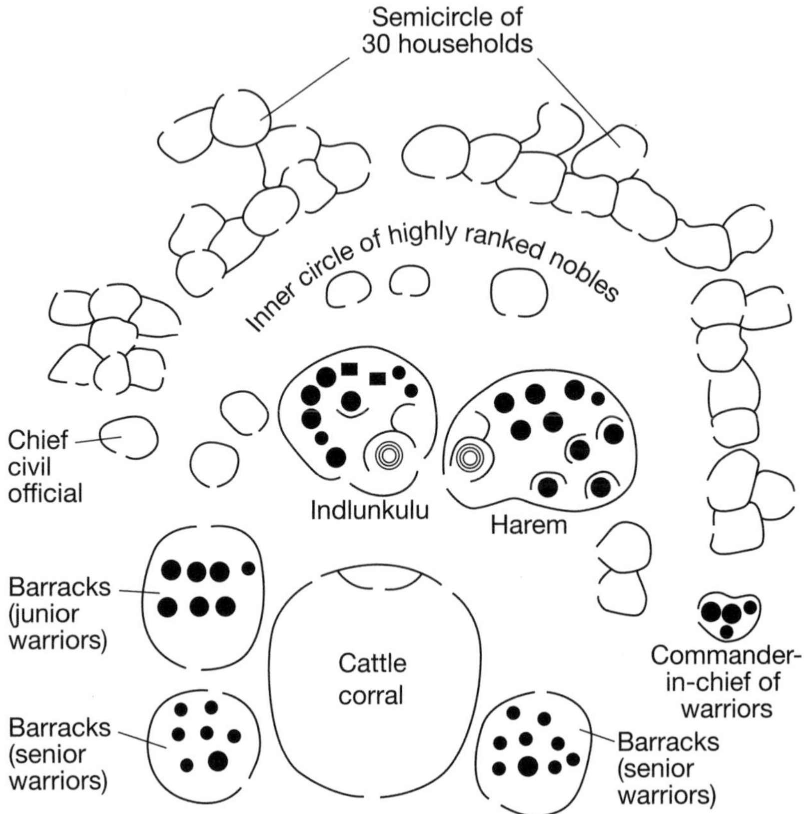

FIGURE 62. Among the Swazi, the queen mother's village was considered the capital of the kingdom. Its cattle corral alone was 180 feet in diameter. The queen mother occupied the *indlunkulu,* a walled enclosure containing 11 huts. The Swazi king's harem was located nearby so his wives could attend the queen mother. The village was defended by warriors from several age regiments.

Flanking the great enclosure and the harem was an inner circle of huts occupied by men of high rank, whose nobility was acknowledged by allowing them to live near the queen mother. To the south lay the household of the chief civil official of the kingdom, the Swazi equivalent of a vizier. To the north lay the household of the military official in charge of age regiments.

Kuper's work shows future archaeologists what a royal homestead at the capital should look like and what to expect to find at each level of the administrative hierarchy. But her potential contributions to archaeology did not end there. She also gave archaeologists a way to study inequality by describing differences in the way Swazi nobles, officials, and ordinary commoners were buried. One of Kuper's discoveries was that the color symbolism of black oxen and goats figured prominently in burial ritual.

Let us begin with a typical commoner homestead. When the headman of such a homestead passed away, he was wrapped in the hide of a black ox and buried in the corral, accompanied by his personal possessions. His main wife was buried either at the entrance to the corral (if she were a native of that settlement) or at the back of the homestead (if she were from a different settlement). His junior wives were buried behind the huts they had occupied in life, accompanied by their personal belongings.

Swazi governmental officials, regardless of clan, were honored by being wrapped in a black ox hide and buried in a royal cave. The tinsila, bonded as they were to the king by transfer of blood, received special treatment. Were an insila to die before his king, the Swazi refused to treat him as a dead person, as this would be considered a threat to the king's well-being. Such tinsila were buried quietly and privately, accompanied only by their personal possessions. Only after the king himself had died could an insila's own relatives begin to mourn him.

As for the princes who ruled the Level 2 royal villages, they were wrapped in black ox hides and buried in one of several royal caves. A black goat was buried alive with each of them; the skeleton of this sacrificed animal would be an archaeological clue to the deceased's high rank.

The queen mother was buried in the cattle corral of her homestead at the capital. She, too, was wrapped in a black ox hide. In addition, a sheep bladder was placed on her forehead, along with other symbolic insignia.

Finally we come to the Swazi king. His body was embalmed so that he could lie in state until spring. At that time he was wrapped in a black ox hide and buried in a special royal grove of trees, accompanied by his personal possessions and insignia of office. Before there were Europeans around to object, the king was buried not only with a live, black goat but also with a number of live men.

Kuper's descriptions of the type of burial associated with the different ranks of Swazi society did not go unnoticed. An archaeology student named Arthur Saxe eventually wrote a doctoral dissertation on burial ritual that featured Kuper's work. While Saxe's dissertation was never published, it became an underground classic, and photocopies of it are still circulating among archaeologists.

There is no evidence that Kuper had any interest in archaeology, nor were there teams of archaeologists working on the origins of the Swazi state during her stay. This is a shame, because the potential for a wonderful collaboration was there.

The Nature of Swazi Inequality

Swazi society had an interesting mix of institutions, some of which had endured from an earlier era of chiefly society and others of which were more typical of kingdoms. Swazi society was organized as a series of ranked clans, which had not yet broken down as they had in many ancient states. The Swazi king controlled numerous homesteads whose occupants filled his storage huts with food, but he owned no private estates. Magic and witchcraft were widespread, but no land was set aside for high gods and temples.

Along with long-lived institutions such as lineages, clans, and ancestor rituals, the Swazi displayed some of the multiethnic aspects of an empire. Rulership was monopolized by the Dlamini clan, and other "true Swazi" clans were relied on to provide the tinsila and the ruler's two principal queens. Ethnic groups already present in the region when the Swazi arrived were treated differently from ethnic latecomers such as the Sotho, Nguni, and Tonga.

In addition to the hereditary differences in Swazi society, commoners had ways of achieving renown. Some rose as military commanders and were rewarded with female captives or shares of plunder. Others managed to accumulate large herds of cattle. In both cases men of renown took as many wives as they could afford. Each aspired to be buried in the black ox hide that symbolized a man of substance.

Archaeologists sometimes infer that they have evidence for a chain reaction in which several kingdoms arose through competitive interaction and strategic marriage alliance. Often this inference cannot be confirmed with historic data. Fortunately, the chain reaction involving the Zulu, Swazi, Mthethwa, Ndwandwe, Sotho, and Nguni is historically well documented. It also shows us that a kingdom does not need settlements larger than 265 people to have the multilevel hierarchy of an archaic state.

THE RISE OF THE ASANTE

One of the most frequent scenarios for the rise of a second-generation state is for an outlying province to break the grip of its overlords and emerge as a monarchy in its own right. The irony of this scenario is that the rebellious province has usually acquired its knowledge of statecraft by studying those very overlords.

Once again, Africa provides us with a historic example. This time the story was played out on the Gold Coast of tropical West Africa, the region known today as Ghana.

Archaeologists are still not sure how many early kingdoms succeeded one another in West Africa. For the purposes of this chapter we need only go back to the eighth or ninth century A.D. By that time the residents of Igbo Ukwu, an archaeological site near the delta of the Niger River, had witnessed the burial of a man who was either a paramount chief or an early king. He was buried in the seated position in a wooden tomb, holding a fan or fly-whisk and wearing a copper crown, breastplate, beaded armbands, and copper anklets. His burial offerings included the tusks of several elephants.

Off to the west of the Niger lay the Oda and Ofin, smaller rivers that flowed south to the Gulf of Guinea. These rivers had a resource the Niger lacked: their alluvium was gold-bearing. Indeed, one stretch of the Ofin floodplain came to be known as Suwiri Sika, "the flood of gold." Here one could mine soil from either bank of the river and then wash gold from the earth.

The region of the Oda and Ofin came to be known as Akan, and its gold was sought by both European and North African traders. By 1482 the Portuguese had established their first trading post on the coast of Akan. Multiple kingdoms arose in the region, basing their wealth not only on gold but on ivory, copper, iron, and slaves.

Between 1660 and 1690, according to historian Thomas McCaskie, a kingdom called the Denkyira was considered the dominant power in Akan. The Denkyira spoke a language called Twi. Their ruler, known as the *denkyirahene,* maintained his capital at Abankeseso on the Oda River. There he welcomed trade emissaries from the Dutch and English, who had by then broken the Portuguese monopoly on Akan gold.

Among the neighboring ethnic groups forced to pay tribute to the Denkyira were the ancestors of today's Asante. They were Twi speakers who lived to the north of Abankeseso, between the Oda and Ofin. The protohistoric Asante grew yams and plantains, raised sheep and chickens, and fished in the local lakes and rivers. They also dug for gold in the alluvium and competed with neighboring groups for control of the trading center of Tafo.

The sixteenth- and seventeenth-century Asante had eight major matrilineal clans and a series of hereditary leaders called *abirempon.* The most tenacious and ambitious abirempon, according to McCaskie, "eventually institutionalized their wealth in chiefship . . . converting their economic clients into a political following of retainers." Sixteenth-century Asante chiefs used three main symbols of authority: a spear, an elephant tail, and a *dwa,* or carved wooden stool.

One strategy used by the Denkyira to ensure the obedience of their tribute-paying neighbors was to require each of those groups to send a person of high rank to the court at Abankeseso. While treated as a guest, this person could become a hostage if his ethnic group rebelled.

During the 1660s or 1670s, according to oral histories, the Asante sent a young man named Osei Tutu to Abankeseso. Osei Tutu was the son of Maanu, a highly ranked woman of the Oyoko clan.

Osei Tutu's stay in Abankeseso took place under the Denkyira ruler Boamponsem, who reigned for some 40 years. Greatly admired by European observers, Boamponsem is considered by many the most successful denkyirahene. His court was the ideal place for a young Asante to learn political and diplomatic skills.

After a few years at Abankeseso, Osei Tutu traveled east to the Volta River and lived in the court of Ansara Sasraku, the ruler of a powerful group called the Akwamu. There, according to political scientist Naomi Chazan, Osei Tutu learned the essential features of military formation. Akwamu armies consisted of the following five elements: the *twifo* (foreguard); *adonten* (center); *nifa* (right flank); *benkum* (left flank); and *kyidom* (rearguard). Osei Tutu now possessed the political and military skills that he would need later. He purchased Western firearms before returning to the Asante.

Two events then changed the course of Akan history. First, the Asante chief Obiri Yeboa was killed by Domaa warriors during one of his battles for the gold trading center of Tafo. Second, Boamponsem's death (in or around 1694) left the Denkyira in the hands of Ntim Gyakari, described by McCaskie as "a capricious young man of uncertain judgment."

Obiri Yeboa owed his high rank to his mother and could not simply be succeeded by his son. An Asante man inherited his *abusua,* or blood, from his mother; a chief could not inherit his father's office, because his blood and his father's came from women of different clans.

A number of men with highly ranked mothers were eager to succeed Obiri Yeboa, and it proved difficult to decide among them. The matter was placed in the hands of Anokye, a "possessed" or ecstatic priest. Anokye selected Osei Tutu and continued to act as his spiritual adviser.

Osei Tutu set for himself the goal of punishing the Domaa who had killed Obiri Yeboa. After defeating them he went on to conquer the neighboring Tafo, Kaase, and Amakom, guaranteeing control of the local gold trade. Osei Tutu established his capital at nearby Kumase and was named the first *asantehene,* or King of the Asante.

There were several keys to Osei Tutu's military success. First, he adopted the five-part formation of the Akwamu army. Second, he strengthened his forces by using subordinate Asante nobles as combat officers, giving his whole corps of administrators a stake in his military campaigns. Third, any defeated rival who agreed to be loyal was allowed to attach his forces to Osei Tutu's. And, finally, at the core of Osei Tutu's army was a contingent of experienced fusiliers on loan from the Akwamu ruler Ansara Sasraku.

While Osei Tutu was rising to power, the Denkyira kingdom was suffering under Ntim Gyakari. Encouraged by the ecstatic priest Anokye, many of Ntim Gyakari's subjects began to switch their loyalties to Osei Tutu. All were welcomed.

Ntim Gyakari was angered by the fact that Osei Tutu had attacked his Domaa vassals without permission. He was further irritated by the fact that Osei Tutu was providing sanctuary to groups fleeing Denkyira rule. In addition, both the Asante and Domaa had now stopped paying tribute to the Denkyira. Ntim Gyakari therefore sent messengers to the Asante with the following list of demands:

1. The Asante were ordered to fill a large brass vessel with gold for the Denkyira.
2. The Asante were ordered to send Ntim Gyakari a long necklace of precious beads, similar to those worn by the asantehene's wives.
3. The Asante king and each of his provincial governors were ordered to surrender their favorite wives to the messengers, who would escort them back to be married to Ntim Gyakari.
4. The Asante king and each of his provincial governors were ordered to surrender their most beloved children to Ntim Gyakari.

Osei Tutu rejected all the Denkyira demands, which was a de facto declaration of war. The Asante and Denkyira then began three tense years of preparation for the final showdown. Anokye advised Osei Tutu that his only hope of victory was to encourage more of Ntim Gyakari's vassals to defect to the Asante. According to legend, Anokye brought about a number of key defections by performing miracles.

One by one the Asabi, Anwianwia, and Awu Dawu threw in their lot with the Asante. In addition, a Denkyira man who had been Osei Tutu's personal servant during his years in Boamponsem's court defected and was made a general in Osei Tutu's army. As the Swazi might have expressed it, Osei Tutu let hundreds of immigrants "flee to the safety of his armpit."

Finally, in 1701, Denyira forces under Ntim Gyakari began advancing north from Abankeseso. At first the Asante fell back, encouraging the Denkyira to pursue them to Feyiase near the Oda River. Here Osei Tutu's main force was waiting. In the ensuing battle of Feyiase the Asante soundly defeated the Denkyira, beheading Ntim Gyakari in the process. Some 166 years later, according to McCaskie, Ntim Gyakari's skull was still kept near Osei Tutu's coffin in the royal Asante mausoleum.

The Asante kingdom was now the most powerful in the region. The Asante had gone from being tribute payers to tribute receivers. Their preeminent position cried out for a miraculous ceremony, one that would reflect the revised ideology of a newly created state. The miracle, orchestrated by Anokye, drew on the principle of the stool as a symbol of authority.

In preparation, Osei Tutu buried the stools of all the kings and chiefs he had vanquished. Anokye then selected a day on which the king, the Asante elders, the provincial governors, and the local chiefs should assemble to see the miracle. Anokye put himself into an ecstatic state and then, according to legend, a fabulous golden stool descended from the sky and landed gently in Osei Tutu's lap.

In this golden stool, or *sika dwa,* the souls of all the Asante people were said to be enshrined (Figure 63). From that point on, the administrative hierarchy of the Asante monarchy would be symbolized as follows. Kingship was associated with the golden stool; the governors of provinces owned silver stools; and local chiefs or headmen owned carved wooden stools.

Asante territory now extended 150 to 200 miles from the capital. The time required to reach its limits depended on whether one was traveling through forest or savanna. To facilitate forest travel, the Asante maintained a series of roads analogous to the ones that connected early Maya cities such as Calakmul to their Level 2 centers.

Unfortunately for Osei Tutu, he did not live to see the Asante kingdom reach its greatest extent. In 1717, sixteen years after his victory over the Denkyira, he died in battle while trying to subdue the Akyem. He was succeeded by Opoku Ware, a great-grandson of Osei Tutu's mother. Opoku Ware continued to expand against the neighbors of the Asante. He beheaded numerous

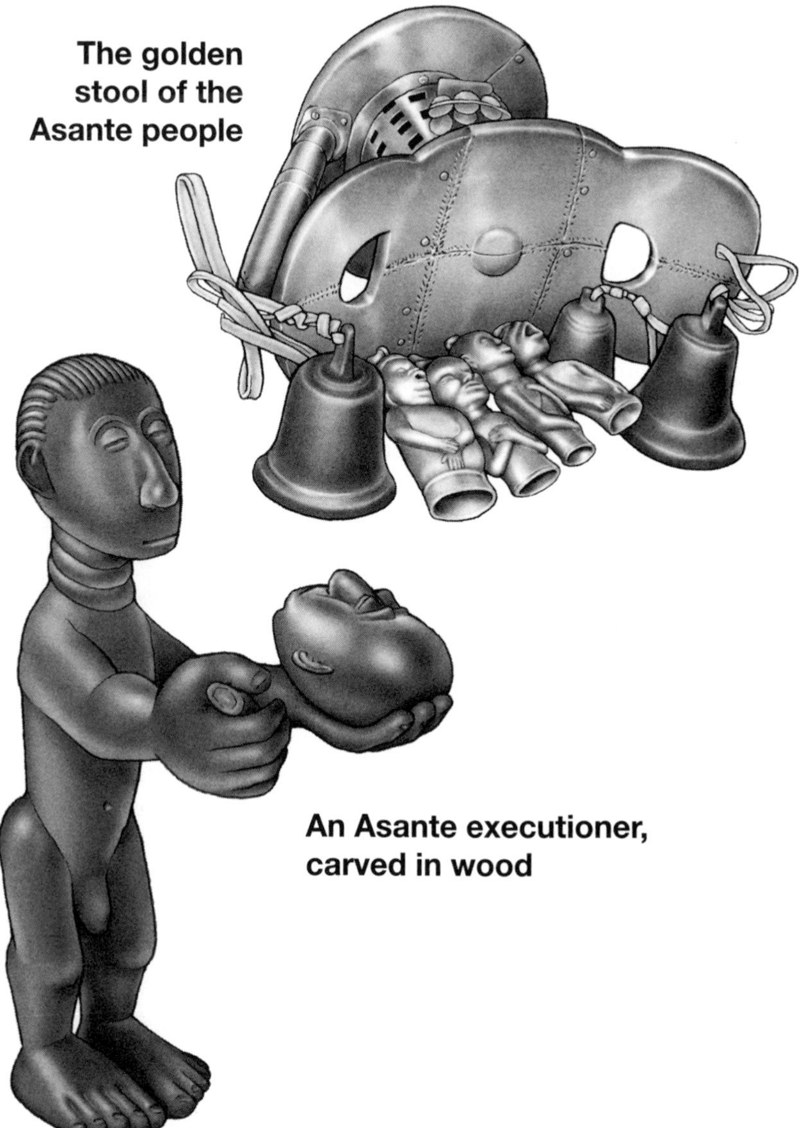

FIGURE 63. The souls of all the Asante people were enshrined in the *sika dwa,* or golden stool. When not in use, the stool was stored on its side, as shown; here it is accompanied by royal bronze bells, as well as golden figurines of the enemies slain by the king. When an Asante king died, many people were sacrificed to serve him in the afterlife. Some of them were beheaded by an *abrafo,* or executioner.

rival leaders—five of whose skulls were still in the royal Asante mausoleum a century later—and increased the size of the kingdom to 100,000 square miles. One of the skulls Opoku Ware added to the royal collection was that of the Akyem ruler responsible for Osei Tutu's death.

The royal Asante mausoleum was a large, mud-walled structure at the village of Bantama, only a mile from Kumase. There the ritually cleansed and reassembled skeletons of past asantehenes were curated so that their *asamanfo,* or ancestral spirits, would continue to protect the Asante people. Near the remains of each ruler were the skulls of the most important enemies he had beheaded. Periodically these skulls were placed in a great brass vessel and displayed to the general populace; afterward, each was carefully returned to its proper place in the mausoleum.

Opoku Ware was succeeded by a distant cousin, Kusi Obodom. The fourth asantehene was Osei Kwadwo, who added more skulls to the mausoleum between 1765 and 1774. The fifth asantehene, Osei Kwame, contributed still others.

By the 1820s the British were attempting to extend their colonial rule to the Asante. The Asante did not take kindly to the idea. The fifth, sixth, seventh, and eighth asantehenes, all of whom were siblings, battled to maintain Asante independence.

In 1824 Osei Yaw, the eighth asantehene, delivered to the Bantama mausoleum a skull so valuable that Asante priests did not risk damaging it by carrying it around. This was the head of the unfortunate Sir Charles McCarthy, who up until the moment of his beheading had been the British governor of the Gold Coast.

Finally, after a half century of bitter fighting, the Asante lost their autonomy. Agyeman Prempe I, the 13th asantehene, surrendered to the British and was exiled in 1896. According to oral history, the Asante felt that if they continued to oppose the well-armed British forces, they risked losing their golden stool. There was little question about Asante priorities: they kept the stool and surrendered their king.

The Elusive Stool

Between 1896 and 1901 a number of Asante nobles figured out how to make the most of the colonial system. A prominent man named Kwabena Kuofor made a fortune from the rubber trade. He then used the colonial courts to establish his right to the golden stool. The British appointed him to the stool in 1901, in part because he was the wealthiest man on the Gold Coast.

Few colonial administrators fully understood that the *sunsum,* or collective soul, of the Asante people was enshrined in the golden stool. Nor did most of them understand that Asante stools had gender. A male stool could never be used by a woman, or vice versa. When word came that Queen Victoria wanted to sit upon the golden stool to symbolize her authority, the Asante were scandalized. They hid that priceless male stool and attacked the British at Kumase. When the dust had settled and the British realized the impropriety of their request, the Asante created a female silver stool for Princess Mary as a gesture of goodwill. "We bind to this stool," the Asante are quoted as saying, "all our love of queen mothers and women."

There were other aspects of Asante society that colonial authorities understood poorly. One was the reckoning of descent in the mother's line. In Europe a king was typically succeeded by his firstborn son. The choosing of his successor did not revert to his mother's clan.

Particularly revealing is a comment made by an Asante elder to Robert Rattray. Rattray, a former British customs officer, had become fluent in Twi. He became the principal investigator of "Ashanti culture" (as it was then spelled) in the Gold Coast colony.

Rattray's knowledge of Twi allowed him to grasp the importance of the queen mother in Asante society. Why, he asked one Asante elder, do you suppose that we British never appreciated her role? "The white man," he was told, "never asked us this; you have dealings with and recognize only the men; we supposed the European considered women of no account, and we know you do not recognize them as we have always done."

Victorian society, Rattray now understood, was more sexist than Asante society—among other things.

Rattray became an advocate for the Asante. He argued for the rehabilitation of Asante queen mothers, kings, chiefs, "and all the wonderful household organization of an Ashanti court, now crumbling into poverty and decay." He urged the Asante not to barter the wealth of their past, "metaphorically and not infrequently in reality, for a coat, a collar, or a tie." The greatest hope for the Asante in the future, he felt, was "to follow and build upon lines with which the national *sunsum* or soul has been familiar since first they were a people."

To be sure, Rattray was a civil servant and a product of Victorian England. Fortunately for historians and anthropologists, however, Rattray's command of Twi and his feeling of common humanity with the Asante challenged him to understand how their society had operated before European interference.

Precolonial Asante Society

The supreme creator of the Asante cosmos was the Sky God. He had a special set of priests committed to him for life, and every residential compound had an altar where prayers could be offered to him. The Sky God was so remote, however, that most tasks were delegated to the *abosom,* a series of lesser supernatural spirits whose actions were more tangible. Many of these lesser spirits, such as Ta Kese and Ta Kora, had mud-walled temples dedicated to them.

There were also the *asamanfo,* or spirits of the ancestors, who protected Asante society and could be ritually contacted for help and advice. Ancestor ritual was pursued at the family or clan level. State-level religion was based on the premise that the collective soul of the Asante people was embodied in the golden stool, and the ghosts of past kings protected Asante society.

The ancestors were honored at ceremonies called *adae,* held every 42 days in the courtyard of the king's residence. At these ceremonies the Asante propitiated the spirits of the departed leaders of their major clans. A dramatic feature of the adae was the oral recitation of Asante history by a specialist who had memorized it. What provided the drama was the fact that an *abrafo,* or executioner, stood behind the man doing the reciting; if the oral historian made a mistake, he was led away by the abrafo. And you think *you* have job-related stress.

The Oyoko clan, headed by the *ohema,* or queen mother, was the social segment from which the kings and nobles of Asante society came. The ohema, in consultation with other members of her clan, typically chose each king's successor. Before the new ruler was allowed to sit on the golden stool, she publicly admonished and advised him. The queen mother also chose each new king's senior wife. When a king left his court to wage war, the ohema served as regent in his absence. Even after his return, she sat next to him in court and served as a check on his power.

The queen mother's personal wooden stool was considered senior to the king's. Neither the king's feet nor his personal wooden stool was allowed to touch the ground directly, as that might bring on famine. The king therefore wore sandals on his feet; he also carried a parasol for protection from the sun.

When a prominent king, queen mother, or provincial governor died, his or her personal wooden stool was blackened with soot and curated in a "stool house." As for the golden stool of the Asante state, which transcended the king's personal property, it was placed on an elephant hide to keep it from touching the ground.

Like the Bemba chief, the Asante king had a group of councillors who inherited their positions. The king was also advised by the Asante elders. His court included spokesmen, heralds, minstrels, and drummers.

Lower-ranking Asante were organized into *oman*, or clans. Each clan controlled its own farmland. The Asante were required to marry outside their clan and expressed preference for what anthropologists call "cross-cousin marriage." For example, a man could marry his father's sister's daughter, since the groom had inherited his blood from his mother, and his bride had inherited her blood from the groom's paternal aunt, who belonged to a different clan.

On the bottom rung of the Asante social ladder were slaves, usually war captives, who belonged to no clan and inherited nothing. Many slaves were kept alive only until they were needed for sacrifice in a major ritual. Other slaves were employed in gold mining, a dangerous activity in which death from cave-ins was always a possibility.

Of all the Asante crafts, goldworking has received the most attention. In the Polynesian terms used by Irving Goldman, goldworking was the ultimate Asante expression of tohunga, or expertise. Goldsmiths were organized into craft guilds or brotherhoods and were the only Asante commoners permitted to wear gold ornaments. Fathers were allowed to bequeath the tools and skills of goldworking to their sons, nephews, or clan mates. This special treatment of goldsmiths was a source of inequality among Asante commoners.

Gold was such an important Asante commodity that a standardized system of weights was created. Experts on the Gold Coast suspect that some units of weight were borrowed from the Portuguese and Arab traders who had been active in the Akan region for centuries. Many Asante weights were cast in bronze, often in geometric shapes or the images of animals. Unfortunately there was so much cheating that it spawned an Asante proverb: "A chief's weights are not the same as a poor man's."

Another Asante skill was long-distance communication with "talking drums." Only chiefs of a certain rank could own the drums, which were made from a hollow log and the membrane of an elephant's ear. According to Rattray, drum communication depended on the fact that Twi is a tonal language, that is, one in which differences in tone are used to distinguish between syllables or short words that would otherwise sound alike. By varying the tone, Asante drummers created sounds that simulated Twi words. The messages frequently involved information about prominent citizens, reports of danger, or calls to war. Rattray devised a way to use English letters to indicate the various tones, allowing him to record messages.

Rattray used his knowledge of both Twi and drum language to give us our most detailed description of an important Asante ceremony, known as the *Odwira*. This ritual, whose name meant "purification," was held each September to honor the Asante kings who had become ancestral spirits.

In its precolonial form Odwira involved the sacrifice of 12 men, usually condemned criminals who had been kept alive for the ceremony. Since the British government had by then banned human sacrifice, Rattray got his information from a senior Asante official who had witnessed the precolonial ritual.

Rattray was told that in the first stage of the ceremony the king and his court, preceded by officials carrying the famous golden stool and the blackened stools of the ancestors, marched to a nearby stream. Sprinkling the golden stool with sacred water, the king asked it to help him behead future enemies, just as his predecessors had done with the Denkyira, Akyem, Domaa, and so on. He prayed that the Asante would find gold to dig and added his hope that "I get some for the upkeep of my kingship."

Some freshly harvested yams were placed on shrines to the abosom, or supernatural spirits; others were offered to royal and noble ancestors. Only after the gods and ghosts had been fed could the king and his people eat.

The procession then moved to the royal mausoleum at Bantama. There, with their arms tied behind them, the 12 men chosen for sacrifice were lined up before the great brass vessel.

The king entered the mausoleum and visited the skeletons of each Asante ruler in chronological order, beginning with Osei Tutu. Each sacrificial victim was assigned the task of serving one past asantehene in the afterlife. From a talking drum came the message, "Osei Tutu! Alas! Alas! Alas! Woe!" This was the signal for the executioner to cry, "Off with you to the land of the ghosts to serve Osei Tutu." He then beheaded the first victim. The king next walked to the coffin of Opoku Ware, where the second victim was given his instructions for the afterlife and decapitated.

After all 12 victims had been sacrificed, their bodies were dragged to the forest behind Bantama. The Asante king then returned to his palace, where he was entertained with singers, drums, and reed pipes. This concert ended the precolonial Odwira.

"I was glad," Rattray's informant told him, "that I still had my head."

Another precolonial context in which the Asante performed human sacrifice was the royal funeral. The venue for these funerals was the mausoleum at Bantama, a place where the royal skeletons and their sumptuary goods were guarded by 1,000 warriors. The funeral of an asantehene was similar to that of the Tattooed Serpent of the Natchez: many wives and members of his retinue expected, indeed, volunteered, to accompany him in the afterlife. Included among the sacrificial victims were the usual condemned criminals and prisoners of war.

As the king lay dying, he might whisper to the queen mother the names of the women he wanted to accompany him in the afterlife. The matriarch was allowed to choose additional women for this privilege; still others volunteered. All of these women dressed in white, the color of celebration, and put on their best gold ornaments. In a ritual reminiscent of chiefly Panamanian funerals, the women were stupefied with palm wine and strangled.

Young boys who had served the king as heralds, pages, or shooers of flies had their necks broken on a large elephant tusk; their bodies were smeared with white pipe clay as a sign of joy. Many noble officeholders, unable to bear the loss of their ruler, volunteered to be garroted. Some servants chose to flee into the forest, but there were always dozens of war captives available as substitutes.

No one dared speak of the asantehene as having died. Common euphemisms were "A mighty tree has been uprooted," or "The king is absent elsewhere." The royal corpse lay for 80 days and nights in a perforated coffin above a pit. As the liquids of decomposition dripped through the perforations and into the pit, a team of funeral attendants fanned away the flies.

By the 80th day, decomposition had reached the point where the king's bones could be removed, cleaned, and rubbed with fat. In precolonial times this fat came from a forest subspecies of the African cape buffalo. Rattray reports, however, that one of the gifts sent by Queen Victoria to the ninth asantehene was a jar of pomade. From that point on this exotic pomade, referred to as "the queen's fat," was used on the royal bones.

Once the fat had been applied, all the king's *suman,* or talismans, were attached to the appropriate bones. The major long bones were then rearticulated with gold wire. The partly reconstructed skeleton was placed in a hexagonal coffin, covered with black velvet, decorated with gold rosettes, and placed in the mausoleum.

A living woman from an important family was then chosen to bring food to the king's remains for the rest of her life. Known as *saman yere,* "the wife of the ghost," this woman dressed totally in white. Wives of the ghosts lived in a special harem, guarded by eunuchs to preserve their virtue for as long as they lived. When one of these women died, she was immediately replaced by another.

The Nature of Asante Inequality

Let us briefly compare the Asante with the Bemba of Zambia and ask this question: Was inequality actually greater under an Asante king than under a Bemba paramount chief?

To begin with, both societies shared what we now recognize as widespread African social institutions. Both retained the matrilineal clans of earlier societies. Both had male leaders whose legitimacy derived from the blood of highly ranked mothers. In both cases a chiefly woman or queen mother continued to supervise the ruler throughout his life. Rulers of both societies shared power with a body of hereditary councillors. Both societies used the public recitation of past leaders' accomplishments as a form of oral history.

There were, however, differences in scale between the two societies. A Bemba chiefdom typically covered only 22,000 square miles, had no elaborate road system, and displayed a three-level administrative hierarchy. At its peak the Asante kingdom covered 100,000 square miles, displayed four administrative levels, and had a concentric pattern of inner and outer provinces linked by carefully maintained roads.

Although Bemba chiefs raided their neighbors for slaves and booty, their subjects were overwhelmingly Bemba by ethnicity. The Asante kingdom contained an empirelike assortment of subordinate ethnic groups such as the Domaa, Tafo, Amakom, Denkyira, Akwamu, Gonja, and so on. Military regiments from these groups were often incorporated into the Asante army.

While the Bemba were content with shrines and informed ritual specialists, the Asante built actual mud-walled temples dedicated to specific deities and staffed by full-time priests. They also had an official state religion, centered on a golden stool.

The Bemba chief had a retinue; the Asante king maintained an actual court, like that of the Denkyira kingdom visited by Osei Tutu. Both the chitimukulu of the Bemba and the asantehene of the Asante had the power of life and death over their subjects. The Bemba chief was often buried with sacrificial victims. Many more people—including wives, officials, servants, slaves, and condemned criminals—were killed at an Asante ruler's funeral. The bones of the Asante king himself, however, were placed in a special coffin in a royal mausoleum, surrounded by the curated remains of earlier kings.

For the vast majority of commoners, inequality under an Asante king was probably only marginally greater than under a Bemba paramount chief. One interesting difference between the two societies lay in the Asante reverence for the golden stool. In 1896 the Asante revealed that they were more concerned with retaining their golden stool than their asantehene. The Asante king, for all his prestige, was not considered the equal of that item of sacred furniture.

This should not surprise us. There are almost certainly U.S. citizens who would rather surrender a sitting president than give up Mount Rushmore or the Statue of Liberty.

To be sure, more than a half century of colonial rule broke down many institutions of Asante society. After an abortive Asante uprising in the late nineteenth century, British authorities burned down the royal mausoleum at Bantama. Fortunately, by that time the Asante priests had quietly removed everything of value. The British then torched the entire village of Bantama. "And a splendid blaze it made," wrote Sir Robert Baden-Powell in his 1896 memoirs.

In 1957 the Gold Coast was awarded its independence from Britain, leaving the Asante in command of their own future. The term Ghana, or "warrior king," was chosen as a name for the new nation. In 1960 it became the Republic of Ghana. The royal Asante mausoleum may be gone, but the golden stool and the collective soul of the Asante people live on.

TWENTY-ONE

The Nursery of Civilization

Of all the world's first-generation states, none were earlier than those of the Near East. They formed at a time when it was still not certain that Hierakonpolis would emerge triumphant in Upper Egypt. They formed at a time when permanent villages had yet to appear in Mexico and Peru.

Thirty years ago, Southern Mesopotamia was considered "the cradle of civilization." Today we know that proto-states were also forming in Northern Mesopotamia and southwest Iran at about the same time (Figure 64). These three regions were all in contact with each other, providing us with another example of a chain reaction: the rise of multiple early states in response to the first aggressive one. The title of this chapter reflects our belief that when you have three cradles, it is a nursery.

The 'Ubaid 4 period ended about 5,700 years ago. During the subsequent Uruk period, 5,700 to 5,200 years ago, states formed in both Iran and Iraq. The first political hierarchy with four administrative levels may have appeared in Iran, but the early state in Iraq was larger. Our use of the generic term "state" reflects the fact that some of these societies were more oligarchy than monarchy.

THE SUSIANA PLAIN

The Susiana plain is southwest Iran's version of the great Mesopotamian plain. It lies between the Karkheh and Karun Rivers at an elevation of 130–550 feet and covers roughly 1,000 square miles.

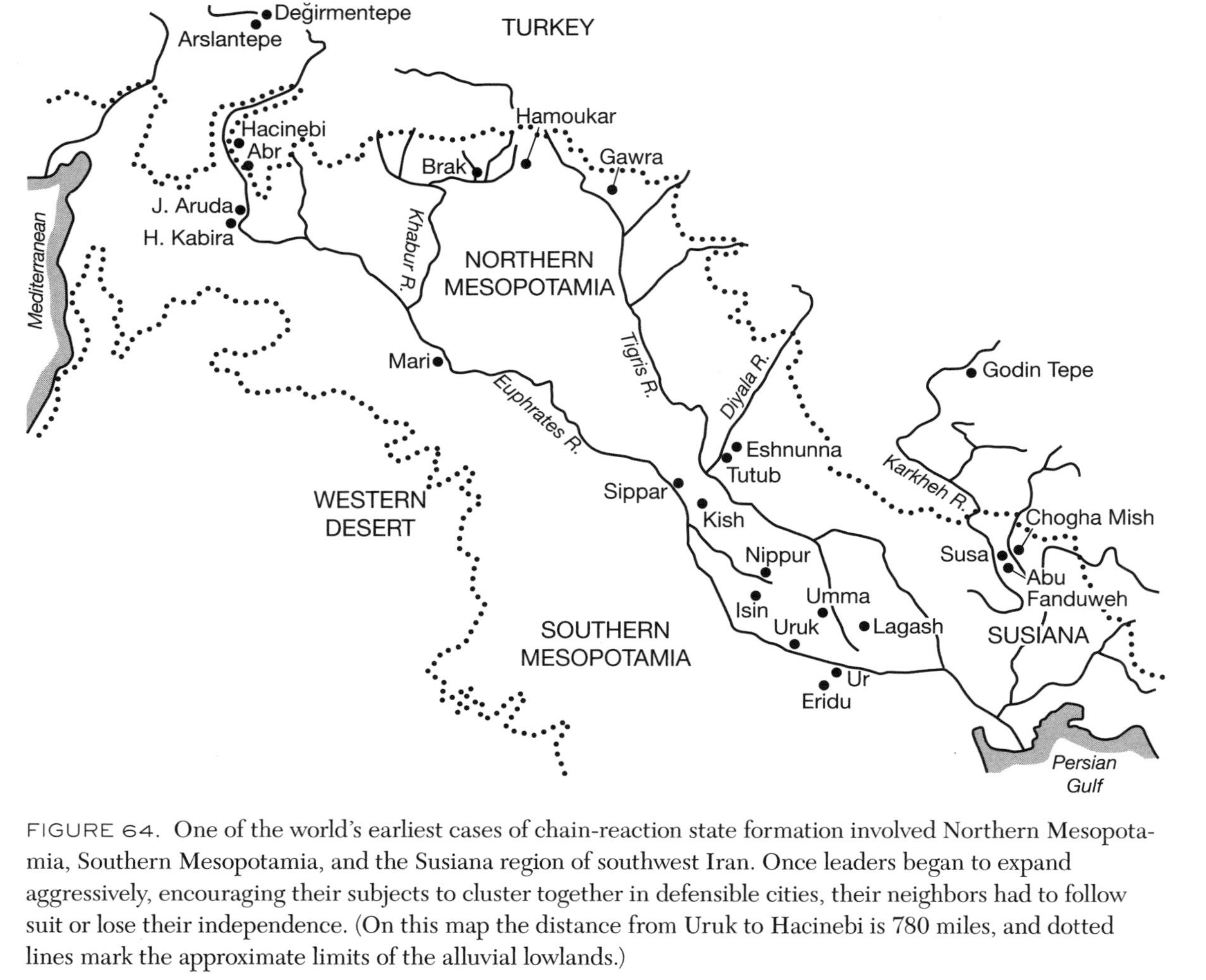

FIGURE 64. One of the world's earliest cases of chain-reaction state formation involved Northern Mesopotamia, Southern Mesopotamia, and the Susiana region of southwest Iran. Once leaders began to expand aggressively, encouraging their subjects to cluster together in defensible cities, their neighbors had to follow suit or lose their independence. (On this map the distance from Uruk to Hacinebi is 780 miles, and dotted lines mark the approximate limits of the alluvial lowlands.)

While Susiana is sometimes described as a smaller version of Mesopotamia, there are two significant differences. Because the Susiana plain was formed by outwash fans from the nearby Zagros Mountains, it has an underlying layer of gravel that reduces the problems of waterlogging and salinization that accompany irrigation. It also receives more rain than Southern Mesopotamia, and this complements canal irrigation from rivers such as the Shaour, Dez, Shur, and Karun.

There has been more than a century of archaeology in Susiana. Some of that work has included surveys for ancient sites, as well as attempts to answer social and political questions with archaeological information. The archaeologists carrying out this work include Robert McC. Adams, Frank Hole, Henry Wright, Gregory Johnson, and James Neely.

Let us pick up the story some 6,400 to 6,200 years ago, a time equivalent to the 'Ubaid 3 period in Southern Mesopotamia. In southwest Iran, this period is known as Susiana d. There were between 85 and 90 villages on the Susiana plain at that time. At least 20 of these villages lay within an easy walk of Chogha Mish, a 27-acre chiefly center on the Shur River floodplain.

Excavations at Chogha Mish by Pinhas Delougaz and Helene Kantor suggest that the family of its *khan* (Persian for "chief") lived in a mud-brick building greater than 48 by 32 feet in size. This building, which may have had a second story, was protected by exterior walls three to six feet thick. One interior room was dedicated to converting flint nodules into blades such as those used for sickles. Another room seems to have been used by potters, who left it filled with carefully stacked storage jars.

The community of Chogha Mish grew bread wheat, barley, oats, peas and lentils, and flax whose seeds were in the size range associated with irrigation. Families at Chogha Mish collected pistachio nuts and caper fruits and harvested clover, perhaps as fodder for their sheep, goats, and cattle.

At least one young woman buried at Chogha Mish displayed cranial deformation. Her skull had been bound shortly after birth, leaving her head elongated as a sign of rank or beauty.

Unfortunately, about 6,200 years ago, the khan's enemies torched his house. The fire preserved the walls of the building to the height of the ceiling and temporarily ended Chogha Mish's role as the dominant chiefly center on the Susiana plain.

The chiefly center that took over from Chogha Mish (and might be implicated in the burning of the khan's house) was Susa on the Shaour River, 18 miles to the west. Between 6,200 and 6,000 years ago, Susa grew to cover

37 acres and seems to have had at least 20 satellite villages. This growth took place during the Susa A period, equivalent in time to ‘Ubaid 4 in Mesopotamia.

During Susa A, the villages of the Susiana plain began to decrease in number and increase in average size. Of the 85 to 90 villages occupied in the previous period, only 60 were still occupied, and that number would drop to 30 over the next few centuries. New villages sprang up, but not in sufficient numbers to counteract the trend, which almost certainly reflects the need to concentrate people in larger settlements for defense against raiding.

By the end of Susa A, the leaders of Susa had erected a mud-brick platform 225 feet on a side and 30 to 35 feet high. Unfortunately, the public buildings on this platform were too eroded to excavate. Near the platform was a cemetery with more than 1,000 burials. These burials reflect a wide range of social ranks. At one extreme were people buried with hordes of copper and masterpieces of painted pottery; at the other extreme were people buried only with cooking pots.

At some time during its period of use, the huge brick platform at Susa suffered the same fate as the khan's house at Chogha Mish: it was so destroyed by fire that its façade collapsed. Susiana, like Northern Mesopotamia, was a region where archaeologists should expect to find evidence for chiefly cycling, endemic warfare, and an individualizing hereditary elite who made use of sumptuary goods.

The Uruk Period

Roughly 5,700 years ago, societies in the Susiana plain entered the Uruk period. The name of this period is borrowed from the ancient city of Uruk in southern Iraq. It is revealing that the pottery of Southern Mesopotamia and southwest Iran was so similar at this time that archaeologists feel comfortable using the term "Uruk" for both areas. When the pottery of two regions displays such similarity, it suggests that their societies were actively involved with each other.

While the people of Susiana and Southern Mesopotamia clearly interacted, they almost certainly belonged to different ethnic groups. We find this likely because of differences in the writing that the two regions later created. The writing in Susiana reflects an early form of the Elamite language. The writing in Southern Mesopotamia reflects an early form of the Sumerian language.

Elamite and Sumerian do not even belong to the same language family. As a result, many community leaders and traders were probably bilingual.

It was during the Uruk period that first-generation states formed in both regions. In order to document all the steps in this process, the archaeologists working in Susiana have subdivided the period into Early Uruk (5,700 to 5,500 years ago); Middle Uruk (5,500 to 5,300 years ago); and Late Uruk (5,300 to 5,100 years ago).

The Early Uruk period was one of social and political reorganization. During the preceding period, both of Susiana's most powerful communities, Chogha Mish and Susa, had been attacked and burned. One result was a temporary loss of population.

Susa rebounded before Chogha Mish did. It grew to 30 acres and emerged as the lone Level 1 community in a political hierarchy of three levels. There were two 15–17-acre villages in Level 2 and more than 45 Level 3 villages covering less than eight acres each.

Gregory Johnson was curious to see how much of the population of the Susiana plain might have been directly or indirectly controlled by Susa. He answered this question by dividing all Early Uruk settlements into three groups: those near Susa, those near the 17-acre community of Abu Fanduweh, and those in the area formerly dominated by Chogha Mish. He then ranked the settlements of each group in order of size.

We will not dwell here on the mathematical details of Johnson's study. To understand what he discovered, we need only consider the following ideas. Geographers have found that many ancient and modern systems of settlement are characterized by a pattern in which the population of the largest settlement is about twice that of the second largest, three times that of the third largest, four times that of the fourth largest, and so on, down to the smallest settlement. Even geographers are not sure why this "ideal" pattern forms, but they suspect that it reflects a well-integrated society.

Geographers, of course, also encounter regions where the rank order of settlements deviates from the ideal. In some cases a region's largest settlement is many times larger than the second, third, and so on. For example, Monte Albán, the capital of Mexico's Zapotec kingdom, was seven to 15 times the size of its Level 2 administrative centers. Such regions tend to be ones in which the capital city has not only integrated its hinterland but also monopolized its region's political and economic interaction with the outside world.

At the opposite extreme are regions in which the second, third, and fourth settlements (and so on) are larger than expected. Geographers suspect that this happens when a region is only weakly integrated, or when the smaller settlements in the region belong to politically independent societies.

What Johnson's discoveries indicated was that Susa was many times larger than the ideal. The region once subject to Chogha Mish, on the other hand, seemed to include a group of independent, or only weakly integrated, societies. Johnson eventually estimated that the Early Uruk population of the Susiana plain was roughly 19,000, of which 9,800 people probably considered themselves subjects of Susa.

The Middle Uruk Period and the Creation of a First-Generation Kingdom

The Middle Uruk period, only two centuries long, was a crucial time for Susiana. At the start of this period Susa grew to cover 60 acres. Taking advantage of Chogha Mish's decline and Abu Fanduweh's smaller size, Susa extended its political control to the entire Susiana plain. The result was a kingdom with a political hierarchy of four levels. Susa remained dominant, even as Abu Fanduweh and Chogha Mish began to grow.

This Middle Uruk state was the first in southwest Iran, and possibly the first in the world. Level 1 of the hierarchy included Susa, followed closely by Chogha Mish and Abu Fanduweh. Level 2 consisted of four administrative centers in the 10- to 17-acre range. Level 3 included 17 villages of five to seven acres. Level 4 consisted of all the remaining villages in the region, most of which covered less than three and a half acres. In 1975 Wright and Johnson, drawing on this Susiana evidence, became the first archaeologists to point out that the appearance of a four-level political hierarchy might be one clue to the creation of a first-generation kingdom.

The Susiana plain of 5,500 years ago thus provides an analogy for what happened in the Oaxaca Valley some 3,500 years later. Both regions had formerly been occupied by rival chiefly societies. The largest of these societies sought to take over the territories of the others, and eventually succeeded. The result in both cases was an early kingdom.

Both of these early kingdoms, to continue the analogy, remained strong until key districts decided to break away from the capital, seeking control of their own territory, tribute, and external relations. This did not happen to Monte Albán until six or seven centuries after it became the capital of a kingdom. It would happen to Susa, however, at the start of the Late Uruk period.

Sacred authority seems to be implicated by the fact that the Middle Uruk leaders built temples at most of their Level 1 and 2 communities. We know this because the Uruk architects had developed a new way of decorating temples. Recall that earlier villages such as Eridu and Tepe Gawra had relieved the monotony of their temple façades by giving them recessed piers, pilasters, and wall niches. The Uruk architects added thousands of ceramic cones that could be set into the wall. The exposed end of each cone was colored white, black, or red. By carefully inserting the cones in the wall before the plaster had set, the masons gave each temple's exterior a series of multicolored geometric designs.

When archaeologists encounter loose ceramic cones on a *tell,* or mounded ruin, they conclude that there is an eroded temple somewhere beneath the surface. The ruins of Susa, Chogha Mish, and at least six other communities had such cones lying on their surfaces.

The Middle Uruk settlements also produced thousands of artifacts related to the administrative duties of the state: seals and seal impressions, bullae and tokens, and beveled-rim bowls. To be sure, seals had already been used 1,500 years earlier at villages of the Halaf period. By Middle Uruk times, however, both the stamps and the clay blobs into which they were pressed had become more varied. Many Uruk seals were shaped like cylinders, carved in such a way that when they were rolled out over wet clay a complex scene was left behind. Seal impressions, no longer limited to the blobs that surrounded knots, now included clay casts of the wooden locks on storeroom doors.

Uruk officials had also invented the prehistoric equivalent of a bill of lading: a lightly baked clay sphere, filled with small tokens. Archaeologists have borrowed the Latin term *bulla* (plural, *bullae*) for these spheres, a reference to the ball of sealing wax attached to a papal bull. It is believed that the tokens, which came in many shapes, reflected the items in a shipment. The recipient would break open the bulla to make sure the tokens matched the items he had received.

Beveled-rim bowls, which were created during the Early Uruk period and became more widespread and numerous with time, may be the least attractive pottery vessels ever made. They were mass-produced by the thousands, the Uruk equivalent of a disposable Styrofoam cup. These bowls appear to be mold-made, and the characteristic beveled rim was produced by trimming excess material from the edge of the mold with a finger. At some Uruk sites in Susiana, beveled-rim bowls were discarded at a rate 11 to 47 times greater than that of any other pottery vessel.

Archaeologist Hans Nissen, drawing on what he knew from later periods in Mesopotamia, has suggested that these eminently disposable vessels might have been ration bowls, used to provide workers with their daily allotment of barley. We know from texts written in the later Akkadian period (roughly 4,200 years ago) that state workers of that era received such rations. The standard Akkadian unit was the *sila,* estimated to be 0.842 liters (0.889 quarts).

Intrigued by Nissen's suggestion, Johnson measured hundreds of beveled-rim bowls from Susiana and found that they came in three modal sizes: 0.90 liters, 0.65 liters, and 0.45 liters. This strongly suggests that they were indeed ration bowls, produced in sizes corresponding to one unit of barley, two-thirds of a unit, and half a unit. The crudeness of the bowls likely reflects the fact that they were made quickly and cheaply, used once, and then discarded.

Here, then, is another analogy with Oaxaca's first kingdom. We have seen evidence that Monte Albán used griddles to mass-produce tortillas for its workers. The Middle Uruk equivalent was a bowl for barley rations. Although both artifacts may have been invented to solve the problem of feeding urban workers, they eventually spread to smaller communities. Some workers took their beveled-rim bowls back to their home villages, where they were put to other uses. For their part, tortilla griddles eventually spread to virtually every household in highland Mexico.

Conflict and Colonization in the Late Uruk Period

As the Susiana plain entered the Late Uruk period, conflict broke out again between Susa and Chogha Mish. Hostilities in the area may have erupted when Chogha Mish decided to break away from Susa, reestablishing its control of the Shur River district.

There were several consequences to this power struggle. In the Late Uruk period, Chogha Mish grew from 20 acres to nearly 45 acres. Susa, on the other hand, shrank from 60 acres to 22 acres, about half the size of Chogha Mish. One of the most dramatic Late Uruk developments was the establishment of a no-man's-land between these two large communities. A buffer zone eight or nine miles wide appeared in the center of the Susiana plain, leaving the productive Dez River floodplain virtually devoid of villages. Some 16 of Susa's satellite communities, representing an estimated 4,500 people, were abandoned. Perhaps 19 villages in the Chogha Mish area, representing an estimated 6,600 people, were also abandoned.

The villagers of the Chogha Mish area may have taken refuge in Chogha Mish itself, accounting for its size increase. On the other hand, exactly where the villagers in the Susa district went is not clear. Johnson wonders if some of them left Susiana and emigrated to Southern Mesopotamia, a possibility to which we return later.

During this period the scenes carved on cylinder seals became increasingly militaristic. One seal impression from Chogha Mish shows a ruler traveling by boat. He holds a mace in one hand; in the other, he holds a cord attached to what may be a pair of prisoners. Another seal impression shows an archer, with his weaponry rendered in great detail, and yet another shows a group of men marching in close formation.

Seal impressions from Susa display similar themes. One shows a line of captives with their hands tied behind their backs. Another depicts a bearded figure armed with a bow and arrow; in front of him are three men with arrows protruding from their bodies.

Johnson suspects that these conflicts reduced Susiana's political hierarchy from four levels to three. He points out that as the result of widespread abandonment, villages measuring five to seven acres (which had constituted Level 3 during the Middle Uruk period) no longer stood out as a separate level. Despite this reduction of administrative levels, other lines of evidence suggest that Chogha Mish's seizure of power had not demoted Susiana from a kingdom to a rank society. Seals and seal impressions, bullae, tokens, and beveled-rim bowls all indicate that the state bureaucracy had survived, and workers continued to receive their rations. In addition, some Late Uruk administrators were beginning to keep their accounts by impressing numbers on clay tablets with a stylus. We would know more about these accounts were it not so difficult to read the earliest writing.

The fact that the apparatus of the state could survive wars (and perhaps the temporary loss of a four-level hierarchy) should not surprise us. We have seen that Maya cities such as Calakmul and Tikal could attack each other, seize each other's Level 2 centers, and even capture royal family members, all without destroying the institutions of the state. Chogha Mish's seizure of power from Susa 5,300 years ago may simply have been analogous to Tikal's seizing of power from Calakmul 4,000 years later. Neither conflict eliminated social stratification and monarchy.

Further evidence of Susa's continued vigor comes from an upper tributary of the Karkheh River. It appears that one of Susa's Late Uruk rulers placed a trade outpost in the Kangavar Valley of the Zagros Mountains, 150 miles north of Susa and more than 4,000 feet above sea level. There, at the community of Godin Tepe, archaeologists T. Cuyler Young and Louis Levine found a fortified enclave of the Late Uruk period. This outpost was surrounded by a much larger community of local Kangavar families. The architecture of the fort was local in style, as was half of its pottery. The other half of its pottery (which included beveled-rim bowls) was Late Uruk in style, as were the cylinder seal impressions. The Godin enclave also included administrators who kept their accounts on clay tablets. The style of notation on the tablets resembles that of Susa.

The fort at Godin Tepe controlled a key pass along one of the Near East's most important trade routes: the Khorasan Road, which led from the Tigris River into the Zagros Mountains and on to the Iranian plateau. This road would lead donkey caravans to sources of copper, turquoise, and lapis lazuli.

SOUTHERN MESOPOTAMIA

When we last looked at Southern Mesopotamia we were trying to reconstruct the society of the 'Ubaid 4 period. We concluded that it was a rank society, but near the group-oriented end of Colin Renfrew's continuum.

Some 5,700 years ago, Southern Mesopotamia entered the Uruk period. It would soon witness the rise of a first-generation state, one supported by irrigation canals from the lower Euphrates River.

Fortunately, the area between the lower Euphrates and Tigris has been the scene of one of archaeology's greatest surveys. During most of three decades, Robert McC. Adams surveyed a strip of Mesopotamia more than 150 miles long. His survey included such ancient Sumerian cities as Sippar, Kish, Nippur, Adab, Shuruppak, Zabalam, Bad-Tibira, and Uruk itself.

Ancient Uruk

Near the modern Iraqi city of Nasiriya, the Euphrates runs east on its way to join the Tigris. Just north of the river lies the huge archaeological mound of Tell Warka. Occupied as long ago as 7,000 years, Warka had once consisted of a pair of mounds called Eanna and Kullaba. These twin mounds were swallowed up as Warka grew to become the ancient Sumerian city of Uruk.

The Euphrates channel at this point is 500 to 650 feet wide and easily navigated. The river has entered its delta and flows only ten feet above sea level, discharging more than 3,000 cubic feet per second during the September dry season. Uruk is surrounded by good alluvial soil that can be irrigated with canals from the left bank of the Euphrates.

Adams's survey of the Uruk region covered some 300 square miles. During the 'Ubaid 1 period (about 7,000 years ago) there were only three villages in the survey area. During 'Ubaid 2 (about 6,500 years ago) the number had increased to seven. The number of villages remained steady at seven through 'Ubaid 3 (6,400 to 6,200 years ago). About 6,000 years ago, during 'Ubaid 4, the number of communities in the survey area rose to 11. The largest were towns in the 25-acre size range; the smallest were two- to three-acre villages.

A significant jump in population took place during the Uruk period. The number of communities rose to 18 in the Early Uruk period and surged to 108 by the Late Uruk period. This increase was so rapid that Adams suspects it included actual immigration from elsewhere. Late Uruk, of course, is exactly the period when Johnson suspects that thousands of families fled Susiana for Southern Mesopotamia. Future archaeologists may be able to use DNA and bone chemistry to determine whether Susiana was indeed the place from which the immigrants came.

Perhaps the most spectacular growth during this period took place at Uruk itself. It grew to 170 acres in the Early Uruk period and reached 250 acres in the Late Uruk period. This growth, too, implies immigration.

Johnson believes that a political hierarchy of four levels appeared in the Late Uruk period. The city of Uruk (250 acres) was all alone in Level 1; Level 2 consisted of eight towns measuring 20 to 35 acres; Level 3 included all villages in the seven- to 15-acre range; and smaller villages constituted Level 4.

Adams is less convinced that a four-level hierarchy was present in the Late Uruk period. He has no doubt that it was present during the subsequent period, known as Jemdet Nasr (5,100 to 5,000 years ago). By that time Uruk surely exceeded 300 acres. Two smaller cities, less than half the size of Uruk, made up Level 2; Level 3 consisted of 20 towns, some of which had reached 50 acres in extent; and roughly 124 villages made up Level 4.

For our purposes here, it is not crucial to know whether Uruk's four-level hierarchy emerged during the Late Uruk or the Jemdet Nasr period. Let us instead consider the following points:

1. As far as one can tell from the archaeological record, the context in which Uruk's leaders created a state does not seem to have involved a local conflict such as the one between Susa and Chogha Mish.

2. The Susiana plain lost thousands of families during the Late Uruk period. The Uruk region gained thousands of families during the Late Uruk and Jemdet Nasr periods.
3. Thus we might not be dealing with two independent cases of state creation, one at Susa and one at Uruk. We might be dealing, instead, with a chain reaction such as the one involving Monte Albán, La Providencia, Monte Negro, Cerro Jasmín, and Huamelulpan in Oaxaca, Mexico.
4. The available information suggests that Susa's crucial growth came first, during the Middle Uruk period, and allowed Susa to dominate both Chogha Mish and Abu Fanduweh. Uruk's crucial growth came slightly later, during the Late Uruk and Jemdet Nasr periods. The scale of every process in the Uruk region, however, was greater than its counterpart in Susiana.
5. Archaeologists, as mentioned earlier, once referred to Uruk as the cradle of civilization. Such Uruk-centrism now appears to be an oversimplification. Mesopotamian civilization is more likely to have been the product of the dynamic competition and alliance-building among several urban societies. One of the best ways to preserve one's autonomy is to become huge. Once communities such as Susa, Chogha Mish, and Uruk had become cities, the chain reaction was on.
6. Examples of inter-city competition include the following. During the Late Uruk period, according to Adams, Uruk seems to have suppressed the growth of other towns within a radius of nine or ten miles. During the Jemdet Nasr period, a newly irrigated area with several large communities appeared about 20 miles northeast of Uruk. Later, however, all these communities were abandoned, possibly because their populations were drawn into the rival city of Umma, 25 miles from Uruk.
7. Some 5,000 years ago, at the start of the Early Dynastic period, Uruk grew to cover an unprecedented one-and-a-half square miles. Adams believes that much of its growth came from rural families taking refuge in the city. One motivation for such immigration may have been widespread violence; it is significant that Uruk built a defensive wall nearly six miles in length. Hans Nissen reveals that this wall had watchtowers at regular intervals and at least two gates into the city.
8. The chain reaction of urban development did not end there. Some 4,700 years ago, perhaps five other cities—Umma, Shuruppak,

> Zabalam, Bad-Tibira, and possibly Larsa—had grown to the point where they could dilute Uruk's influence over Southern Mesopotamia. From that point on, Mesopotamian cities would experience cycles of dominance and decline such as those already described for the Maya.

The Monumental Building Program at Uruk

The ruins of Uruk have been excavated repeatedly since the 1850s. No city has produced a more spectacular series of early public buildings. In fact, the pace at which the city's architects worked makes it hard to determine the function of a given building, so often were they torn down and replaced.

Recall that Uruk once consisted of two mounds, Kullaba on the west and Eanna on the east. Hans Nissen considers Kullaba, buried today beneath the remains of later epochs, "the kernel of the whole great settlement of Uruk."

The oldest recognizable public building on Kullaba was an 'Ubaid 4 temple, resembling those found at Eridu and Tepe Gawra. This temple stood on a high artificial terrace. During the Uruk period this terrace was continually enlarged and raised, eventually becoming a truncated pyramid more than 30 feet high. The White Temple on its summit was visible from a great distance.

The White Temple measured 72 by 57 feet and had the typical central cella, podium, burnt offerings, rows of smaller rooms, and multiple entrances (Figure 65). The temple's nickname was inspired by the layers of gypsum whitewash that covered its walls. The corners of the White Temple faced the cardinal directions; the building was dedicated by burying a leopard and a lion below its eastern corner.

The sequence of buildings from Eanna, east of Kullaba, was even more complex. According to Nissen, the oldest levels in the Eanna sector included the remains of mud-plastered reed huts. This discovery suggests that prehistoric Uruk, like Eridu, had both neighborhoods of mud-brick houses and neighborhoods of reed buildings.

By the start of the Late Uruk period, Eanna had become the scene of a prolonged and ambitious public building program, set off from the rest of the city by an enclosure wall. If we knew the function of every structure, we would have a better picture of the secular and religious hierarchies of Uruk society. Some examples include the following:

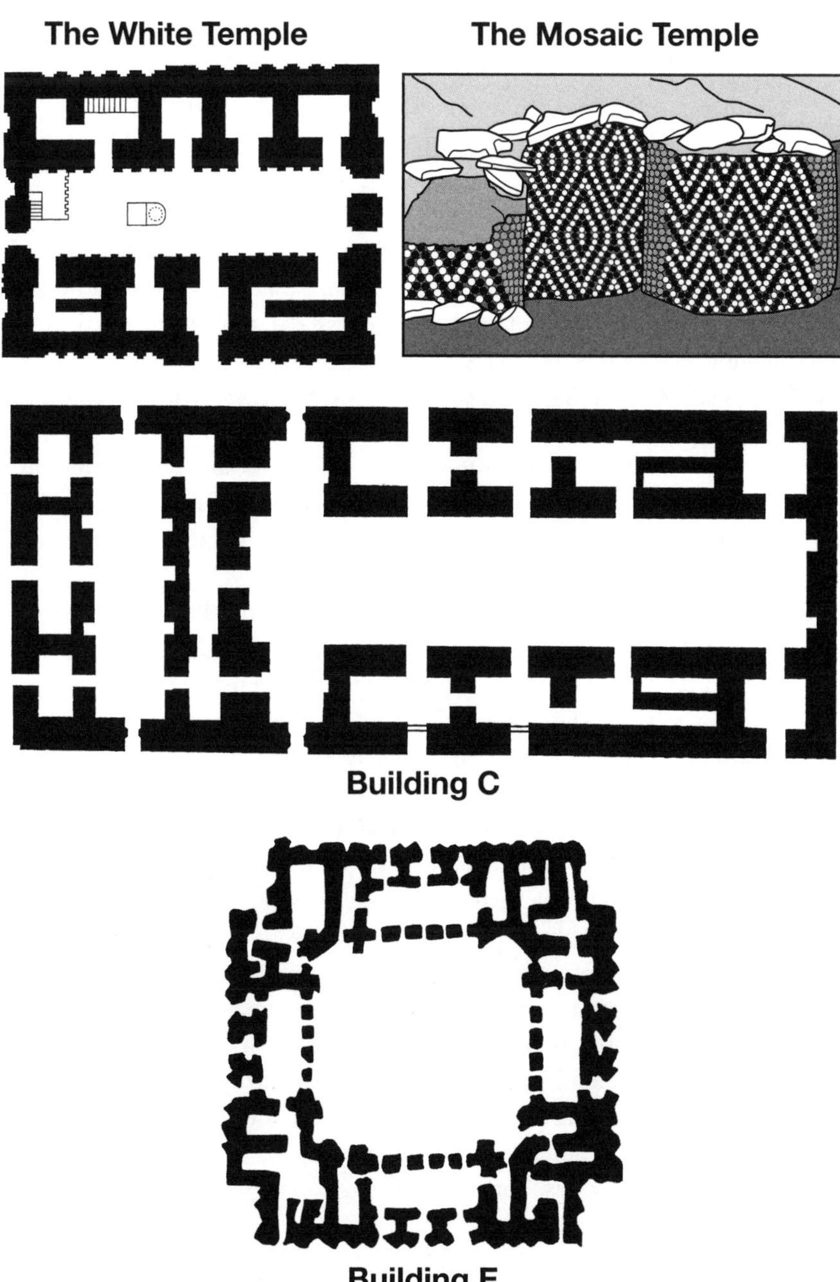

FIGURE 65. The variety of public buildings at the city of Uruk was impressive. Here we see a white-plastered temple, a temple covered with colored cone mosaics, a nave-and-apse temple, and a building that may have been a secular place of assembly. The dimensions of these buildings are given in the text.

1. *Easily recognized temples.* One of the largest of the Eanna structures was Building D. It had the temple layout with which we are now familiar: a cella, flanking rooms, wall niches, and pilasters. Unfortunately, Building D was incomplete when discovered; its original length would have been greater than 175 feet.

   Some distance away lay another temple, within its own separate enclosure. Its walls were of limestone blocks. The floor had been paved with a layer of white gypsum, laid over limestone slabs set in natural asphalt. Similar to other temples, its recessed and pilastered walls were decorated with colored cones.

   These were not, however, the usual pottery cones. They were labor-intensive cones of cut and polished stone, whose colors came from the raw material used: red limestone, black limestone, and white alabaster. The colored cones had been set in gypsum plaster to form mosaic zigzags, chevrons, diamonds, or lozenges. This decoration gave the building its nickname, the Mosaic Temple.
2. *"Nave-and-apse" temples.* Several of Eanna's public buildings, while sharing the standard temple's central cella and rows of smaller rooms, had a different overall shape. Building C is the most complete example. An impressive 175 feet long and 72 feet wide, Building C had the look of two conjoined temples. Two-thirds of the building had its cella oriented northwest-southeast. One-third had its cella running northeast-southwest. Analogous to a cathedral, two-thirds of Building C was a nave and one-third an apse. We wonder if this building's floor plan might not reflect a small, highly sacred sanctuary reached by a longer, and less highly sacred, ritual space. Everyone in a cathedral is allowed in the nave, but only the priests get to use the apse.

   Building C was not unique. An earlier and larger version, called the Limestone Temple, also had an apselike section whose long axis was at right angles to a longer, navelike section. While this building had been destroyed down to its floor of limestone slabs, the excavators estimate it to have measured at least 247 by 98 feet.
3. *Colonnaded halls.* Just north of Building C was the Hall of Pillars, a structure from which little but the façade had survived. The entrance to this building was a portico with two rows of columns eight or nine feet in diameter. The columns were made of small bricks, set like the radii of a circle, and had been decorated with red, white, and black cones. This building looks more like an audience hall than a temple.

4. *Assembly halls.* In the same part of Eanna lay Building E, which looks to us like a secular place of assembly. It consisted of an open court 100 feet wide, flanked by four complexes of large and small rooms that gave it the shape of a plus sign.

The sheer number and variety of Uruk buildings invite us to reconstruct the society that created them, but we need help. That help comes from two sources. The first is a series of sociopolitical terms written on clay tablets of the Late Uruk and Jemdet Nasr periods. The second source is information from a later period called the Early Dynastic.

Let us begin with the Early Dynastic period (5,000 to 4,350 years ago). Early Dynastic society was an oligarchy whose rulers shared power with other aristocrats. One power-sharing institution was an Assembly of Elders, analogous to the noble councillors of the Bemba, the 70 aristocratic advisers of the Merina king, or the marika of the Hunza. Another Early Dynastic institution was a Public Assembly, where commoners could air their concerns.

Although not divine, Early Dynastic rulers were supposed to be pious. Their dreams were considered encrypted instructions from their city's patron deity. That god needed his own temple, and so did the goddess who was his wife or consort. Add a few minor deities and it becomes clear why the cities needed multiple temples.

Let us turn next to the clay tablets of the Late Uruk and Jemdet Nasr periods. Because writing was still in an early stage, these tablets are difficult to read. Some of the signs on them, however, can be recognized as prototypes for Early Dynastic terms.

Among the recognizable words are *en,* "lord" or "spiritual leader"; *nun,* "great nobleman"; *ab-ba,* "elder"; and *ukkin,* "assembly." It sounds as if early versions of an oligarchy, Council of Elders, and Public Assembly may have existed in the Late Uruk and Jemdet Nasr periods. Building E and the Hall of Pillars represent the kinds of buildings in which councils and assemblies might have met. Standard temples such as Building D and the Mosaic Temple, as well as "nave-and-apse" temples such as Building D and the Limestone Temple, suggest that various gods and goddesses were honored at Uruk. Thus the best interpretive guide to Late Uruk society, not surprisingly, may be Early Dynastic society.

Offices and Professions in Late Uruk and Jemdet Nasr Society

The vocabulary of the Early Dynastic period reflected both bureaucratic offices and craft specialties. Many of these terms, as Adams points out, were already present on Late Uruk and Jemdet Nasr tablets. Among the bureaucratic terms were *sanga,* "accountant"; *lagar,* "servant/official"; *dub-sar,* "scribe"; *ugula,* "steward"; *sukkal,* "messenger"; and *nubanda,* "overseer."

Another important term on the early tablets was *dam-gar,* meaning "temple agent in charge of procurement." It is significant that this term eventually came to mean "merchant," suggesting that entrepreneurial trade branched off from the more supervised economy of the temple. As we have seen, something similar happened in Egypt.

There are hints that Uruk/Jemdet Nasr society already had a wide range of professions and craft specialties for commoners. Recognizable terms on the clay tablets include *simug,* "smith" or "metal caster," as well as *simug-gal,* "head smith" or "foreman of the smiths." The presence of smiths is confirmed by a metal foundry and piles of copper ore in the ruins of Uruk.

Additionally, there are a number of occupational terms in Early Dynastic texts that may be a legacy from earlier times. Benno Landsberger, an expert on Early Dynastic writing, felt that the terms for urban craftsmen such as potters, masons, carpenters, weavers, leatherworkers, launderers, and cooks might fall into this category. For the rural professions he listed fishermen, shepherds, plowmen, gardeners, and "fatteners of oxen."

The Uruk/Jemdet Nasr tablets also have terms for servants *(zur)* and slaves. The signs for both male and female slaves indicate that they were "from the mountains" (that is, from a foreign land). This fact suggests that, as in so many other societies, slavery in Mesopotamia began with captives from other regions.

STRATIFIED SOCIETIES TO THE NORTH

We have now looked at two cradles of civilization: one in southwest Iran and the other in southern Iraq. Today many archaeologists would argue that there was a third cradle, this one in northern Iraq and adjacent parts of Syria and Turkey.

Between 6,000 and 5,500 years ago, a number of societies in Turkey, eastern Syria, and northern Iraq began to show signs of social stratification, urban life, and administrative bureaucracy. Some of these stirrings in the north took place before the formation of the Late Uruk state in the south and were clearly indigenous developments. One could make the case, in other words, that if left alone the north was on course to develop its own cities and multilevel hierarchies.

As it happens, the north was not left alone. At the start of the Late Uruk period, Southern Mesopotamian societies began to interfere with a number of northern societies. Archaeologist Marcella Frangipane believes that she can recognize four alternative scenarios for this period, which we paraphrase as follows.

1. Some northern communities continued to develop on their own terms.
2. Others borrowed individual strategies from Southern Mesopotamia (including accounting practices) but essentially created their own distinctive political centers.
3. Some northern communities were actual colonies of Southern Mesopotamian people, founded from scratch in formerly unoccupied places.
4. In some cases people from Southern Mesopotamia directly interfered in the lives of established northern communities. This interference varied from placing a trade enclave in the midst of a settlement to taking over by military force a northern community.

We are struck by how similar these alternatives are to the scenarios we saw in the highlands of Oaxaca. Early Zapotec rulers sent colonists into sparsely occupied valleys, conquered some neighboring regions by military force, and annexed other regions through peaceful alliance. Throughout this process a series of Mixtec rulers, who were already on a course toward monarchy, borrowed individual strategies from the Zapotec but essentially created their own powerful urban centers. The people of Northern Mesopotamia, like the Mixtec, were not the passive recipients of someone else's civilization. They were part of a chain reaction in which no ruler wanted to be somebody else's subordinate.

Northern Societies That Marched to the Beat of Their Own Drummer

Tepe Gawra, Levels XI–VIII. When we last looked at Tepe Gawra, the village of Level XII had been attacked and burned. After a period of abandonment, the summit of the mound was reoccupied. By that time, northern Iraq had entered a period known as the Gawran, or northern Uruk.

An estimated five to seven villages were built, one above the other, at Gawra during this period. We cannot be more precise than that, because some buildings were renovations of preexisting structures. We limit our comments to Levels XI–VIII.

The community of Level XI began as a village of large, extended-family houses, accompanied by a temple 27 feet on a side. At some point the occupants of Level XI began to perceive an external threat. While the commoners arranged their houses so as to present blank walls to the outside world, the community's elite ordered the construction of a circular stronghold, more than 60 feet in diameter, in the center of the village. Included within the defensive wall were granaries large enough to allow the occupants to survive a siege. The Level XI village had watchtowers, heaps of sling missiles, and mud-brick tombs with sumptuary goods.

Levels X and IX were closely related. Gawra at this time lacked defensive works but had streets, a large, secular public building, and at least one centrally located temple. Many of its estimated 185 to 198 residents were involved in craft activity, or the marking of trade shipments with stamp seals.

In addition to multiple temples, Gawra's Level VIII had several important secular buildings. One, described by Ann Perkins as a "large vaulted hall," may have been a place of assembly. Another is reconstructed by Mitchell Rothman as an eight-room warehouse of some kind. In addition to the usual stamp sealings and evidence for craft activity, Gawra seems to have served as a point of transshipment for volcanic glass from sources in Turkey.

The population of Gawra may have dropped below 100 at this point, arguably too low for a community with so many important public buildings. We therefore suspect that Gawra was serving the needs of a wider region. Level VIII had no defensive works, but it probably should have, because it was eventually attacked and burned.

Rothman was impressed by the escalating richness of Gawra's elite burials. By Level VIII they looked like the burials of a stratified society, with gradations of rank within each stratum.

The simplest burials in Levels XI–VIII (85 out of a total of 301) had been placed directly in the earth. Some of these individuals had little or nothing in the way of grave goods. Others wore bracelets or necklaces of rock crystal, obsidian, turquoise, mother-of-pearl, carnelian, or even gold.

Another 78 burials, however, had been placed in mud-brick tombs. Many of these individuals were buried in garments from which only the golden studs, ribbons, and rosettes had survived. Some had ivory combs or ivory-inlaid pins for their hair. Meals for their afterlife had been placed in vessels carved from marble, serpentine, or obsidian. Some people in the tombs wore necklaces, bangles, and other ornaments of gold, silver, copper, turquoise, and lapis lazuli. Among their possible symbols of office were maceheads and stamp seals.

Among those given their own mud-brick tombs were youths, children, and infants. One such infant, Burial 12, was accompanied by 331 beads and other ornaments. Among the raw materials used were gold, turquoise, lapis lazuli, ivory, and carnelian.

While the individuals in brick tombs at Gawra received the most impressive sets of stone vessels and symbols of office, personal ornaments of exotic material were not restricted to the tombs. A possible analogy for this situation can be found in sixteenth-century Colombia and Panama, where the wearing of gold was not restricted to members of the most highly ranked families.

Gawra knew about the Southern Uruk cities but seems to have created its own stratified society. It did just fine without cylinder seals, cone mosaic temples, or clay accounting tablets. Most importantly, it continued Northern Mesopotamia's tradition of individualizing leaders who made flamboyant use of prestige goods.

Arslantepe. On Turkey's Malatya plain, a well-watered region in the headwaters of the Euphrates, lies the archaeological mound of Arslantepe. Founded more than 6,000 years ago, Arslantepe went on to become the civic and ritual center of an indigenous highland society.

According to excavator Marcella Frangipane, Arslantepe had several different kinds of public buildings. Building XXIX, perhaps contemporary with the Middle Uruk period, appears to have been a hall for public assembly. It stood on a platform of huge stone slabs and mud-bricks; its walls were five feet thick; and its main hall was almost 60 feet in length. Not far away lay another massive building whose rooms contained scores of seal impressions, mass-produced ceramics, and other traces of craft activity.

At a time equivalent to the Late Uruk period, Arslantepe featured several temples, each accompanied by a series of storage units. Temples A and B, despite having been built between 5,300 and 5,100 years ago, do not resemble typical Uruk temples. Instead of a central cella, each has an inner room 30 by

15 feet in size, entered from a much smaller outer room. Lacking the complex pilasters and wall cones of Uruk temples, these buildings were clearly the product of a local highland tradition.

Frangipane reconstructs Arslantepe society as stratified, with a ruling class and commoners. The rulers occupied a palatial residence with polychrome wall paintings. Their staff made use of corvée labor, oversaw the flow of commodities, and intensified wool and mutton production. Their temple staff could draw on storerooms full of grain. Arslantepe was in contact with Uruk peoples but had become a proto-state in its own right rather than an Uruk enclave.

Northern Communities Clearly Founded by Immigrants from Southern Mesopotamia

We last looked at the Great Bend of the Euphrates while describing the early village of Abu Hureyra. The Great Bend was also home to two later settlements, Jebel Aruda and Habuba Kabira South, which look as if they had been built by people from Southern Mesopotamia.

Jebel Aruda. The community of Jebel Aruda covered seven or eight acres of a steep bluff overlooking the Euphrates. The center of the community was a walled precinct with at least two Late Uruk temples. To either side of this precinct a Dutch archaeological team found extensive residential neighborhoods that included elite families. In one storeroom at Jebel Aruda the excavators found eight copper axes of roughly equal weight which, according to Guillermo Algaze, almost certainly served as ingots.

Habuba Kabira South. Only five miles south of Jebel Aruda was an even larger Late Uruk settlement. This was Habuba Kabira South, which may have covered more than 50 acres. Habuba Kabira was naturally defended on the east by the 25-foot bluffs of the Euphrates. On the west it was defended by a wall with regularly spaced watchtowers.

Its German and Belgian excavators discovered that Habuba Kabira was a fortified city with streets, residential neighborhoods, artisans' wards, bureaucrats who kept accounts on clay tablets, and an acropolis with public buildings. The metalsmiths at Habuba Kabira had facilities for extracting both lead and silver from the same mineral ore.

Jebel Aruda and Habuba Kabira were not embedded in a preexisting city. They were newly founded in the Late Uruk period, lasted 100 to 150 years, and were abandoned at the end of the Late Uruk period. According to archaeologist Joan Oates, the "identity of material culture, ideology, accounting practices, use of space and building techniques render inconceivable any interpretation other than that the settlements at both Habuba and Jebel Aruda were built and lived in by south Mesopotamians."

That having been said, archaeologists are not in agreement regarding the motivation behind these Southern Mesopotamian forays into the Great Bend. Were they placed there by Uruk itself, to serve as middlemen in its trade with Turkey and the Mediterranean coast? Or were they founded by members of noble Uruk lineages who saw no political future for themselves in the south? We have seen that both primogeniture and ultimogeniture could force some elite sons and their followers to seek new territories.

Resolving these issues will be difficult, but not hopeless. It requires the tracing of any obviously imported pottery vessels to their Southern Mesopotamian clay sources and the use of DNA and bone chemistry to find out from which urban center of Southern Mesopotamia the immigrants came. Archaeologists can then ask whether that urban center was growing, losing population, or experiencing upheaval during the Late Uruk period.

Northern Societies Whose Lives Were Changed by Southern Mesopotamian Immigrants

Hacinebi. The village of Hacinebi occupied a defensible limestone bluff above the Euphrates in southern Turkey. To the north lay the Taurus Mountains; to the south lay the Northern Mesopotamian steppe.

Founded more than 6,000 years ago, Hacinebi grew to cover eight acres. During its first two or three centuries of occupation, Hacinebi belonged to a society in which large villages were surrounded by small satellite villages. Excavator Gil Stein recovered modest evidence of inherited rank; one infant at Hacinebi, buried in a jar, was accompanied by two silver earrings and a copper ring.

Between 5,500 and 5,300 years ago, at a time equivalent to the Middle Uruk period in Southern Mesopotamia, an apparent Uruk enclave established itself in the northern part of Hacinebi. In light of the earlier 'Ubaid 4 trade enclaves at places like Değirman Tepe and Tell Abr, this Uruk enclave was not without precedent. It did, however, lie an impressive 780 miles north of the city of Uruk.

For hundreds of years the families of the Uruk enclave continued to maintain a lifestyle as much like that of their homeland as possible. For example, even though flint was available near Hacinebi, they used overfired clay sickles like those of Southern Mesopotamia. They ate more sheep and goats and fewer cattle and pigs than the local Hacinebi families.

The Uruk enclave used both stamp seals and cylinder seals. They received their own shipments of goods, accompanied by clay bullae covered with seal impressions. They made their own beveled-rim bowls from local clay. Scattered through their refuse were ceramic cones like those used to decorate Uruk temples.

Relations between the Uruk enclave and the Hacinebi people seem to have been peaceful; Stein found no defensive wall, such as the one encircling the Uruk outpost at Godin Tepe. Hacinebi, therefore, continues the tradition of conflict-free enclaves that began in the ʻUbaid period.

Tell Brak. The Khabur River is the last major tributary feeding the Euphrates on its journey south. Its upper tributaries cross the Northern Mesopotamian steppe, homeland of the rank societies of the Halaf period. At least two communities in this region may have grown to the size of cities before Uruk developed its four-level hierarchy. They were undoubtedly part of a widespread chain reaction, but their initial growth cannot be explained by immigration from Southern Mesopotamia.

One of those early cities was Tell Brak on the Jaghjagh tributary of the Khabur. First excavated by Max Mallowan in the 1930s, Brak was already occupied in Halaf times. It had grown to more than 100 acres by the Northern Mesopotamian equivalent of the Middle Uruk period, at which time it was surrounded by satellite communities. During the late twentieth and early twenty-first centuries Joan and David Oates carried out new excavations at Tell Brak, sometimes collaborating with Geoffrey Emberling, Henry Wright, and others.

It appears that during a time equivalent to the Early and Middle Uruk periods, Brak took a backseat to no one. Some 5,800 to 5,500 years ago, it already had a wall with a monumental city gate. Within the next few centuries it achieved its maximum urban size of 106 acres. The rulers of Brak commissioned a huge temple that would undergo three subsequent rebuildings; its final stage was completed approximately 5,300 to 5,100 years ago.

The final stage of this temple is estimated to have measured 97 by 81 feet. There is no question that its builders were familiar with Uruk temples, since they gave it the usual cone mosaic decoration, niches, and pilasters.

Look closely at the temple, however, and you will see local features, almost suggesting a kind of ethnic resistance to Southern Mesopotamia. To begin with, the temple's sides, rather than its corners, were aligned to the cardinal directions. This situation reminds us of Tilcajete's choice of an astronomical orientation different from Monte Albán's. Second, the temple's central cella is shaped like a Latin cross rather than a long, narrow rectangle. Third, the builders of the temple made abundant use of metals from the nearby mountains. The walls of the cella were given copper paneling impressed with a human eye motif. This motif has given the building its nickname, the Eye Temple.

During a time equivalent to the Late Uruk period, Tell Brak began to show more signs of Southern Mesopotamian interference in the lives of its occupants. Cylinder seals, seal impressions, bullae, and tokens became more and more common, as did houses built with the small, distinctive type of brick used extensively at Uruk itself. This period of increased Uruk contact, however, does not seem to have been beneficial to Brak. Some 5,300 to 5,100 years ago it shrank steadily, as if its population, resources, and tribute were being siphoned off.

Tell Hamoukar. On an eastern tributary of the Khabur River, barely five miles from the modern border between Syria and Iraq, lay Tell Hamoukar. Already occupied in Early Uruk times, Hamoukar had grown to 32 acres and erected a defensive wall by the Middle Uruk period.

Excavators McGuire Gibson and Muhammad Maktash found that during the Middle Uruk period, Hamoukar had abundant stamp seals and seal impressions but no tablets with early writing. Within Hamoukar's city walls they found evidence for beveled-rim bowls, domed ovens, a bakery, and a brewery. Apparently wheat, barley, and oats were being converted into meals for large work crews.

Then suddenly, at the end of the Middle Uruk period, Hamoukar was the scene of a massive attack. Thousands of sling missiles, many of them blunted by impact, were found in a layer of debris from burning and destruction. By Late Uruk times, a colony of people using Southern Mesopotamian pottery and artifacts had settled into Hamoukar.

The Hamoukar case allows us to make two points. First, indigenous processes of urbanization and state creation were under way in the north by Middle Uruk times. Second, by the Late Uruk period, Southern Mesopotamia evidently had a military advantage over the north. The south did not create the cities of the north, but it had the power to destroy them. In the words of Gibson and Maktash, this "was not a case of a more developed [southern] core expanding its influence into an underdeveloped [northern] periphery, but of equally matched areas in cooperation and competition over a long time, with the south eventually colonizing parts of the north."

WHY CLUSTER TOGETHER IN CITIES?

Some 5,000 to 4,750 years ago, as pointed out by Guillermo Algaze, none of the surviving settlements in Northern Mesopotamia were large enough to be called cities. Some archaeologists take this as a sign of promise unfulfilled, as if the creation of cities was a lofty goal to which all societies should aspire. We cannot agree.

Our earliest ancestors lived in small-scale societies where everyone knew his or her relationship to everyone else. Nothing could be further from Rousseau's State of Nature than a city. Short of living in a space station, one could hardly imagine a more artificially created environment.

Why, then, would people cluster together in cities? We favor Robert McC. Adams's explanation of urbanization in Susiana, which involves "the drawing together of the population into larger, more defensible political units." Many rural populations of the Uruk period felt exposed and vulnerable; they left their fields and corporate kin groups for the security of the city wall. Still other people fled wounded cities such as Susa, emigrating to regions where they had no traditional right to farm the land. Those with craft skills found work in the city. Those without skills became sharecroppers on the estates of temples or wealthy families. Still others performed manual labor in return for rations of barley and beer.

The rulers of emerging cities were evidently willing to accept as many refugees as they could get. The larger their labor pool and military force, the grander their buildings and the smaller the likelihood that they would lose their autonomy to another urban society.

For every commoner who found security and employment in the Uruk city, however, there were probably several who considered it the lesser of two evils. Many Mesopotamian commoners had traded village life in Mayberry, where people never lock their doors, for life behind three deadbolts in a South Bronx tenement.

THE DYNAMICS OF COMPETITIVE INTERACTION

Previous generations of Mesopotamian archaeologists left us a legacy of brilliant, large-scale excavations. They also left us a lot of folkloric beliefs. One is the notion that "civilization" began in one spot, like an oil spill in the Gulf of Mexico, and spread until it was washing up on distant shores. Another is the notion that the first city was the creation of visionaries who somehow knew in advance that urban life would be superior, a more efficient way to organize crafts and labor, a more exciting place to live, and a magnet for rural people. How you gonna keep 'em down on the farm, after they've seen Uruk?

The fact is that there is nothing inherently superior about urbanization. In Mesopotamia it was all about power building and responding to real or perceived threats. The khan of Chogha Mish did not want his house burned by his rival at Susa. The leaders of Uruk did not want to see Umma grow at their expense, drawing away the rural populations that grew barley for them. Tell Hamoukar did not want to be taken over by people from the south. One way to prevent those things from happening was to get bigger.

Competitive interaction is one of the most important forces driving social and biological evolution. It determines which species leaves behind more offspring; which chimpanzee becomes the troop's alpha; which of the chief's sons succeeds in unifying Hawai'i; which company gets the biggest market share; and which team wins the World Cup.

Many of the ingredients of Mesopotamian city life preceded the Uruk period. Tell Maghzaliyah had a defensive wall 8,500 years ago. Tell es-Sawwan had walls, ditches, sling missiles, and irrigated fields 7,300 years ago. During roughly the same period, Chogha Mami had residential wards walled off from the rest of the village. Samarran pottery was produced by artisans who signed their work. Arpachiyah had streets 7,000 years ago and, like many Halaf sites, monitored shipments of goods by pressing seals into clay.

As early as the 'Ubaid 1 period, Eridu had created a ritual precinct where temples would be built for centuries. Between 6,000 and 5,600 years ago, Tells 'Oueili and Uqair had secular public buildings, one of which had the capacity to store tons of grain. Tell Abada had two-story houses for highly ranked families; Eridu had reed-and-clay houses for fishermen. By 'Ubaid 4 times, there were already trade enclaves embedded in the Euphrates headwaters.

Whatever its timing, we doubt that city life began at one community and spread like an oil slick. It likely grew out of long-term competitive interaction, not only between neighbors such as Susa and Chogha Mish but among regions such as Susiana, Southern Mesopotamia, and Northern Mesopotamia. Competitive interaction drives ambitious leaders to take unprecedented measures. In addition to transforming whole societies, of course, it produces winners and losers. We flock to the winners like paparazzi, forgetting that the competition itself was the real engine of change.

TWENTY-TWO

Graft and Imperialism

They called themselves "the black-headed people," a likely reference to raven hair. During the Early Dynastic period, 5,000 to 4,350 years ago, they dominated Southern Mesopotamia. For two centuries, 4,350 to 4,150 years ago, they lost their autonomy to people speaking a different language. During the Third Dynasty of Ur, 4,150 to 4,000 years ago, they returned to power, only to be ravaged by invaders and internal revolt.

We call the land of the black-headed people Sumer. While that land has seen extensive archaeological survey and excavation, much of what we know about the Sumerians is the product of epigraphy, the meticulous translation of their own written texts. Some of these texts allow us to assign the reigns of Sumerian rulers to specific years in our twenty-first century calendars.

Sumerian society was, in the words of epigrapher Igor Diakonoff, an aristocratic oligarchy in which both the ruler and the oligarchs struggled for supremacy. The Sumerians may have been the first people on earth to privatize land. The relentless purchase of land by noble families, combined with the charging of high interest on loans, created both private wealth and a body of landless serfs.

THE BUREAUCRATIC STATE

While the Sumerians are usually credited with creating the first bureaucratic state, a great deal of the groundwork was laid by their Late Uruk/Jemdet Nasr ancestors. We know a lot about inequality in the Early Dynastic state because the Sumerians wrote so much, and because the writing of that period, referred to as cuneiform because of its wedge-shaped stylus marks, is easier to read than that of the Uruk period.

Many early states had strong, highly centralized governments with a professional ruling class. Politically based social units began to replace the clans and ancestor-based descent groups of earlier societies. One can still detect clanlike units in Sumerian society, but many people in the cities were beginning to live in residential wards based on shared occupation or social class.

One of the most dramatic innovations of states is that the central government monopolizes the use of force, dispensing justice according to rules of law. Achievement-based and rank societies tended to respond to theft or assault at the level of the individual, family, clan, or village. For the Sumerians, most crimes were treated as crimes against the state. It then became the state's responsibility to implement one of a series of punishments, which were codified in order to give the appearance of fairness. This required a system of judges and bailiffs, who were also called upon to decide disputes.

While individuals in Sumerian society were constrained from violence and revenge, the state had the right to draft soldiers and wage war. During the Early Dynastic period, commoners were rounded up to serve as foot soldiers when needed. The artists of that time depicted rulers driving war chariots, followed by soldiers with helmets, spears, and bows and arrows. By the time of the Third Dynasty of Ur, the horse, first domesticated on the steppes of central Asia, was replacing the donkey as a puller of chariots.

Bureaucracies are expensive to maintain, and one Sumerian solution was to levy taxes. Every official transaction had to be witnessed and archived, and an official took his cut. While rations of barley, wool, and beer were still supplied to state employees, the Sumerians turned to standardized units of silver for many taxes, fees, and fines.

A lot of lower-level officials were commoners, and as we will see later, some abused their offices. The distinction between officials and nonofficials enhanced inequality within the commoner stratum: I can now overcharge you, and your only recourse is to bitch.

Finally, the Sumerian state supported what amounted to an official religion. Each city had a patron deity whose temple was larger than that of any other. Temple activities and staff were supported by an estate on which crops were grown, livestock was raised, and artisans labored. The wealth of the largest estates was staggering.

Many economic historians see in the temple estates the germ of a capitalist society. Early Dynastic temples were profit-making, surplus-accumulating, money-lending, interest-charging corporations, and foreclosure on loans may have driven thousands of needy farmers into servitude. Temple managers unwittingly showed the Sumerian aristocrats how to do the same thing.

One other aspect of Sumerian religion deserves mention. On a continuum from tolerant to authoritarian, Sumer lay toward the authoritarian end. Hundreds of rules of social behavior had allegedly been established by the gods; human priests, judges, and bailiffs were there to make sure that they were followed. The state decided what men were allowed to do, what women were allowed to do, who could marry, who could divorce, who could strike whom, and so on. The penalties included fines, corporal punishment, and even death by stoning.

COSMOLOGY

The Sumerians themselves, of course, did not know that their state had been created out of less complex societies, because for them the world had begun in the chaos of a mythological past. Out of the clouds and mist of the cosmos appeared Ki, the earth, floating on a great freshwater sea called Abzu, the Mesopotamian counterpart to Egypt's Nun. The mating of Abzu with Ti'amat, the ocean, produced the high god Anu (Sky). Celestial bodies such as Utu (Sun) and Nanna (Moon), as well as powerful forces such as Enlil (Lord Wind), crossed the sky from horizon to horizon.

The mating of these high gods, some of which were incestuous, gave rise to other deities. Among these were Enki (Lord Earth), Ninhursaga (Mother Earth), and Inanna (Queen of Heaven). For his part, Enlil gave rise to *enten*, the farmer, and *emesh*, the shepherd, the workhorses of Sumerian society.

Obedience was a prime Sumerian virtue. The human ruler obeyed his city's patron deity. In the case of Eridu, that was Enki; at Nippur, Enlil; at Ur of the Chaldees, Nanna; at Larsa, Utu; and at Uruk, Anu and Inanna. The rest of human society obeyed the ruler. Only a ruler was powerful enough to have a direct relationship with his city's patron deity. A commoner interacted only with his tutelary god, a lesser deity who had taken an interest in him. As with the Big Man of Bougainville, who benefited from the love of a demon, a Sumerian's success was explained by his having acquired a deity who would intervene on his behalf.

Deities loved gifts, and a lot of rituals called for pouring libations of beer and making offerings of food. Wealthy citizens could commission small statues that portrayed them praying. Each statue would be stored at the temple and, for a fee, brought out at the right time to stand in for the worshipper.

The gods were the alphas of two dominance hierarchies, one human and one divine. In the city of Lagash, for example, there was a great temple called Eninnu, considered the manor house of the god Ningirsu. It had two temple staffs: one visible and one invisible.

The invisible staff began with a doorkeeper and butler, both minor deities. Below them were a divine chamberlain, counselor, and bailiff, and still further down the list a divine charioteer, gamekeeper, inspector of fisheries, and goatherd, as well as musicians, singers, and errand boys.

The visible staff began with a high priest and continued with human counterparts for all the divine officials. The city's ruler was ex officio head of the church and determined what Ningirsu wanted by having his dreams interpreted. For example, a long narrative poem from Lagash tells us that Gudea, who ruled that city from 2141 to 2122 B.C., was troubled by a vivid dream. He knew that the goddess Nanshe, patron deity of Nina (a Level 2 center in the hierarchy below Lagash), was skilled at interpreting dreams.

Gudea made a pilgrimage to Nina, praying for guidance at other temples along the way. Nanshe revealed the dream to be a sign that Ningirsu wanted Eninnu rebuilt. Upon his return to Lagash, Gudea ordered the work done. This act of piety was described hyperbolically in the aforementioned poem, which was inscribed on two clay cylinders found at Lagash.

SOCIAL CLASS AND LAND

So closely tied were land and people in Sumer that social classes can almost be inferred from the archives of land use. Rulers, upper-level administrators, high priests, and judges of the supreme court were drawn from the hereditary aristocracy. Individual aristocrats (or their families) owned large estates whose fields were worked for them by commoners and slaves. Considerable land had been privatized, and wealthy families could acquire more over time, leaving less and less for everyone else.

Many free commoners still belonged to corporate social units called *imru-a* ("clans"). These units owned communal land that was supposedly inalienable, but it appears that parts of it could be sold, as long as everyone in the unit agreed.

Robert McC. Adams has concluded that most free commoners lived in nuclear family households. There are suggestions, however, that families might have been grouped into larger units called *dumu-dumu* (extended families or lineages?), which in turn were grouped into the "clans" mentioned earlier.

For example, one Early Dynastic text mentions 539 dumu-dumu grouped into seven im-ru-a, some of which were named for deities, animals, or professions. It appears that, over time, these traditional clanlike segments gave way to politically organized units, based on residence or profession within the city.

Sumerian descent was reckoned in the male line, although elite women were mentioned in the genealogies of aristocrats, and women could hold high office. Sumerian kings, like the monarchs of other societies, were allowed multiple wives. Royal polygamy was not just a perquisite of office but a diplomatic strategy, allowing rulers to forge marriage alliances with the aristocracy of other cities.

Commoner marriage, with few exceptions, was limited to one man and one woman. Divorce was allowed, but bigamy and adultery were punished, often severely. One inscription discovered at Lagash states that "the women of former days used to take two husbands, [but] the women of today [if they attempted this] were stoned with stones [upon which was inscribed their evil] intent."

What evil intent? Most of the societies discussed in earlier chapters saw no harm in polygamous marriage. For societies that believed in reincarnation, paternity was not a concern. Babies were seen as recycled ancestors, and all children born into a polygamous marriage were considered full siblings.

The logic of Sumer was different. Men were seen as "planting a seed" in the woman, and because of the male-oriented system of inheritance, the origin of this seed was a major concern. A woman who lost her virginity before marriage, committed adultery, or took two husbands had created intolerable doubt about paternity. The state intervened to protect what it saw as a husband's rights but phrased it in terms of good and evil to make it appear that it was carrying out the will of a deity.

The term for "father's brother" appears in Sumerian cuneiform texts. This suggests to Adams that one of the preferred types of marriage might have been between a man and his father's brother's daughter. Anthropologists call this "patrilateral parallel cousin marriage," and it is still common today in parts of the Near East.

Sumerian marriages, like those of the less complex societies seen in earlier chapters, required gifts between the bride's and groom's relatives. Exchanges of gifts could go on for months. Marriage was considered a legally binding contract, and divorce could cost the husband a fee in silver. Owing to sexism, it was harder for women to get a divorce.

It is probably from the Sumerians that later Near Eastern societies, including the Aramaic-speaking authors of the Old Testament, got the notion that marriage should be restricted to one man and one woman. The flexible marriage partnerships of egalitarian societies, which came in six or seven varieties, had been arbitrarily reduced to a legal contract between a man and a woman. Nothing could be allowed to make a man worry that his male heir was the result of someone else's "seed."

Let us return now to the relationship between social groups and land. Among the wealthiest landowners were the temples, of which there were several in each city. One category of temple land, called *níg-en-na,* was cultivated by temple employees. Its products were (1) distributed as rations or wages to the plowmen, millers, weavers, cooks, and brewers who worked for the temple; (2) stored as a safeguard against droughts or famines; (3) traded by the *dam-gar,* or temple agents, for imported goods; and (4) used to feed the priests, scribes, and other officials of the temple.

Two other categories of temple land were *gán-shukura,* or prebend (a British term for the land set aside to support the staff of a cathedral) and *gán-apin-lá,* or tenant fields. The latter were lands sharecropped by commoners, who turned over to the temple a percentage of the harvest.

Thousands of people in Sumer, of course, were landless, and their numbers grew as time went on. Sometimes referred to as *gurush,* or serfs, at least some of these people were fugitives from other districts or regions. Others had simply lost their land through debt or misfortune. As aristocratic families acquired more and more land, serfs increasingly sharecropped or worked for standard rations.

Finally, there were the slaves owned by rulers, temples, and private citizens. Slave women, more numerous than their male counterparts, worked mainly at spinning, weaving, cooking, and brewing. Male slaves were used as farm laborers and burden carriers. Most slaves were war captives, but late in Sumerian history, some impoverished families resorted to selling their children into slavery.

Slaves could engage in business, borrow money, and even buy their freedom. On the other hand, if they tried to escape they could be branded, flogged, or even killed. Some, in fact, had already been blinded when they were captured in combat.

RULES, ORDER, AND RITUAL PURITY

In the language of every society there are abstract terms that underlie many of the logical premises. The Polynesians had mana, the Merina hasina, and the Egyptians ma'at. For their part, the Sumerians had *me* and *nam*.

Me, sometimes translated as "order," referred to the rules that the gods had established so that society would run smoothly. In the words of epigrapher Benno Landsberger, me "emanated from gods and temples in a mystic manner, was imagined as a substance, was symbolized by emblems, and could be transferred from one god to another." The task of a human ruler was to make sure that the rules of his city's god were obediently carried out, and that the society he commanded was sufficiently orderly. Much of the order was achieved by appointing overseers for every activity and keeping extensive written documents.

Nam has been translated as "fate," but its meaning was more subtle than that. In earlier chapters we learned that names could be magic. In Sumer, Landsberger explains, the name defining the essence of a thing determined its life trajectory and destiny. Temples, people, animals, plants, and bodies of water had names and, ultimately, fates pronounced by the gods.

In contrast to Egypt, where rulers were divine, the Early Dynastic ruler was essentially an aristocratic mortal who did his god's bidding. Whether or not he was a relative of the previous ruler, he found it difficult to succeed without the support of the Council of Elders and the other aristocrats. Even at the peak of his power, an early Sumerian ruler did not claim descent from a deity. His inscriptions might portray him as "beloved" by a series of major and minor deities; he might even claim that the city's patron god had chosen him to rule. But not until the reign of a ruler named Naram-Sin did Mesopotamian kings begin regularly to portray themselves as divine.

The need to please the gods made ritual purity a major concern in Sumer. As early as the Uruk period, some temple precincts had been walled off from the secular parts of the city. Before entering the temple, even a Sumerian ruler had to perform ritual ablution, washing away the pollution of the secular world.

Perhaps no archaeological discovery reveals more clearly the importance of ritual purity than the oval temple enclosure at the ancient city of Tutub. During Early Dynastic II, Tutub was one of two major cities on the Diyala River between the Tigris and the Zagros Mountains. Its ruins today are known as Tell Khafajah.

Excavations by Pinhas Delougaz determined that occupation at Tutub began at least 5,100 years ago. Over the centuries, house upon house, street upon street, the remains of secular human settlement accumulated. By Early Dynastic II, the mounded debris stood 26 feet high.

At this point the ruler of Tutub, perhaps in response to a divine order encrypted in a dream, decided to build a great temple. This "mansion of the god" was to occupy an oval precinct, walled off from the secular part of the city. There was just one problem: the place chosen for the temple had been polluted by centuries of secular houses and human waste.

The ruler's solution was to have his workers dig down 26 feet to the underlying sterile soil, removing all traces of human settlement over an area of 7.4 acres. This excavation was then filled with 64,000 cubic meters (2,260,160 cubic feet) of clean sand. Now it was sufficiently free of pollution to support a temple.

The Temple Oval of Tutub was given two concentric walls (Figure 66). The high priest's residence, tucked into a corner between the inner and outer walls, resembled a palace. Roughly 130 by 98 feet in extent, it was entered by a small door that led the visitor past a guard room to a narrow corridor. Off this corridor were two antechambers; one was flanked by a bath and toilet, where the priest and his visitors could purify themselves before proceeding further. Once their ablutions were complete, they could enter the building's central court, perhaps pouring libations at its offering table.

The central court was a hub for traffic within the residence. To its south lay the priest's reception room, complete with a divan on which he could receive his visitors. Behind the reception room were his archive for cuneiform tablets and his dressing/sleeping room. East of the court was a dining room and behind it a pantry with access to the servants' quarters. To the north of the central court lay a storage room and the priest's private chapel.

Just as the British monarch is titular head of the Church of England, Sumerian rulers were keepers of the faith. The me, or divine rules of society, however, were established by gods and not by kings. The ruler's duty was to see that a pious and orderly society was maintained.

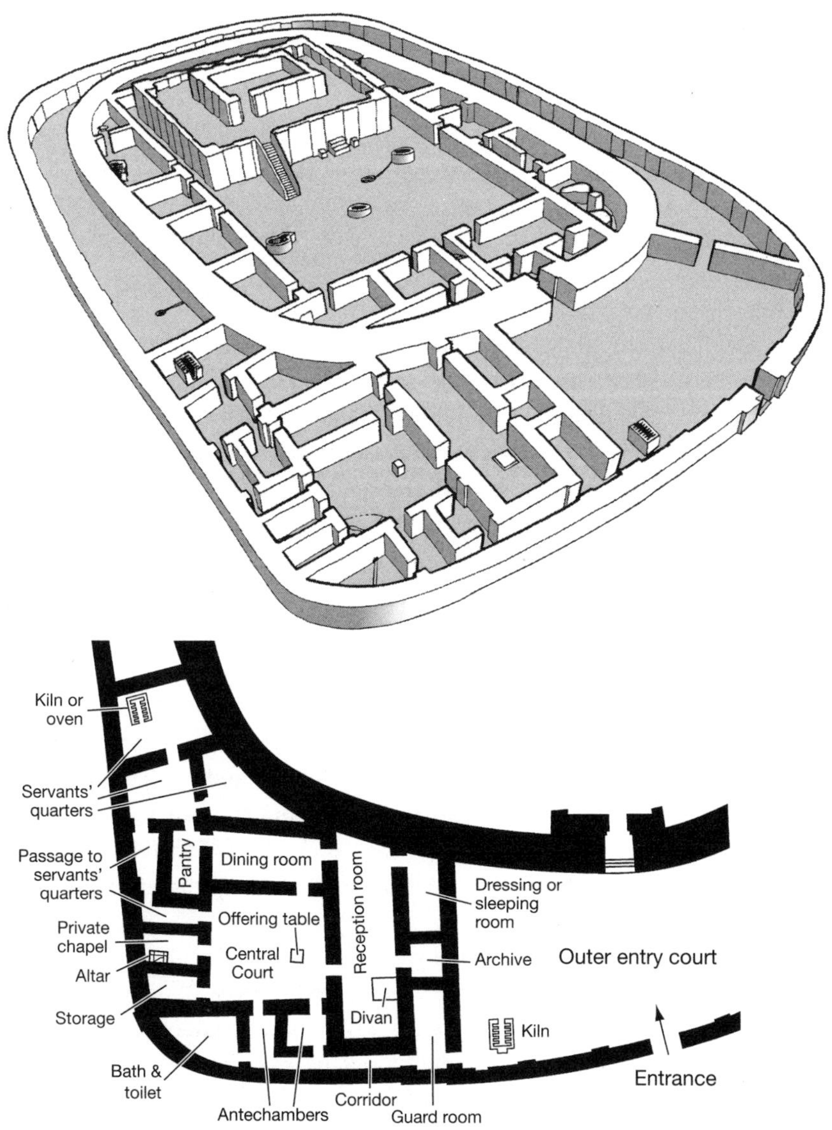

FIGURE 66. Ritual purity was very important in Early Dynastic Sumer. Before this oval temple enclosure at ancient Tutub could be built, workers had to remove the impure debris of human settlement from an area of 7.4 acres and replace it with clean sand. The high priest's residence (shown in detail below) was tucked into a space between the inner and outer enclosure walls.

POLITICAL HIERARCHY

Like ancient Egypt, Sumer was made up of numerous provinces. These provinces, each of which had a capital city and a hierarchy of towns, large villages, and small villages, have been compared by Diakonoff to *nomes* (the Greek word for the hesps of Egypt). The comparison is apt in the sense that each Sumerian province had a governor, like the *nomarchs* of Egyptian nomes.

While the nome is not a perfect analogy, we prefer it to the term "city-state," which has often been applied to Sumerian provinces. This term strikes us as an inappropriate comparison to the Classical Greek city-state, or *polis*. We are not convinced that the polis, whose leaders were elected by the populace, closely resembles any other society of the ancient world.

During the Early Dynastic period, the capital cities of Sumerian provinces fell into three clusters. In the south were Ur, Eridu, Larsa, Bad-Tibira, Uruk, Umma, and Lagash. Farther upstream were Nippur, Adab, and Shuruppak. And still farther upstream, where the Tigris and Euphrates more closely approach each other, lay Kish and Akshak.

While the Sumerian language was dominant from Nippur to Ur, there are words in the cuneiform texts of Kish that reflect a second language, Akkadian. Early epigraphers recognized Akkadian as a Semitic language, part of the family to which later languages such as Hebrew and Arabic belong. Semitic languages apparently extended from the Mediterranean Sea to northern Iraq. Some speakers of Semitic languages lived in settled communities; others were pastoral people, who spread their language widely while traveling with their herds.

We have one long, largely mythological list of early Sumerian kings, plus shorter king lists from individual provinces. Obviously we know much more about those provinces where thousands of cuneiform tablets are available. One of the best-documented provinces was headed by the city of Lagash, which lay not far from the Persian Gulf.

The population of the province of Lagash has been estimated at 100,000 "free citizens" (that is, excluding slaves). Some 36,000 of those free citizens may have lived at Lagash itself. The ruins of Lagash, known today as Tell al-Hiba, cover 1,284 acres.

In Level 2 of the administrative hierarchy below Lagash were two smaller cities, Girsu and Nina. Girsu, whose ruins are known as Tell Luh, covered 914 acres and may have been home to 19,000 free citizens. Nina, whose ruins are known as Tell Shurgal, covered 370 acres. Both Girsu and Nina were already occupied 6,000 years ago. Lagash became large in the Early Dynastic period and eventually subordinated the two smaller cities. In Level 3 of the hierarchy were settlements with names like Urú, E-Ninmar, Kinunir, and Guaba. We do not know the names of the Level 4 villages. We also do not know the full extent of the province of Lagash, but its irrigated fields alone covered 772 square miles. A 28-mile-long stretch of irrigated land on the border between the provinces of Lagash and Umma was, as we shall see, under dispute for centuries.

Lagash featured at least ten temple estates, the largest of which belonged to the patron deity Ningirsu and his divine wife, Bau. Among the smaller temple estates were those dedicated to Utu (Sun) and Nanshe (that skilled interpreter of dreams). Temple estates may have covered more than 200 square miles of the province and employed 5,000 to 12,000 free citizens.

At the apex of a province's political hierarchy was its king, for whom two Sumerian words existed. One of those words, *ensí,* is the older, and it incorporates the word *en,* "lord" or "spiritual leader." This derivation of the word may reflect the fact that early Sumerian rulers had a degree of ritual authority.

A second word, *lugal* (from *lu,* "man," and *gal,* "big"), appeared later, and in several contexts it seems to outrank ensí. Diakonoff notes that one ensí of Lagash changed his title to lugal when he embarked on an ambitious campaign of conquest. Rulers claiming control of more than one province sometimes referred to themselves as "lugal of the land" or "lugal of the universe." Few rulers, Diakonoff feels, would dare to assume the title of lugal if their own province was claimed by a "lugal of the universe."

Unlike a typical ensí, a powerful lugal may have felt that he could ignore the Council of Elders and the Popular Assembly. This, according to Diakonoff, made him a forerunner of the more despotic kings of later times.

A lugal's Level 2 cities were usually run by ensís. For his part, the ensí delegated many tasks to a vizier like that of Egypt. The temple estates of the province were run by overseers called *sanga.* In the complex hierarchy of Sumer, aristocratic administrators supervised commoner foremen, who in turn supervised gangs of plowmen, weavers, and burden carriers. Wages and products were listed by scribes on clay tablets, to be stored eventually in archives.

KINGS, PALACES, AND ROYAL TOMBS

The Sumerians believed that kingship had descended from heaven during mythological times. The first two kings of Eridu are said to have ruled for a total of 64,800 years. Three later kings of Bad-Tibira ruled for 108,000. In Shuruppak one king ruled for 18,600.

After a total of eight mythical rulers came a giant flood that covered the earth. Kingship then had to descend from heaven for a second time, and it now centered on Kish. They evidently were not making rulers the way they had before the flood, because the first king of Kish ruled for a mere 1,200 years.

Even the kings of Early Dynastic I remain shadowy figures. One monarch, Etana of Kish, is described as "he who stabilized all the lands," implying that his influence extended beyond his own province. Finally, during Early Dynastic II, inscriptions from places as widely separated as Nippur, Adab, Girsu, and the Diyala River basin began to mention rulers of Kish who are likely to have been flesh and blood.

One of these Early Dynastic rulers was Mesalim of Kish. His inscriptions indicate that he controlled provinces beyond his own, and his political influence extended further still. As we shall see later, Mesalim was once called upon to mediate a border dispute between the rival provinces of Umma and Lagash.

Given the importance of the kings of Kish, it is no surprise that archaeologists have found two impressive palaces there. Palace A, built at no great distance from the city's temples, consists of at least two architectural units covering an area 300 by 200 feet. The larger of the two units was surrounded by a massive, buttressed defensive wall. Its royal residential quarters were embedded deep in the western portion of the building. The monumental entrance to the palace lay to the southeast and led to offices and archives that had only indirect access to the royal apartments. The smaller of the two architectural units was separated from the larger by a narrow corridor and had the appearance of an annex. Deep in its interior was a decorated reception hall with columns.

A second Early Dynastic palace at Kish, known as the Plano-Convex Building, had a triangular ground plan like the well-known "flat iron building" in New York City. It lay more than a mile from Palace A, underscoring the fact that each Mesopotamian king preferred to build his palace in a new area, designed to his own specifications.

During Early Dynastic III, a greater number of Sumerian kings made the transition from legend to history. One such ruler was Mesannepadda, alleged founder of the first royal dynasty of Ur. This dynasty is of interest because the most spectacular Early Dynastic tombs ever discovered come from the city of Ur.

During the period 1927–1928, archaeologist C. Leonard Woolley discovered an Early Dynastic III cemetery at Tell al-Muqayyar, the ruins of the ancient city of Ur. The 1,800 graves he excavated almost certainly include the remains of well-to-do commoners, government officials, minor nobles, and members of royal families. The 16 graves Woolley considered royal provided a contrast to the graves of commoners. Many commoners were simply wrapped in matting or given a coffin of basketwork, wood, or clay. They were accompanied by their personal belongings, which in the case of some bureaucrats included their administrative seals.

The most spectacular pair of tombs (Graves 789 and 800) belonged to a king, possibly named A-bara-gi, and his queen, Pu-abi. The king's tomb had been broken open and plundered, but the queen's was intact (Figure 67).

The queen of Ur had been laid to rest on a raised platform, or bier, inside a limestone and mud-brick tomb measuring 14 by 9 feet. She wore an elaborate headdress of gold leaves and ribbons, carnelian rings, lapis lazuli beads, and a golden comb decorated with lapis lazuli flowers. A huge pair of crescent-shaped golden earrings adorned her ears, and the entire upper part of her body was covered with gold jewelry and semiprecious stones. Three lapis lazuli cylinder seals were found at the queen's right shoulder. One was inscribed with the name Pu-abi and the title *nin,* "Lady." Another seal bore the name A-bara-gi and is presumed to refer to her husband.

Two additional skeletons were found on the floor near the queen's bier; their headdresses suggested that they had been ladies-in-waiting. A third skeleton is believed to have been that of a male attendant.

Just outside the queen's tomb began Grave 1237, an associated chamber covering more than 500 square feet and described by Woolley as "the great Death Pit." In it were the remains of an estimated 74 sacrificial victims, mostly young women. There were also two wagons, drawn by oxen and attended by drivers and grooms. All the animals, drivers, and grooms had apparently been sacrificed in place.

Along the wall of the Death Pit, closest to the king's tomb, were the skeletons of nine women with headdresses of gold, carnelian, and lapis lazuli. These women were accompanied by the disintegrating remains of two harps;

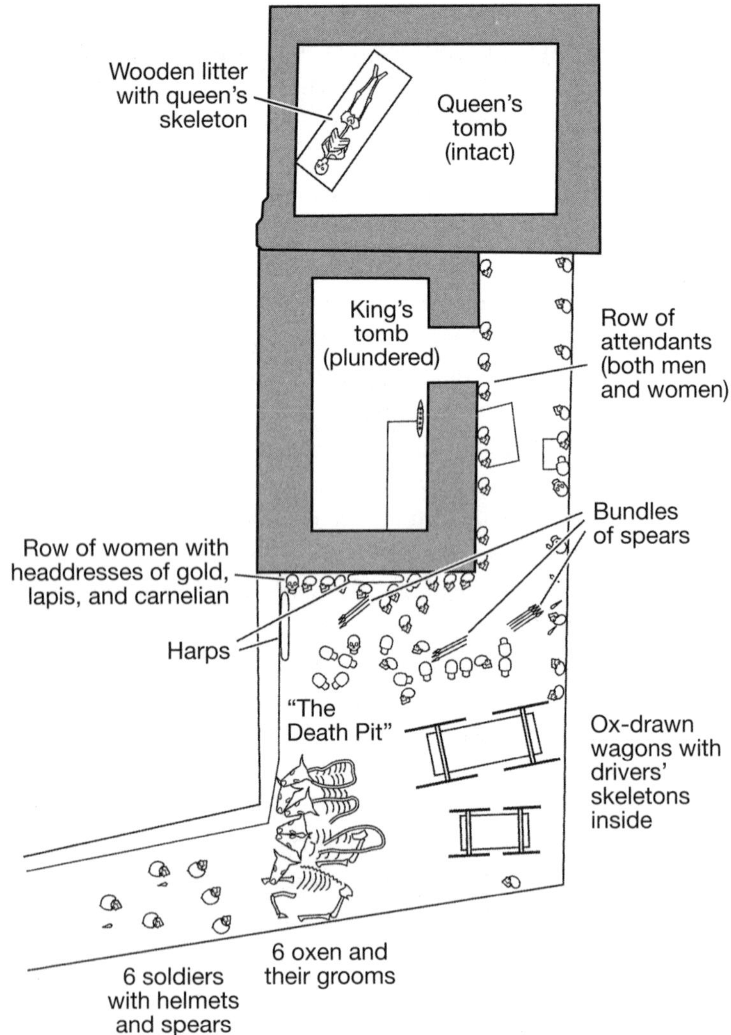

FIGURE 67. During the Early Dynastic period, some funerals of kings and queens at the city of Ur included human sacrifice. This drawing shows the tombs of King A-bara-gi and Queen Pu-abi. More than 70 attendants, including soldiers, ladies-in-waiting, grooms, ox-cart drivers, and musicians, had apparently been sacrificed to accompany their rulers in the afterlife.

elsewhere Woolley found the remains of lyres. These musical instruments had likely been used to accompany the laments that we know were sung at Sumerian funerals.

Between the skeletons of these women and the two wagons were the remains of men with bundles of spears. Leading upward from the Death Pit was the ramp down which the wagons had been led; this ramp was still guarded by the skeletons of six soldiers with helmets and spears.

Woolley saw no evidence for violent death in this part of the cemetery. In his scenario all the soldiers, musicians, grooms, attendants, and ladies-in-waiting went to their death willingly, perhaps by taking poison. Woolley's scenario is plausible, but it is currently being reevaluated by Mesopotamian scholars.

What is intriguing about the royal tombs of Ur is that they show us a level of human sacrifice as great as that of Peru's Moche tombs, or the burials of Panama's Coclé chiefs. While these sacrifices allow us to compare Early Dynastic Sumer to Moche and Coclé societies, such behavior is considered atypical for the Sumerians and apparently did not continue into later periods.

CORRUPTION AND MALFEASANCE IN OFFICE

The Sumerians left the world an amazing legacy of urban civilization. Unfortunately, they also created a legacy of bureaucratic corruption that even today's politicians must work hard to equal.

The chiefs of rank societies expected to receive tribute from their subjects. The "thigh-eating chiefs" of the Kachin, for example, accepted a hind limb from every animal sacrificed. Such tribute was rationalized by the belief that Kachin chiefs were descended from celestial spirits, who would consume the animal's essence. In Sumer, however, even officials with no celestial ancestors began to demand exorbitant fees for every bureaucratic transaction. As for the rulers themselves, some began to covet the wealthy temple estates.

The Early Dynastic texts of Lagash describe growing corruption and malfeasance, interrupted by occasional reform. For some examples, let us look at the period covered by the reigns of Entemena, Enanatum II, Enentarzi, Lugalanda, and Urukagina, who ruled Lagash between 2404 and 2342 B.C.

We have seen that the two largest temple estates in Lagash were those of the city's patron deity, Ningirsu, and his wife, Bau. Bau's estate alone has been estimated at 25 square miles, some 17 of which consisted of agricultural fields.

Every temple estate, while considered the property of a deity, was run by a human overseer known as a *sanga*. Under the sanga the estate was treated like a profit-making corporation, producing a surplus, engaging in foreign trade, extending loans to private citizens, calculating the long-term impact of interest rates, and foreclosing on debts. At the start of the Early Dynastic period, there were checks on the avarice of the sanga; his accounts were kept by scribes, and he had to answer to his community.

Little by little, however, the ensís of Lagash began to confer the title of sanga on their eldest son and heir. The ruler Entemena, for example, made his son Enentarzi the sanga of the temple estate of Ningirsu. After the brief reign of his uncle, Enanatum II, Enentarzi acceded to the throne of Lagash.

What Enentarzi then did reveals two interesting changes in the logic of Lagash society. First, in the words of Diakonoff, "the temple estate [of Ningirsu] came to be regarded as the property of the ensí." Enentarzi, who had overseen the estate for years, simply retained control of it when he left for the palace. Second, by logical extension, the temple estate of the god's wife came to be regarded as the property of the ensí's wife. Enentarzi's wife, therefore, assumed control of the Bau estate. This privatization of what had been the gods' land continued under the ruler Lugalanda and his wife, Barnamtarra.

Epigrapher A. I. Tyumenev's description of the Bau estate makes it clear why a ruler's wife might find the property desirable. At one point in time the estate employed an estimated 1,200 persons, 250 to 300 of whom were slaves. The products of its 17 square miles of fields were kept in 30 storehouses, one of which held 9,450 tons of barley. At least 205 female slaves worked in a centralized weaving establishment, while others brewed beer and cooked for the work gangs. Administrators and professional plowmen at Bau received rations, while other residents of the estate were sharecroppers. The temple estate of Ningirsu was presumably even larger than Bau's, but we cannot estimate its size because fewer texts from its archives have come to light.

Other aristocrats, observing the embezzlement of the temple estates, used the ensís as role models. They possessed themselves of more and more land, often by making loans on which they eventually foreclosed. For their part, the commoners who held bureaucratic positions began to line their pockets as well.

Eventually the citizens of Lagash began to complain that the ensí and his wife had appropriated the temple estates. They also complained that the ensí was using the temple's oxen to plow his personal onion fields. The sanga, they added, was raiding the orchards set aside to support indigent mothers. As if that were not enough, rich men were stealing fish from poor men's ponds.

Corruption was rampant among appointed officials. The overseer of boatmen claimed the best boats for himself. The overseer of fisheries preempted the best fishing locations. The ensí's officials hired blind men to draw water from wells and then fed them only table scraps.

When shepherds arrived at the shearing station with valuable white sheep, they were charged an exorbitant five shekels of silver to have them shorn. Men who wanted to divorce their wives also had to pay five shekels. The official whose job it was to deliver a corpse to the cemetery was charging the deceased's family 420 loaves of bread and seven pitchers of beer. Priests were often shortchanged on their barley rations. The *gish-kin-ti,* or temple craftsmen, reported having to beg for the bread they were owed. Many of these abuses drove ordinary Sumerian families further into debt.

And finally, epigrapher Samuel Noah Kramer reports, there was this widespread complaint: "From the borders of [the estate of] Ningirsu to the sea, there was the tax collector."

In this atmosphere of corruption, a noble named Urukagina began to curry favor with influential priests, promising reforms. Many other aristocrats, aware of the commoners' complaints, agreed that reforms would be necessary.

Urukagina became ensí of Lagash in a coup in 2351 B.C., promising to return "the house of the ruler [Ningirsu]," "the house of the woman [Bau]," and "the house of the children [the divine offspring of Ningirsu and Bau]" to their rightful owners. Urukagina also freed priests from taxation and canceled many of the commoners' outstanding debts. His may have been the first government bailout.

Urukagina claimed in his royal inscriptions that he had been given the kingship of Lagash by Ningirsu himself. He prohibited officials from shortchanging priests' rations, seizing the best boats, and occupying the best fisheries. He prohibited bailiffs from charging five shekels to shear white sheep. He lowered the fee for delivering a corpse to 80 loaves of bread and three pitchers of beer. Aristocrats were forbidden to take fruit from the orchards set aside for indigent mothers. No longer would temple craftsmen have to beg for their rations. No longer would the wealthy take advantage of widows, orphans, and the blind.

Unfortunately, Urukagina's return of the temple estates was largely cosmetic. Records from the Bau estate show that it was still being managed for Urukagina's wife, Shag-Shag, by an overseer named Eniggal, who had previously managed it for Lugalanda's wife, Barnamtarra.

Who knew that the politicians of 4,350 years ago would not fulfill their campaign promises?

CONFLICT BETWEEN PROVINCES

Along the border between the provinces of Umma and Lagash lay a 28-mile-long tract of land called Gu'edena, irrigated by a canal from the Euphrates. For 150 years, through the reigns of at least ten rulers of Lagash, both provinces quarreled over Gu'edena.

During the reign of Lugal-sha-engur of Lagash, the rulers of both provinces called upon the great Mesalim of Kish to adjudicate the dispute. Mesalim made the 90-mile trip and erected a stela, or freestanding stone monument, at the disputed border. According to epigrapher Jerrold Cooper, Mesalim's version of events was that the god Enlil himself had established the boundary between Ningirsu (patron deity of Lagash) and Shara (patron deity of Umma). Mesalim was thus simply carrying out divine orders.

Umma felt that the settlement favored Lagash, so the dispute continued. During the reigns of Ur-Nanshe and Akurgal of Lagash, there were acts of defiance by Umma. At one point the border stela was ripped out, and Umma began to grow barley on land claimed by Lagash.

Eannatum of Lagash (2454–2425 B.C.) attacked and defeated Umma and established a new border treaty with its ruler Enakale. He improved the canal that irrigated Gu'edena and, to lessen the likelihood of war, established a no-man's-land on Umma's side of the frontier. To sanctify the new agreement, Eannatum built chapels to the gods Enlil, Ninhursaga, Ningirsu, and Utu. He also forced Enakale to swear oaths on several deities, agreeing that any barley Umma had managed to grow at Gu'edena would be considered an interest-bearing loan from Lagash. The ensí of Umma also had to swear that his people would not trespass on Gu'edena, destroy the new border stela, or modify the course of the canal.

At the city of Girsu, a Level 2 center in the province of Lagash, Eannatum erected a great stone stela to commemorate his victory over Umma. One side depicted Eannatum and his troops marching over the bodies of his enemies, while vultures made off with portions of the victims' corpses. (This scene has given the monument its name, the Stela of the Vultures.) The opposite side of the stela featured a metaphoric scene in which Eannatum cast the Great Net of Enlil over the men of Umma.

Despite the fact that Eannatum was wounded by an arrow, he claimed that his army killed 3,600 enemies, so many that their heaped-up corpses required 20 funerary mounds to cover. It should be noted that calculations in Sumerian math were based on units of 60, so any claim of 3,600 (60 × 60) is probably an idealized number.

Eannatum's victory did not end the dispute over Gu'edena. Urluma, a later ruler of Umma, began to divert water from the Gu'edena canal in violation of his predecessor's oaths. Urluma is said to have recruited foreign mercenaries to smash the boundary stela, destroy the chapels Eannatum had built, and invade what Lagash regarded as "Ningirsu's land." The new ruler of Lagash, Enanatum I (2424–2405 B.C.), went to war against Urluma; the latter fled the battlefield but was tracked down and killed in Umma.

Finally, after decades of being soundly thumped by Lagash, Umma had its moment in the sun. While Urukagina of Lagash (2351–2342 B.C.) was busy implementing his reforms, new rulers arose in Umma. The son of one ensí of Umma, a man named Lugal-zagesi, had ambitions beyond his native city.

Lugal-zagesi came to power in 2340 B.C. He managed to conquer Girsu, which allowed him to seize Gu'edena and move the frontier closer to Lagash. One by-product of Umma's victory was that its engineers were able to divert water from the canal watering Gu'edena.

According to Robert McC. Adams, this diversion of water interrupted the flow to the city of Lagash itself, which was an even greater tragedy than the loss of barley. In desperation, Lagash tried to dig an alternative canal from the Tigris, which was not a viable long-term solution.

For his part, Lugal-zagesi extended his conquests. He soon added Ur and Uruk to his possessions, virtually isolating Lagash, and went on to be considered "irresistible in all lands." An inscription at Nippur, far to the north, claims that the god Enlil had made Lugal-zagesi king of all Mesopotamia, from the Persian Gulf to the Mediterranean.

Here, then, is another legacy of the Sumerians. Once a piece of land is disputed by two Near Eastern groups, the conflict never really ends. No oath taken in the name of a deity, no cease-fire, no mediation by a third party, and no amount of bloodshed is enough to convince either party to let the matter drop.

STEPS IN THE CREATION OF EMPIRE

See if this story sounds familiar. No one knows who his parents were, though his mother is rumored to have been a priestess. To conceal her pregnancy, she gave birth to her son in secret. She placed him in a wicker basket, waterproofed it with pitch, and set it afloat in a river. A gardener drawing water from the river noticed the basket, rescued the little boy, and raised him as his own. Working his way up the social ladder, the boy rose to become a major historic figure.

An alternative version of Moses's birth? No, it is the legendary origin of Sargon of Akkad, the ruler most often credited with unifying all of Mesopotamia. Just as the Sumerians had the myth of a Great Flood long before the authors of the Old Testament, they also had the legend of the Boy in the Basket before it was applied to Moses.

The fact that we have no plausible account of Sargon's birth suggests that he may have been a usurper. Epigraphers are sure that his native tongue was Akkadian, but they cannot tell us his actual name. "Sargon" is simply our version of the title Sharru-kin, "the true king." He claimed to be from Azupiranu, a city on the banks of the Euphrates, but no archaeologist knows exactly where that is. Because much of Sargon's early life was spent in Kish, we assume that his home city must lie somewhere nearby.

It was allegedly a gardener named Akki who fished baby Sargon from the Euphrates, sometime around 2300 B.C. Following in his adoptive father's footsteps, the boy became an apprentice gardener. He showed enough talent and intelligence to lift himself up by his own sandal straps. Eventually he was appointed cupbearer to Ur-Zababa, the king of Kish.

Cupbearer was a position of trust, one that placed young Sargon in close proximity to the ruler. One night, or so the legend goes, Sargon had a terrifying dream in which he saw the goddess Inanna (or Ishtar, as she was called in Akkadian) drowning Ur-Zababa in a river of blood. Ur-Zababa heard Sargon cry out in his sleep and asked him to describe the dream the next day.

A cuneiform tablet from Uruk, translated by Jerrold Cooper and Wolfgang Heimpel, explains what happened next. Ur-Zababa's interpretation of Sargon's dream was that the goddess Inanna was planning to replace him with Sargon. The author of the tablet describes Ur-Zababa's fear in colorful terms:

> King Ur-Zababa . . . he was frightened in that residence,
> Like a lion, he was dribbling urine, filled with blood and pus, down his legs,
> He struggled like a floundering salt-water fish, he was terrified there.

Distressed that Inanna was planning to replace him, Ur-Zababa asked his chief metalsmith to assassinate Sargon. But Inanna protected Sargon from harm, forcing Ur-Zababa to try an alternative strategy. He dispatched Sargon to Uruk with a sealed letter to Lugal-zagesi, the mightiest king of Mesopotamia. The letter implored Lugal-zagesi to kill Sargon. Once again, the goddess intervened and Sargon was spared.

Legend aside, it does appear that Sargon usurped the throne of Ur-Zababa at roughly 2270 B.C. Sargon later moved his capital to a city called Akkad, which has given its name to his native language. Unfortunately, archaeologists do not know which ancient mound represents the ruins of Akkad, although they suspect that it lay not far from Kish.

With all due respect to the goddess Inanna, it is unlikely that the adopted son of a gardener could have usurped the throne of Kish without support from the high priests and many other influential aristocrats. No currently available inscription reveals what Sargon did to deserve such support; the 20 years leading up to his usurpation are a blank.

Some anthropologists, however, are willing to bet that Sargon actually rose through the military, and that the tales of support from Inanna were simply an attempt to legitimize him after the fact. Sargon's later conquests display a military expertise uncharacteristic of gardeners and cupbearers. Ur-Zababa's fear of Sargon makes the latter seem more like a renowned warrior than a palace attendant.

Whatever the case, we know that Sargon set out to bring all of Mesopotamia under his control. He was not, of course, the first ruler to attempt this. Whoever pulled off Southern Mesopotamia's conquest of Tell Hamoukar in the Late Uruk period was attempting to unify several provinces. The Early Dynastic ruler Mesalim of Kish exercised hegemony beyond his own province. Royal inscriptions were left by the kings of Kish at distant places such as Adab, Girsu, Nippur, and the Diyala basin. Ur-Nanshe, who ruled Lagash between 2494 and 2465 B.C., defeated the ruler of Ur and captured Pabilgaltuk, the ruler of Umma. According to Cooper, Eannatum of Lagash also assumed the kingship of Kish.

Then came two rulers who desired nothing less than the total unification of Mesopotamia. A man named Enshakushana, calling himself "King of Sumer," came close to unifying the entire region between 2432 and 2403 B.C. He conquered Kish and Akshak and dedicated war booty to the god Enlil at Nippur. Some 63 years later, as we have already seen, Lugal-zagesi claimed to have controlled all of Mesopotamia from the Persian Gulf to the Mediterranean. Truth be told, his may have been Mesopotamia's first empire.

Had Lugal-zagesi put Sargon to death, as Ur-Zababa is said to have requested, there might have been no Akkadian empire. Instead, Sargon wound up leading an army against Lugal-zagesi, whose capital was now Uruk. Sargon conquered Uruk and claimed in his inscriptions that he brought Lugal-zagesi back with his neck in a stock. Now it was Sargon's turn to expand his territory.

Most accounts of Sargon's imperialism were written long after his death. While we do not have many details, we know what some of his policies were. For one thing, he made Akkadian the official language of his realm. For another, he sent an Akkadian governor to rule each province of Sumer after he had conquered it.

Another of Sargon's policies was to create a council of diplomats and military officers, described as numbering 5,400, who "ate bread before him." The figure 5,400 may be another exaggeration based on units of 60, but archaeologist J. Nicholas Postgate argues that it constitutes Mesopotamia's first mention of a permanent military establishment. Sargon's armies moved south through Uruk, Ur, Umma, and Lagash. He then made a point of washing his weapons in the waters of the Persian Gulf, symbolizing his total conquest of Sumer.

Sargon, however, was just getting warmed up. By the 11th year of his reign, according to one account, his conquests had reached the Mediterranean coast. Akkadian armies marched to the cedar forests of Lebanon and the copper and silver mines of Turkey. Sargon expanded east into Elam, making the king of Susa his vassal. He boasted that he was now "the Lord of the Four Quarters": Subartu (north), Sumer (south), Elam (east), and Martu (west).

Sargon allegedly ruled until 2215 B.C., at which point some scholars estimate that he would have been 85 years old. In the 55th year of his life, when his enemies assumed that his grip would be weakening, many territories rebelled against him. Sargon, however, put down every revolt. Toward the end of his life, he took stock of his career and asked whether any future ruler could equal his conquests:

> Whatsoever king shall be exalted after me . . . let him govern the black-headed peoples; mighty mountains with axes of bronze let him destroy; let him ascend the upper mountains, let him break through the lower mountains; the country of the sea let him besiege three times; Dilmun [Bahrain] let him capture. . . .

Clearly, Sargon considered himself the greatest ruler of all time. About the only thing he did not claim is that he could float like a butterfly and sting like a bee.

Sargon is generally credited with creating Mesopotamia's first empire—that is, a macro-state, each province of which had once been a kingdom in its own right. It would be wrong, however, to ignore the possibility that earlier rulers such as Mesalim, Enshakushana, and especially Lugal-zagesi established the agenda for Sargon.

We have previously seen that Kamehameha received credit for the unification of Hawai'i, although it was begun by 'Umi and Alapai. For his part, Shaka received credit for the unification of Natal begun by Dingiswayo. The story of Sargon is analogous. His unification of Mesopotamia was spectacular and significant, but it may simply have been the most successful attempt in a long sequence that began with the assault on Tell Hamoukar.

CYCLING IN MESOPOTAMIAN STATES

The dynasty established by Sargon of Akkad lasted nearly 200 years. Sargon was succeeded by his sons Rimush and Manishtu and then by his grandson Naram-Sin, perhaps the first Mesopotamian ruler whose monuments portray him as divine. As so often happens, however, later Akkadian rulers lacked the talent and motivation of the dynastic founder and found it impossible to deal with recurrent famines and rebellious provinces. Eventually many of Sargon's territories were overrun by swarms of Gutians from the Zagros Mountains, people regarded by the Akkadians as barbarians. What followed was a half-century "dark age" for which written documentation is inadequate.

If the states created in the Late Uruk period were first-generation states, any new state of the Early Dynastic period could be considered a second-generation state, making Sargon's empire a third-generation state. Sargon's realm differed from a first-generation state in that it was created not from a group of rank societies but from a group of preexisting kingdoms, including Lugal-zagesi's expansionist state. And there would be a fourth-generation state: out of the ashes of the dark age brought on by the Gutians, a Sumerian-speaking ruler named Ur-Nammu would rise to power in 2112 B.C. His state, modeled on those of earlier rulers, would be known as the Third Dynasty of Ur. Unfortunately, this fourth-generation state would suffer a fate similar to that of its predecessor: its later kings would have trouble hanging on to the territory put together by Ur-Nammu.

The problems facing Ur were both external and internal. External pressure came from a foreign ethnic group known as the Amurru, or "Amorites." The Amorites spoke a Semitic language, and their homeland seems to have been in the arid region west of the Euphrates. Ur-Nammu's successor, King Shulgi, invested in a 150-mile wall to keep out the Amurru. This wall proved ineffective, giving rise to prophecies of doom for the people of Ur.

The fifth and last king of the dynasty was a man named Ibbi-Sin, who ruled from 2028 to 2004 B.C. His many problems are reflected in the cuneiform tablets of his reign, including actual palace documents. Ibbi-Sin faced not only Amorite invasions but disloyalty and usurpation among his own subjects.

Marc van de Mieroop, who has analyzed the available texts, points out that the Amorites were far more than a horde of nomadic barbarians. Many Amorites were urban; they lived in a number of Sumerian and Akkadian cities while retaining their ethnic identity. Some had even worked their way up as bureaucratic officials in Ur's Third Dynasty state. In van de Mieroop's words, however, some cuneiform texts portray the Amorites as a "loathed" ethnic group. This may be another case of pejorative stereotyping, the kind we witnessed in kingdoms such as Egypt and the Zulu.

In the fifth year of his reign, Ibbi-Sin realized that Ur was running out of grain. He needed shipments from the northern fields of his kingdom, a task for which he would have to dispatch a royal official.

The man chosen for the journey was Ishbi-Erra, an Amorite born in Mari, a city far up the Euphrates from Ur. We do not know how Ishbi-Erra came to be a royal official. Had he been a person of some rank in Mari? Or was he a talented commoner who had worked his way up the bureaucratic ladder like Uni had in ancient Egypt? Either scenario is possible since, according to van de Mieroop, individuals with Amorite names were present in all classes of Third Dynasty society.

Ishbi-Erra collected the grain. He then claimed, however, that marauding nomads made it impossible to deliver the shipment to Ur. As a result, he stored the grain at the city of Isin and then suggested that he be put in charge of defending both Isin and the neighboring city of Nippur. According to van de Mieroop, Ibbi-Sin sensed treason coming but felt compelled to grant Ishbi-Erra's request.

Ibbi-Sin's worries were well-founded. Ishbi-Erra soon established his own dynasty at Isin, a feat suggesting that his background may indeed have been aristocratic. His control of Nippur, the religious capital of Mesopotamia, gave him the clout to establish links to Uruk and Larsa. Ishbi-Erra's army conquered Kish and penetrated the Diyala River region as far as Eshnunna. While he was unable to conquer Ur, he received what amounted to "protection money" from that city.

Impoverished by his loss of land and tribute, Ibbi-Sin was forced to raid the temple treasuries of Ur to buy provisions from other cities at inflated prices. He cut all barley rations to his palace staff, but famines became more frequent and thus many laborers fled his kingdom. In the final decade of Ibbi-Sin's reign, his kingdom was under attack from both Amorites and Elamites.

Archaeologist J. Nicholas Postgate has described the growing famine that gripped Ur in its final days, as its enemies cut off its supplies of barley, fish, and oil, and inflation soared from fivefold to sixtyfold. It was a cruel way for prices to begin reflecting supply and demand instead of bureaucratic guidelines, and it sped the collapse of the venerable old city.

As dust blew in the empty streets of Ur, an anonymous poet wrote a lament that has survived more than 4,000 years. Elamites from Susiana, taking advantage of the chaos brought on by Amorite raids and internal revolts, had just sacked Ur and carried off King Ibbi-Sin. Many of Ur's temples lay in ruins, and it was to the goddess Ningal that the poet began his lament.

> Oh Ningal, how has your heart led you on? How can you stay alive?
> Your house has become a house of tears. How has your heart led you on?
> Your city has been made into ruins. How can you exist?
> Ur, the shrine, has been given over to the wind. How now can you exist?

Obviously we must credit the Sumerians with one additional legacy. They gave birth to the blues.

The collapse of Ur was followed by a fifth-generation state, led by the cities of Isin and Larsa. The sixth-generation state for the region would be the one created by Hammurabi of Babylon (1792–1750 B.C.). The latter was an empire virtually as large as Sargon's, and by that time, Sumerian was well on its way to becoming a dead language.

Some authors choose to portray Mesopotamia as a land of petty kingdoms or "city-states," only briefly consolidated into expansionist states or empires. Other authors, including Postgate, portray that part of the world as going through repetitive cycles of strong centralization, separated by political breakdown and regional autonomy.

We find the latter portrayal more convincing. Rather than making Mesopotamia unique, it makes it comparable both to Egypt (with its cycles of centralized Kingdoms and decentralized Intermediate periods) and to ancient Mexico and Peru (whose cycles we describe later in this book). At the heart of all these cycles is a principle with which we are already familiar: For every leader seeking greater territory and power, there are others seeking to bring him down.

THE SOCIAL CONTRACT

In Rousseau's scenario for mankind's past, the archaic state represented the final stage. It was in societies like that of Sumer that the poor had signed the Social Contract, agreeing to inequality forever.

Rousseau did not know as much about ancient Mesopotamia as today's archaeologists and epigraphers, but Sumerian society comes as close to matching his hypothetical final stage as any we know of. Sumer was, in fact, a society of signed contracts. The Sumerian ideal was order, and the method for achieving order was obedience to hundreds of me, or rules, interpreted for the poor by the aristocracy and the priests.

All but the lowliest Sumerians signed multiple contracts in the course of their lives, and all understood the penalties for default. Many signers, who put up their personal liberty as collateral for a loan, were in the long run signing a contract for inequality. Rousseau could not possibly have known that in 1753, but his instincts were correct.

Bureaucratic micromanagement of Sumerian society resulted in huge archives of cuneiform tablets, equivalent to today's "red tape" or "paperwork." The intervention of an authoritarian government into commoners' private lives reduced the variety of marriages to one. The Sumerians created harsh punishments for behavior that other societies would merely have gossiped about.

Many societies had hereditary aristocracies. Prior to Sumer, however, few created such economic inequality within the commoner stratum. The growing privatization of land, combined with the charging of high interest on loans, undercut the safety net provided by the traditional descent groups of earlier societies. The Sumerian government seems to have been aware of what was happening and thus periodically slowed the process by canceling debts. Unfortunately, it could not reverse it.

We have seen that Hawai'i's paramount chiefs eliminated the landed gentry by placing all garden land under chiefly control. In contrast, the Sumerians swelled the ranks of the landed gentry and turned thousands of commoners into sharecroppers.

Debt slavery was widespread in rank societies. Private ownership of land was not. It was a defining feature of Mesopotamian society, one that might even have begun during the Late Uruk period. The first step was to remove land from circulation and assign it to the city's most important temple. Late Uruk temples were impressive; Early Dynastic temples were grander still.

Sumerian kings often claimed to be loved by a patron deity. Appointing their heirs to take care of the gods' estates was thus only logical. After spending years as sanga of such an estate, the royal heir knew just how valuable a property it was. On his ascent to the throne, he held on to the god's estate and then provided his wife with the estate of the god's wife.

It was on such temple estates that Sumerian record-keeping, standard weights and measures, the renting of land at interest, and the accumulation of capital for use by temple merchants were perfected. The lessons were not lost on the aristocracy, which soon established its own private estates.

We have seen that the Chumash of California learned how to create a fourfold increase in the value of shell beads. Sumerian accountants learned how to calculate the wealth-producing effects of long-term interest rates. And the skilled merchants who carried out trade for the Sumerian temples inspired others to become private entrepreneurs.

Economic historian Michael Hudson considers the Sumerian temple the forerunner of the corporation. The wealth of the palace estates encouraged other aristocrats to see how much land they could pry away from commoner descent groups. Lacking the clout of the royal family, the aristocrats relied on extending loans at interest rates of up to 33⅓ percent. As we have seen, many commoners used their personal freedom as collateral and wound up becoming serfs. Reform-minded rulers tried to prevent this from happening, but as Hudson points out, private wealth eventually grew strong enough to undermine royal power. One outcome was real estate as we know it.

Although Mesopotamian economic behavior looks capitalistic, it was not yet laissez-faire capitalism. As late as the period of the Isin-Larsa Dynasties and the First Dynasty of Babylon, the government was still attempting to micromanage prices. While the economy had some elements of a market, bureaucrats set guidelines for the exchange of goods. Silver was used as a standard of value, but it was not yet an actual medium of exchange.

The state was especially interested in long-distance trade, such as the movement of copper from Turkey to Mesopotamia. The price of copper, however, was not allowed to fluctuate with supply and demand. Instead, guilds of merchants traveled to the highlands of Turkey and, by negotiating with the local princes, established a long-term price for copper. The merchants then purchased specific amounts of copper on consignment and received a commission on their return to Mesopotamia. Their profits depended on the high volume of trade, and their business was low-risk because the price of copper had been fixed.

The rarity of laissez-faire market systems in early civilizations has fueled a long-standing debate between two kinds of economists: formalists and substantivists. Formalists believe that the laws of supply and demand usually determine what societies do. Substantivists, as exemplified by economic historian Karl Polanyi, believe that, on the contrary, the economy is embedded in society and constitutes a special form of social relations. Indeed, many substantivists would argue that economics began with the reciprocal gifts exchanged by hunters and gatherers and grew from there.

We once had the pleasure of dining with a formal economist who spent most of the evening telling us how silly archaeologists were for believing that prehistoric behavior was determined by anything but supply and demand. As he finished his dessert, he took out a fine cigar and asked if we minded his lighting up. We knew better than to deny him that pleasure. He then complained that it was no longer possible to get his favorite brand of Havana cigar. It never occurred to him that the cultural values and social policies of the United States had prevented Cuba's supply from getting together with his demand.

Perhaps the best way to leave the debate is this: Substantivists can cite dozens of anecdotal cases in which cosmology, religion, or cultural values restrict the operation of supply and demand. The formalists, however, have produced all the sexy equations that might win you a Nobel Prize.

TWENTY-THREE

How New Empires Learn from Old

An empire is a kind of macro-state and has its own social and political logic. In many parts of the world we can point to multiple generations of kingdoms and empires. This allows us to observe third- and fourth-generation states borrowing strategies from their predecessors.

Two New World societies can serve as examples. The Aztec belonged to the fifth generation of states in central Mexico. The Inca constituted the fourth-generation empire in the Andes. Both used the logic of their predecessors as templates.

CENTRAL MEXICO'S FIRST STATE

The Basin of Mexico lies 7,200 feet above sea level and occupies 3,700 square miles. When the Spaniards arrived in 1519, they found a series of interconnected lakes covering 400 square miles. The most productive farmland lay in the southern part of the lake system, where the annual rainfall exceeded 40 inches and the lake margin was swampy. The central part of the lake system was brackish. The northern basin received less than 24 inches of rain and required irrigation.

This dry northern region, however, had a unique environmental feature. At a place now called San Juan Teotihuacan, 80 permanent springs brought more than a billion cubic feet of water to the surface annually. Today this water is collected by a canal system irrigating more than 10,000 acres. Two thousand years ago, the water supported Teotihuacan, one of Mexico's earliest and largest cities. Archaeologists are not yet sure whether the state headed by Teotihuacan was a monarchy or an oligarchy.

Fifteen hundred years ago, Teotihuacan had an estimated population of 125,000. One of the ways that it grew so large was by drawing most of the rural population of the Basin of Mexico into the city. This deliberate relocation of the rural population was so extensive that no serious candidates for Level 2 administrative centers were left within the basin. Teotihuacan's impact can be seen at distant settlements, from the Mexican states of Hidalgo in the north and Veracruz in the east to the Republic of Guatemala in the south. Let us look at some of the Teotihuacan behaviors that were emulated by later states.

1. The capital was divided into quadrants and had major roads leading into and out of the city.
2. Craft specialists of various kinds lived in their own residential wards. Teotihuacan had more than 2,000 large, multifamily apartment compounds, surrounded by high walls and separated from other compounds by narrow alleys. At least 500 of these compounds were involved in craft activity. Some produced artifacts of obsidian, some made pottery of a specific type, some produced mold-made figurines, others produced masks for rituals or funeral bundles, and so on.
3. Some compounds may have been occupied by hundreds of people, suggesting that a social segment with clanlike properties may have been involved. These large, possibly corporate social segments may have served as an archetype for the *calpulli* of the later Aztec.
4. There were enclaves of people from foreign ethnic groups. For example, Zapotec immigrants from Oaxaca lived in one part of the city and traders from the Gulf Coast in another. Some of these enclaves may have supplied Teotihuacan's craftsmen with raw materials from distant regions.
5. At least two of the supernatural beings represented in art at Teotihuacan were the forerunners of Aztec deities. These were the Feathered Serpent (called Quetzalcoatl by the Aztec) and a goggle-eyed personification of Lightning or Rain (called Tlaloc by the Aztec).
6. Beneath a temple pyramid with depictions of these supernatural beings, the officials of Teotihuacan sacrificed and buried people who appear to have been military captives.

Several centuries after reaching its peak population, Teotihuacan began to decline. By A.D. 800 it had lost many of its craftspeople. By 1000 it was barely a city at all.

THE SECOND GENERATION OF STATES

It appears that during the height of its power Teotihuacan was able to inhibit the growth of nearby urban centers, much as Uruk did for a time in early Mesopotamia. Once Teotihuacan began to decline, however, its hinterland broke up into a series of kingdoms or political confederacies. It is likely that many of these small kingdoms had once been part of an inner ring of subject provinces, extending out 75 to 100 miles from urban Teotihuacan.

Some of the second-generation kingdoms that took advantage of Teotihuacan's decline were Cantona to the east (in the state of Puebla), Cacaxtla to the southeast (in the state of Tlaxcala), and Xochicalco to the south (in the state of Morelos). The capitals of these kingdoms achieved their greatest growth between A.D. 600 and 900.

Many second-generation kingdoms were preoccupied with defense from hostile neighbors. Xochicalco, for example, was set on a rugged mountaintop. The city was defended by a series of dry moats and walls and could only be entered by three narrow causeways. Its summit had a plaza with several temples, a royal acropolis with storage rooms, residential areas for lesser nobles, facilities for sweat baths, and several courts for playing ritual ball games.

One temple platform was decorated both with feathered serpents, like those of Teotihuacan, and hieroglyphs referring to a series of subject territories. Some of the hieroglyphs depict open jaws holding an ancient symbol for tribute: a circular cake of cacao or chocolate, divided by incisions into four quadrants.

Archaeologists suspect that Xochicalco's eventual collapse was brought on by factional or ethnic rivalry. In its last days, Xochicalco's royal lineage turned the acropolis into a mini-fortress by dismantling its access stairways. During a final conflagration, women and children were trapped under falling roof beams along the escape route.

In the pine-forested highlands east of the Basin of Mexico lies another city whose concern with defense was obvious. Cantona occupied the summit of a lava hill so rugged and abrasive as to shred the sandals of anyone attempting to scale it. The builders of Cantona added a dry moat and restricted traffic to a series of narrow causeways monitored by guard rooms.

Archaeologists believe that Cantona was built by a confederacy of petty kingdoms that, by pooling their manpower, created a virtually impregnable city. One reflection of this confederacy can be seen in the 24 ball courts scattered throughout Cantona. These courts varied significantly in size, architectural style, and astronomical orientation, as if each participating group had its own version of the ball game.

Still another second-generation city was Cacaxtla, which occupied a defensible hill in Tlaxcala. Cacaxtla lacked the impressive moats and walls of Xochicalco, but its murals depicted battles, captive taking, and the names of subjugated towns. One prominent mural, more than 60 feet long, shows a battle scene in which nobles wearing bird helmets are menaced by warriors wearing jaguar pelts and carrying spears.

One stairway at Cacaxtla, called the Captive Stair, was given several coats of stucco. On the tread, Cacaxtla's artists painted images of prisoners whose skin-and-bone corpses leave little doubt that they had been deliberately starved (Figure 68). On the riser of the same step they painted the hieroglyphic names of subjugated towns, presumably those from which the captives had come.

Another community founded during this period was Tula in the state of Hidalgo. There can be little doubt that the region of Tula had once been sub-

FIGURE 68. The Captive Stair at Cacaxtla, a hilltop citadel in Tlaxcala, Mexico, was painted with polychrome images. On the tread were the corpses of prisoners who had been starved until they were literally skin and bone. On the riser were hieroglyphs referring to places subjugated by Cacaxtla. Such militarism was typical of second-generation states in central Mexico.

ject to Teotihuacan: the earlier administrative center for the region, an archaeological site called Chingú, featured an unmistakable Teotihuacan architectural style. Chingú's abandonment coincided with Tula's growth. By A.D. 900 Tula had become a city covering more than a square mile.

Many of central Mexico's second-generation states declined after 900, often because the confederacies that built them had dissolved. Tula was an exception; its greatest days still lay ahead.

THE THIRD GENERATION OF STATES

In 1577 King Philip II of Spain asked every colonial administrator in Mexico to fill out a questionnaire on the province under his command. The result was a series of documents called *Relaciones Geográficas,* kept in an archive in Seville. These documents are a gold mine of information on the Indian societies of Mexico, but they are only the tip of the iceberg. In addition to the authors of the *Relaciones,* highly motivated Spanish missionaries, soldiers, and officials interviewed Indian leaders about their history, customs, religious beliefs, kings, and conquests.

It is from such documents, augmented by archaeological data, that we learn much of what we know about the Aztec. But the legendary histories go back farther than that. They speak of a pre-Aztec people called the Toltec, who ruled central Mexico between A.D. 900 and 1200. The Toltec spoke Nahuatl like the later Aztec.

Thanks to historian Wigberto Jiménez Moreno, we know that the archaeological site of Tula, roughly 35 miles north of Teotihuacan, was the Toltec capital. We also know that the Toltec created not merely a third-generation state but a multiethnic empire.

Tula was already occupied in 700, but its influence at that time did not extend far outside its immediate region. The city lay along both banks of the Río Tula, the main source of irrigation water for a dry basin 7,000 feet above sea level. According to Jiménez Moreno, Tula's later growth reflected an influx of at least two major immigrant groups. From the arid north and west came the Toltec proper. From the south and east came the Nonoalca, a collection of ethnic groups whose emigration was prompted by the decline of earlier central Mexican cities.

Now take a large grain of salt and listen to one of the romanticized native accounts of the Toltec rise to power.

The story begins with a leader named Mixcoatl ("Cloud Serpent"). He led a great horde of people from the north into the Basin of Mexico, where he battled with an ethnic group called the Otomí. In the process, Mixcoatl was assassinated. His son, Ce Acatl ("One Reed"), avenged his father's death and then led his people north to the less bitterly contested region of Tula. The date given for his arrival at Tula corresponds to A.D. 968 in our calendar.

One Reed then assumed two honorific titles: Topiltzin (the equivalent of "lord" or "sir") and Quetzalcoatl ("Feathered Serpent"). The latter title suggests an attempt to legitimize his rulership by associating himself with an important deity.

Tula eventually grew to cover five square miles. While it lacked the large traffic arteries of Teotihuacan, its layout suggests that much of its growth was planned. Tula's artisans did not live in large compounds like those of Teotihuacan, but their crafts were just as well developed. Two huge workshops turned volcanic glass into thousands of lancets, blades, and knives. The enormous numbers of spindle whorls, or flywheels for spinning fiber, suggest large-scale production of cotton textiles. Since cotton cannot be grown at 7,000 feet, it must have been imported from the lowlands on a grand scale.

In fact the evidence for long-distance trade is so great as to suggest that Toltec society included forerunners of the Aztec *pochteca,* a guild of special entrepreneurs that led trade missions to far-off regions. One residence at Tula had storerooms with Plumbate pottery imported from the Pacific coast of Guatemala and Papagayo Polychrome pottery from Costa Rica or Nicaragua.

The temple precinct at Tula was separated from the secular parts of the city by a wall, analogous to the walls that surrounded the sacred precincts of some Sumerian cities. In Tula this structure was a *coatepantli,* or "serpent wall," decorated with undulating rattlesnakes. The later Aztec would borrow the concept of the serpent wall from the Toltec.

Another Toltec creation adopted by the Aztec was the *chac mool.* This was a sculpture depicting a reclining man holding a receptacle on his abdomen (Figure 69). According to oral histories, the receptacle's purpose was for the placement of offerings, including the hearts of sacrificial victims. The final resting place for the heads of many victims was a *tzompantli,* or skull rack, many layers high.

The peak of Toltec influence occurred in the twelfth century A.D. The extent of their trade network was impressive indeed. To the south they had access to the products of Nicaragua and Costa Rica. To the north they had access to turquoise, mined either in the U.S. Southwest or in northwest Mexico.

Archaeologist Patricia Crown and chemist W. Jeffrey Hurst have found residue from chocolate inside a series of painted beakers at Pueblo Bonito in New Mexico. The vessels were locally made, but the chocolate must have come from Mexico. The beakers date to the period when the Toltec were importing a lot of turquoise from regions to the north, perhaps offering chocolate beans in return.

After centuries of expansion, the Toltec succumbed to internal conflict. Once again their oral histories romanticize the story, attributing it to competition between two deities. According to legend, conflict arose between Quetzalcoatl (a deity associated with creativity, arts, and crafts) and Tezcatlipoca ("Smoking Mirror," a deity associated with militarism and human sacrifice).

FIGURE 69. This Toltec sculpture, called a *chac mool,* occupied a place of honor at Tula in the state of Hidalgo, Mexico. The sculpture is a bit over two feet tall and depicts a priestly attendant, with a sacrificial knife tucked into his armband, holding the basin in which a victim's heart would be placed. The carving of such figures was one of the Toltec practices borrowed by the later Aztec.

Tezcatlipoca is alleged to have tricked Quetzalcoatl into public drunkenness, an act so scandalous that the latter was forced to leave Tula.

This legend is probably the romantic version of a conflict between two royal families or political factions, each with a different patron deity. Tula eventually went into a downward spiral from which it never recovered. Huemac, the last Toltec ruler, is said to have left the city in A.D. 1156 or 1168, moving to the Basin of Mexico. There the old Toltec elite took up residence on the lake system, occupying places with Nahuatl names like Azcapotzalco, Texcoco, Tlacopan, and Colhuacan.

As it shrank in population, Tula became vulnerable to attack. A series of ethnic groups from the north and west, known to the Toltec by the derisive term *Chichimec*, or "Dog People," entered and burned parts of Tula. One of these Chichimec groups claimed to have come from an island within a lagoon, a place called Aztlan ("Place of the Heron"). People from such a place would be referred to as "Azteca," which is the origin of the word Aztec. During their migrations, however, these people changed their name to "Mexica," from which we get the word Mexico.

THE FOURTH GENERATION OF STATES

Between A.D. 1200 and 1300, a fourth generation of states arose in central Mexico. None reached the status of empire. Most were petty kingdoms, forced to form alliances with their neighbors in order to avoid being taken over by ambitious rivals.

Archaeological surveys suggest that some small kingdoms of this era had an administrative hierarchy of no more than three levels. Having previously crossed the rubicon to monarchy, however, these societies had no intention of giving up the trappings of kingship. Each attempted to maintain its own royal lineage, however modest its territory.

Each of these petty kingdoms was referred to as an *altepetl,* a word that combined the Nahuatl terms for land and water. On average, an altepetl had an estimated 10,000 to 20,000 people, several thousand of whom resided at the capital.

While legendary histories describe the ethnic groups of this era as "arriving" in the Basin of Mexico and "settling" in specific localities, the archaeological record shows that many of those localities had already been occupied for centuries. We suspect that such long-term occupations reflect the presence of Nahuatl-speaking commoners, who farmed the land and provided long-term stability for each community. The migrations referred to in the legends were probably those of royal lineages who, like the great Ang families of the Konyak Naga, moved from place to place as older lineages declined and communities were left leaderless. Such a scenario is supported by the later actions of the Mexica, who, as we shall see, repeatedly asked other communities to send them a leader of royal blood.

According to their own oral history, Mexica leaders were not sufficiently elite to rule their own altepetl. From 1250 to 1298 they lived as vassals of Azcapotzalco. Then, from 1299 to 1323, they became vassals of Culhuacan. Eager to establish their own royal lineage, the Mexica asked the ruler of Culhuacan to give them his daughter, claiming that she would be both their sovereign and the bride of their main deity. Keeping one's vassals happy often involved sending them a noble marriage partner, so the Mexica got their princess.

If the oral histories are to be believed, however, the Mexica then committed an incredible faux pas: they decided to honor the princess by deifying her. This ritual involved dressing the princess as a goddess, sacrificing her, skinning her corpse, and having a priest dance in her skin.

The ruler of Culhuacan was invited to the dance, recognized his daughter, and was horrified. Soon the Mexica were forced to flee, taking refuge on a pair of swampy islands in the central lake. These islands lay in a buffer zone between the territories of Azcapotzalco, Texcoco, and Culhuacan, a familiar venue for the founding of a new rank society. The Mexica named one island Tenochtitlan, "Place of the Prickly Pear Cactus," and the other Tlatelolco, "Where There Are Earthen Mounds."

Would this be the end of the Mexica? Not a chance.

THE AZTEC: A FIFTH-GENERATION STATE

According to Mexica legend, the most important moment in their migration from Aztlan was their discovery of an idol in a cave. The idol was that of Huitzilopochtli, "Hummingbird on the Left," the patron deity who told them to call themselves Mexica.

Huitzilopochtli's mother was the widowed goddess Coatlicue, "She of the Serpent Skirt." One day as she swept the earth on Coatepec, a mythical "Serpent Hill" near Tula, she was miraculously impregnated by a ball of feathers.

Her daughter Coyolxauhqui (the embodiment of the moon) was angered by her mother's licentious behavior. Coyolxauhqui encouraged her 400 brothers (the stars of the southern sky) to decapitate their mother.

This event would later be commemorated in a colossal statue of the beheaded Coatlicue, with blood gushing from her neck in the form of serpents. The statue portrays Coatlicue as a bruiser, an offensive tackle in a rattlesnake skirt, a goddess only a ball of feathers could love. Her most winsome accessory was a necklace of severed hands and hearts.

Despite her beheading, Coatlicue gave birth to a warrior son, Huitzilopochtli, who emerged from his mother's womb fully armed. He chopped his sister Coyolxauhqui into pieces, hurled her remains to the base of Coatepec hill, and drove his 400 brothers from the sky. This myth is believed to symbolize the sun's daily banishment of the moon and stars.

The Mexica survived through deal-making and hard work, including the reclamation of farmland from swampy lakeshore. By 1376, enough time had elapsed so that their sacrifice of the Culhuacan princess had been forgiven. The occupants of Tenochtitlan petitioned for, and received, a prince from Culhuacan named Acamapichtli (1376–1395). Tlatelolco, for its part, received a prince from Azcapotzalco. The two new royal lineages thus created were, of course, considered junior (and therefore subordinate) to the ones from which they had been derived.

One of the major political trends of this period was the growing power of Tezozomoc, the king of Azcapotzalco. Soon he moved aggressively on Texcoco and drove its ruler, Nezahualcoyotl, into exile. Nezahualcoyotl, considered the most eminent sage and poet of his time, sought refuge with allies in Puebla and Tlaxcala. As he fled, he composed a poem as touching as the lament written by the Sumerians in response to the destruction of Ur:

> I am bent over, I live with my head bowed beside the people.
> For this I am weeping, I am wretched!
> I have remained alone beside the people on earth.
> How has Your heart decided, Giver of Life?
> Dismiss Your displeasure! Extend Your compassion!
> I am at Your side, You are God.
> Perhaps You would bring death to me?

Sometime between 1426 and 1428, Tezozomoc of Azcapotzalco was succeeded by Maxtla, who apparently had no love for either Tenochtitlan or Tlatelolco. One of his first acts was to arrange the murders of both islands' rulers.

These political assassinations brought to a boil years of simmering resentment of Azcapotzalcan despotism. The leaders of Tenochtitlan and Tlatelolco sent messengers to Nezahualcoyotl, Texcoco's ruler-in-exile, plotting revenge. They were joined by Tlacopan, an altepetl just south of Azcapotzalco, whose people felt special antipathy for Maxtla. Soon the conspiracy spread to kingdoms in Puebla and Tlaxcala, well outside the Basin of Mexico.

One of the first acts of the rebels was to restore Texcoco's exiled ruler to power. The allies then began taking away some of Azcapotzalco's subject territories and encouraging others to defect. By 1428 they had effectively isolated Azcapotzalco and defeated Maxtla.

While different in detail, the overthrow of Azcapotzalco was analogous to the overthrow of the Denkyira by Osei Tutu's Asante-led alliance. The Mexica ruler who played Osei Tutu's role was Itzcoatl ("Obsidian Serpent"), who succeeded the murdered ruler of Tenochtitlan. Itzcoatl did not create a golden stool to celebrate his winning of independence for Tenochtitlan. He did, however, discard his official seat of reed bundles for a throne made of woven mats, and he directed his prime minister to burn all the old Mexica picture-writing so that he could give his people a more glorious (albeit revisionist) history.

The large towns of Tenochtitlan, Texcoco, and Tlacopan now decided that as long as they maintained their political and military alliance, no other altepetl could resist them. They therefore embarked on their own campaign of political expansion. In the course of their conquests, the spoils of war were divided into five equal portions. Tenochtitlan and Texcoco provided the bulk of the warriors and received two portions each; Tlacopan received one portion for transporting provisions to the battlefield.

It was only at the level of this Triple Alliance between Tenochtitlan, Texcoco, and Tlacopan that an Aztec empire could have been created. No single altepetl had the political and military power to succeed on its own; it would have suffered the same fate as Azcapotzalco. In order to cement their alliance, the royal houses of Tenochtitlan, Texcoco, and Tlacopan began to intermarry in such a way that their rulers would be related as uncles, nephews, or cousins.

Over the next century the population of the Basin of Mexico grew to an estimated 1.5 million. This estimate is based on two sources: colonial Spanish documents and a fine-scale archaeological survey of the Basin of Mexico by William Sanders, Jeffrey Parsons, and Robert Santley.

By the time of the Spanish Conquest, the Basin of Mexico had at least 60 *altepemeh* (the plural of altepetl) with average populations of 15,000 to

30,000. Each altepetl included a town of about 3,000 people and a series of smaller rural communities. These towns and villages provided the second, third, and fourth levels of the Aztec administrative hierarchy; the Triple Alliance constituted Level 1. Texcoco and Tlacopan had populations of about 25,000 each. Estimates for the population of Tenochtitlan range from a low of 60,000 to a high of 300,000. The islands of Tenochtitlan and Tlatelolco, once thought of as refuge areas, were now connected to the mainland by three major causeways and thousands of canoes.

Aztec Society

By the time the Spaniards arrived, the Mexica had possessed their own royal house for 152 years, and their society was as highly stratified as that of their Toltec predecessors. Anyone born into the ruling stratum was known as a *pilli,* or hereditary noble. The *pipiltin* (the plural of pilli) received special education in an elite school called a *calmecac,* where they learned how to behave as aristocrats. Pipiltin wore sandals in public and were allowed to wear cotton mantles extending below the knee (Figure 70).

Some pipiltin rose by achievement and public service to become *tecuhtin,* or major nobles. Judges, governors, the rulers of conquered cities, generals of the army, and highly ranked civil officials were all tecuhtin. They paid no taxes and were given official residences, subsisting on income from lands set aside for their office. By right of membership in a noble family, the tecuhtin also had access to the products of other fields.

The Mexica ruler was known as *tlatoani,* "he who speaks [for us]." He was chosen from among the eligible tecuhtin by a council of 100 noble electors. In theory the council could choose the best person available. As time went on, however, rulers tended, increasingly, to be brothers, cousins, or nephews from the same family.

The tlatoani of Tenochtitlan was the de facto commander in chief of the Triple Alliance. The Mexica ruler built his palace in downtown Tenochtitlan, where he maintained guest quarters for allied rulers. The tlatoani appeared in public only on special occasions and traveled in a litter carried by other nobles.

Below the tlatoani was a prime minister who, like the vizier of ancient Egypt, attended to the everyday running of the state. This minister was deputy ruler in the king's absence, chief justice of the supreme court, and chair of the council of noble electors.

The Aztec ruler was not considered a god. He was simply the most powerful and respected of all the tecuhtin, sometimes described as a "great tree" whose branches sheltered all the Aztec people. Nor did the tlatoani have to be descended from his father's most highly ranked wife; oral history suggests that Itzcoatl, who won independence for the Mexica, was the politically savvy offspring of a nobleman and a woman of lowly rank.

The *macehualtin* were commoners who belonged to corporate groups called *calpultin* (the plural of calpulli). Each of the calpultin claimed descent from a remote ancestor; some of these ancestors were alleged to have lived during the time of the Toltec empire.

Calpultin might consist of 150 to 200 families. Each calpulli held corporate rights to specific resources, which could include agricultural land in rural settings or the raw material for crafts in urban settings. Some calpultin were more prestigious than others, as were some families within each calpulli. People

FIGURE 70. The Codex Mendoza is a sixteenth-century picture book, painted by Aztec artists at the request of their Spanish conquerors. Its images include people from all levels of Aztec society. On the left we see a *pilli,* or hereditary noble. In the center is a commoner at work. On the right we see two slaves with their necks in stocks; the hairdo of the slave at upper right identifies her as a woman.

maintained their rights to resources by marrying within their calpulli, and many positions of authority descended through family lines.

The head of a calpulli was elected for life. He took several wives, enjoyed numerous privileges, and represented his calpulli to the outside world. It was his duty to collect taxes from the families in his calpulli and pass it on to the ruler of his altepetl. He himself paid no taxes, because he was expected to entertain visitors and provide food and *pulque* (agave cider) at ceremonies. Early in Aztec history, the tlatoani was advised by councils of calpulli heads; as time went on, however, such power-sharing institutions were bypassed.

Not all commoners belonged to calpultin. An estimated 30 percent were *mayeque,* or landless serfs. Mayeque could be foreign immigrants, freed slaves, or commoners who had lost their land as a result of crime or debt. Some documents suggest that just as the Sumerians created debtors by charging high interest rates, Aztec rulers sometimes created debtors by overtaxing their subjects. Mayeque worked the lands of others and could easily be distinguished from nobles because they were permitted only knee-length mantles of agave fiber. It was not only the difference between cotton and agave garments, of course, that separated nobles from commoners. When the Aztec organized hunting parties, the venison went to the nobles; the commoners, who usually beat the brush to drive out the game, contented themselves with rabbits, pack rats, and lizards.

Some subject provinces paid their tribute to the Aztec in slaves, in lieu of any other desirable resource. Other slaves were prisoners of war. An average slave could be purchased for 20 cotton mantles; an exceptional dancer might be worth 40. Male slaves were used as field laborers, domestic servants, or burden carriers; female slaves worked in kitchens and textile workshops. Slaves were allowed to acquire land, property, and even slaves of their own. Some worked their way up to positions of responsibility or married free citizens.

Some neighborhoods at Tenochtitlan, like those seen earlier at Teotihuacan, featured immigrants from other ethnic groups. Many were artisans who transformed gold, copper, silver, jade, turquoise, and rock crystal into sumptuary goods; made cloaks from the feathers of macaws, cardinals, quetzals, and hummingbirds; produced polychrome ceramics for the tables of nobles; and wove multicolored mantles that the tlatoani, like a film star in a Versace gown, is said to have worn in public only once.

Long-distance trade was in the hands of a special guild of wealthy commoners called *pochteca.* The pochteca lived in more than a dozen altepemeh, concealing their wealth behind high walls and coming and going by night. They used slaves as porters and, because they often traveled through hostile territory, had their own armed guards. One of their favorite destinations was Xicalango, a port-of-trade on a coastal lagoon in the Mexican state of Tabasco. There the pochteca exchanged highland products such as gold, rock crystal, obsidian, polychrome pottery, and fine textiles for lowland products such as quetzal feathers, jaguar pelts, chocolate, coral, and seashells.

Many of these lowland items were considered sumptuary goods by the tlatoani and tecuhtin, who frequently invested in the pochteca for personal gain. They also encouraged the pochteca to act as spies, debriefing them on the defenses and military strength of foreign peoples.

Xicalango, of course, was not the only important market of that period. Tenochtitlan's sister island, Tlatelolco, maintained an open market whose size impressed even the sixteenth-century Spaniards. There, thousands of market women displayed their wares, while vigilant officials settled disputes and collected the equivalent of sales taxes.

The Aztec market, or *tianguis,* was an institution for which the Inca had no counterpart. Along with tribute, slave labor, and the efforts of pochteca entrepreneurs, it was one of the pillars of the Aztec economy. Another of those pillars was intensive agriculture, including the system of swamp reclamation known as *chinampa* gardening.

Over thousands of acres of the southern lake shore, organic mud was brought to the surface and piled in long parallel ridges. Each fertile ridge was bracketed by narrow canals that allowed canoes to enter and depart. Some chinampas were so productive that three or more vegetable crops could be harvested every year. Add thousands of square miles of irrigated fields and hillside terraces producing corn, and it is no surprise that the Spaniards were impressed.

Downtown Tenochtitlan

Spanish eyewitnesses described a sacred precinct in downtown Tenochtitlan, part of which featured a truncated pyramid whose twin temples were dedicated to Huitzilopochtli and Tlaloc. This pyramid represented Coatepec, the sacred hill where Coatlicue was impregnated, and a stone monument found near its base depicted her dismembered daughter, Coyolxauhqui. One temple was equipped with a chac mool to receive the hearts of sacrificial victims. Archaeologists such as Eduardo Matos Moctezuma and Leonardo López

Luján have labored for years to expose the great temple complex, which had been enlarged and renovated many times.

To the south was a temple to Tezcatlipoca; to the west lay a temple to Quetzalcoatl. Other institutions borrowed from the Toltec included a serpent wall and a giant skull rack, built from the heads of sacrificed captives. There was also a court for the ritual ball game and a school for the children of nobles. It is significant that the palace of the ruler was built on secular ground, outside the sacred precinct.

The Logic of Aztec Imperialism

For roughly 90 years the Triple Alliance worked to acquire new territory, exact tribute from it, and keep it from breaking away. We list here some of the principles the Aztec followed.

1. Military prowess became an increasingly important criterion for selecting the tlatoani. Some candidates were asked to demonstrate their skills by bringing back 40 captives to be sacrificed at their inaugurations.
2. Exceptional warriors were rewarded with special costumes, reflecting the number of captives they had taken.
3. Like the Zapotec before them, the Aztec maintained détente with their most powerful neighbors while conquering weaker ethnic groups. They subjugated people on all sides of Tlaxcala but never the Tlaxcalans themselves. The Tarascan people of west Mexico put up such resistance that their frontier with the Aztec became a buffer zone between two lines of forts.
4. When a community on the border of their empire proved rebellious, the Aztec pacified the area by slaughtering the adults and replacing them with loyal, Nahuatl-speaking immigrants. The children of the slaughtered rebels were then brought back to the Basin of Mexico and raised to be Aztec.
5. What the Aztec sought as tribute were goods, including corn, cotton, chocolate, vanilla beans, tropical fruits, gold and precious stones, rubber, the pelts and plumage of exotic animals, and slaves. This stands in contrast to the Inca, who preferred corvée labor as tribute.
6. The Aztec wanted to integrate other ethnic groups into their society, so they brought back foreign idols and built temples to foreign gods in

Tenochtitlan. One of the reasons the Aztec had such an extensive pantheon, in fact, was that it grew by accretion as more and more deities were added. It was only logical to the Aztec that each society would have its own deities and ancestors. There was no forced conversion in Mexico until the Spaniards arrived, bringing with them the Inquisition.

7. Aztec war had its protocols, its pre-Hispanic equivalent of the Geneva Convention. For example, major campaigns were never fought until the corn harvest was in and the dry season had begun. The logic behind this rule was that commoner foot soldiers, who were drafted as needed, should not be prevented from producing their crops, as this would impoverish even the victors.

The Later Aztec Kings

At the request of the Spaniards the Aztec recounted the exploits of their past kings, both through oral histories and through carefully painted picture-writing. From these histories we learn that Itzcoatl (1427–1440), first head of the Triple Alliance, sent his armies southwest to subjugate towns in what is now the Mexican state of Guerrero. He was succeeded by Motecuhzoma I (1440–1469), who had been a highly ranked general. Motecuhzoma solidified Aztec control of Morelos and Guerrero and then began a series of eastern campaigns against the Huastec people to the east. During this campaign the Aztec never strayed more than a one-day or two-day march from a friendly town that could supply them with food. This contrasted with the strategy of the Inca, who created their own extensive system of roads and imperial storehouses.

Motecuhzoma I was succeeded by his son, Axayacatl (1469–1481). The latter's preinaugural campaigns had taken him to the Isthmus of Tehuantepec, and he went on to conquer Huatulco on the Pacific coast of Oaxaca. He was forced to reconquer Guerrero and the Huastec region.

Axayacatl was succeeded by Tizoc (1481–1486), the first Aztec king to be considered a military failure. Tizoc's problems began during his preinaugural campaign. In a region called Metztitlan, Tizoc's troops were defeated but took the requisite 40 prisoners to be sacrificed at his inauguration. Upon his return to Tenochtitlan, Tizoc therefore commissioned a stone monument that depicted him taking captives. This monument was Tizoc's version of a "mission accomplished" banner; it failed to mention that the battle had cost him 300 of his noble officers.

Tizoc continued to disappoint his supporters; two years later, he proved unable to prevent the people of Metztitlan from killing an Aztec tribute collector. After less than five years on the throne, Tizoc died, allegedly poisoned by people he trusted.

Tizoc was quickly replaced by his more aggressive brother Ahuitzotl (1486–1502). Ahuitzotl put down regions that had resisted Tizoc and conquered (or reconquered) parts of Guerrero, Puebla, and Veracruz. According to one account, he wiped out several rebellious communities and repopulated them with 9,000 loyal, Nahuatl-speaking married couples, 600 of whom came from the cities of the Triple Alliance. In the process, the thousands of orphans he created were relocated to other parts of the empire.

One of Ahuitzotl's major accomplishments was his annexation of Xoconochco, a chocolate-growing region on the Pacific coast of Guatemala. The most direct route to Xoconochco was through Zapotec territory, but the Zapotec refused to allow the Aztec free passage through the Oaxaca Valley. Ahuitzotl therefore attempted to open a route by conquering Zapotec communities.

Under pressure from the Aztec, the Zapotec ruler Cociyoeza (1487–1529) moved his army from the Oaxaca Valley to a fortified mountaintop in the tropical Isthmus of Tehuantepec. The Zapotec troops were joined at Tehuantepec by a contingent of Mixtec-speaking allies from a kingdom called Achiutla. Soon Ahuitzotl's army was under attack from two sides.

The Aztec received reinforcements three times in seven months but made no inroads into the Zapotec and Mixtec defenses. Finally, Ahuitzotl, his troops weakened by casualties and demoralized by the tropical heat, realized that his best strategy was to arrange a truce.

The key ingredient of the truce was a political marriage: the Aztec princess Coyolicatzin ("Cotton Flake"), daughter of Ahuitzotl, was betrothed to the Zapotec ruler. To make this union palatable to both ethnic groups, a romantic legend was created. It was alleged that Cociyoeza came upon Cotton Flake bathing in a mountain pool and was instantly smitten. Note how similar this is to the Hawai'ian legend of Liloa and Akahi, the parents of 'Umi.

Ahuitzotl was succeeded by Motecuhzoma II (1502–1520), who solidified the conquests of his predecessors and added new territories. Unfortunately for Motecuhzoma, he happened to be on the throne when the Spaniards arrived in 1519.

Although outnumbered, the Spaniards had cannons, blunderbusses, crossbows, horses, and armor. They were joined by troops from Tlaxcala, skilled warriors who had a vested interest in seeing the Aztec defeated. The Spaniards also took the Aztec by surprise by attacking during the agricultural season, a breach of pre-Hispanic protocol. European diseases for which the Aztec had no immunity preceded each Spanish advance. In a relatively short time, Mexico's last indigenous empire had collapsed.

PERU'S SECOND-GENERATION EMPIRES

When we last turned to the Andes it was to describe the Moche empire, which spread over 15 valleys on Peru's north coast. The Moche were one of Peru's earliest monarchies. Another was the Nasca kingdom of Peru's south coast.

Even before the decline of the Moche and Nasca, two new expansionist states had begun to form. In contrast to the Moche and Nasca, both of which were centered on the Pacific coast, these second-generation states arose in the Andean highlands. Both would create institutions that were adopted by the later Inca.

Wari

We have identified the southern highlands of Peru as a place where the long-term gathering of tubers and hunting of guanacos led to farming and herding. The rugged Ayacucho basin, averaging more than 9,000 feet above sea level, is embedded in this region.

At the moment we do not understand the early history of the Ayacucho basin well enough to explain why the capital of a new expansionist state would arise there. We do know that between 200 B.C. and A.D. 200, perhaps half a dozen communities in the basin had temples and elite residences.

Archaeological surveys show that a place called Wari was already a large village with public buildings at that time, but it was not the most important community in the basin. That distinction belonged to Ñawinpukyo, a hilltop civic-ceremonial center that continued to grow for the next 500 years. There is no evidence, however, that Ñawinpukyo's influence extended beyond the southern part of the basin.

Sometime between A.D. 500 and 700, Wari began to grow. It eventually came to cover a square mile of volcanic plateau, swallowing up smaller communities as it expanded. At least five towns that may have been Level 2 centers in Wari's administrative hierarchy arose to the west.

At Wari's peak (A.D. 600–900) many of its residents lived in large rectangular compounds, surrounded by stone walls 20 to 40 feet high. Some of these compounds had three stories and measured more than 900 by 400 feet. Craft activities were well developed, with mass production of mold-made pottery. The bones of llamas (raised as burden carriers) and alpacas (raised for their wool) have been found in the refuse at Wari.

Water was brought to Wari by a long canal that tapped into high-altitude sources. This main canal fed secondary canals that irrigated thousands of hillside terraces. The style of agriculture used at Wari anticipated, and perhaps provided a model for, later societies such as the Inca. While potatoes and other Andean tubers were among the staple crops, hundreds of terraces at lower elevations were used to grow corn for *chicha,* or maize beer.

Chicha was used by the Wari both as a ceremonial beverage and as a reward for labor gangs. The Wari drank it from special beakers called *keros,* a tradition carried on by later Andean peoples. Wari administrators kept elaborate accounts, using a system of knotted cords called *khipu* or *quipu.* This technology was also adopted by later societies such as the Inca.

The number and spacing of knots in the cords of a khipu allowed its owner to keep count of numbers of animals or units of commodities. It was, in other words, analogous to an abacus. There have been attempts to argue that khipu information was a kind of writing, but we find this unconvincing. In writing, there is a relationship between a set of symbols and the grammar of a spoken language. The mathematical relationships of knots do not meet this definition, and the information being recorded on one person's khipu would not necessarily have been clear even to another person speaking the same dialect.

Wari Imperialism

Wari extended its political control over the entire Ayacucho basin, bringing about the disappearance of earlier population centers such as Ñawinpukyo. It then began establishing colonies in more distant regions. Wari affected places as far away as Viracochapampa (480 miles to the northwest) and Cerro Baúl

(420 miles to the southeast). At the same time, it is not always clear whether Wari controlled, or merely influenced, these distant places.

Wari colonists often built high-walled rectangular enclosures with a central patio and long, narrow galleries. These enclosures, along with Wari-style pottery, provide clues to Wari expansion. In some places the expanding empire co-opted existing settlements; in other cases, it built brand-new settlements from scratch.

One of Wari's newly created colonies was the 495-acre settlement of Pikillaqta. Pikillaqta was built only 20 miles from the Cusco basin, the region that later gave rise to the Inca. The Cusco basin was occupied by a distinctive local society at that time, and the Wari chose not to confront it directly. Instead, Wari colonists chose a large mountain shelf above Lake Lucre, near the confluence of the Vilcanota and Huatanay Rivers. The centerpiece of Pikillaqta was a walled enclosure with a series of multiroom buildings that, when seen from the air, resemble giant ice cube trays. The colonists also built irrigation canals and terraces, bringing large tracts of previously marginal land into agricultural production.

It is possible that Wari administrators found the Cusco society of that period too uncooperative or underdeveloped for their purposes, so they built their own administrative center and staffed it with Wari officials. Archaeologists are convinced, however, that the inhabitants of the nearby Cusco basin watched every move the Wari colonists made and learned a great deal of statecraft from them.

As we saw earlier, Moche public constructions were built by rotating labor gangs. The construction of Pikillaqta seems to have been similar. The great walls enclosing the centerpiece of the community were made up of sections, built in slightly different styles. This architectural variation reflects a pattern of construction by multiple gangs. Such a pattern conforms to the Andean ideal of rotating responsibility, but it is less than ideal from the standpoint of architectural strength; the seams between sections eventually become weak points.

After occupying the Lucre basin for an estimated 200 years, Pikillaqta was eventually abandoned. Sometime later it was burned, presumably by the local inhabitants of the region. Pikillaqta's abandonment was part of the general decline of the Wari empire, whose causes are not fully understood. With this imperial presence removed, the societies of the Cusco region could now begin their own trajectory toward statehood and empire.

Tiwanaku

While Wari was expanding over the central highlands of Peru, a rival second-generation empire was emerging in the Titicaca basin on the Peru-Bolivia border. One of the highest large bodies of water in the world, Lake Titicaca lies more than 12,000 feet above sea level. This altitude rules out many frost-sensitive crops. However, fields of quinoa and root crops such as potatoes, *oca* (*Oxalis* sp.), *mashwa* (*Tropaeolum* sp.), and *ollucu* (*Ullucus* sp.) could be grown on the slopes near the lake. The region could also support herds of llamas and alpacas.

Archaeologist Charles Stanish has provided us with a scenario for the consolidation of the Lake Titicaca basin into one centralized, state-level society. This scenario begins around 500 B.C. and involves the emergence of an urban kingdom out of a group of competing chiefly societies.

At 500 B.C. the Titicaca basin, stretching more than 200 miles from northwest to southeast, was occupied by at least six or seven rank societies. The largest of these were Qaluyu in the northwest and Chiripa in the southeast. Within 500 years the northwestern society, with its paramount center at a place called Pukara, had expanded to control an area 90 miles in diameter. Its relations with the smaller southeastern society, whose center was Tiwanaku, were hostile.

Between A.D. 200 and 300, Pukara society underwent a rapid decline, perhaps owing to depredations by its aggressive neighbors. This gave Tiwanaku an opening to grow unimpeded, and by A.D. 600 it had no rival in the basin. At its height it had an estimated population of 30,000 to 60,000, living in a city covering two square miles. The city's core was a planned complex of plazas and public buildings; beyond this were commoner neighborhoods of artisans and laborers. Agriculture was intensified by the deliberate construction of raised field systems, and communities of llama and alpaca herders extended well into the mountains.

The residences of Tiwanaku's rulers covered the summit of the Akapana, a terraced pyramid whose summit rises to 55 feet and whose base measures 835 by 640 feet. Adjacent to the north face of the Akapana was the Kalasasaya, a ritual enclosure 400 feet on a side. Multiple generations of archaeologists, from Carlos Ponce Sanginés to Alan Kolata and Juan Albarracín-Jordan, have worked to increase our understanding of urban Tiwanaku.

The Tiwanaku state expanded far beyond Lake Titicaca, though it is not always clear which regions it controlled and which ones it merely influenced.

Tiwanaku-style pottery has been found within 20 miles of the Wari colony at Pikillaqta. To the south, settlements with Tiwanaku pottery were placed within sight of Cerro Baúl, a fortified Wari outpost in the Moquegua Valley (Figure 71). In other words, the Wari and Tiwanaku empires expanded until they literally came within a few miles of each other. At such meeting places they seem to have coexisted without bloodshed, perhaps realizing that an all-out war would not have been good for either empire.

Like the Wari, Tiwanaku established a number of patterns that were borrowed by the later Inca. Tiwanaku public structures were often built of stones so tightly fitted together that one could not have inserted a razor blade between them. Tiwanaku also built roads through the highlands that anticipated the later Inca imperial roads. Like the Wari, Tiwanaku nobles drank chicha from keros.

A number of imperial strategies served to communicate Tiwanaku's subjugation of neighboring peoples. Its architects decorated a sunken court in the Kalasasaya enclosure with carved stone versions of trophy heads. Tiwanaku also captured and removed the stone monuments of the foreign groups it subjugated. For example, the so-called Thunderbolt Stela at Tiwanaku has turned out to be the upper half of a stone monument from Arapa, a community 150 miles to the northwest. This monument, carved before the rise of Tiwanaku, had been deliberately broken; half was left at Arapa and the other half taken to Tiwanaku. Such "monument capture" was a strategy borrowed by the later Inca.

After centuries of expansion, Tiwanaku finally declined. By 1200 the Titicaca basin was decentralized, broken down into a dozen or more small societies whose elites sought refuge in *pukaras*, or fortified settlements. Between 1450 and 1475, the Inca moved into the Titicaca basin.

THE THIRD-GENERATION ANDEAN EMPIRE

In Peru, just as in Mexico, Spanish officials produced manuscripts on the Native American societies they encountered. While a high percentage of these documents deal with the Inca, some record tales of a pre-Inca people called the Chimu, who ruled the north coast between A.D. 850 and 1460.

The Chimu are named for the Kingdom of Chimor, whose capital lay in the same valley that gave rise to the earlier Moche. The Chimu state eventually became an empire stretching 600 miles along the coast.

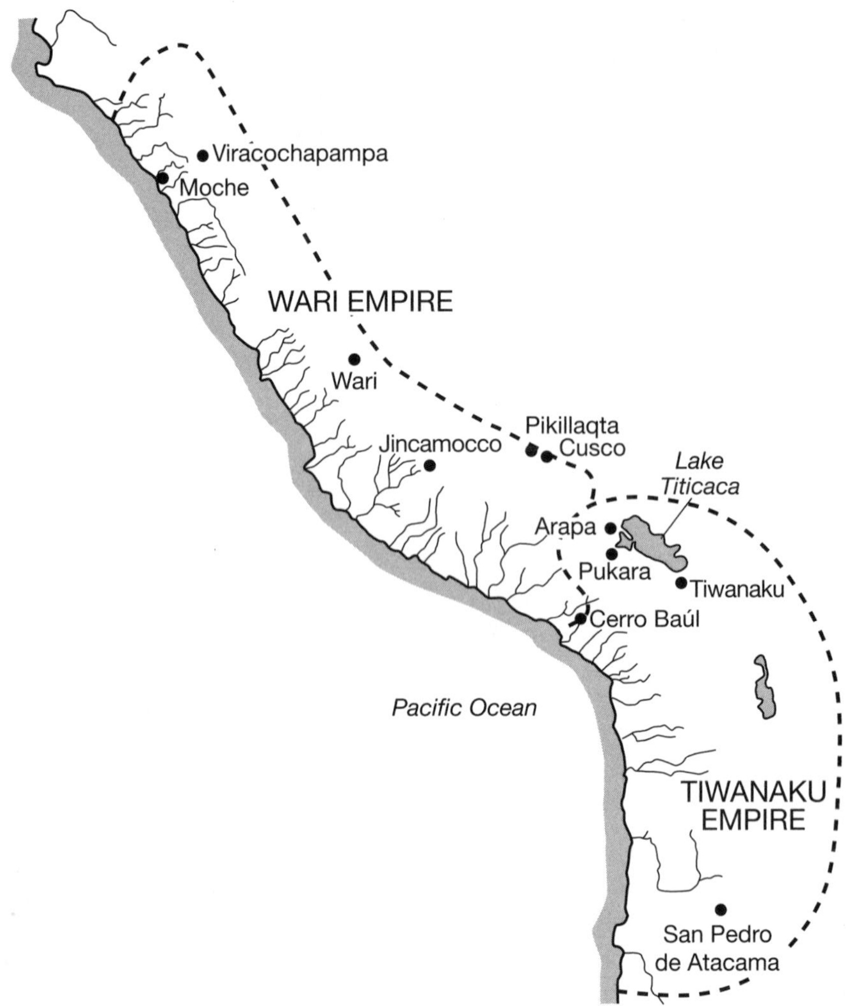

FIGURE 71. Some 1,400 years ago, two empires spread over the Andean highlands. The Wari empire's capital was in the Ayacucho basin of Peru; the Tiwanaku empire's capital was in the Lake Titicaca basin of Bolivia. The two empires met at Cerro Baúl in southern Peru. There, Tiwanaku-influenced villages coexisted with a fortified Wari colony in what seems to have been an atmosphere of détente. (On this map, the distance from Moche to San Pedro de Atacama is 1,250 miles.)

The rise of the Chimu was facilitated by the gradual collapse of the Moche empire. At its peak, this empire had been divided into northern and southern regions. Collapse began in the southern region, where valleys such as Virú, Santa, and Nepeña achieved independence from their Moche overlords.

In the northern region the Moche shifted their capital from the Huacas de Moche to the site of Pampa Grande in the Lambayeque Valley. The task of administrating the Moche Valley was left to Galindo, 12 miles inland from Huacas de Moche. Galindo was a planned urban center covering more than two square miles. Some of its architecture anticipated that of the later Chimu.

Chan Chan

By 1000, Galindo was in decline, and the center of political power in the Moche Valley had shifted coastward to a place called Chan Chan. There the water table was so high that communities could create extensive *mahamaes*, or sunken fields, for gardening, as well as walk-in wells for drinking water. Chan Chan became the capital of the Kingdom of Chimor.

In the world of legend, the first Chimu king is said to have been a noble named Tacaynamo, who sailed to the Moche Valley on a balsa raft. In the world of political reality, the Chimu state is likely to have been created by a junior royal lineage that split off from a senior lineage somewhere on the north coast.

At its peak, the urban core of Chan Chan covered 2.3 square miles and had an estimated 60,000 inhabitants. Excavator Michael Moseley estimates that 6,000 of these inhabitants were nobles who lived in large adobe-walled residential compounds. The ten largest of these compounds, known as *ciudadelas*, are thought to have been built by a succession of royal families, while the lesser nobles lived in 30 smaller compounds. An estimated 26,000 craftspeople lived in their own *kincha*, or cane-and-clay houses. Three thousand commoners lived immediately adjacent to the royal compounds.

Oral histories report that the Chimu kings practiced a strategy called "split inheritance." Upon the death of a ruler, his residential compound and any territory he had conquered were retained in his name and administered in perpetuity by a special bureaucracy. The new ruler inherited his office but not his predecessor's property; he was therefore forced to build his own compound and conquer new territory, which would in turn be administered in his name. Just as the need to take captives for his inauguration forced an Aztec ruler to

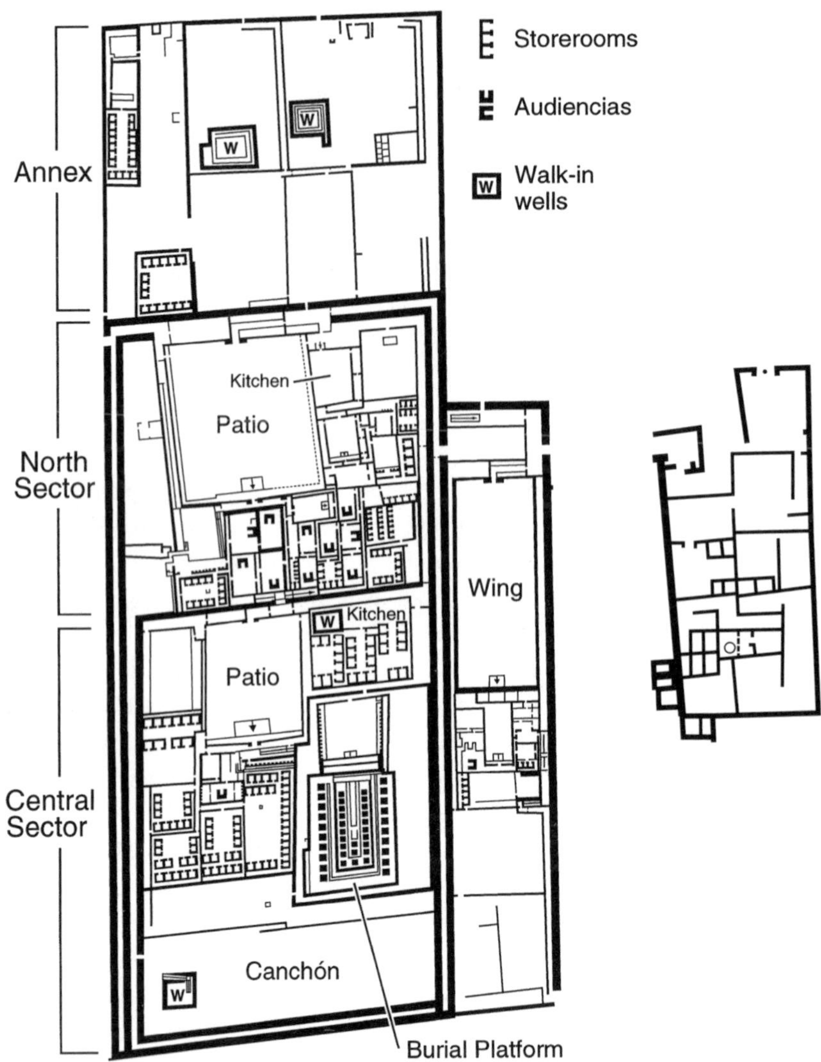

FIGURE 72. Chimu emperors lived in huge royal compounds with patios, kitchens, servants' quarters, storerooms, walk-in wells, royal burial platforms, and overseers' rooms called *audiencias*. Lesser nobles lived in smaller versions of these compounds. Chimu commoners lived in neighborhoods of small room complexes. On the left we see a royal compound from Chan Chan, roughly 1,770 feet in length. On the right we see two adjoining complexes of rooms for commoners, totaling 75 feet in length.

extend his conquests, split inheritance forced a Chimu ruler to add lands to Chimor.

No two ciudadelas were identical in plan, though all shared a number of elements. Walls more than 30 feet high ensured privacy for those who lived and worked there. Traffic flow inside the compound was strictly controlled. Workers, who entered during the day and went home at night, were monitored by officials sitting in U-shaped rooms, referred to by Spanish speakers as *audiencias*. There were open courts for craft activity and innumerable storage units to be filled. Thick walls separated the residence of the royal family from the rooms where their commoner staff worked (Figure 72).

One part of each compound was set aside for a massive adobe platform, where the ruler's body would be hidden after death. The royal burial chamber was surrounded by subterranean cells for family members, sacrificed humans and llamas, and offerings of precious metals, pottery, fine textiles, and spiny oyster shells.

The Logic of Chimu Imperialism

Archaeologist Carol Mackey has provided us with insight into the way the Chimu turned their kingdom into an empire. For roughly 250 years, from A.D. 1050 to 1300, the Chimu consolidated power in their heartland, which consisted of the adjacent valleys of Moche, Chicama, and Virú. Then, from roughly 1300 to 1450, they expanded south to the Casma Valley and north to the Leche Valley.

During their expansion the Chimu established three large, Level 2 administrative centers. One was Farfán in the Jequetepeque Valley, 70 miles north of Chan Chan. According to oral history, a Chimu general named Pacatnamu subdued this valley after a fierce battle.

A second Level 2 administrative center was Manchán in the Casma Valley, 180 miles south of Chan Chan. The last of the Level 2 centers to be established was Túcume in the Leche Valley, 150 miles north of Chan Chan. This 370-acre city had been occupied by a rival group since 1100, and it proved difficult to conquer.

In addition to Farfán, Manchán, and Túcume, the Chimu maintained Level 3 centers, such as Talambo in the Jequetepeque Valley and Quebrada Katuay in the Moche Valley. Below these smaller centers were thousands of villages, some of which Mackey believes had been newly created to produce agricultural staples for the Chimu state. This suggestion of an economy planned from the top down anticipates the later Inca.

All of the Level 2 centers of the Chimu empire were communities that had been previously occupied. At each place, in Mackey's words, "the Chimu altered the infrastructure, either by constructing anew or rebuilding, with the result that [Chimu] state presence was highly visible." She identifies the following four Chimu imperial policies, from which we can infer the logic of subjugation and administration:

1. Level 2 centers were provided with large public areas and stored enough food to host state-sponsored rituals and feasts. Such feasts were designed to communicate the generosity of Chimu overlords.
2. Level 2 centers were also located near long-established routes for the acquisition of valued resources, including the ores needed for making metal tools and sumptuary goods, as well as the spiny oyster shells used in so many rituals.
3. In two of the Level 2 centers—Manchán and Túcume—local nobles and Chimu provincial lords seem to have lived side by side, using residential compounds built in different styles. This strategy of joint rule was borrowed by the later Inca.
4. Level 3 centers show much less imperial interference, perhaps because the Chimu believed that indirect rule was less likely to disrupt the productivity of the local population.

The Chimu empire lasted until roughly A.D. 1460. Unlike the earlier empires discussed in this chapter, it did not collapse as the result of internal factionalism or the revolt of its colonies. It was conquered by Peru's fourth-generation empire, the Inca, who expanded out of the southern highlands.

THE RISE OF THE INCA

The heartland of the Inca state was a network of mountain valleys inhabited by speakers of the Quechua language. Lying at an average elevation of 11,000 feet was the basin of Qosko (Hispanicized Cusco). To the north was the valley of the Vilcanota River, some 9,000 feet above sea level. To the east was the basin of Lake Lucre, where the Wari empire had built Pikillaqta. The Wari eventually withdrew from Pikillaqta, having given local leaders no end of ideas about how to create their own empire.

Our two main sources of information on post-Wari developments consist of indigenous historical accounts and intensive archaeological surveys. Because Peru did not have hieroglyphic texts or picture-writing like the Aztec of Mexico, the accounts are purely oral histories, like those memorized and recited by Asante specialists. The intensive surveys are the work of archaeologists Brian Bauer, R. Alan Covey, and their associates.

Surveys show that from A.D. 400 to 1000, even during the peak of the Wari empire, there was no settlement in the Vilcanota Valley larger than a village. The Wari colony at Pikillaqta seems to have had minimal impact on this area. During the post-Wari era, however, the Vilcanota Valley gradually developed its own four-level hierarchy of settlements. Many of these settlements seem to have chosen defensible localities.

According to Covey, the site of Pukara Pantillijlla in the Vilcanota Valley grew from village to civic-ceremonial center by 1400. An emerging ruling class built royal estates at Larapa in the Cusco basin and Qhapaqkancha on the southern rim of the Vilcanota Valley. Eventually Cusco became the capital for the region, while Pukara Pantillijlla was reduced to a Level 2 center.

The Inca took advantage of altitude differences that allowed them to produce corn at lower elevations and potatoes at higher elevations, accompanied by the herding of llamas and alpacas. Like the Wari before them, they dug irrigation canals and turned mountain slopes into thousands of agricultural terraces. The royal estates mentioned earlier were used to produce surplus food for the rulers, enabling them to host neighboring elites and impress workers with their generosity.

In the Cusco dialect of Quechua, the word *Inka* was used for the ruler. To avoid confusion, we use Inka (with a k) to refer to the emperor and Inca (with a c) to refer to the society.

Oral histories have given us the names of 12 Inkas, the earliest of whom may be partly legendary. The list begins with Manco Qhapaq, whose principal wife is said to have been his sister or half sister. If true, this claim suggests that the rulers of Cusco had suspended the usual incest taboos in order to ensure that their heirs would have the bluest bloodlines. We note that sibling marriage seems to have been more common in societies whose rulers were considered divine (Egypt and the Inca) or filled with the life force called mana (Tonga and Hawai'i).

As the expanding Cusco kingdom began to incorporate more and more of its neighbors, however, rulers turned frequently to marriage alliance as a way of solidifying their takeovers. Sinchi Roca, the second Inka, and Lloque Yupanki, the third, married noblewomen from other regions. Neighboring royal houses who accepted Cusco's domination were declared "Inca by privilege," making them feel less subordinate and giving them a stake in future imperial expansion.

Oral histories attribute a major escalation in territorial expansion to the ninth ruler, Pachacuti Inka Yupanki, who in 1438 is said to have defeated a powerful rival kingdom called the Chanka. This defeat removed a major roadblock to westward expansion and helped the Inca become the largest empire in the New World. Pachacuti Inka Yupanki returned to an earlier strategy by marrying his own sister. He also created a new set of royal estates, one of which—the spectacular mountaintop community of Machu Picchu—is now a major tourist attraction.

By conquest, intimidation, and alliance the Inca created an empire stretching 2,000 miles, from Ecuador in the north to Chile and Argentina in the south. From the Wari they borrowed the khipu, the kero for drinking chicha, and the practice of building many imperial installations from scratch. From the Tiwanaku they learned to connect the capital to its provinces with impressive roads, capture the monuments of defeated peoples, and use very tightly fitted stones for major buildings. From the Chimu they borrowed the policy of joint rule, which they used wherever it worked. To all these preexisting institutions, of course, they added policies of their own.

Inca Society

An important unit of Inca society was the *ayllu,* a corporate social segment analogous to the calpulli of the Aztec. The ayllu probably began as a large kin group that reckoned descent in the male line. In today's Quechua-speaking communities, however, an ayllu can include unrelated extended families that live in the same region and share the same system of crop rotation.

At some point in the past, members of certain ayllus came to be considered nobles, while members of most ayllus were commoners. At the time of the Spanish conquest there were 11 noble ayllus in Cusco, six in one moiety and five in another. High-level officials were chosen from these ayllus.

Commoner ayllus had corporate land, and each married couple was allowed to cultivate as much of it as they needed. Relatives exchanged labor with each other on a day-for-day, person-for-person basis, a system of reciprocal aid known as *ayni.* Each ayllu had a leader whose lands were cultivated for him by his fellow members. Under Inca rule, state officials ordered the exchange of family plots each year to ensure the proper rotation of crops.

Contrasting with ayni was *mit'a,* or unreciprocated labor. Each citizen of the empire owed the government a period of labor service each year. Buildings, agricultural terraces, irrigation canals, and roads were all built with mit'a labor, which the Inca preferred over tribute in goods.

Not every commoner belonged to an ayllu. There were also landless serfs called *yanakuna,* the Andean counterpart of the Aztec mayeque. Many yanakuna were male laborers who worked full time for the state. Caring for royal herds of llamas and alpacas was a typical assignment for yanakuna.

While reciprocal exchange and unreciprocated labor service were important in the Inca economy, markets were not. This was one of two major economic differences between the Aztec and the Inca. The other economic difference, already mentioned, was that the Aztec preferred goods as tribute, while the Inca preferred labor.

Stratification

The hereditary ruling stratum of Inca society was divided into lineages of major nobles *(inka)* and lesser nobles *(kuraka).* The Inka claimed lineal descent from Inti, the Sun, and was considered divine. He had his hair trimmed short and wore large golden spools in his earlobes (Figure 73). He might wear a headdress of multicolored braid, wound four times around his head and decorated with golden tubes and crimson tassels. At times he carried a mace with a golden head or sat on a stool covered with fine cloth. As his kingdom grew, he came to be referred to as *Qhapaq Apu* (Emperor). Later emperors added titles such as *Sapa Inka* (Unique Inka), *Intip Cori* (Son of the Sun), and *Wakca Khoyaq* (Lover of the Poor). An emperor had many wives and concubines, but his principal spouse was referred to as *Qoya* (Empress) or *Mamancik* (Our Mother).

Anyone coming to see the Inka had to remove his sandals and place a burden on his back as a sign of subservience. The emperor ate from gold and silver plates held by female servants. All leftovers, including his used clothing, were burned in an annual ceremony. When the Inka traveled it was on a litter, carried slowly and with such dignity that his bearers rarely moved more than 12 miles a day. Oral histories claim that the route ahead of him was swept by members of an ethnic group called the Rucana.

FIGURE 73. Felipe Guamán Poma de Ayala was the offspring of Inca nobles from the province of Rucanas. During the sixteenth century he served as interpreter for the Spanish priest Cristóbal de Albornoz. Guamán Poma de Ayala later wrote a critique of Spanish colonial rule that included drawings of indigenous life. On the left we see two Inca commoners engaged in agriculture. On the right is an Inca emperor in a checkerboard tunic, brandishing a lance and shield. (The artwork in Guamán Poma de Ayala's book shows a mixture of Inca and Spanish elements.)

One Spanish eyewitness claims to have seen the Inka stop his litter bearers when he saw a blanketless commoner shivering in the mountain air. After ordering that the man be issued a blanket, the Inka berated the local administrator for not taking proper care of his subjects. This story illustrates the paternalistic, top-down attitude of the Inca state.

Because the Inka could not be in two places at the same time, he commissioned a statue referred to as his *huauque,* or "brother." This statue stood in for the ruler when he was unavailable. Even in death, the emperor's mummy would continue to counsel the living through an oracle. On occasion, a bundle that contained his hair and fingernail clippings was brought out to legitimize the rights of his descendants or the caretakers of his estates. This practice reminds us of the Tongan belief that a chief's hair and nail clippings had powerful mana.

In a funerary ritual reminiscent of an Egyptian king's mummification, the dead Inka's intestines were removed and placed in a special container, while his body was dried and treated with preservatives. And just as among the Natchez and the ancient Panamanians, many of the Inka's favorite women and servants volunteered to be stupefied and strangled so that they could accompany their ruler in the afterlife.

The Spaniards were surprised to find that each Inka's mummy was served by royal attendants and paraded in public on a regular basis. Whenever a mummy, huauque statue, or hair/fingernail bundle was brought out, its caretakers sang about the history of the emperor's reign, his conquests and his accomplishments. Each emperor's oracle sat next to his mummy, passing along the dead Inka's advice and his ongoing need for food and drink.

While Inca nobles kept long genealogies and venerated many generations of ancestors, the memories of commoners rarely extended beyond their grandparents. Understandably, commoners who belonged to an ayllu were more concerned with their ancestors than were the yanakuna, whose ancestral ties had been disrupted.

Rulers and other nobles had multiple wives. Since most commoner men could afford only one wife, polygamy was seen as a sign of wealth. Women participated in agricultural tasks, such as hoeing potato fields, so multiple wives could enrich a family. Sometimes a man of modest means was given an extra wife by the emperor for services rendered, or captured a foreign second wife while serving in war.

People taken in war could be considered slaves, but the differences between slaves and yanakuna were subtle at best. While Aztec slaves were always potential victims of sacrifice, the Inca were much more interested in them as a source of labor.

Downtown Cusco

Even as the Inca empire expanded beyond those of the Wari, Tiwanaku, and Chimu, the Cusco Valley remained its dynastic capital. At its apogee, Cusco was home to more than 20,000 people; thousands more lived elsewhere in the valley.

Cusco grew up between two rivers, the Saphy and Tullumayu. The banks of these rivers were walled and canalized, and the Inca made annual offerings at their confluence. At the heart of Cusco was a Great Plaza divided in half by the Río Saphy, its western half called Cusipata and its eastern half Aucaypata. Thousands gathered in Aucaypata on the June and December solstices, the August planting ceremony, and the May harvest celebration. On such occasions the mummies of previous Inkas were brought to the plaza and lined up in the order of their reigns.

In 1559 the conquering Spaniards broke ground for a cathedral in Aucaypata. They were surprised to find that the entire foundation for this eastern plaza was a thick layer of sand, brought hundreds of miles from the Pacific coast. This massive use of sand as a clean foundation for ritual space recalls the 7.4 acres of that material placed beneath the Temple Oval of Tutub in ancient Sumer. In Aucaypata the sand layer had also been filled with golden images and vessels of precious metals, a windfall for the Spaniards.

The Spaniards were equally excited to find the Ushnu, a sacred stone altar covered with gold. The Inca regularly poured offerings of chicha on this altar, watching the stream of corn beer disappear down a canal leading to the Coricancha, or Golden Enclosure. The latter was a Temple to the Sun, much of it covered with fine sheets of gold.

The Coricancha was the ritual epicenter of Cusco, and from it a series of long sight lines, called *ceques,* radiated in many directions. Among other things, these sight lines divided the Inca empire into four *suyus,* or quadrants, corresponding to the four world directions. In addition, the ceques served to align a series of *huacas,* or shrines, built at increasing distances from the Coricancha. There were nearly 400 of these shrines, the most distant of which lay beyond the limits of the Cusco Valley.

Many huacas were rock outcrops, springs, caves, or places associated with sacred visions. In other words, despite the fact that they were the New World's greatest empire, the Inca continued to share several principles with smaller-scale societies. Like the hunter-gatherers of Australia, they considered springs and rock outcrops sacred places; like the Tewa of San Juan Pueblo, they envisioned a sacred landscape extending far beyond the limits of human settlement.

Several other buildings in downtown Cusco are worthy of mention. One was the Casana, a palace allegedly built by the 11th Inka, Huayna Capac (1493–1527). Another was the Aklla Wasi, or "House of the Chosen Women." The Aklla Wasi housed hundreds of women whose lives were dedicated to the Inca state. These women wove cloth and brewed chicha for the emperor, and sometimes they served as priestesses in temples.

The Logic of Inca Imperialism

In the course of creating and maintaining their empire, the Inca borrowed no end of principles from Wari, Tiwanaku, and Chimor. These principles provide us with a framework from which we can infer some of the logic of Inca imperialism. There were multiple layers of imperial strategy, depending on whether the Inca were dealing with their capital city, their heartland in the southern highlands of Peru, their more distant provinces, or the outer frontiers to which they expanded.

At the level of Cusco, a number of behaviors maintained the social distance between members of noble ayllus and everyone else. Royals and nobles were exempt from labor service, and the ruler, as we have seen, was allowed to marry his sister or half sister to maximize the rank of his offspring.

Within their heartland the Inca acted upon the ethnocentric belief that their neighbors longed to emulate them. Inca rulers married the sisters and daughters of noble allies and then declared their in-laws' ethnic groups "Inca by privilege." Many "Inca by privilege" were later given positions of trust within the imperial bureaucracy.

Inca rulers also selected thousands of young girls to become the aforementioned "chosen women." Called *mamakuna,* these women were used to weave textiles and brew beer, a role analogous to that of many women who labored for the Sumerian state. Much of the beer was used to host the allies of the Inca. In some cases a key alliance might be cemented by the betrothal of one of the mamakuna to a neighboring leader.

When it came to adding more distant provinces, the Inca (like so many pre-Hispanic empires) chose the path of least resistance. They lavished gifts upon those who accepted Inca rule, using force only on enemies they felt they could defeat.

We have seen that one widespread principle of Andean logic was ayni, or balanced reciprocity. If, for example, members of Group A assisted Group B in the harvest of their crops, they could count on Group B to provide them with comparable help in the future. The Inca convinced many of their provinces to accept an asymmetrical version of this reciprocity: in return for long-term manual labor, the Inca would "balance the account" with a burst of feasting and drinking.

Like their Tiwanaku predecessors, the Inca improved access to distant provinces with a road system. At intervals along the roads the Inca created *tampus,* or way stations, for resting and provisioning their troops. Near each tampu they built long lines of *collcas,* or storage rooms, many filled with foodstuffs. The Inca pattern was to maintain hundreds of small collcas rather than a few large ones. If insects or disease attacked one of these small units, it could quickly be burned before the problem spread.

Before resorting to an all-out war, the Inca used pressure to bring in line resistant provinces. Having borrowed "monument capture" from the Tiwanaku state, they gave this practice an additional twist: the captured statue of a foreign god or ruler might be set up in a public square and flogged for days. Some provinces, distraught over this torture of an icon they considered a living being, eventually capitulated.

When all else failed, the Inca relied on their *sinchi,* or war leader, to subjugate a resistant province. The well-organized Inca army was based on a decimal system and had units of 10, 50, 1,000, 2,500, and 5,000 soldiers. Sometimes resistant groups were obliterated, moved en masse to another region, or replaced with loyal Quechua speakers.

Sinchis were under great pressure to succeed. According to the sixteenth-century Spaniard Miguel Cabello Balboa, an Inca general who had lost several battles was sent women's clothing and ordered to wear it upon his return to Cusco.

The Administration of Provinces

The Inca had three basic policies for the administration of distant provinces. Two were borrowed from the Chimu: the negotiation of joint rule over some district capitals, and the establishment of direct rule by Inca administrators over others. The third strategy, borrowed from Wari, was to build a brand-new administrative center from scratch. In this section we look at one example of each strategy.

The Chincha Valley. The valley of the Río Chincha lies on the south coast of Peru, 110 miles from the city of Lima. Before the Inca rose to power, Chincha was the scene of a kingdom that divided its time between fishing, irrigation agriculture, and long-distance trade. Oral histories suggest that traders from Chincha, hugging the coast in rafts made from balsa logs, traveled to Ecuador's Gulf of Guayaquil and returned with *Spondylus,* or spiny oyster. The shells of this sacred mollusk were reduced to powder and used to cover the floors of temples, or to create sparkling footpaths for rulers and priests.

In addition to its economic success, Chincha was also the seat of an important oracle. The Oracle of Chincha was allegedly associated with a complex of pyramids at La Centinela, one of two major archaeological sites at the mouth of the Chincha River. The other major site, Tambo de Mora, was part of the same urban sprawl, and the area between the two complexes of pyramids and palaces was filled with the cane-and-clay houses of commoners. Archaeologists have found extensive evidence for the working of spiny oyster and metal at Tambo de Mora.

Archaeologist Craig Morris discovered signs that the Inca takeover of Chincha was bloodless and that it involved joint rule. On the main ceremonial plaza at La Centinela, the Inca built a pair of palaces, one for the local lord of Chincha and one for the Inca administrator. These palaces were not built in the local Chincha style, which involved the pouring of clay between wooden molds; instead, they were built of adobe bricks in typical Inca style. Nearby the Inca built a Temple to the Sun. They also co-opted the Oracle of Chincha, changing the access to the shrine so that it could only be entered from the Inca administrator's palace.

In return for the Chincha lord's acquiescence to joint rule, the Inca allowed him to be carried around in a litter, an honor normally reserved for Inca nobles. The Inca also gave him gifts of gold, fine clothing, and pottery vessels that Morris instantly recognized as having been imported from Cusco. The Inca benefited from joint rule at Chincha, since the navigational skills of the local traders kept the supply of spiny oyster shells coming from Ecuador.

The Cañete Valley. The valley of the Río Cañete lies 80 miles south of Lima and 30 miles north of Chincha. During the Inca rise to power the Cañete Valley was occupied by two small kingdoms, Huarco on the coastal plain and Lunahuaná in the piedmont.

The Kingdom of Huarco was encircled by a major irrigation canal that took off from the Cañete River and ended at Cerro Azul Bay. The kuraka, or ruler, lived in a palace at the hilltop settlement of Cancharí. At Cerro Azul Bay he established a specialized fishing community that produced industrial quantities of dried fish, shipped inland in exchange for corn, potatoes, sweet potatoes, and other products of irrigation agriculture.

In contrast to Chincha, the Kingdom of Huarco refused to capitulate or agree to joint rule. The Inca army therefore took over Lunahuaná, using it as their staging point for an invasion of Huarco.

In 1470 representatives of the Inca requested a truce with Huarco. Thrilled by this proposal, the people of Huarco entered their watercraft and sailed out for a joyous ritual offshore. Unfortunately, the Inca had lied. Once the nobles of Huarco were at sea, the Inca troops rushed the coast and took them by surprise.

The Inca massacred the Huarco elite and then built two Cusco-style buildings at Cerro Azul. One of these was an adobe brick structure with typical Inca trapezoidal niches in its walls. The other was an oval building resembling an Inca ushnu, set on the brink of a cliff so that it could be seen far out at sea. This building was constructed of volcanic stones imported from the distant highlands. In typical Inca style the stones were so tightly fitted that a razor blade could not be inserted between them. From this building a stairway descended the cliff so that offerings could be made to the sea.

Archaeologists found no evidence of joint rule at Cerro Azul, no twin palaces, and no gifts of gold or pottery sent from Cusco. The Inca had simply wiped out the local elite and installed their own administrators.

Huánuco Pampa. One of the most important stretches of the imperial road system was the highland route connecting Cusco with Quito, Ecuador. This route passed through the Huánuco area of north-central Peru, a region mentioned during our discussion of the site of Kotosh.

In one part of the Huánuco area the Inca road traversed a high and sparsely inhabited plain, some 12,350 feet above sea level. To either side lived ethnic groups whom the Inca regarded as "warlike" and "uncivilized." Included were the peoples of the upper Marañon River and more distant groups called Chupaychu and Yacha.

The Inca decided to build a major city from scratch on the high plain, known as Huánuco Pampa. Their strategy was twofold. First, they would be such generous hosts that local groups would be attracted to the city. Second, they would obligate their guests to repay their generosity with labor. At the heart of their strategy was the belief that periodic feasting and beer drinking more than compensated for long stretches of hard labor.

We have two main sources of information on Huánuco Pampa. One is a sixteenth-century Spanish eyewitness, Pedro Cieza de León. The other is archaeological information from the city of Huánuco Pampa, excavated by Craig Morris and analyzed by Morris and R. Alan Covey.

Huánuco Pampa covered more than a square mile and its east-west axis consisted of three plazas, each with its own complex of public buildings. The imperial road passed directly through the largest plaza, which was 500 yards in length. The centerpiece of this plaza was an ushnu in typical Inca style, adorned with images of pumas.

This largest plaza appears to have been dedicated to huge assemblies of commoners, many invited from neighboring groups. The Inca administrators lived in a smaller and more private plaza. As Cieza de León wrote in 1553:

> There was an admirably built royal palace, made of very large stones artfully joined. This palace . . . was the capital of the province, and beside it there was a temple to the Sun with many priests.

Cieza de León claims that at its height Huánuco Pampa was served by 30,000 people. Morris was able to map the foundations of more than 4,000 structures, and on a hill south of the city he found nearly 500 storage units arranged in neat rows.

One of the notable buildings at Huánuco Pampa was an Aklla Wasi that may have housed 200 chosen women. In addition to weaving fine textiles, these women brewed the vast quantities of corn beer with which the Inca entertained representatives of the Chupaychu, Yacha, and other neighboring societies. Thousands of laborers from visiting groups came to reside seasonally at Huánuco Pampa, performing labor for their Inca hosts in return for food and drink.

As for the leaders of the local ethnic groups, they did not become "Inca by privilege" like the Quechua-speaking allies of the Cusco region. The Inca did, however, assign them an intermediate status, somewhere between nobles and commoners.

Three terms for social rank appear in Spanish documents for the region. The word *collana* was used for rulers and important officials who had kinsmen in the Cusco region. The term *cayao* was used for commoners from all non-Inca ethnic groups. A third word, *payan,* was created for people of intermediate status, who had connections both to the collana and cayao by intermarriage or fictive kin relations.

Morris and Covey believe that by turning local leaders into payan, the Inca were responding to a growing need for a stratum of intermediate-level elites who could manage parts of their empire for them. The payan provided the Inca with thousands of workers in return for lavish Inca hospitality. And, like the Kachin chiefs who accepted Shan brides, the payan received collana women in marriage, which raised their prestige.

To be sure, the collana did not look upon the payan as equals. They were willing, however, to flatter the payan, treating them as honored guests and betrothing women to them. The Inca knew that this status-enhancing treatment would obligate the payan to provide them with thousands of cayao workers.

The End of Empire

One of the proud moments of Inca history, as we have seen, was the conquest of the Chanka by Pachacuti Inka Yupanki (1438–1471). In the tradition of Tiwanaku monument capture, Pachacuti took possession of the most important Chanka statue, a sculpture of the first Chanka ruler's mummy. After Pachacuti's death this Chanka statue was kept on one of his royal estates, next to his own mummified remains.

Pachacuti was succeeded by his son Topa Inka Yupanki (1471–1493), who became the tenth Inka. He was in turn succeeded by Huayna Capac (1493–1527), the last Inka to rule over a unified empire. It was during this period that Columbus discovered the New World, setting in motion a European colonization that would doom the Inca.

Huayna Capac died near Quito, the northern limit of his empire, leaving behind a bitterly divided realm. Two half brothers, Huascar and Atahualpa, each claimed to be Huayna Capac's true successor. Huascar managed to get himself installed as emperor, but civil war between the royal siblings eventually broke out.

Atahualpa's forces invaded Cusco, seeking to kill all nobles loyal to Huascar. Included among the latter were many relatives of Topa Inka Yupanki. Atahualpa's men hanged countless Huascar supporters and then tracked down Topa Inka Yupanki's mummy and reduced it to ashes.

The Spaniards arrived in 1532 and quickly took advantage of the civil war. They captured Atahualpa in Cajamarca, determined to ransom him for gold. Despite Atahualpa's captive status, his supporters overtook and assassinated Huascar. The conquistador Francisco Pizarro then ordered the execution of Atahualpa, leaving the Inca without a ruler.

The Spaniards harvested all the gold and silver objects from the sand layer in Cusco's Aucaypata and stripped the gold from the Great Ushnu and the Coricancha. They then began to search for the mummies of past Inkas, whose mystique made them a threat to Spanish rule. By 1559 they had located and burned the majority of the royal mummies.

In the logic of the Inca the mummies, huauque statues, and hair/fingernail bundles of rulers were not only alive but continued to advise their heirs, legitimize lower-level officials, and govern their subjects. Their destruction severed the divine chain of command and left the administrative hierarchy with no Level 1.

INEQUALITY IN EMPIRES

The empires described in this book were significantly larger than most kingdoms. One wonders, however, if social inequality in empires was significantly greater than in kingdoms. It seems unlikely that slaves and landless serfs would have noticed much difference.

One new source of inequality in empires was the stripping of autonomy and authority from conquered elites. Empires swallowed up rival kingdoms the way kingdoms had once swallowed up rival rank societies. Many a monarch whose word had been law in his own kingdom was now forced to take orders from an emperor.

Subjugated monarchs responded to their newly created inequality in a variety of ways. In the Basin of Mexico, the ruler of Texcoco went into exile until he could reclaim his throne. In Peru's Chincha Valley the local lord agreed to joint rule in order to preserve some of his privileges, such as being carried in a litter. Similar promises of joint rule convinced many an ambitious prince from a junior lineage to cooperate with the Inca ruler, if the latter helped him usurp a senior ruler's position. The kings of Huarco and the Chanka, in contrast, refused to submit to the Inca and had to be defeated militarily.

As powerful as they were, both kings and emperors often had to resort to generosity to get what they wanted. The Aztec are reported to have thrown feasts for the Tlaxcalan nobles, whom they were never able to subdue. Culhuacan overlords provided their Mexica vassals with a prince. The Inca of Huánuco Pampa gained access to Yacha workers by wining and dining their hereditary leaders. So deeply ingrained are the first principles of generosity and reciprocity that even emperors learned to manipulate them.

THE LEGACY OF EARLIER KINGDOMS AND EMPIRES

There is a reason we have emphasized the generation to which each kingdom or empire belonged. Fourth- and fifth-generation kingdoms were not created in the same way as first-generation kingdoms. All later generations of kingdoms and empires were able to borrow strategies and institutions from their predecessors.

The creators of first-generation kingdoms had no template to follow. They did not know that they were creating a new type of society; they simply thought that they were eliminating rivals and adding subordinates. Only later did they discover that they had created a realm so large that they would need new ways to administer it.

Once a template existed, however, there were many alternative routes to the creation of the next kingdom. We doubt that the founders of the first kingdom in each region had as many options, and our suspicions are supported by some remarkable similarities in the way that first-generation states were created.

Archaeologists and social anthropologists sometimes ignore generational differences by lumping together all ancient kingdoms. We hope that in the future they will isolate first-generation kingdoms and investigate them as a special case. Only by doing so will they learn why certain groups of rank societies could be consolidated into kingdoms and others could not.

V

Resisting Inequality

TWENTY-FOUR

Inequality and Natural Law

Our earliest ancestors were all born equal, but the Ice Age had barely thawed when some of them began surrendering bits of equality.

The rise of complex human societies, which began with hereditary rank and peaked with empires, has been compared to hypertrophic growth in biology. Social complexity, however, was not caused by genes. It grew out of perceived differences in life force, virtue, intellectual property, generosity, debt, and prowess in combat.

In biological evolution, population increase is considered a measure of success. One species grows at the expense of others. Either brand-new genes made it more successful, or a change of environment favored its preexisting genes. Social evolution was different. Some of humanity's largest increases followed the adoption of agriculture, a change that had nothing to do with our genes. The decision to live in permanent villages, the rise of aggressive rank societies, and the creation of expansionist kingdoms were frequently accompanied by population growth.

Despite their obvious differences, one can point to useful analogies between biological and social evolution. Biologists used to rely largely on anatomical similarities and differences to infer how animals had changed over time. Now that we can consult their DNA, we often learn that many outwardly similar species are unrelated, while others that look different have a common ancestor. Many of today's biologists would say that our knowledge of any species is therefore incomplete until we have worked out its genetic code.

This is the reason we have chosen to focus on social logic. For social anthropologists and archaeologists, the printout of any society's logic would be analogous to having its DNA profile. When we do not understand society's changing premises, we are left with unanswered questions. Did states with divine kings arise from rank societies where sacred authority was preeminent? Did secular kingdoms arise from rank societies where military force was uppermost? Or could any type of monarchy arise from any type of rank society?

UPDATING ROUSSEAU

Rousseau held that our ancestors were born without sovereign masters, governments, or laws, and that the only differences among them lay in their strength, agility, and intelligence. Those inequalities were authorized by Natural Law. Most later inequalities resulted not from nature but from the actions of society itself.

Today we suspect that our Ice Age forebears were not wholly without masters or laws. They almost certainly believed themselves to have been the creations of celestial spirits, powerful masters who gave men laws of social behavior. Most likely our ancestors also believed that the first humans had abilities beyond ours. Those "old ones" had taken on the role of betas in society's dominance hierarchy and, when treated properly, would intercede on their descendants' behalf with the alphas of the spirit world.

Ice Age people lived on foods whose pursuit tended to keep societies small and mobile. Because fluctuations in the food supply might force some families to forage in the territories of others, our ancestors could not afford to have hostile neighbors. Foragers, we have seen, are not only diplomatic, but actually make neighbors into honorary kinsmen. They do this by creating partners with whom they exchange such things as magical names, food, or gifts. Such partnerships allow one family to host another in times of need, just as if they had been related by blood or marriage.

The logic of small-scale foragers has its own first principles. The following would be typical:

There is an invisible life force within us.
Certain spirits, places, and objects are sacred.
Individuals differ in virtue.
Generosity is one of those virtues.
Older, initiated people tend to be more virtuous than younger, uninitiated people.

Later arrivals in a territory are obliged to defer to earlier arrivals.
Our way of life is inherently superior to that of our neighbors.

Despite the widespread nature of such first principles, most anthropologists would not argue that they are encoded in our genes. Generosity is a widespread principle among hunters and gatherers, yet constant social pressure must be applied to ensure that individuals continue to be generous. Such pressure would not be necessary if there were genes for generosity.

The secondary premises that grew out of the first principles were not as widely shared as the latter. For example, most foragers agree that humans differ in virtue, but they frequently disagree on which specific behaviors make individuals more virtuous. Such variations are the raw material for ethnic diversity, long-term social change, and greater inequality.

Some foragers, for example, considered sharing so important that they declined to store food lest they be accused of hoarding. Such behavior is often associated with immediate-return economic strategies. Other foragers had delayed-return strategies that allowed for drying, smoking, and storing food, and even some modest engineering of the environment. To avoid accusations of hoarding, they threw feasts at which foods were shared.

In some parts of the world delayed-return foraging probably set the stage for agriculture. For its part, feasting conferred increased respect on the host. A commitment to reciprocity meant that unreciprocated generosity could translate into debt. Gift-giving could either keep the playing field level, or be manipulated to achieve the opposite result.

It would be useful to know the circumstances under which such manipulation occurred. A decline in sharing may be indicated by privatized storage units, which archaeologists have detected in Near Eastern villages occupied 8,000 years ago.

In parts of the Ice Age world foragers went beyond exchange and food sharing. The archaeological evidence suggests that some of them created large, permanent groups of people who considered themselves related, whether it was true or not. Early clans may have made use of the templates for patrilineal or matrilineal society that, as we saw earlier, sometimes could be found in the gender makeup of forager camps.

Most clanless foragers worked hard to treat everyone as equals. This ethic usually persisted within one's clan but did not always extend to other clans. Some clans, for example, felt a sense of intellectual property and sought to keep their rituals secret from others. This need to protect secrets may have inspired the first attempts to have leadership pass from father to son. In other cases Clan A was willing to let Clan B perform one of its rituals in exchange for valuables.

Even clanless foragers have been known to save the bones of deceased relatives. The enhanced importance of the ancestors in clans increased this curation of skeletal parts. Some clanless foragers built sweat houses or bachelors' huts; clan-based societies sometimes built men's houses or charnel houses. Such buildings appeared in the Near East 9,000 years ago.

Even clanless foragers practiced initiation and bride service. In societies with larger social units, such rituals came to include even greater exchanges of valuables, not only between families but also between descent groups. The addition of so many levels of ritual behavior helped escalate the archaeological evidence for art, music, and dance.

In some regions having corporate groups created new logical premises. The alleged difference in virtue between bride-givers and bride-takers is one example. Here is a second case where formerly reciprocal exchanges could be converted to sources of inequality.

Finally, the "us versus them" mentality of clans justified raiding. The principle of social substitutability meant that anyone from another group was fair game. Some raiding parties returned with trophy heads. Others returned with captive women and children, turning them into slaves. The groundwork had been laid for larger-scale war.

What clues lead archaeologists to suspect that a prehistoric society possessed clans or ancestor-based descent groups? The clues are multigenerational cemeteries, wooden palisades or masonry defensive walls, men's houses, charnel houses, trophy heads, the saving of skulls from burials, and an increase in the circulation of valuables such as those used in bride-price exchange. The first clues in the Nile Valley appeared even before farming and herding had begun. The evidence was strong in the early agricultural villages of the Near East, Mexico, and Peru.

We cannot assume, however, that clanless foragers represent some kind of "original" society. There are hints that some clanless foragers (the Basarwa, for example) may have had descent groups or clans in the past, only to lose them when they were driven into marginal environments. At the same time, societies such as that of the Andaman Islanders show us that even if a group lives in a relatively lush environment, there is no guarantee that it will develop clans. For all these reasons we should probably view clans or descent groups as one of several alternative social networking strategies rather than as an inevitable second stage of foraging society.

BALANCING PERSONAL AMBITION AND THE PUBLIC GOOD

Rousseau considered the replacement of self-respect with self-love an important moment in the creation of inequality. It now seems obvious, however, that both self-respect and self-love were there from the beginning. The tug-of-war between them may have been one of Ice Age society's most significant logical contradictions.

With the rise of agricultural villages 9,000 years ago in the Near East, 7,000 years ago in Egypt, and 4,000 years ago in Mexico, the environment for self-love had improved. In many parts of the world, however, the adoption of agriculture did not lead immediately to inequality. Lots of societies struck a balance between personal ambition and the public good, and in some regions that balance lasted well into the twentieth century. There are archaeological hints, to be sure, that many of today's achievement-based societies once flirted with greater inequality. Most of those flirtations, however, ended with a return to egalitarian behavior.

What achievement-based societies excelled at was providing ambitious individuals (those who, in Rousseau's words, "desired to be thought of as superior") with acceptable ways of increasing their prestige. Those ways included prowess in raiding or head-taking, skill in entrepreneurial exchange, or sponsorship of increasingly important rituals. While all these paths could lead to renown, prominent individuals were not allowed to become a hereditary elite. They could serve as role models for their children but could not guarantee them the same prestige.

Let us look first at the taking of scalps or heads. Some idealistic anthropologists have chosen to downplay such violence as a path to renown, but it was often celebrated in native memory. "Once we had leaders who lined the walls of our men's house with enemy skulls," some tribes lamented, "but now we are reduced to squabbling like girly men."

An interesting aspect of achievement-based society is the not-infrequent link between raiding and exchange. The tee cycle of the Enga shows us that war could be changed from blood feuds to a means of profiting from war reparations. The escalation of mokas, potlatches, and feasts of merit shows us that competitive exchange could fill the vacuum left by the colonial suppression of raiding.

Exchange, to be sure, is unlikely to produce captives that one can turn into slaves. Sometimes, however, it produces debtors that one can force into servitude. Differences in expertise at accumulating and giving away valuables can also divide communities into Big Men, ordinary men, rubbish men, and "legs."

One of the most common paths to renown involved climbing a ladder of ritual achievement. A Tewa man could rise from Warm Clown to Fully Made Person. A Mandan woman could rise from Goose society to White Buffalo Cow society. An Angami Naga could rise to the position of holy man. What none of these high achievers could do was bequeath their renown to their children.

Many Americans will find familiar the logic of achievement-based societies. All men are created equal. Work hard, play by the rules, and anyone can grow up to be prominent. If one provides one's children with privileges they have not really earned, they will be so spoiled that they will get their own reality TV show.

The difference is this: the United States had to fight a Revolutionary War to get rid of hereditary aristocracy and never did figure out how to reduce disparities in wealth. Achievement-based societies, on the other hand, usually pressured all of their members to give away the valuables they had accumulated.

By what date did societies first show signs of achievement-based leadership? Perhaps 9,000 years ago in the Near East, 4,000 years ago in the Andes, and 3,500 years ago in Mexico. And what would be some of the clues? Archaeologists look for the building of men's houses, either the larger and more inclusive type or the smaller and more exclusive type. They also look for accumulations of trade items that might be used in entrepreneurial exchange. They analyze residences and burials carefully, and unless they find convincing evidence that certain families' children were entitled to sumptuary goods, they are likely to conclude that any obvious differences in prestige were achieved, not inherited.

Archaeologists examine as many of a society's villages as they can, looking for any evidence that hamlets were obliged to contribute tribute or labor to a larger village nearby. When no such evidence appears, an achievement-based society is indicated. Archaeologists also try to evaluate any evidence for monument building, with the caveat that an occasional plaza, stone monument, or massive slit-gong might be evidence for achievement rather than hereditary leadership.

How did the old hunter-gatherer logic come to be changed, creating routes to renown? Even foragers considered some individuals more virtuous than others and believed that one could increase one's virtue over a lifetime. Building on this principle, many village societies created a series of formal steps to increase one's virtue through the learning of sacred lore.

Another route, using entrepreneurial exchange, was created by manipulating three principles we saw among foragers: (1) Generosity is good; (2) Exchanges of gifts create social bonds; and (3) The farther away one's trade goods come from, the more impressed one's peers will be. Some achievement-based societies, such as the Enga, tried to keep exchanges equal, using principles such as "Give one pig and one pig only." Others, such as the Melpa, decided that giving one's neighbors more pigs than they could repay made one more generous (and hence more virtuous) than they.

Once the latter principle was accepted, embarrassing one's rivals with spectacular gifts became an acceptable path to renown. An unanticipated consequence of competitive exchange was that whole families and clans might be pressured into bankrolling an aspiring Big Man. If he were defeated by a rival, they could kiss their investment good-bye.

The loss of face created by asymmetrical exchange could lead to blood feuds, and blood feuds could increase the scalping and head-hunting. Many societies believed that the taking of a head could add to one's life force. Leading warriors into combat, counting coup, or returning with captives or body parts thus became another route to prestige.

Achievement-based societies had great stability. At various times and places in the ancient world, however, self-love persisted until a hereditary elite arose. We have seen that this phenomenon was not the inevitable outcome of population growth, intensive agriculture, or climatic improvement, even though all those factors could create a favorable environment for inequality. The key process involved one group of human agents battling for greater privilege, while other agents resisted with all the strength they could muster.

Even when one segment of society succeeded in achieving elite status, the struggle was not necessarily over. Some societies, such as the Kachin and the Konyak Naga, cycled between hereditary rank and achievement-based society for decades.

Archaeologists have proposed several scenarios for the creation of hereditary rank. Most take as their starting point a society that already had a history of achieved inequality, but we do not consider this a prerequisite. At least a few societies might have gone from egalitarian to ranked through the use of debt slavery, without spending much time in a phase of achievement-based villages. If that is the case, it will one day be confirmed by archaeologists.

In those cases where rank society did develop out of achievement-based society, there were many preexisting inequities that could serve as raw material. Included were the differences in prestige between Big Men and rubbish men; between people who had climbed the ritual ladder and those who had not; between the clan that arrived first and everyone else; and between the man chosen for success by a demon and lesser men.

Another strategy for achieving rank was the aforementioned use of debt, which turned needy clan members into servants and neighbors into slaves. Debt could result from exorbitant bride-price, loans to aspiring Big Men, excessive war reparations, or the desperate cries of impoverished kinsmen. It was a route built on the principle that failure to repay a gift or loan made one less virtuous.

One of the interesting facts of hereditary rank was that it could be created even by hunters and gatherers such as the Nootka. Neither slavery nor aristocracy, in other words, had to wait until agriculture had arisen.

What are the archaeological clues for the appearance of rank society? That is not as easy a question as it sounds, because rank came in so many forms. One clan might be ranked above others. One lineage within each clan might be considered a chiefly lineage. There might be a continuum of rank, based on genealogical distance from the chief. There might be stepwise gradations of nobility, a landed gentry, and commoners. And, as if this diversity were not enough, there is also Renfrew's continuum from individualizing to group-oriented rank societies.

Archaeologists should thank their lucky stars for individualizing rank societies, the ones in which the children of the elite get buried with sumptuary goods, and the chief's corpse gets bundled, smoked, or surrounded by sacrificed servants. They should also be grateful for all the symbolically charged pottery, goldwork, and jade exchanged by noble families. At the regional level, they should be thankful for archaeological evidence that chiefly centers grew by attracting new followers, or were surrounded by satellite villages to whom they sent brides.

In rank societies, temples dedicated to celestial spirits often replaced the men's house. Even in group-oriented rank societies, where elites generally refrained from flamboyant displays, chiefly families often lived in bigger houses with greater storage facilities and more evidence of trade goods.

Rank clearly represents a loss of equality, but let us play the devil's advocate. Was rank really such a bad thing? Don't lots of species have a dominance hierarchy, and doesn't it provide stability to their society? In fact, don't our closest primate relatives have pecking orders?

They do, but with an important difference. It is not predestined from the moment of birth that a given chimpanzee will become an alpha or a beta. Having an alpha parent may increase the likelihood, but in the end an individual's position in the hierarchy is the result of his or her interactions with other individuals. And any chimp's position can rise or fall over time.

Human rank societies are different. The child of great Ang parents is born to be a great Ang, no matter how short of talent he or she may be. The child of commoner parents will never become a great Ang, no matter how clever he or she is. The ability to negotiate one's position in rank society is much more limited than in a chimpanzee troop.

There are confrontational interactions in rank society, to be sure, but they are usually between rivals of high rank. Chiefly polygamy leads to situations in which a number of heirs have roughly similar ranks. Some of the bitterest competition is between noble siblings, half siblings, and first cousins.

Another set of violent confrontations involves territorial expansion. Both chimpanzee troops and chiefly human societies like to take territory away from their neighbors. Both also prefer ambushes and numerical superiority. Some aggressive chiefs, however, dare to take on larger enemy forces if they feel that their military tactics are superior. A number of Shaka's greatest victories came when his troops were outnumbered.

Among rank societies, war became a tool for chiefly aggrandizement. When that aggrandizement simply meant the acquisition of titles (as in parts of Samoa), it did not necessarily change the basic principles of society. When aggrandizement meant the acquisition of land (as in Madagascar and Hawai'i), it could produce territories too large for the management principles of rank society. That set the stage for the political hierarchy characteristic of kingdoms.

Many of the earliest kings, in the course of changing the way they administered their territories, created new strategies. Instead of continuing to move his residence so that all provinces could share in his support, the Hawai'ian king appointed a trusted governor for each province. Instead of letting each ethnic group provide its own age regiments, Shaka created state-level regiments that were loyal only to him. Instead of appointing their brothers to administer parts of their realm, some Egyptian kings chose talented commoners who were less likely to usurp the throne.

The first kingdoms or oligarchic states appeared 5,000 years ago in Egypt and Mesopotamia and 2,000 years ago in Mexico and Peru. We find it hard to date the moment of state formation, because the creation of a state often required several generations of aggressive rulers. And despite all the similarities we have seen in first-generation states, they were neither common nor inevitable. As late as the twentieth century, many parts of the world still displayed nothing more complex than rank societies.

What are the clues that a kingdom has been created? At the regional scale, archaeologists look for signs that the political hierarchy had at least four levels, the upper three of which featured administrators. They look for the standardized temples of a state religion, as well as for secular buildings whose ground plans reflect councils or assemblies. At the capital they look for palaces built by corvée labor and tombs with sumptuary goods appropriate for royalty. At Level 2 administrative centers there may be smaller versions of such residences and tombs, often displaying the standardized architecture of a top-down administration. Another clue would be workers' receipt of rations doled out with standardized bowls, griddles, or redeemable tokens. Sometimes the archaeologist's task is made easier by a kingdom's use of writing or art to convey the agenda of its leaders.

Few of the rulers who created kingdoms were content with the territories they controlled. Whenever a new state was surrounded by weaker neighbors, the temptation to expand was great. Sometimes, as in the Mexican state of Oaxaca, this expansion set off a chain reaction that created multiple fortified kingdoms. In other cases, as on the north coast of Peru, expansion created a multiethnic empire. The key to expansion lay in knowing which neighbors were vulnerable and which were best left alone.

Who created the world's first empire? While many archaeologists would point to Sargon of Akkad, he may have received more credit than he deserves. An earlier king, Lugal-zagesi, claims to have held sway from Mesopotamia to the Mediterranean. And even before Lugal-zagesi rose to power, some Egyptian kings may have subjugated the whole region from Nubia to the Southern Levant.

Empires, in other words, are probably more than 4,300 years old. And along with empires came ethnic stereotyping, an escalation of simpler societies' long-standing ethnocentrism. The precedent for racial, religious, and ethnic intolerance had been set.

Early kingdoms and empires did more than this, of course. Many state regimes took away whatever vestiges of equality the individual commoner had left. In the Aztec state, even commoners who cultivated cotton were forbidden to wear cotton mantles. Sumerian law restricted commoner marriage to one man and one woman, giving later societies the impression that monogamy was a divinely sanctioned norm. The Sumerians also strengthened economic inequality among commoners, increasing the likelihood that it would endure even if hereditary privilege were to disappear.

Finally, empires took away the freedom of other societies by turning them into subject colonies. To be sure, the commoners in those societies had been treated as an underclass even before they were colonized; it was their elite who wound up losing the most. Sometimes conquered leaders were mollified with gifts, or they were allowed to participate in the joint rule of their former territories.

We have left the topic of colonialism until now because few subjects evoke more passion from today's anthropologists. That field has had a long-standing love affair with political correctness, and many anthropology courses preach that colonialism is evil and that resistance to colonialism is good. So pervasive is this mantra that many of today's professors refuse to assign the anthropological literature written in Queen Victoria's era, even those works considered classics. Some go so far as to accuse the nineteenth-century social anthropologists of being complicit in colonialism, since few of them vigorously denounced it.

This is political correctness times ten. Colonialism was created neither by anthropologists nor by Queen Victoria. It is at least 4,300 years old, the product of kings who sought to add land and tribute to their realms. The Sumerians, Akkadians, Assyrians, Hittites, Greeks, Romans, Moors, Aztec, and Inca did not learn their craft from anthropologists, and most of their leaders make Queen Victoria sound like Mother Teresa.

Roman archaeologists do not refuse to read Caesar's commentaries on the grounds that he was "the tool of a colonial power." Latin Americanists do not ignore the 1580 *Relaciones Geográficas* on the grounds that the Spaniards writing them had colonized Mexico. One can thus oppose the phenomenon of colonialism without trashing every author who lived in an empire.

WHAT IF FORAGERS WERE IN CHARGE?

Archaeologists are frequently asked two questions about inequality. One, which we have tried to answer, is: How did it arise in the first place? The second is: How can we get rid of it?

Rousseau had his own ideas about the second question. He believed that people could only be happy and free in a community simple enough to be intelligible to them and small enough to enable them to take a full and equal part in its government. In a huge society with a complex economy, there would, out of necessity, be hierarchy and inequality; the majority of what Rousseau called "passive citizens" would be controlled and exploited by the "active few." Some of Rousseau's readers took this to mean that hereditary privilege in eighteenth-century France could only be overturned by a bloody revolution.

The perspective taken in this book, however, allows for alternatives to bloody revolution. If inequality is the result of incremental changes in social logic—and if those changes can be reconstructed—might we not be able to return society to equality just as incrementally, beginning with the most recent changes and working back?

If inequality could be reversed by identifying and retracing its steps, at least some of the information would need to come from archaeology and social anthropology. That fact should provide both fields with incentives to work together.

We once broached this subject with Scotty MacNeish, an archaeologist who had spent 40 years studying social evolution. How, we wondered, could society be made more egalitarian? After briefly consulting his old friend Jack Daniels, MacNeish replied, "Put hunters and gatherers in charge."

We are not sure whether the suggestion came from Jack or Scotty, but it gave us something to think about. Putting hunter-gatherers in charge would reduce inequality overnight. It would, to be sure, require a bit of getting used to, because modern society has eliminated many behaviors that foragers took for granted.

Let us briefly consider what our life would be like if we were to leave it in the hands of egalitarian hunter-gatherers or achievement-based farmers. To begin with, there are a few things that probably would not change. Even after our society had been turned over to the people mentioned earlier, a certain degree of sexism and age-based discrimination would remain. Not all egalitarian societies believed that women had the capacity to be as virtuous as men. And few of them considered young men to possess the virtue of older men.

Our society would also retain its ethnocentrism. Our treatment of other groups, however, would no longer include religious proselytizing. Foragers and achievement-based farmers believed that each ethnic group had been created by different celestial spirits, received its own instructions for living, honored its own ancestors, and could not expect other groups to share its beliefs. If our neighbors' dress, religion, and behavior were different from ours, it would not be because they were wicked but because their origins were different.

Our society's tolerance of variation would extend to marriage. A man with two or more wives, a wife with two or more husbands, or even a foursome such as the one in *Bob and Carol and Ted and Alice* would be accepted. We would permit same-sex weddings, such as those involving Native American "two-spirit" people. Marriage would not be seen as a match made in heaven but as an economic partnership in which maximum flexibility was desirable.

Since many foragers practiced infanticide, our new leaders would not outlaw abortion. Because of their belief in reincarnation (a view that survives even among twenty-first-century Americans), foragers felt that every "spirit child" would have multiple opportunities to be born.

Tribal societies had no laws preventing child labor. For our teenagers there would not be as many hours of video games and hanging out at the mall, just lots of chores. It is likely, however, that our teenagers' frenzied music and dancing would bring on the same awe-inspiring high that tribal societies experienced.

Foragers believed that success depended partly on skill and partly on magic. If you think that anything has changed, watch a dugout full of baseball players putting on their "rally caps" to influence the outcome of a game.

In fact, despite their pragmatism, hunters and gatherers saw no contradiction in combining magic, science, and religion. Our belief in the separation of church and state would surprise them. At the same time, whenever their cosmology interfered with the adoption of a useful scientific or technological innovation, they would change the cosmology.

Foragers had an ethic of sharing that would alter business as we know it. They would never allow CEOs to earn thousands of times what assembly-line workers earn. Achievement-based villagers, for their part, would pressure management into throwing huge feasts for the workers and their families. They would also insist on a safety net for the less fortunate, such as the Tewa distribution of food to poor families.

Hunters and gatherers would admire philanthropists. At the same time, they would keep those generous millionaires from getting too pleased with themselves. They would rely on sarcastic comments such as, "You call that a charitable donation? The check was hardly worth cashing."

As for people who have the opposite problem—those who have accepted so much from others that they cannot pay it back—achievement-based villagers would have a solution. Such people would be turned into servants or slaves, forced to work off their debt through hard labor. Don't tell Master Card.

Then there are thieves who take others' property with no intention of returning it. Traditional foragers reacted angrily to theft and had little patience with repeat offenders. They believed in capital punishment and had no concept whatsoever of long-term imprisonment. If it were left up to the Basarwa, Bernie Madoff would simply have been lured into the wilderness and shot with poisoned arrows.

How would foragers handle the problem of illegal immigration? They would establish hxaro exchanges or namesake partnerships with as many families on the other side of the border as possible. When times were hard, they would allow those partners to share in the bounty of their territory. On the other hand, strangers who showed up without having established a prior relationship might be driven away.

Our drug policies would change. Many foragers and small-scale horticulturalists used narcotic and hallucinogenic plants, so they would not believe in criminalizing them. At the same time, they would not want to see drugs used merely for "recreation." They considered them sacred plants, because they possessed the power to open a window into the spirit world. Such drugs would therefore be used exclusively in the context of ritual.

While many achievement-based villagers were willing to massacre their enemies, burn their villages, poison their wells, and turn them into slaves, they never engaged in anything resembling nation building. They found it implausible that an enemy society, with its different supernatural ancestors and social logic, could be turned into a replica of their own.

Consider, for example, Mesopotamia. We have seen that it developed rank societies more than 7,000 years ago. It has had monarchies or oligarchic states for at least 5,000 years. Never once in all its millennia of ensís, lugals, sheikhs, emirs, sultans, warlords, and military dictators has Mesopotamia voluntarily created a democracy.

Societies do not embrace forms of governance that are incompatible with their social logic, especially when that governance is imposed from the outside. The aggressor doing the imposing usually finds that maintaining the illusion of democracy requires an effort that makes head-hunting and pincushioning seem rational by comparison. That is why so many empires relied on joint rule instead.

In addition to the high cost of forcing every other society to be just like ours, there is a compelling reason not to do so. One day we may discover that preserving the world's reservoir of diverse social logic was just as important as preserving its biodiversity.

THE MYSTERIOUS FIRST PRINCIPLES

Imagine, for the sake of argument, that we have just reversed the premises that led to social inequality. Still looming before us would be the first principles of social logic. Does their widespread nature mean that they are innate to our species? Or is it simply the case that the limitations and biases of human logic are widely shared?

From time to time one prominent scientist or another has argued that our society would work better if we grounded our values in science and logic instead of religion. This idea appeals to anyone fed up with religion's intolerance of diversity, or its frequent disdain for science.

The problem with the idea is its underlying assumption that our ancestors began with logic and acquired religion later. The fact is that even the first principles of hunter-gatherer logic include notions of the sacred. And when we search for the source of those first principles, we do not uncover an earlier, even more primordial logic; instead we encounter a cosmology filled with the instructions of celestial spirits. Cosmologies are built on sacred propositions, unchallenged despite the fact that there is no empirical evidence to support them.

Even among the most pragmatic hunters and gatherers, this is where logic ends. Cosmological propositions can be validated only by strong emotions, because they defy validation by logic or evidence. And while anthropologists doubt the existence of genes for religion, no one doubts that we have genes for emotions.

Our emotions also play a role in the subordination of our self-interest for the good of the group. Some evolutionary biologists have a problem when individual humans subordinate their self-interest in this way. After all, subordination of self-interest fits poorly with the notion that natural selection operates at the individual, rather than the group, level. What is not clear is the degree to which individuals handicap themselves when they subordinate self-interest by giving away food or valuables. Their generosity will almost certainly result in their being considered superior in virtue, and this superiority can result in more mates and offspring than if they behaved selfishly.

In other words, as long as our behavior results in leaving behind more offspring, it may not matter whether that behavior was directed by genes, by logic, or by unverifiable sacred propositions.

INEQUALITY AND RESISTANCE

There can be no more exciting story for an archaeologist than the way new societies were created from old. A system based on arbitrary premises, in theory, has the potential to give rise to thousands of different societies, and so it did. As we have seen, however, five or six ways of organizing people work so well that strikingly similar societies have appeared in different regions of the world. We recognize those societies in the archaeological record, whether they arose in Africa, Asia, or the Americas.

The similarities among societies in different parts of the world were not lost on early anthropologists. Some even assumed that those societies constituted an inevitable sequence of stages, through which all human groups had passed on their way from foraging to civilization. No one believes such a thing today. In fact, some of today's anthropologists would even deny that recognizable types of societies exist. Such denials are every bit as misguided as our predecessors' belief in a monolithic sequence of stages.

Today we know that even when two regions happened to go through similar stages, their social history did not proceed at the same rate. Just look, for example, at the Near East and Mexico. Both regions began to domesticate plants at the end of the Ice Age, perhaps 10,000 years ago. The Near East gave rise to villages with ritual houses 9,000 years ago. The process took longer in Mexico, in part because early corn was not as productive as wheat and barley. There the first villages with ritual houses did not appear until 3,500 years ago.

Once Mexico had developed achievement-based village societies, however, the transition to stratified societies and kingdoms was much more rapid. The first monarchies or oligarchic states in Mesopotamia arose between 5,500 and 5,000 years ago, some 4,000 to 3,500 years after the first villages. The first monarchies or oligarchic states in Mexico arose 2,000 years ago, barely 1,500 years after the first villages.

Why did it take states more than twice as long to develop in the Near East? Did military force play a greater role in Mexico, hastening the shift from "traditional" to "stratified" society in Goldman's terms? Were the efforts to preserve a level playing field more successful in Southern Mesopotamia, prolonging the period of achievement-based leadership? What roles did sacred authority, expertise, and military prowess play in speeding or slowing social change? Were societies with exclusionary ritual houses more likely to give rise to hereditary elites than those whose ritual houses were open to all?

Archaeologists will not be able to answer these questions until they have better ways of reconstructing the logic of ancient societies. We would like to be able to work out scenarios for a wide variety of societies, providing plausible explanations for why certain varieties appeared so frequently and lasted so long. We suspect, for example, that complex societies could only arise after changes in logic had reduced the pressure to suppress self-interest. Some families or descent groups were then free to place their less successful neighbors in a position of disadvantage. They justified their superiority by claiming special relationships with the very beings who had given humans their laws of behavior in the first place.

We are struck, however, by the fact that each escalation of inequality required the overcoming of resistance. There seems to have been an ongoing struggle between those who desired to be superior and those who objected. That is undoubtedly why some of our most complex and stratified societies formed in a crucible of intense competition among clans, chiefly lineages, and ethnic groups.

Man is born free, Rousseau declared, yet we see him everywhere in chains. We have our ancestors to thank for that. They had dozens of chances to resist inequality, but they did not always have the resolve. We can forgive them for admiring virtue, entrepreneurial skill, and bravery. We simply wish they had not accepted the idea that those qualities were hereditary.

American society, of course, has abolished hereditary privilege, but today we make entertainers and professional athletes into an aristocracy. Many of us go deeply in debt to emulate them, buying countless toys that we do not need. Our celebrities surround themselves with the equivalent of chiefly entourages; we make do with yardmen and undocumented nannies.

Forbidden from mutilating their subjects like Bemba chiefs, American aristocrats settle for hitting their servants with cell phones and coat hangers. Celebrities ease into rehab for crimes that would land most of us in jail. Prevented from practicing chiefly polygamy, they accumulate multiple cocktail waitresses instead.

What can the rest of us do to avoid becoming an underclass? We can remember that Natural Law permits inequality only in strength, agility, and intelligence, and we can resist. The Maliyaw subclan could not become Avatip's elite as long as the other subclans fought back. The Bear clan could not become a Hopi aristocracy as long as the other clans objected. The Mandan refused to let certain families accumulate all the tribe's valuables. The Kachin periodically told their thigh-eating chiefs to get lost. And, once in a while, a civilization's passive majority takes back the privileges of the active few.

We may never be entitled to sumptuary goods, but we can work to increase our virtue. And it is no one's fault but our own if we allow our society to create "nobles by wealth." We can resist just as surely as any self-respecting !Kung would do. So the next time a pampered star tells you that his last film made him $20 million, tell him which charity to give it to.

Then explain that you have not actually seen the film, but that you and your dog have discovered that the DVD makes a great Frisbee.

NOTES

SOURCES OF ILLUSTRATIONS

INDEX

Notes

PREFACE

ix Numerous editions of Jean-Jacques Rousseau's essays are available; two examples are *A Discourse on Inequality,* with an introduction by Maurice Cranston (Penguin, New York, 1984) and *The Social Contract,* with an introduction by Charles Frankel (Hafner, New York, 1951).

x Christopher Boehm, *Hierarchy in the Forest: The Evolution of Egalitarian Behavior* (Harvard University Press, 1999).

x Edward O. Wilson, *On Human Nature* (Harvard University Press, 1978).

xii Robin Fox, "One World Archaeology: An Appraisal," *Anthropology Today* 9 (1993): 6–10.

CHAPTER 1: GENESIS AND EXODUS

3 In 2009 the *Proceedings of the National Academy of Sciences* devoted part of its volume 106, number 38, to the special feature "Out of Africa: Modern Human Origins." This consisted of nine articles by Richard G. Klein, Ian Tattersall, Timothy D. Weaver, J. J. Hublin, Michael P. Richards, Erik Trinkaus, John F. Hoffecker, G. Philip Rightmire, and other leading experts on the origins of modern humans. See also Paul Mellars, "Why Did Modern Human Populations Disperse from Africa ca. 60,000 Years Ago? A New Model," *Proceedings of the National Academy of Sciences* 103 (2006): 9381–9386. Early dispersals of humans from Africa are also discussed by Ofer Bar-Yosef and Anna Belfer-Cohen in "From Africa to Eurasia—Early Dispersals," *Quaternary International* 75 (2001): 19–28, and Michael Bolus and Nicholas J. Conrad in "The Late Middle Paleolithic and Earliest Upper

Paleolithic in Central Europe and Their Relevance for the Out of Africa Hypothesis," *Quaternary International* 75 (2001): 29–40.

4 Neanderthal DNA has been analyzed by an international team led by geneticist Svante Pääbo. See Richard E. Green et al., "A Draft Sequence of the Neandertal Genome," *Science* 328 (2010): 710–722.

4 Paul Mellars, *The Neanderthal Legacy* (Princeton University Press, 1996).

4 C. Loring Brace, "'Neutral Theory' and the Dynamics of the Evolution of 'Modern' Human Morphology," *Human Evolution* 20 (2005): 19–38.

4 Dorothy A. E. Garrod and Dorothea M. A. Bate, *The Stone Age of Mount Carmel,* vol. 1 (Clarendon Press, Oxford, 1937); Theodore D. McCown and Arthur Keith, *The Stone Age of Mount Carmel,* vol. 2 (Clarendon Press, Oxford, 1932–1934); Erella Hovers, Shimon Ilani, Ofer Bar-Yosef, and Bernard Vandermeersch, "An Early Case of Color Symbolism: Ochre Use by Modern Humans in Qafzeh Cave," *Current Anthropology* 44 (2003): 491–522. See also Paul Mellars, "Why Did Modern Human Populations Disperse from Africa ca. 60,000 Years Ago? A New Model."

5 Ice Age cooling is discussed by Paul Mellars in *The Neanderthal Legacy* and by Miryam Bar-Mathews and Avner Ayalon in "Climatic Conditions in the Eastern Mediterranean during the Last Glacial (60–10 ky) and Their Relations to the Upper Paleolithic in the Levant as Inferred from Oxygen and Carbon Isotope Systematics of Cave Deposits," in Nigel Goring-Morris and Anna Belfer-Cohen, eds., *More than Meets the Eye: Studies on Upper Paleolithic Diversity in the Near East* (Oxbow Books, Oxford, 2003), 13–18.

5 Neanderthal extinction is discussed by John F. Hoffecker in "The Spread of Modern Humans into Europe," *Proceedings of the National Academy of Sciences* 106 (2009): 16040–16045. Ecological release is discussed by Edward O. Wilson in *On Human Nature* (Harvard University Press, 1978).

6 Kristen Hawkes, "Grandmothers and the Evolution of Human Longevity," *American Journal of Human Biology* 15 (2003): 380–400.

6 Richard G. Klein, "Fully Modern Humans," in Gary M. Feinman and T. Douglas Price, eds., *Archaeology at the Millennium* (Kluwer-Plenum, New York, 2001), 109–135.

6 Hilary J. Deacon and Janette Deacon, in *Human Beginnings in South Africa: Uncovering the Secrets of the Stone Age* (AltaMira Press, Walnut Creek, Calif., 1999), discuss Blombos Cave and Klasies River Mouth.

6 The origins of personal ornamentation are discussed by Francesco d'Errico et al. in "Additional Evidence on the Use of Personal Ornaments in the Middle Paleolithic of North Africa," *Proceedings of the National Academy of Sciences* 106 (2009): 16051–16056.

7 For the burning of *Watsonia* and other fynbos plants, see Hilary J. Deacon and Janette Deacon, *Human Beginnings in South Africa* (1999).

8 Fred Wendorf, Romuald Schild, and Angela Close, eds., *The Prehistory of Wadi Kubbaniya,* vols. 1 and 2 (Southern Methodist University Press, Dallas, 1989). The Wadi Kubbaniya skeleton is described in vol. 1.

9 The Sahul Shelf, the Sunda Shelf, and the colonization of New Guinea, Australia, and Tasmania are described in James F. O'Connell and Jim Allen, "Dating the Colonization of Sahul (Pleistocene Australia-New Guinea): A Review of Recent Research," *Journal of Archaeological Science* 31 (2004): 835–853. See also Andrew S. Fairbairn, Geoffrey S. Hope, and Glenn R. Summerhayes, "Pleistocene Occupation of New Guinea's Highland and Subalpine Environments," *World Archaeology* 38 (2006): 371–386.

10 Elizabeth Culotta, "Ancient DNA Reveals Neanderthals with Red Hair, Fair Complexions," *Science* 318 (2007): 546–547; Rebecca L. Lamason et al., "Slc24a5, a Putative Cation Exchanger, Affects Pigmentation in Zebrafish and Humans," *Science* 310 (2005): 1782–1786.

10 For human adaptation to high altitudes, see Cynthia M. Beall, "Two Routes to Functional Adaptation: Tibetan and Andean High-Altitude Natives," in "In the Light of Evolution I: Adaptation and Complex Design," *Supplement 1 of the Proceedings of the National Academy of Sciences* 104 (2007): 8655–8660; Mark Aldenderfer, "Modelling Plateau Peoples: The Early Human Use of the World's High Plateaux," *World Archaeology* 38 (2006): 357–370.

11 David J. Meltzer, *First Peoples in a New World: Colonizing Ice Age America* (University of California Press, 2009).

11 Edward Vajda, "A Siberian Link with the Na-Dené," *Anthropological Papers of the University of Alaska* 6 (2009): 75–156.

11 The date of 15,000 years ago has recently been confirmed by Michael R. Waters et al., in "The Buttermilk Creek Complex and the Origins of Clovis at the Debra L. Friedkin Site, Texas," *Science* 331 (2011): 1599–1603.

12 John F. Hoffecker, in *A Prehistory of the North: Human Settlement of the Higher Latitudes* (Rutgers University Press, 2005), describes the Gravettians and the site of Gagarino.

13 Ludmilla Iakovleva, in "Les Habitats en Os de Mammouths du Paléolithique Superieur d'Europe Orientale: Les Données et leurs Interpretations," in S. A. Vasil'ev, Olga Soffer, and J. Kozlowski, eds., "Perceived Landscapes and Built Environments," *BAR International Series* 1122 (Archaeopress, Oxford, 2003), 47–57, discusses the sites of Kostienki and Sungir.

13 Olga Soffer, in *The Upper Paleolithic of the Central Russian Plain* (Academic Press, 1985), discusses the site of Mezhirich.

14 J. G. D. Clark, *Prehistoric Europe: The Economic Basis* (Stanford University Press, 1966); Paul G. Bahn, *Cave Art: A Guide to the Decorated Ice Age Caves of Europe* (Frances Lincoln, London, 2007).

15 Raymond C. Kelly, *Warless Societies and the Origin of War* (University of Michigan Press, 2000).

17 Leslie G. Freeman, "Caves and Art: Rites of Initiation and Transcendence," in *Anthropology without Informants: Collected Works in Paleoanthropology* (University Press of Colorado, Boulder, 2009), 329–341.

CHAPTER 2: ROUSSEAU'S "STATE OF NATURE"

20 Commonsense opinions concerning the usefulness of living hunter-gathers for understanding the past can be found in Ernest S. Burch Jr., "The Future of Hunter-Gatherer Research," in Ernest S. Burch Jr. and Linda J. Ellanna, eds., *Key Issues in Hunter-Gatherer Research* (Berg, Oxford, 1994), 441–455; Richard B. Lee, "Art, Science or Politics? The Crisis in Hunter-Gatherer Studies," *American Anthropologist* 94 (1992): 31–54; Susan Kent, *Cultural Diversity among Twentieth-Century Foragers: An African Perspective* (Cambridge University Press, 1996).

21 Overviews of the peopling of the American Arctic can be found in Don E. Dumond, *The Eskimos and Aleuts: Revised Edition* (Thames and Hudson, 1987); David Damas, ed., *Handbook of American Indians, vol. 5: Arctic* (Smithsonian Institution Press, Washington, D.C., 1984).

21 Overviews of Eskimo society can be found in Ernest S. Burch Jr.'s books, *The Eskimos* (Macdonald, London, 1988), and *Alliance and Conflict: The World System of the Iñupiaq Eskimos* (University of Nebraska Press, 2005). (The references in both books include the pioneering works of Knud Rasmussen.)

23 Kaj Birket-Smith, *The Caribou Eskimos: Material and Social Life and Their Cultural Position* (Gyldendal, Copenhagen, 1929).

24 Asen Balikci, *The Netsilik Eskimo* (The Natural History Press, Garden City, N.Y., 1970). Netsilik seal-sharing was originally described by Frans Van de Velde in "Les Règles du Partage des Phoques pris par la Chasse aux Aglus," *Anthropologica* 3 (Wilfred Laurier University Press, Waterloo, Ontario, 1956), 5–14.

27 Mark Stiger, "A Folsom Structure in the Colorado Mountains," *American Antiquity* 71 (2006): 321–351; Edwin N. Wilmsen, *Lindenmeier: A Pleistocene Hunting Society* (Harper & Row, New York, 1974); Edwin N. Wilmsen and Frank H. H. Roberts Jr., "Lindenmeier, 1934–1974: Concluding Report on Investigations," *Smithsonian Contributions to Anthropology* 24 (1978).

30 Early studies of the Basarwa are found in Lorna Marshall, "The Kin Terminology of the !Kung Bushmen," *Africa* 27 (1957): 1–25, and "!Kung Bushmen Bands," *Africa* 30 (1960): 325–354; George B. Silberbauer, *Bushman Survey Report* (Bechuanaland Government Press, Gaborone, 1965); Rich-

ard B. Lee, "What Hunters Do for a Living, or, How to Make Out on Scarce Resources," in Richard B. Lee and Irven DeVore, eds., *Man the Hunter* (Aldine, Chicago, 1968), 30–48, and "!Kung Bushmen Subsistence: An Input-Output Analysis," in Andrew P. Vayda, ed., *Environment and Cultural Behavior* (Natural History Press, New York, 1969), 47–79. A classic overview of the !Kung is found in Richard B. Lee, *The !Kung San* (Cambridge University Press, 1979), which includes a description of arrow exchange. Hxaro exchange is described in Pauline Wiessner, "Hxaro: A Regional System of Reciprocity for Reducing Risk among the !Kung San" (PhD diss., University of Michigan, 1977). The anthropology and archaeology of Basarwa camps are discussed in John E. Yellen, *Archaeological Approaches to the Present: Models for Reconstructing the Past* (Academic Press, New York, 1977). An overview of !Kung cosmology can be found in Lorna J. Marshall, "Nyae Nyae !Kung Beliefs and Rites," *Peabody Museum Monographs* 8 (Harvard University, 1999).

35 See James Woodburn's "An Introduction to Hadza Ecology," in Richard B. Lee and Irven DeVore, eds., *Man the Hunter* (Aldine, Chicago, 1968), 49–55; "Stability and Flexibility in Hadza Residential Groupings," in Richard B. Lee and Irven DeVore, *Man the Hunter,* 103–110; "Ecology, Nomadic Movement and the Composition of the Local Group among Hunters and Gatherers: An East African Example and Its Implications," in Peter J. Ucko, Ruth Tringham, and Geoffrey W. Dimbleby, eds., *Man, Settlement and Urbanism* (Duckworth, London, 1972), 193–206; "African Hunter-Gatherer Social Organization: Is It Best Understood as a Product of Encapsulization?" in Tim Ingold, David Riches, and James Woodburn, eds., *Hunters and Gatherers, vol. 1: History, Evolution and Social Change* (Berg, Oxford, 1988), 31–64. See, also, two papers by Kristen Hawkes, James F. O'Connell, and Nicholas Blurton Jones, "Hadza Women's Time Allocation, Offspring Provisioning, and the Evolution of Long Postmenopausal Life Spans," *Current Anthropology* 38 (1997): 551–577, and "Hadza Meat Sharing," *Evolution and Human Behavior* 22 (2001): 113–142.

37 For interesting discussions of the early evolution of human social groups and kinship systems, see Nicholas J. Allen, Hillary Callan, Robin Dunbar, and Wendy James, eds., *Early Human Kinship: From Sex to Social Reproduction* (Blackwell, Oxford, 2008).

38 Marshall D. Sahlins, "The Social Life of Monkeys, Apes, and Primitive Men," in Morton H. Fried, ed., *Readings in Anthropology,* vol. 2 (Thomas Y. Crowell, New York, 1959), 186–199.

38 William S. Laughlin, "Hunting: An Integrating Behavioral System and Its Evolutionary Importance," in Richard B. Lee and Irven DeVore, eds., *Man the Hunter* (Aldine, Chicago, 1968), 304–320.

CHAPTER 3: ANCESTORS AND ENEMIES

40 Raymond C. Kelly, *Warless Societies and the Origin of War* (University of Michigan Press, 2000). For additional reading, see Lawrence H. Keeley, *War Before Civilization* (Oxford University Press, 1996); Steven A. LeBlanc, *Constant Battles* (St. Martin's, New York, 2003); Keith F. Otterbein, *The Anthropology of War* (Waveland Press, Long Grove, Ill., 2009).

41 Fred Wendorf, "Site 117: A Nubian Final Paleolithic Graveyard Near Jebel Sahaba, Sudan," in Fred Wendorf, ed., *The Prehistory of Nubia,* vol. 2 (Southern Methodist University Press, 1968), 954–995; Fred Wendorf and Romuald Schild, "Late Paleolithic Warfare in Nubia: The Evidence and Causes," *Adumatu: A Semi-Annual Archaeological Refereed Journal on the Arab World* 10 (2004): 7–28.

42 A. R. Radcliffe-Brown, *The Andaman Islanders* (Cambridge University Press, 1922).

46 Adolphus P. Elkin, *The Australian Aborigines* (Doubleday-Anchor, Garden City, N.Y., 1964); Ian Keen, *Aboriginal Economy and Society* (Oxford University Press, 2004); M. J. Meggitt, *Desert People: A Study of the Walbiri Aborigines of Central Australia* (Angus & Robertson, Sydney, 1962).

47 H. Ling Roth, *The Aborigines of Tasmania: Second Edition* (F. King & Sons, Halifax, UK, 1899).

48 Baldwin Spencer and F. J. Gillen, *The Native Tribes of Central Australia* (Macmillan & Co., London, 1899), and *The Northern Tribes of Central Australia* (Macmillan & Co., London, 1904).

53 W. Lloyd Warner, *A Black Civilization* (Harper & Bros., New York, 1937).

CHAPTER 4: WHY OUR ANCESTORS HAD RELIGION AND THE ARTS

55 Donald E. Brown, *Human Universals* (McGraw-Hill, New York, 1991).

56 Nicholas Wade, *The Faith Instinct: How Religion Evolved and Why It Endures* (Penguin Press, New York, 2009).

56 Joyce Marcus and Kent V. Flannery, "Ethnoscience of the Sixteenth-Century Valley Zapotec," in Richard I. Ford, ed., "The Nature and Status of Ethnobotany," *Anthropological Papers* 67 (Museum of Anthropology, University of Michigan, 1978), 51–79.

57 Roy A. Rappaport, "The Sacred in Human Evolution," *Annual Review of Ecology and Systematics* 2 (1971): 23–44, and *Ritual and Religion in the Making of Humanity* (Cambridge University Press, 1999).

58 John C. Mitani, David P. Watts, and Martin N. Miller, "Recent Developments in the Study of Wild Chimpanzee Behavior," *Evolutionary Anthropology* 11 (2002): 9–25; Richard Wrangham and Dale Peterson, *Demonic*

Males: Apes and the Origins of Human Violence (Houghton Mifflin, Boston, 1996).

59 Christopher Boehm, *Hierarchy in the Forest: The Evolution of Egalitarian Behavior* (Harvard University Press, 1999).

62 Yosef Garfinkel, in *Dancing at the Dawn of Agriculture* (University of Texas Press, 2003), confirms the link between art and dance by showing how often dancing was represented in prehistoric art.

62 The churinga ilpintira is described by Baldwin Spencer and F. J. Gillen in *The Northern Tribes of Central Australia* (Macmillan & Co., London, 1904).

63 Edward O. Wilson, *On Human Nature* (Harvard University Press, 1978).

CHAPTER 5: INEQUALITY WITHOUT AGRICULTURE

67 Useful introductions to California's Native Americans can be found in A. L. Kroeber, "Handbook of California Indians," *Bulletin 78* (Bureau of American Ethnology, Smithsonian Institution, 1925), and Robert F. Heizer, ed., *Handbook of North America Indians, vol. 8: California* (Smithsonian Institution Press, Washington, D.C., 1978).

68 Jeanne E. Arnold, ed., "Foundations of Chumash Complexity," *Perspectives in California Archaeology* 7 (Cotsen Institute of UCLA, Los Angeles, 2004); Jeanne E. Arnold, "Credit Where Credit Is Due: The History of the Chumash Oceangoing Plank Canoe," *American Antiquity* 72 (2007): 196–209.

70 H. E. Bolton, ed., "Expedition to San Francisco Bay in 1770: Diary of Pedro Fagés," *Publications* 2, no. 3 (1911): 141–159 (University of California Academy of Pacific Coast History); Pedro Fagés, "The Chumash Indians of Santa Barbara," in Robert F. Heizer and M. A. Whipple, eds., *The California Indians: A Sourcebook* (University of California Press, Berkeley, 1951), 255–261.

71 Useful introductions to the Native Americans of the Pacific Northwest can be found in Wayne Suttles, ed., *Handbook of North American Indians, vol. 7: Northwest Coast* (Smithsonian Institution Press, Washington, D.C., 1990).

74 Philip Drucker, "The Northern and Central Nootkan Tribes," *Bulletin 114* (Bureau of American Ethnology, Smithsonian Institution, 1951); Eugene Arima and John Dewhirst, "Nootkans of Vancouver Island," in Wayne Suttles, ed., *Handbook of North American Indians, vol. 7*, 391–411.

76 The Nootka whaling shrine at Jewitt's Lake is described in Peter Nabokov and Robert Easton, *Native American Architecture* (Oxford University Press, 1989), and Eugene Arima and John Dewhirst, "Nootkans of Vancouver Island" (see previous reference).

77 Gary Coupland, Terence Clark, and Amanda Palmer, "Hierarchy, Communalism, and the Spatial Order of Northwest Coast Plank Houses: A Comparative Study," *American Antiquity* 74 (2009): 77–106; Brian Hayden, "The Emergence of Large Villages and Large Residential Corporate Group Structures among Complex Hunter-Gatherers at Keatley Creek," *American Antiquity* 70 (2005): 169–174; Anna Marie Prentiss et al., "The Emergence of Status Inequality in Intermediate Scale Societies: A Demographic and Socio-Economic History of the Keatley Creek Site, British Columbia," *Journal of Anthropological Archaeology* 26 (2007): 299–327; Anna Marie Prentiss et al., "Evolution of a Late Prehistoric Winter Village on the Interior Plateau of British Columbia: Geophysical Investigations, Radiocarbon Dating, and Spatial Analysis of the Bridge River Site," *American Antiquity* 73 (2008): 59–81.

80 Frederica de Laguna, "Tlingit," in Wayne Suttles, ed., *Handbook of North American Indians, vol. 7,* 203–228; George T. Emmons, "The Tlingit Indians," *Anthropological Papers of the American Museum of Natural History* 70 (1991); Aurel Krause, *The Tlingit Indians* (University of Washington Press, Seattle, 1970); Kalervo Oberg, *The Social Economy of the Tlingit Indians* (University of Washington Press, Seattle, 1973).

84 The Tutchone, Tagish, and Teslin are described in June Helm, ed., *Handbook of North American Indians, vol. 6: Subarctic* (Smithsonian Institution Press, Washington, D.C., 1981).

85 Catherine McClellan, "The Inland Tlingit," in Marian W. Smith, ed., "Asia and North America: Transpacific Contacts," *Memoirs of the Society for American Archaeology* 9 (1953): 47–51, and "Inland Tlingit," in June Helm, ed., *Handbook of North American Indians, vol. 6,* 469–480.

CHAPTER 6: AGRICULTURE AND ACHIEVED RENOWN

91 The worldwide literature on the origins of plant and animal domestication is vast. Introductions to that literature can be found in Bruce D. Smith, *The Emergence of Agriculture* (Scientific American Library and W. H. Freeman, New York, 1995); C. Wesley Cowan and Patty Jo Watson, eds., *The Origins of Agriculture: An International Perspective* (Smithsonian Institution Press, Washington, D.C., 1992); Melinda A. Zeder, Daniel G. Bradley, Eve Emshwiller, and Bruce D. Smith, eds., *Documenting Domestication: New Genetic and Archaeological Paradigms* (University of California Press, 2006).

93 Elizabeth A. Cashdan, "Egalitarianism among Hunters and Gatherers," *American Anthropologist* 82 (1980): 116–120.

93 Jacques Barrau, in "L'Humide et le Sec: An Essay on Ethnobiological Adaptation to Contrastive Environments in the Indo-Pacific Area," *Journal of the*

Polynesian Society 74 (1965): 329–346, introduced us to the key plants in New Guinea agriculture. Chapter 7 in Peter Bellwood, *First Farmers: The Origins of Agricultural Societies* (Blackwell, Oxford, 2005), summarizes what is now known of the origins of agriculture in New Guinea.

94 Raymond C. Kelly, *Constructing Inequality: The Fabrication of a Hierarchy of Virtue among the Etoro* (University of Michigan Press, Ann Arbor, 1993). See also Raymond C. Kelly, *Etoro Social Structure: A Study in Structural Contradiction* (University of Michigan Press, 1974).

95 Paula Brown, *The Chimbu: A Study of Change in the New Guinea Highlands* (Schenkmen, Cambridge, Mass., 1972). See also Paula Brown, "Chimbu Tribes: Political Organization in the Eastern Highlands of New Guinea," *Southwestern Journal of Anthropology* 16 (1960): 22–35; Harold C. Brookfield and Paula Brown, *Struggle for Land: Agriculture and Group Territories among the Chimbu of the New Guinea Highlands* (Oxford University Press, 1963).

97 Bruce M. Knauft, *South Coast New Guinea Cultures: History, Comparison, Dialectic* (Cambridge University Press, 1993).

99 Pauline Wiessner and Akii Tumu, *Historical Vines: Enga Networks of Exchange, Ritual, and Warfare* (Smithsonian Institution Press, Washington, D.C., 1998). See, also, Mervyn J. Meggitt, "System and Subsystem: The 'Te' Exchange Cycle among the Mae Enga," *Human Ecology* 1 (1972): 111–123.

101 Marilyn Strathern, *Women in Between: Female Roles in a Male World, Mount Hagen, New Guinea* (Seminar Press, London, 1972).

101 Andrew Strathern, *The Rope of Moka: Big-Men and Ceremonial Exchange in Mount Hagen, New Guinea* (Cambridge University Press, 1971).

105 John H. Hutton, *The Angami Nagas* (Macmillan & Co., London, 1921).

CHAPTER 7: THE RITUAL BUILDINGS OF ACHIEVEMENT-BASED SOCIETIES

111 James P. Mills, *The Rengma Nagas* (Macmillan & Co., London, 1937).

113 James P. Mills, *The Ao Nagas* (Macmillan & Co., London, 1926).

113 Maureen Anne MacKenzie, *Androgynous Objects: String Bags and Gender in New Guinea* (Harwood Academic Publishers, Melbourne, Australia, 1991).

115 Fredrik Barth, *Cosmologies in the Making: A Generative Approach to Cultural Variation in Inner New Guinea* (Cambridge University Press, 1987).

115 Igor Kopytoff, "Ancestors as Elders in Africa," *Africa* 41 (1971): 129–141.

117 Douglas L. Oliver, *A Solomon Island Society* (Harvard University Press, 1955).

CHAPTER 8: THE PREHISTORY OF THE RITUAL HOUSE

122 For a summary of scholarly debates on the origins of agriculture in the Near East, see Michael Balter, "Seeking Agriculture's Ancient Roots," *Science* 316 (2007): 1830–1835.

122 For discussions of Ohalo II, see Nigel Goring-Morris and Anna Belfer-Cohen, "Structures and Dwellings in the Upper and Epi-Paleolithic (ca. 42–10 k BP) Levant: Profane and Symbolic Uses," in S. A. Vasil'ev, Olga Soffer, and J. Kozlowski, eds., "Perceived Landscapes and Built Environments," *BAR International Series* 1122 (Archaeopress, Oxford, 2003), 65–81; Dani Nadel et al., "Stone Age Hut in Israel Yields World's Oldest Evidence of Bedding," *Proceedings of the National Academy of Sciences* 101 (2004): 6821–6826; Ehud Weiss et al., "The Broad Spectrum Revisited: Evidence from Plant Remains," *Proceedings of the National Academy of Sciences* 101 (2004): 9551–9555; Dolores R. Piperno, Ehud Weiss, and Dani Nadel, "Processing of Wild Cereal Grains in the Upper Paleolithic Revealed by Starch Grain Analysis," *Nature* 430 (2004): 670–673.

123 Stefan Karol Kozlowski, ed., "M'lefaat: Early Neolithic Site in Northern Iraq," *Cahiers de l'Euphrate* 8 (1998): 179–273.

123 El-Wad Cave is described in Dorothy A. E. Garrod and Dorothea M. A. Bate, *The Stone Age of Mt. Carmel*, vol. 1 (Clarendon Press, Oxford, 1937). Natufian use of dentalium is discussed by Daniella E. Bar-Yosef Mayer, "The Exploitation of Shells as Beads in the Paleolithic and Neolithic of the Levant," *Paléorient* 31 (2005): 176–185.

126 Ofer Bar-Yosef, B. Arensburg, and Eitan Tchernov, *Hayonim Cave: Natufian Cemetery and Habitation* (Bema'aravo Shel Galil, Haifa, 1974); Anna Belfer-Cohen, "The Natufian Settlement at Hayonim Cave" (PhD diss., Hebrew University, Jerusalem, 1988); Patricia Smith, "Family Burials at Hayonim," *Paléorient* 1 (1973): 69–71.

126 Dorothy A. E. Garrod and Dorothea M. A. Bate, "Excavations at the Cave of Shukbah, Palestine, 1928," *Proceedings of the Prehistoric Society for 1942*, n.s., vol. 8 (1942): 1–20.

127 Jean Perrot, "Le Gisement Natoufien de Mallaha (Eynan), Israël," *L'Anthropologie* 70 (1966): 437–483; François R. Valla, "Les Natoufiens de Mallaha et l'Espace," in Ofer Bar-Yosef and François R. Valla, eds., *The Natufian Culture in the Levant* (International Monographs in Prehistory, Ann Arbor, Mich., 1991), 111–122.

127 Wadi Hammeh 27 is discussed in Nigel Goring-Morris and Anna Belfer-Cohen, "Structures and Dwellings in the Upper and Epi-Paleolithic (ca. 42–10 k BP) Levant" (see previous reference).

127 Natalie D. Munro and Leore Grosman, "Early Evidence (ca. 12,000 B.P.) for Feasting at a Burial Cave in Israel," *Proceedings of the National Academy of Sciences* 107 (2010): 15362–15366.

128 Klaus Schmidt, "Göbekli Tepe, Southeastern Turkey: A Preliminary Report on the 1995–1999 Excavations," *Paléorient* 26 (2001): 45–54; also see Klaus Schmidt, *Sie Bauten die Ersten Tempel: Das Rätselhafte Heiligtum der Steinzeitjäger—Die Archaeologische Entdeckung am Göbekli Tepe* (Verlag C. H. Beck, Munich, 2006).

131 Harald Hauptmann, "Ein Kultgebäude in Nevali Çori," in Marcella Frangipane et al., eds., *Between the Rivers and Over the Mountains* (Università di Roma "La Sapienza," Rome, 1993), 37–69.

131 Andrew M. T. Moore, Gordon C. Hillman, and Anthony J. Legge, *Village on the Euphrates: From Foraging to Farming at Abu Hureyra* (Oxford University Press, 2000).

132 Nikolai O. Bader, "Tell Maghzaliyah: An Early Neolithic Site in Northern Iraq," in Norman Yoffee and Jeffrey J. Clark, eds., *Early Stages in the Evolution of Mesopotamian Civilization* (University of Arizona Press, 1993), 7–40.

134 Kathleen Kenyon's comments on the plastered skulls from Jericho are taken from her book *Archaeology in the Holy Land, Third Edition* (Praeger, New York, 1970).

134 Gary O. Rollefson, Alan H. Simmons, and Zeidan Kafafi, "Neolithic Cultures at 'Ain Ghazal, Jordan," *Journal of Field Archaeology* 19 (1992): 443–470.

136 The ritual buildings of Çayönü are discussed in Mehmet Özdoğan and A. Özdoğan, "Çayönü: A Conspectus of Recent Work," *Paléorient* 15 (1989): 65–74, and in Wulf Schirmer, "Some Aspects of Building at the 'Aceramic-Neolithic' Settlement of Çayönü Tepesi," *World Archaeology* 21 (1990): 363–387. The first excavators of Çayönü were Robert Braidwood, Halet Çambel, Charles Redman, and Patty Jo Watson. See their "Beginnings of Village-Farming Communities in Southeastern Turkey," *Proceedings of the National Academy of Sciences* 68 (1971): 1236–1240.

139 Overviews of the sites of Gheo-Shih and Guilá Naquitz Cave can be found in Joyce Marcus and Kent V. Flannery, *Zapotec Civilization: How Urban Society Evolved in Mexico's Oaxaca Valley* (Thames and Hudson, London, 1996).

141 Richard S. MacNeish et al., eds., *The Prehistory of the Tehuacán Valley, vol. 5: Excavations and Reconnaissance* (University of Texas Press, Austin, 1972).

141 Jane E. Dorweiler and John Doebley, "Developmental Analysis of Teosinte Glume Architecture 1: A Key Locus in the Evolution of Maize (Poaceae)," *American Journal of Botany* 84 (1997): 1313–1322; Adam Eyre-Walker

et al., "Investigation of the Bottleneck Leading to the Domestication of Maize," *Proceedings of the National Academy of Sciences* 95 (1998): 4441–4446; Yoshiro Matsuoka et al., "A Single Domestication for Maize Shown by Multilocus Microsatellite Genotyping," *Proceedings of the National Academy of Sciences* 99 (2002): 6080–6084; Viviane Jaenicke-Després et al., "Early Allelic Selection in Maize as Revealed by Ancient DNA," *Science* 302 (2003): 1206–1208.

141 Dolores R. Piperno and Kent V. Flannery, "The Earliest Archaeological Maize (*Zea mays* L.) from Highland Mexico: New Accelerator Mass Spectrometry Dates and Their Implications," *Proceedings of the National Academy of Sciences* 98 (2001): 2101–2103; Bruce F. Benz, "Archaeological Evidence of Teosinte Domestication from Guilá Naquitz, Oaxaca," *Proceedings of the National Academy of Sciences* 98 (2001): 2104–2106. For additional evidence of early maize, see Dolores R. Piperno et al., "Late Pleistocene and Holocene Environmental History of the Iguala Valley, Central Balsas Watershed of Mexico," *Proceedings of the National Academy of Sciences* 104 (2007): 11874–11881.

142 The Atexcala Canyon site is reported in Richard S. MacNeish and Angel García Cook, "Excavations in the San Marcos Locality in the Travertine Slopes," in Richard S. MacNeish et al., eds., *Prehistory of the Tehuacán Valley, vol. 5: Excavations and Reconnaissance* (University of Texas Press, 1972), 137–160.

142 Atlatl point exchange at Cueva Blanca is described in Joyce Marcus and Kent V. Flannery, *Zapotec Civilization.*

143 The men's houses at San José Mogote are described in Joyce Marcus and Kent V. Flannery, *Zapotec Civilization.*

144 Daniel H. Sandweiss et al., "Early Maritime Adaptations in the Andes: Preliminary Studies at the Ring Site, Peru," in Don S. Rice, Charles Stanish, and Phillip R. Scarr, eds., "Ecology, Settlement and History in the Osmore Drainage, Peru," *BAR International Series* 545 (Archaeopress, Oxford, 1989), 35–84.

145 Melinda A. Zeder, Daniel G. Bradley, Eve Emshwiller, and Bruce D. Smith, eds., *Documenting Domestication: New Genetic and Archaeological Paradigms* (University of California Press, 2006); Tom D. Dillehay et al., "Preceramic Adoption of Peanut, Squash, and Cotton in Northern Peru," *Science* 316 (2007): 1890–1893.

146 C. A. Aschero and Hugo D. Yacobaccio, "20 Años Después: Inca Cueva 7 Reinterpretado," *Cuadernos del Instituto Nacional de Antropología y Pensamiento Latinoamericano* 18 (1998–1999): 7–18. See, also, Guillermo L. Mengoni Goñalons and Hugo D. Yacobaccio, "The Domestication of South American Camelids: A View from the South-Central Andes," in Melinda A.

Zeder, Daniel G. Bradley, Eve Emshwiller, and Bruce D. Smith, eds., *Documenting Domestication: New Genetic and Archaeological Paradigms* (University of California Press, 2006), 228–244.

146 Mark S. Aldenderfer, *Montane Foragers: Asana and the South-Central Andean Archaic* (University of Iowa Press, 1998).

147 Jane C. Wheeler, "La Domesticación de la Alpaca (*Lama pacos* L.) y la Llama (*Lama glama* L.) y el Desarrollo Temprano de la Ganadería Autóctona en los Andes Centrales," *Boletín de Lima* 36 (1984): 74–84.

147 Jane C. Wheeler, Lounès Chikhi, and Michael W. Bruford, "Genetic Analysis of the Origins of Domestic South American Camelids," in Melinda A. Zeder, Daniel G. Bradley, Eve Emshwiller, and Bruce D. Smith, eds., *Documenting Domestication,* 329–341. See also M. Kadwell et al., "Genetic Analysis Reveals the Wild Ancestors of the Llama and Alpaca," *Proceedings of the Royal Society of London* 268 (2001): 2575–2584.

147 For useful overviews of the earliest Andean societies, see Danièle Lavallée, *The First South Americans* (University of Utah Press, 2000); Michael E. Moseley, *The Incas and Their Ancestors* (Thames and Hudson, London, 1992).

148 Robert A. Benfer, "The Challenges and Rewards of Sedentism: The Preceramic Village of Paloma, Peru," in Mark Nathan Cohen and George Armelagos, eds., *Paleopathology at the Origin of Agriculture* (Academic Press, New York, 1984), 531–558; Jeffrey Quilter, *Life and Death at Paloma: Society and Mortuary Practices in a Preceramic Peruvian Village* (University of Iowa Press, 1989).

148 Christopher B. Donnan, "An Early House from Chilca, Peru," *American Antiquity* 30 (1964): 137–144.

149 Terence Grieder, Alberto Bueno Mendoza, C. Earle Smith Jr., and Robert M. Malina, *La Galgada, Peru: A Preceramic Culture in Transition* (University of Texas Press, 1988).

151 Richard L. Burger and Lucy Salazar-Burger, "The Early Ceremonial Center of Huaricoto," in Christopher B. Donnan, ed., *Early Ceremonial Architecture in the Andes* (Dumbarton Oaks, Washington, D.C., 1985), 111–138.

151 Seiichi Izumi and Toshihiko Sono, *Andes 2: Excavations at Kotosh, Peru, 1960* (Kadokawa Press, Tokyo, 1963).

CHAPTER 9: PRESTIGE AND EQUALITY IN FOUR NATIVE AMERICAN SOCIETIES

154 W. H. Wills, in *Early Prehistoric Agriculture in the American Southwest* (School of American Research Press, Santa Fe, NM, 1988), describes Bat Cave.

154 Shabik'eschee village and the SU site are discussed by W. H. Wills, "Plant Cultivation and the Evolution of Risk-Prone Economies in the Prehistoric American Southwest," in A. B. Gebauer and T. D. Price, eds., "Transitions to Agriculture in Prehistory," *Monographs in World Archaeology* 4 (Prehistory Press, Madison, Wisc., 1992), 153–176, and by Stephen Plog, *Ancient Peoples of the American Southwest* (Thames and Hudson, London, 1997).

155 Steven A. LeBlanc, *Prehistoric Warfare in the American Southwest* (University of Utah Press, 1999).

155 Tim D. White, *Prehistoric Cannibalism at Mancos 5MTUMR-2346* (Princeton University Press, 1992).

155 G. T. Gross, "Subsistence Change and Architecture: Anasazi Storerooms in the Dolores Region, Colorado," *Research in Economic Anthropology, Supplement* 6 (1992): 241–265. For the transition from circular to rectangular houses, also see Timothy A. Kohler, "News from the Northern American Southwest: Prehistory of the Edge of Chaos," *Journal of Archaeological Research* 1 (1993): 267–321.

156 For thoughtful discussions of the level of inequality reached in the Southwest, see Stephen Plog, *Ancient Peoples of the American Southwest*, and two books by Linda S. Cordell: *Prehistory of the Southwest* (Academic Press, New York, 1984); *Ancient Pueblo Peoples* (St. Remy Press, Montreal, 1994).

156 Winifred Creamer, *The Architecture of Arroyo Hondo Pueblo, New Mexico* (School of American Research, Santa Fe, 1993).

157 R. Gwinn Vivian, "An Inquiry into Prehistoric Social Organization in Chaco Canyon, New Mexico," in William A. Longacre, ed., *Reconstructing Prehistoric Pueblo Societies* (School of American Research, Santa Fe, 1970), 59–83.

158 Larry Benson et al., "Ancient Maize from Chacoan Great Houses: Where Was It Grown?" *Proceedings of the National Academy of Sciences* 100 (2003): 13111–13115.

158 George H. Pepper's discoveries have been reanalyzed by Stephen Plog and Carrie Heitman in "Hierarchy and Social Inequality in the American Southwest, A.D. 800–1200," *Proceedings of the National Academy of Sciences* 107 (2010): 19619–19626.

159 James W. Judge, "Chaco Canyon—San Juan Basin," in Linda S. Cordell and George J. Gumerman, eds., *Dynamics of Southwest Prehistory* (Smithsonian Institution Press, Washington, D.C., 1989), 209–261.

160 Edward P. Dozier, "The Pueblos of the Southwestern United States," *Journal of the Royal Anthropological Institute of Great Britain and Ireland* 90 (1960): 146–160; Fred Eggan, *Social Organization of the Western Pueblos* (University of Chicago Press, 1950).

161 Images showing the diversity of kivas can be found in Peter Nabokov and Robert Easton, *Native American Architecture* (Oxford University Press, 1989).

163 Alfonso Ortiz, *The Tewa World: Space, Time, Being, and Becoming in a Pueblo Society* (University of Chicago Press, 1969).

167 Mischa Titiev, "Old Oraibi: A Study of the Hopi Indians of Third Mesa," *Papers of the Peabody Museum of American Archaeology and Ethnology* 22 (Harvard University, 1944).

170 Jerrold Levy, *Orayvi Revisited: Social Stratification in an "Egalitarian" Society* (School of American Research Press, Santa Fe, N.Mex., 1992).

171 For a perspective on the introduction of Northern Flint corn, see David S. Brose, "Early Mississippian Connections at the Late Woodland Mill Hollow Site in Lorain County, Ohio," *Midcontinent Journal of Archaeology* 18 (1993): 97–110, plus the Appendix on pp. 111–130 by Robert P. Mensforth and Stephanie J. Belovich.

171 W. Raymond Wood, "Plains Village Tradition: Middle Missouri," in Raymond J. DeMallie, ed., *Handbook of North American Indians, vol. 13: Plains* (Smithsonian Institution Press, Washington, D.C., 2001), 186–195.

172 The contributions of Prince Maximilian and Charbonneau are discussed in Robert H. Lowie, *The Crow Indians* (Farrar & Rinehart, New York, 1935).

173 For an overview of sacred bundles and the concept of xo'pini, see Frank Henderson Stewart, "Hidatsa," in Raymond J. DeMallie, ed., *Handbook of North American Indians, vol. 13*, 329–348.

173 For the story of the Water Buster clan's long-lost bundle, see Patrick Springer, "Medicine Bundles Help Keep Stories from 'Dream Time,'" *The Forum* (Forum Communications, Fargo, N.Dak., 2003), 1–3.

176 Alfred W. Bowers, *Mandan Social and Ceremonial Organization* (University of Chicago Press, 1950); W. Raymond Wood and Lee Irwin, "Mandan," in Raymond J. DeMallie, ed., *Handbook of North American Indians, vol. 13*, 349–364.

180 Frank Henderson Stewart, "Hidatsa," in Raymond J. DeMallie, ed., *Handbook of North American Indians, vol. 13*; Alfred W. Bowers, "Hidatsa Social and Ceremonial Organization," *Bulletin* 194 (Bureau of American Ethnology, Smithsonian Institution, Washington, D.C., 1965).

181 References to "two-spirit people" can be found in Alfred W. Bowers, "Hidatsa Social and Ceremonial Organization"; W. Raymond Wood and Lee Irwin, "Mandan," in Raymond J. DeMallie, ed., *Handbook of North American Indians, vol. 13*; and throughout Raymond J. DeMallie, ed., *Handbook of North American Indians, vol. 13.*

CHAPTER 10: THE RISE AND FALL OF HEREDITARY INEQUALITY IN FARMING SOCIETIES

188 Simon J. Harrison, *Stealing People's Names: History and Politics in a Sepik River Cosmology* (Cambridge University Press, 1990).

192 Edmund R. Leach, *Political Systems of Highland Burma* (G. Bell & Sons, London, 1954).

198 Jonathan Friedman, *System, Structure and Contradiction: The Evolution of "Asiatic" Social Formations* (National Museum of Denmark, Copenhagen, 1979).

201 Christoph von Fürer-Haimendorf, *The Konyak Nagas: An Indian Frontier Tribe* (Holt, Rinehart and Winston, New York, 1969).

CHAPTER 11: THREE SOURCES OF POWER IN CHIEFLY SOCIETIES

208 Irving Goldman, *Ancient Polynesian Society* (University of Chicago Press, 1970).

210 Raymond Firth, *We the Tikopia* (George Allen & Unwin, London, 1936); *Social Change in Tikopia: Re-Study of a Polynesian Community After a Generation* (George Allen & Unwin, London, 1959); *History and Traditions of Tikopia* (The Polynesian Society, Wellington, New Zealand, 1961); *Tikopia Ritual and Belief* (George Allen & Unwin, London, 1967).

211 Patrick V. Kirch and Douglas E. Yen, "Tikopia: The Prehistory and Ecology of a Polynesian Outlier," *Bernice P. Bishop Museum Bulletin* 238 (Honolulu, 1982).

216 Robert L. Carneiro, "The Nature of the Chiefdom as Revealed by Evidence from the Cauca Valley of Colombia," in A. Terry Rambo and Kathleen Gillogly, eds., "Profiles in Cultural Evolution," *Anthropological Papers* 85 (Museum of Anthropology, University of Michigan, Ann Arbor, 1991), 167–190. Also see Hermann Trimborn, *Señorío y Barbarie en el Valle de Cauca* (Consejo Superior de Investigaciones Científicas, Instituto Gonzalo Fernández de Oviedo, Madrid, 1949).

220 Mary W. Helms, *Ancient Panama: Chiefs in Search of Power* (University of Texas Press, 1979).

220 Richard G. Cooke et al., "Who Crafted, Exchanged, and Displayed Gold in Pre-Columbian Panama?" in Jeffrey Quilter and John W. Hoopes, eds.,

Gold and Power in Ancient Costa Rica, Panama, and Colombia (Dumbarton Oaks, Washington, D.C., 2003), 91–158.

222 Samuel Kirkland Lothrop, "Coclé: An Archaeological Study of Central Panama," *Memoirs VII, Part I* (Peabody Museum of Archaeology and Ethnology, Harvard University, 1937). Lothrop not only cites the Spanish colonial manuscripts of Gaspar de Espinoza and Gonzalo Fernández de Oviedo, but he also describes chiefly burials at the Sitio Conte archaeological site.

222 For an iconographic analysis of Coclé polychrome sumptuary vessels, see Olga F. Linares, "Ecology and the Arts in Ancient Panama: On the Development of Social Rank and Symbolism in the Central Provinces," *Studies in Pre-Columbian Art and Archaeology* 17 (Dumbarton Oaks, Washington, D.C., 1977).

223 David Phillipson, in *The Later Prehistory of Eastern and South Africa* (Heinemann, London, 1977), discusses the Bantu migration. Andrew D. Roberts, in *A History of Zambia* (Africana, New York, 1976) and in *A History of the Bemba: Political Growth and Change in North-Eastern Zambia Before 1900* (The Longman Group, London, 1973), discusses early evidence for rank in Luba country.

224 Audrey I. Richards, "The Political System of the Bemba Tribe—North-Eastern Rhodesia," in Meyer Fortes and E. E. Evans-Pritchard, eds., *African Political Systems* (Oxford University Press, 1940), 83–120.

CHAPTER 12: FROM RITUAL HOUSE TO TEMPLE IN THE AMERICAS

230 Joyce Marcus and Kent V. Flannery, *Zapotec Civilization: How Urban Society Evolved in Mexico's Oaxaca Valley* (Thames and Hudson, London, 1996); Michael E. Whalen, "Excavations at Tomaltepec: Evolution of a Formative Community in the Valley of Oaxaca, Mexico," *Memoir* 12 (Museum of Anthropology, University of Michigan, Ann Arbor, 1981); Robert D. Drennan, "Fábrica San José and Middle Formative Society in the Valley of Oaxaca," *Memoir* 8 (Museum of Anthropology, University of Michigan, Ann Arbor, 1976).

235 Elsa M. Redmond and Charles S. Spencer, "Rituals of Sanctification and the Development of Standardized Temples in Oaxaca, Mexico," *Cambridge Archaeological Journal* 18 (2008): 230–266.

238 Ruth Shady, Camilo Dolorier, Fanny Montesinos, and Lyda Casas, "Los Orígenes de la Civilización en el Perú: El Área Norcentral y el Valle de Supe Durante el Arcaico Tardío," *Arqueología y Sociedad* 13 (2000): 13–48.

239 Robert A. Feldman, "Áspero, Peru: Architecture, Subsistence Economy, and other Artifacts of a Preceramic Maritime Chiefdom" (PhD diss., Harvard University, 1980).

239 Ruth Shady, *La Ciudad Sagrada de Caral-Supe en los Albores de la Civilización en el Perú* (Universidad Nacional Mayor de San Marcos, Lima, Peru, 1997); "Sustento Socioeconómico del Estado Prístino de Supe-Perú: Las Evidencias del Caral-Supe," *Arqueología y Sociedad* 13 (2000): 49–66; *La Civilización de Caral-Supe: 5000 Años de Identidad Cultural en el Perú* (Instituto Nacional de Cultura, Lima, Peru, 2005); "Caral-Supe y su Entorno Natural y Social en los Orígenes de la Civilización," in Joyce Marcus and Patrick Ryan Williams, eds., *Andean Civilization: A Tribute to Michael E. Moseley* (Cotsen Institute of Archaeology Press, UCLA, Los Angeles, 2009), 99–120. The fish remains from Caral were analyzed by Philippe Beárez and Luís Miranda. See "Análisis Arqueo-Ictiológico del Sector Residencial del Sitio Arqueológico de Caral-Supe, Costa Central del Perú," *Arqueología y Sociedad* 13 (2000): 67–77.

243 Frederic Engel, "A Preceramic Settlement on the Central Coast of Peru: Asia, Unit 1," *Transactions of the American Philosophical Society,* n.s., vol. 53, part 3 (1963).

244 Lorenzo Samaniego, Enrique Vergara, and Henning Bischof, "New Evidence on Cerro Sechín, Casma Valley, Peru," in Christopher B. Donnan, ed., *Early Ceremonial Architecture of the Andes* (Dumbarton Oaks, Washington, D.C., 1985), 165–190; Elena Maldonado, *Arqueología de Cerro Sechín, vol. 1: Arquitectura* (Pontificia Universidad Católica del Perú, Lima and Fundación Volkswagenwerk-Alemania, 1992).

246 The sumptuary goods found at Kuntur Wasi are described in Richard L. Burger, "Current Research in Andean South America," *American Antiquity* 56 (1991): 151–156.

247 An excellent overview of Chavín de Huántar can be found in Richard L. Burger, *Chavín and the Origins of Andean Civilization* (Thames and Hudson, London, 1992). See also Luis G. Lumbreras and Hernán Amat, "Informe Preliminar Sobre las Galerías Interiores de Chavín (Primera Temporada de Trabajos)," *Revista del Museo Nacional* 34 (1965–1966): 143–197.

247 Craig Morris's suggestion that Chavín possessed an oracle can be found in Craig Morris and Adriana von Hagen, *The Inka Empire and Its Andean Origins* (Abbeville Press, for the American Museum of Natural History, New York, 1993). Michael Moseley's interpretation of Chavín can be found in Michael Moseley, *The Incas and Their Ancestors* (Thames and Hudson, London, 1992).

CHAPTER 13: ARISTOCRACY WITHOUT CHIEFS

251 Christoph von Fürer-Haimendorf, *The Apa Tanis and Their Neighbours* (Routledge & Kegan Paul, London, 1962).

CHAPTER 14: TEMPLES AND INEQUALITY IN EARLY MESOPOTAMIA

263 Seton Lloyd and Fuad Safar, "Tell Hassuna," *Journal of Near Eastern Studies* 4 (1945): 255–289.

263 Joan Oates, "Choga Mami 1967–68: A Preliminary Report," *Iraq* 31 (1969): 115–152.

263 Faisal El-Wailly and Behnam Abu al-Soof, "The Excavations at Tell es-Sawwan: First Preliminary Report (1964)," *Sumer* 21 (1965): 17–32; K. H. al-A'dami, "Excavations at Tell es-Sawwan (Second Season)," *Sumer* 24 (1968): 57–98; Ghanim Wahida, "The Excavations of the Third Season at Tell es-Sawwan, 1966," *Sumer* 23 (1967): 167–178; Behnam Abu al-Soof, "Tell es-Sawwan Excavations of the Fourth Season," *Sumer* 24 (1968): 3–16.

265 Manfred Korfmann, "The Sling as a Weapon," *Scientific American* 229 (1973): 34–42.

267 Nikolai Y. Merpert and Rauf M. Munchaev, "Burial Practices of the Halaf Culture," in Norman Yoffee and Jeffrey J. Clark, eds., *Early Mesopotamian Civilization: Soviet Excavations in Northern Iraq* (University of Arizona Press, 1993), 207–223.

268 Ismail Hijara, "The Halaf Period in Northern Mesopotamia" (PhD diss., University of London, 1980).

270 Patty Jo Watson, "The Halafian Culture: A Review and Synthesis," in T. Cuyler Young Jr., Philip E. L. Smith, and Peder Mortensen, eds., "The Hilly Flanks and Beyond: Essays on the Prehistory of Southwestern Asia," *Studies in Ancient Oriental Civilization* 36 (1983): 231–250 (University of Chicago).

270 Steven A. LeBlanc, "Computerized, Conjunctive Archaeology and the Near Eastern Halafian" (PhD diss., Washington University, St. Louis, Mo., 1971). See also Steven A. LeBlanc and Patty Jo Watson, "A Comparative Statistical Analysis of Painted Pottery from Seven Halafian Sites," *Paléorient* 1 (1973): 117–133.

271 Patty Jo Watson, "The Halafian Culture" (see previous reference).

271 Max E. L. Mallowan, in "Excavations in the Balih Valley (1938)," *Iraq* 8 (1946): 111–156, discusses the building at Tell Aswad that he believes to be a Halaf temple.

271 Nikolai Y. Merpert and Rauf M. Munchaev, "Burial Practices of the Halaf Culture" (see previous reference).

272 Max E. L. Mallowan and J. Cruikshank Rose, "Excavations at Tall Arpachiyah, 1933," *Iraq* 2 (1935): 1–178; Ismail Hijara, "The Halaf Period in Northern Mesopotamia" (see previous reference); Ismail Hijara, "Three New Graves at Arpachiyah," *World Archaeology* 10 (1978): 125–128.

275 Ephraim A. Speiser, *Excavations at Tepe Gawra,* vol. 1 (University of Pennsylvania Museum, Philadelphia, 1935); Arthur Tobler, *Excavations at Tepe Gawra,* vol. 2 (University of Pennsylvania Museum, Philadelphia, 1950); Ann Louise Perkins, "The Comparative Archaeology of Early Mesopotamia," *Studies in Ancient Oriental Civilization* 25 (University of Chicago Press, 1949); Mitchell S. Rothman, *Tepe Gawra: The Evolution of a Small, Prehistoric Center in Northern Iraq* (University of Pennsylvania Museum, Philadelphia, 2002).

282 The *British Naval Intelligence Handbook* BR 524 (1944) provides useful data on the regimes of the Tigris and Euphrates.

283 Wright's study of 'Ubaid sickles is reported in Henry T. Wright and Susan Pollock, "Regional Socio-Economic Organization in Southern Mesopotamia: The Middle and Later Fifth Millennium," *Colloques Internationaux CNRS: Préhistoire de la Mésopotamie* (Editions du CNRS, Paris, 1986), 317–329.

284 Seton Lloyd, "The Oldest City of Sumeria: Establishing the Origins of Eridu," *The Illustrated London News,* September 11, 1948, 303–305; Fuad Safar, Mohammad Ali Mustafa, and Seton Lloyd, *Eridu* (Iraqi Ministry of Culture and Information, Baghdad, 1981).

284 Joan Oates, "Ur and Eridu, the Prehistory," *Iraq* 22 (1960): 32–50, and "Ubaid Chronology," in O. Aurenche, J. Évin, and F. Hours, eds., "Chronologies du Proche Orient," *BAR-Maison de l'Orient Archaeological Series* 3 (Lyon-Oxford, 1987), 473–482.

287 Fuad Safar, Mohammad Ali Mustafa, and Seton Lloyd, in chapter 7 of *Eridu* (see previous reference), discuss the possible fishermen's ward.

287 Fuad Safar, Mohammad Ali Mustafa, and Seton Lloyd, in chapter 4 of *Eridu,* discuss the 'Ubaid cemetery, which is reanalyzed by Henry T. Wright and Susan Pollock in "Regional Socio-Economic Organization in Southern Mesopotamia" (see previous reference).

289 Seton Lloyd and Fuad Safar, "Tell Uqair: Excavations by the Iraq Government Directorate of Antiquities in 1940–1941," *Journal of Near Eastern Studies* 2 (1943): 131–155.

290 Sabah Abboud Jasim, "Excavations at Tell Abada, Iraq," *Paléorient* 7 (1981): 101–104, and "Excavations at Tell Abada: A Preliminary Report," *Iraq* 45 (1983): 165–186.

291 Joan Oates, "Ubaid Mesopotamia Reconsidered," in T. Cuyler Young Jr., Philip E. L. Smith, and Peder Mortensen, eds., "The Hilly Flanks and Beyond," 251–281 (see previous reference).

291 Jean-Louis Huot et al., "Larsa et 'Oueli: Travaux de 1978–1981," *Mémoire* 26 (Éditions Recherche sur les Civilisations, Paris, 1983).

293 'Ubaid trading colonies in Syria and Turkey are discussed by Joan Oates, "Trade and Power in the Fifth and Fourth Millennia B.C.: New Evidence from Northern Mesopotamia," *World Archaeology* 24 (1993): 403–422.

295 Colin Renfrew, "Beyond a Subsistence Economy: The Evolution of Social Organization in Prehistoric Europe," in Charlotte B. Moore, ed., "Reconstructing Complex Societies: An Archaeological Colloquium," *Supplement to the Bulletin of the American Schools of Oriental Research* 20 (Cambridge, Mass., 1974), 69–85.

CHAPTER 15: THE CHIEFLY SOCIETIES IN OUR BACKYARD

298 Peter Nabokov and Robert Easton, *Native American Architecture* (Oxford University Press, 1989), 96–97.

299 Robert S. Neitzel, "Archaeology of the Fatherland Site: The Grand Village of the Natchez," *Anthropological Papers* 51, part 1 (American Museum of Natural History, New York, 1965); "The Grand Village of the Natchez Revisited: Excavations at the Fatherland Site, Adams County, Mississippi, 1972," *Archaeological Report* no. 12 (Mississippi Department of Archives and History, Jackson, 1983).

300 Charles Hudson, *The Southeastern Indians* (University of Tennessee Press, 1976).

302 Antoine le Page du Pratz's account of the funeral of Tattooed Serpent was translated into English by John R. Swanton, "The Indians of the Southeastern United States," *Bulletin* 137 (Bureau of American Ethnology, Smithsonian Institution, 1946), 728.

303 Frank G. Speck, "Notes on Chickasaw Ethnology and Folk-Lore," *Journal of American Folk-Lore* 20 (1907): 50–58.

304 Vernon James Knight Jr., "Moundville as a Diagrammatic Ceremonial Center," in Vernon James Knight Jr. and Vincas P. Steponaitis, eds., *Archaeology of the Moundville Chiefdom* (Smithsonian Institution Press, Washington, D.C., 1998), 44–62. Also see Vernon James Knight Jr. and Vincas P. Steponaitis, "A New History of Moundville," in Vernon James Knight Jr. and Vincas P. Steponaitis, eds., *Archaeology of the Moundville Chiefdom*, 1–25.

307 David J. Hally, "The Settlement Patterns of Mississippian Chiefdoms in Northern Georgia," in Brian R. Billman and Gary M. Feinman, eds., *Settlement Pattern Studies in the Americas: Fifty Years since Virú* (Smithsonian Institution Press, Washington, D.C., 1999), 96–115.

307 Igor Kopytoff, "Internal African Frontier: The Making of African Political Culture," in Igor Kopytoff, ed., *The African Frontier: The Reproduction of Traditional African Societies* (Indiana University Press, 1987), 3–84.

309 Adam King, *Etowah: The Political History of a Chiefdom Capital* (University of Alabama Press, 2003).

309 Lewis H. Larson Jr., "Archaeological Implications of Social Stratification at the Etowah Site, Georgia," in James A. Brown, ed., "Approaches to Social Dimensions of Mortuary Practices," *Memoirs of the Society for American Archaeology* 25 (1971): 58–67.

311 Charles Hudson et al., "Coosa: A Chiefdom in the Sixteenth-Century Southeastern United States," *American Antiquity* 50 (1985): 723–737. Also see Charles Hudson, *The Southeastern Indians,* 112–118 (see previous reference).

311 David J. Hally, *King: The Social Archaeology of a Late Mississippian Town in Northwestern Georgia* (University of Alabama Press, 2008).

312 Helen C. Rountree, *The Powhatan Indians of Virginia: Their Traditional Culture* (University of Oklahoma Press, 1989); Martin D. Gallivan, "Powhatan's Werowocomoco: Constructing Place, Polity, and Personhood in the Chesapeake C.E. 1200–C.E. 1609," *American Anthropologist* 109 (2007): 85–100.

CHAPTER 16: HOW TO TURN RANK INTO STRATIFICATION: TALES OF THE SOUTH PACIFIC

314 The literature on Polynesian rank societies is vast. A good place to begin is with three theoretical overviews, written from three different perspectives. Marshall D. Sahlins, in *Social Stratification in Polynesia* (University of Washington Press, Seattle, 1958), relates levels of inequality to differences in the adaptation of societies to their environments. Irving Goldman, in *Ancient Polynesian Society* (University of Chicago Press, 1970), shows how certain basic principles of rank can be used to explain the variation among island societies. Patrick V. Kirch, in *The Evolution of the Polynesian Chiefdoms* (Cambridge University Press, 1984), uses archaeological data to document the evolution of diverse island societies from a common ancestral culture.

314 Patrick V. Kirch and Roger C. Green, *Hawaiki, Ancestral Polynesia: An Essay in Historical Anthropology* (Cambridge University Press, 2001); also see chapter 3 of Patrick V. Kirch, *The Evolution of the Polynesian Chiefdoms* (see previous reference).

315 See chapter 11 of Irving Goldman, *Ancient Polynesian Society,* and Appendix III of Marshall D. Sahlins, *Social Stratification in Polynesia* (both previously referenced).

317 Edward Winslow Gifford, "Tongan Society," *Bulletin* 61 (Bernice P. Bishop Museum, Honolulu, 1971); Will Carleton McKern, "Archaeology of Tonga,"

Bulletin 60 (Bernice P. Bishop Museum, Honolulu, 1929); chapter 12 of Irving Goldman, *Ancient Polynesian Society*; chapter 9 of Patrick V. Kirch, *The Evolution of the Polynesian Chiefdoms*; chapter 2 of Marshall D. Sahlins, *Social Stratification in Polynesia.*

324 Lapaha is described by Will Carleton McKern in "Archaeology of Tonga," 92–101 (see previous reference).

328 Robert F. Heizer, "Agriculture and the Theocratic State in Lowland Southeastern Mexico," *American Antiquity* 26 (1960): 215–222; Philip Drucker, Robert F. Heizer, and Robert J. Squier, "Excavations at La Venta, Tabasco, 1955," *Bulletin* 170 (Bureau of American Ethnology, Smithsonian Institution, Washington, D.C., 1959); Rebecca González Lauck, "La Venta: An Olmec Capital," in Elizabeth P. Benson and Beatriz de la Fuente, eds., *Olmec Art of Ancient Mexico* (National Gallery of Art, Washington, D.C., 1996), 73–81; Rebecca B. González Lauck, "La Venta (Tabasco, Mexico)," in Susan Toby Evans and David Webster, eds., *Archaeology of Ancient Mexico and Central America* (Garland, N.Y., 2001), 798–801.

332 See chapter 10 in Patrick V. Kirch, *The Evolution of the Polynesian Chiefdoms* (see previous reference).

334 Samuel M. Kamakau, *Ruling Chiefs of Hawaii* (The Kamehameha Schools Press, Honolulu, 1961); Marshall D. Sahlins, *Historical Metaphors and Mythical Realities: Structure in the Early History of the Sandwich Islands Kingdom* (University of Michigan Press, 1981); Valerio Valeri, *Kingship and Sacrifice: Ritual and Society in Ancient Hawaii* (University of Chicago Press, 1985); chapter 10 in Irving Goldman, *Ancient Polynesian Society* (see previous reference).

CHAPTER 17: HOW TO CREATE A KINGDOM

341 Samuel M. Kamakau, *Ruling Chiefs of Hawaii* (The Kamehameha Schools Press, Honolulu, 1961); also see chapter 10 in Irving Goldman, *Ancient Polynesian Society* (University of Chicago Press, 1970); Valerio Valeri, *Kingship and Sacrifice: Ritual and Society in Ancient Hawaii* (University of Chicago Press, 1985); Patrick Vinton Kirch, *How Chiefs Became Kings: Divine Kingship and the Rise of Archaic States in Ancient Hawai'i* (University of California Press, 2010).

345 Kathleen Dickenson Mellen, *The Lonely Warrior: The Life and Times of Kamehameha the Great of Hawaii* (Hastings House, New York, 1949); Herbert H. Gowen, *The Napoleon of the Pacific: Kamehameha the Great* (Fleming H. Revell, New York, 1919); Ralph S. Kuykendall, *The Hawaiian Kingdom, 1778–1854: Foundation and Transformation* (University of Hawaii Press, 1938).

346 Patrick V. Kirch and Marshall Sahlins, *Anahulu: The Anthropology of History in the Kingdom of Hawaii* (University of Chicago Press, 1992). Marshall Sahlins, in vol. 1, *Historical Ethnography,* discusses Hawai'ian society from the 1770s to the mid-nineteenth century. Patrick V. Kirch, in vol. 2, *The Archaeology of History,* describes the archaeology of that same time span in the Anahulu Valley of Oahu. This is one of the most successful collaborations ever carried out by a social anthropologist and an archaeologist.

348 Dorothy B. Barrère, in "Kamehameha in Kona: Two Documentary Studies," *Pacific Anthropological Records* 23 (Bernice P. Bishop Museum, Honolulu, 1975), discusses the burial of Kamehameha.

348 William H. Davenport, "The 'Hawaiian Cultural Revolution': Some Political and Economic Considerations," *American Anthropologist* 71 (1969): 1–20; Ralph S. Kuykendall, *The Hawaiian Kingdom, 1778–1854* (see previous reference).

348 The Bantu migration is discussed by David Phillipson, *The Later Prehistory of Eastern and South Africa* (Heinemann, London, 1977).

349 Late Iron Age Natal is discussed by Tim Maggs in "The Iron Age Farming Communities," in Andrew Duminy and Bill Guest, eds., *Natal and Zululand: From Earliest Times to 1910* (University of Natal Press, Pietermaritzburg, 1989), 28–48 (see previous reference).

349 Early rank societies of Natal are discussed by E. A. Ritter in *Shaka Zulu: The Rise of the Zulu Empire* (G. P. Putnam's Sons, New York, 1957) and by John Wright and Carolyn Hamilton in "Traditions and Transformations: The Phongolo-Mzimkhulu Region in the Late Eighteenth and Early Nineteenth Centuries," in Andrew Duminy and Bill Guest, eds., *Natal and Zululand,* 49–82 (see previous reference).

349 Shaka's rise to power has been described in two articles by Max Gluckman, "The Kingdom of the Zulu of South Africa," in Meyer Fortes and E. E. Evans-Pritchard, eds., *African Political Systems* (Oxford University Press, 1940), 25–55, and in "The Rise of a Zulu Empire," *Scientific American* 202 (1960): 157–168. His exploits are detailed by E. A. Ritter in *Shaka Zulu* and by John M. Selby in *Shaka's Heirs* (Allen & Unwin, London, 1971).

353 Both colonial powers and Zulu nationalists manipulated the legend of Shaka for their own purposes. See Carolyn Hamilton, *Terrific Majesty: The Powers of Shaka Zulu and the Limits of Historical Invention* (Harvard University Press, 1998).

355 Homayun Sidky, *Irrigation and State Formation in Hunza: The Anthropology of a Hydraulic Kingdom* (University Press of America, New York, 1996).

356 Irmtraud Müller-Stellrecht, *Hunza und China 1761–1891* (Franz Steiner Verlag, Wiesbaden, Germany, 1978).

359 Mervyn Brown, in *Madagascar Rediscovered: A History from Early Times to Independence* (Anchor Books, Hamden, Conn., 1979), draws on the Tantàran 'ny Andrìana.

359 Conrad P. Kottak, *The Past in the Present: History, Ecology, and Cultural Variation in Highland Madagascar* (University of Michigan Press, 1980); Henry T. Wright and Susan Kus, "An Archaeological Reconnaissance of Ancient Imerina," in Raymond K. Kent, ed., *Madagascar in History* (Foundation for Malagasy Studies, Berkeley, Calif., 1979), 1–31; Robert E. Dewar and Henry T. Wright, "The Culture History of Madagascar," *Journal of World Prehistory* 7 (1993): 417–466; Zoe Crossland, "Ny Tani sy ny Fanjakana, the Land and the State: Archaeological Landscape Survey in the Andrantsay Region of Madagascar" (PhD diss., University of Michigan, 2001); Henry T. Wright, "Early State Formation in Central Madagascar: An Archaeological Survey of Western Avaradrano," *Memoir* 43 (University of Michigan Museum of Anthropology, Ann Arbor, 2007).

361 John Mack, *Madagascar: Island of the Ancestors* (British Museum Publications, London, 1986); Alain Delivré, *L'Histoire des Rois d'Imerina: Interprétation d'une Tradition Orale* (Klincksieck, Paris, 1974); Conrad P. Kottak, *The Past in the Present.*

364 Robert L. Carneiro, *Evolutionism in Cultural Anthropology: A Critical History* (Westview Press, Boulder, Colo., 2003).

364 Charles S. Spencer, "A Mathematical Model of Primary State Formation," *Cultural Dynamics* 10 (1998): 5–20.

365 Herbert S. Lewis, *A Galla Monarchy: Jimma Abba Jifar, Ethiopia, 1830–1932* (University of Wisconsin Press, 1965).

CHAPTER 18: THREE OF THE NEW WORLD'S FIRST-GENERATION KINGDOMS

368 See chapters 11–13 in Joyce Marcus and Kent V. Flannery, *Zapotec Civilization: How Urban Society Evolved in Mexico's Oaxaca Valley* (Thames and Hudson, London, 1996); also see Richard E. Blanton, *Monte Alban: Settlement Patterns at the Ancient Zapotec Capital* (Academic Press, New York, 1978); Joyce Marcus, *Monte Albán* (Fondo de Cultura Económica, Mexico City, 2008).

369 The political importance of Tilcajete is discussed by Charles S. Spencer and Elsa M. Redmond in "Multilevel Selection and Political Evolution in the Valley of Oaxaca, 500–100 B.C.," *Journal of Anthropological Archaeology* 20 (2001): 195–229, and in "Militarism, Resistance and Early State Development in Oaxaca, Mexico," *Social Evolution and History* 2 (2003): 25–70. Kent V. Flannery and Joyce Marcus, in "The Origin of War: New ^{14}C Dates

from Ancient Mexico," *Proceedings of the National Academy of Sciences* 100 (2003): 11801–11805, discuss the role of warfare in creating kingdoms.

369 Charles S. Spencer and Elsa M. Redmond, "A Late Monte Albán I Phase (300–100 B.C.) Palace in the Valley of Oaxaca," *Latin American Antiquity* 15 (2004): 441–455.

370 See chapters 11 and 12 in Joyce Marcus and Kent V. Flannery, *Zapotec Civilization* (see previous reference). Also see chapter 11 in Joyce Marcus, *Mesoamerican Writing Systems* (Princeton University Press, 1992).

371 Richard E. Blanton, *Monte Alban* (see previous reference).

372 Stephen A. Kowalewski et al., "Monte Albán's Hinterland, Part II, Vols. 1 and 2," *Memoir* 23 (Museum of Anthropology, University of Michigan, Ann Arbor, 1989).

372 Andrew K. Balkansky, "The Sola Valley and the Monte Albán State: A Study of Zapotec Imperial Expansion," *Memoir* 36 (Museum of Anthropology, University of Michigan, Ann Arbor, 2002); Gary M. Feinman and Linda M. Nicholas, "At the Margins of the Monte Albán State: Settlement Patterns in the Ejutla Valley, Oaxaca, Mexico," *Latin American Antiquity* 1 (1990): 216–246.

372 Charles S. Spencer and Elsa M. Redmond, "The Chronology of Conquest: Implications of New Radiocarbon Analyses from the Cañada de Cuicatlán, Oaxaca," *Latin American Antiquity* 12 (2001): 182–202. Also see chapter 14 in Joyce Marcus and Kent V. Flannery, *Zapotec Civilization.*

373 Andrew K. Balkansky, Verónica Pérez Rodríguez, and Stephen A. Kowalewski, "Monte Negro and the Urban Revolution in Oaxaca, Mexico," *Latin American Antiquity* 15 (2004): 33–60; Andrew K. Balkansky, "Origin and Collapse of Complex Societies in Oaxaca (Mexico): Evaluating the Era from 1965 to the Present," *Journal of World Prehistory* 12 (1998): 451–493.

376 See chapter 1 in Joyce Marcus and Kent V. Flannery, *Zapotec Civilization;* also see Joseph W. Whitecotton, *The Zapotecs: Princes, Priests, and Peasants* (University of Oklahoma Press, 1977).

378 The buildup to the Moche state is described by Brian R. Billman in "Reconstructing Prehistoric Political Economies and Cycles of Political Power in the Moche Valley, Peru," in Brian R. Billman and Gary M. Feinman, eds., *Settlement Pattern Studies in the Americas: Fifty Years Since Virú* (Smithsonian Institution Press, Washington, D.C., 1999), 131–159, and in "How Moche Rulers Came to Power: Investigating the Emergence of the Moche Political Economy," in Jeffrey Quilter and Luis Jaime Castillo, *New Perspectives on Moche Political Organization* (Dumbarton Oaks, Washington, D.C., 2010), 181–200. A four-level hierarchy in the Casma Valley was detected by David J. Wilson; see his *Prehispanic Settlement Patterns in the*

Casma Valley, North Coast of Peru (Report to the Committee for Research and Exploration, National Geographic Society, Washington, D.C., 1995).

378 Curtiss T. Brennan, "Cerro Arena: Early Cultural Complexity and Nucleation in North Coastal Peru," *Journal of Field Archaeology* 7 (1980): 1–22. Also see Brian R. Billman, "Reconstructing Prehistoric Political Economies and Cycles of Political Power in the Moche Valley, Peru," in Brian R. Billman and Gary M. Feinman, eds., *Settlement Pattern Studies in the Americas* (see previous reference).

380 Early Moche work effort has been estimated by Brian R. Billman in "Reconstructing Prehistoric Political Economies and Cycles of Political Power in the Moche Valley, Peru," in Brian R. Billman and Gary M. Feinman, eds., *Settlement Pattern Studies in the Americas.* The placement of makers' marks on adobe bricks is discussed by C. M. Hastings and M. E. Moseley in "The Adobes of Huaca del Sol and Huaca de la Luna," *American Antiquity* 40 (1975): 196–203.

380 Walter Alva and Christopher B. Donnan, *Royal Tombs of Sipán* (Fowler Museum of Culture History, UCLA, Los Angeles, 1993).

383 Christopher B. Donnan and Donna McClelland, *Moche Fineline Painting: Its Evolution and Its Artists* (Fowler Museum of Cultural History, UCLA, Los Angeles, 1999); Christopher B. Donnan, "Archaeological Confirmation of a Moche Ceremony," *Indiana* 10 (1985): 371–381; Christopher B. Donnan and Luis Jaime Castillo, "Finding the Tomb of a Moche Priestess," *Archaeology* 45 (1992): 38–42.

384 Norman Hammond, ed., in *Cuello: An Early Maya Community in Belize* (Cambridge University Press, 1991), discusses mass graves of Maya warriors.

384 Nakbe is discussed by Richard D. Hansen in "The First Cities—The Beginnings of Urbanization and State Formation in the Maya Lowlands," in Nikolai Grube, ed., *Maya: Divine Kings of the Rain Forest* (Könemann, Cologne, Germany, 2001), 50–65, and in "Continuity and Disjunction: The Preclassic Antecedents of Classic Maya Architecture," in Stephen D. Houston, ed., *Function and Meaning in Classic Maya Architecture* (Dumbarton Oaks, Washington, D.C., 1998), 49–122. See also Joyce Marcus, "Recent Advances in Maya Archaeology," *Journal of Archaeological Research* 11 (2003): 71–148.

385 The road system linking El Mirador to other centers is discussed by William J. Folan, Joyce Marcus, and W. Frank Miller in "Verification of a Maya Settlement Model through Remote Sensing," *Cambridge Archaeological Journal* 5 (1995): 277–283, and in William J. Folan et al., "Los Caminos de Calakmul, Campeche," *Ancient Mesoamerica* 12 (2001): 293–298.

385 El Mirador's monumental buildings are discussed in Richard D. Hansen, "The First Cities—The Beginnings of Urbanization and State Formation in

the Maya Lowlands," in Nikolai Grube, ed., *Maya: Divine Kings of the Rain Forest,* 50–65, and in Robert J. Sharer and Loa Traxler, *The Ancient Maya* (Stanford University Press, 2006).

386 William J. Folan et al., "Calakmul: New Data from an Ancient Maya Capital in Campeche, Mexico," *Latin American Antiquity* 6 (1995): 310–334; Joyce Marcus, "The Inscriptions of Calakmul: Royal Marriage at a Maya City in Campeche, Mexico," *Technical Report* 21 (Museum of Anthropology, University of Michigan, Ann Arbor, 1987); William J. Folan, Joyce Marcus, and W. Frank Miller, "Verification of a Maya Settlement Model through Remote Sensing" (see previous reference); Joyce Marcus, "Recent Advances in Maya Archaeology"; Joyce Marcus, "Maya Political Cycling and the Story of the Kaan Polity," in *The Ancient Maya of Mexico: Reinterpreting the Past of the Northern Maya Lowlands,* edited by Geoffrey E. Braswell (Equinox Press, London, England, 2011); Richard D. Hansen, Wayne K. Howell, and Stanley P. Guenter, "Forgotten Structures, Haunted Houses, and Occupied Hearts," in Travis W. Stanton and Aline Magnoni, eds., *Ruins of the Past: The Use and Perception of Abandoned Structures in the Maya Lowlands* (University Press of Colorado, Boulder, 2008), 25–64.

388 William J. Folan et al., in "Calakmul: New Data from an Ancient Maya Capital in Campeche, Mexico," *Latin American Antiquity* 6 (1995):310-334, discuss the tomb in Calakmul's palace. See also Sophia Pincemin, "Entierro en el Palacio: La Tumba de la Estructura III de Calakmul, Campeche," *Colección Arqueología* 5 (Universidad Autónoma de Campeche, Mexico, 1994).

389 The rivalry of Calakmul and Tikal is discussed in Joyce Marcus, "Calakmul y su Papel en el Origen del Estado Maya," *Los Investigadores de la Cultura Maya* 12: 14–31 (Universidad Autónoma de Campeche, Mexico, 2004); Simon Martin and Nikolai Grube, *Chronicle of the Maya Kings and Queens: Deciphering the Dynasties of the Ancient Maya, Second Edition* (Thames and Hudson, London, 2008); Simon Martin, "In Line of the Founder: A View of Dynastic Politics at Tikal," in Jeremy A. Sabloff, ed., *Tikal: Dynasties, Foreigners & Affairs of State* (SAR Press, Santa Fe, N.Mex., 2003), 3–45; Joyce Marcus, "Recent Advances in Maya Archaeology."

390 Spanish eyewitness accounts of Maya society are discussed in Ralph L. Roys, "The Indian Background of Colonial Yucatan," *Publication* 548 (Carnegie Institution of Washington, Washington, D.C., 1943); Ralph L. Roys, "Lowland Maya Society at Spanish Contact," in Robert Wauchope and Gordon R. Willey, eds., *Handbook of Middle American Indians, vol. 3* (University of Texas Press, 1965), 659–678; Matthew Restall, *The Maya World: Yucatec Culture and Society 1550–1850* (Stanford University Press, 1997); Alfred M. Tozzer, "Landa's Relación de Las Cosas de Yucatán," *Pa-*

per 18 (Peabody Museum of American Archaeology and Ethnology, Harvard University, 1941).

390 Royal Maya women are discussed by Tatiana Proskouriakoff in "Portraits of Women in Maya Art," in Samuel K. Lothrop et al., eds., *Essays in Precolumbian Art and Archaeology* (Harvard University Press, 1961), 81–99; also see Joyce Marcus, "Breaking the Glass Ceiling: The Strategies of Royal Women in Ancient States," in Cecelia F. Klein, ed., *Gender in Pre-Hispanic America* (Dumbarton Oaks, Washington, D.C., 2001), 305–340; Carolyn E. Tate, *Yaxchilan: The Design of a Maya Ceremonial City* (University of Texas Press, 1992).

390 Prisoner taking by Maya nobles is discussed by Tatiana Proskouriakoff in "Historical Data in the Inscriptions of Yaxchilan, Part II," *Estudios de Cultura Maya* 4 (1964): 177–201; also see Joyce Marcus, "Mesoamerica: Scripts," in Peter T. Daniels, ed., *Encyclopedia of Language and Linguistics, Second Edition*, vol. 8 (Elsevier Press, San Diego, 2006), 16–25; Silvia Trejo, ed., *La Guerra entre Los Antiguos Mayas: Memoria de la Primera Mesa Redonda de Palenque* (CONACULTA and INAH, Mexico City, 2000).

392 Oral exams for Maya officeholders are discussed in Daniel G. Brinton, ed., *The Maya Chronicles* (Library of Aboriginal American Literature, Philadelphia, 1882); also see Ralph L. Roys, *The Book of Chilam Balam of Chumayel* (University of Oklahoma Press, 1967); Joyce Marcus, *Mesoamerican Writing Systems* (Princeton University Press, 1992).

CHAPTER 19: THE LAND OF THE SCORPION KING

394 George B. Cressey, in *Crossroads: Land and Life in Southwest Asia* (J. B. Lippincott, New York, 1960), provides a useful description of Egypt before the Aswan High Dam changed the Nile environment.

395 Wadi Or is discussed by W. E. Wendt in "Two Prehistoric Archeological Sites in Egyptian Nubia," *Postilla* 102: 1–46 (Peabody Museum of Natural History, Yale University, 1966).

395 Jack R. Harlan, "The Tropical African Cereals," in David R. Harris and Gordon C. Hillman, eds., *Foraging and Farming: The Evolution of Plant Exploitation* (Unwin Hyman, London, 1989), 335–343; Wilma Wetterstrom, "Foraging and Farming in Egypt: The Transition from Hunting and Gathering to Horticulture in the Egyptian Nile Valley," in Thurstan Shaw et al., eds., *The Archaeology of Africa: Food, Metals and Towns* (Routledge, London, 1993), 165–226.

395 Gertrude Caton-Thompson and Elinor W. Gardner, *The Desert Fayum* (The Royal Anthropological Institute, London, 1934); Robert J. Wenke, Janet

E. Long, and Paul E. Buck, "Epipaleolithic and Neolithic Subsistence and Settlement in the Fayyum Oasis of Egypt," *Journal of Field Archaeology* 15 (1988): 29–51. New work on the Fayum sites by Willeke Wendrich and René Cappers was reported by John Noble Wilford, "5200 B.C. Is New Date for Farms in Egypt," *New York Times,* February 12, 2008.

396 Fred Wendorf, Romuald Schild, and Angela E. Close, eds., *Cattle-Keepers of the Eastern Sahara: The Neolithic of Bir Kiseiba* (Department of Anthropology and Center for the Study of Earth and Man, Southern Methodist University, Dallas, 1984).

396 Fred Wendorf, Angela E. Close, and Romuald Schild, "Prehistoric Settlements in the Nubian Desert," *American Scientist* 73 (1985): 132–141; Fred Wendorf and Romuald Schild, "Nabta Playa and Its Role in Northeastern African Prehistory," *Journal of Anthropological Archaeology* 17 (1998): 97–123.

397 Daniel G. Bradley and David A. Magee, "Genetics and the Origins of Domestic Cattle," in Melinda A. Zeder, Daniel G. Bradley, Eve Emshwiller, and Bruce D. Smith, eds., *Documenting Domestication: New Genetic and Archaeological Paradigms* (University of California Press, 2006), 317–328.

397 Herman Kees, in *Ancient Egypt* (Phoenix Books/University of Chicago Press, 1977), discusses Nun, Sothis, nilometers, and the Egyptian cosmos.

397 Pliny the Elder, *The Natural History of Pliny, vol. 5,* translated by John Bostock and Henry T. Riley (H. G. Bohn, London, 1856).

398 William C. Hayes, in *Most Ancient Egypt* (University of Chicago Press, 1965), provides a pioneering comparison of early Upper and Lower Egypt. See more recent papers by Kathryn A. Bard, "The Egyptian Predynastic: A Review of the Evidence," *Journal of Field Archaeology* 21 (1994): 265–288; Robert J. Wenke, "The Evolution of Early Egyptian Civilization: Issues and Evidence," *Journal of World Prehistory* 5 (1991): 279–329; Stephen H. Savage, "Some Recent Trends in the Archaeology of Predynastic Egypt," *Journal of Archaeological Research* 9 (2001): 101–155.

399 Hermann Junker, "Bericht über die von der Akademie der Wissenschaften in Wien nach dem Westdelta entsendete Expedition," *Denkschrift Akademie Wissenschaft Philosophische-Historische Klasse* 3 (1928): 14–24; Josef Eiwanger, *Merimde-Benisalâme,* vols. 1–3 (Archäologische Veröffentlichungen 59, Mainz am Rhein, Germany, 1984–1992).

399 Oswald Menghin and Moustafa Amer, *Excavations of the Egyptian University in the Neolithic Site at Maadi: First Preliminary Report* (Cairo University, Cairo, 1932); Oswald Menghin, *Excavations of the Egyptian University in the Neolithic Site at Maadi: Second Preliminary Report* (Cairo University, Cairo, 1936); Isabella Caneva, Marcella Frangipane, and Alba Palmieri,

"Predynastic Egypt: New Data from Maadi," *African Archaeological Review* 5 (1987): 105–114.

400 The subterranean houses of Shiqmim are described in Thomas E. Levy, ed., *The Archaeology of Society in the Holy Land* (Leicester University Press, London, 1995), 226–244, and in Thomas E. Levy, "Shiqmim 1," *BAR International Series* 356 (Archaeopress, Oxford, 1987).

400 Edwin C. M. van den Brink and Thomas E. Levy, eds., in *Egypt and the Levant: Interrelations from the 4th through the Early 3rd Millennium* B.C.E. (Leicester University Press, London, 2002), describe Egypt's interactions with Canaan (Israel-Jordan-Palestine), beginning as early as Ma'adi and Shiqmim.

400 Site H in Gaza is discussed by Ram Gophna in "The Contacts between Besor Oasis and Southern Canaan and Egypt during the Late Predynastic and the Threshold of the First Dynasty: A Further Assessment," in Edwin C. M. van den Brink, ed., *The Archaeology of the Nile Delta: Problems and Priorities* (Netherlands Foundation for Archaeological Research in Egypt, Amsterdam), 385–394.

400 The copper source used by Ma'adi is discussed by Andreas Hauptmann in *The Archaeometallurgy of Copper: Evidence from Faynan, Jordan* (Springer, New York, 2007), and in "The Earliest Periods of Copper Metallurgy in Feinan, Jordan," in Andreas Hauptmann, Ernst Pernicka, and Günther A. Wagner, eds., *Old World Archaeometallurgy: Proceedings of the International Symposium Held in Heidelberg 1987* (Selbstverlag des Deutschen Bergbau-Museums, Bochum, Germany, 1989), 119–135.

400 Stine Rossel et al., "Domestication of the Donkey: Timing, Processes, and Indicators," *Proceedings of the National Academy of Sciences* 105 (2008): 3715–3720.

400 The cemeteries at Ma'adi are discussed in William C. Hayes, *Most Ancient Egypt* (1965), and in Michael A. Hoffman, *Egypt before the Pharaohs: The Prehistoric Foundations of Egyptian Civilization* (Michael O'Mara Books, London, 1991).

401 Guy Brunton and Gertrude Caton-Thompson, in *The Badarian Civilisation and Prehistoric Remains near Badari* (British School of Archaeology in Egypt, London, 1928), describe Hemamieh.

401 Fekri A. Hassan, "Predynastic of Egypt," *Journal of World Prehistory* 2 (1988): 135–185.

402 Sir William Matthew Flinders Petrie and James E. Quibell, *Naqada and Ballas* (British School of Archaeology in Egypt, London, 1896); Fekri A. Hassan et al., "Agricultural Developments in the Naqada Region during the Predynastic Period," *Nyame Akuma* 17 (1980): 28–33.

402 The cemeteries at Naqada were dug by Sir William Matthew Flinders Petrie and James E. Quibell (see their *Naqada and Ballas*) and have been restudied by Kathryn A. Bard, *From Farmers to Pharaohs: Mortuary Evidence for the Rise of Complex Society in Egypt* (Sheffield Academic Press, Sheffield, 1994).

402 See chapter 1 in Barry J. Kemp, *Ancient Egypt: Anatomy of a Civilization* (Routledge, London, 1989).

404 Michael A. Hoffman, Hany A. Hamroush, and Ralph O. Allen, "A Model of Urban Development for the Hierakonpolis Region from Predynastic through Old Kingdom Times," *Journal of the American Research Center in Egypt* 23 (1986): 175–187.

405 James E. Quibell and Frederick W. Green, in *Hierakonpolis, vols. I, II* (Bernard Quaritch, London, 1900–1902), discuss the Rosette Scorpion macehead found at Hierakonpolis.

405 Tomb 11 at Hierakonpolis is discussed in Barbara Adams, "Excavations in the Locality 6 Cemetery at Hierakonpolis 1979–1985," *BAR International Series* 903 (Archaeopress, Oxford, 2000).

405 Günter Dreyer describes Tomb U-j at Abydos in "Recent Discoveries at Abydos Cemetery U," in Edwin C. M. van den Brink, ed., *The Nile Delta in Transition, 4th–3rd Millennium BC* (Israel Exploration Society, Tel Aviv, 1992), 293–299, and in "Umm el-Qaab: Nachuntersuchungen im Frühzeitlichen Königsfriedhof 5./6. Bericht," *Mitteilungen des Deutschen Archäeologischen Instituts Abteilung Kairo* 49 (1993): 23–62.

406 James E. Quibell and Frederick W. Green, in *Hierakonpolis,* discuss the Narmer Palette found at Hierakonpolis. See also Toby A. H. Wilkinson, "What a King Is This: Narmer and the Concept of the Ruler," *Journal of Egyptian Archaeology* 86 (2000): 23–32.

408 Thomas E. Levy, Edwin C. M. van den Brink, Yuval Goren, and David Alon, "New Light on King Narmer and the Protodynastic Egyptian Presence in Canaan," *Biblical Archaeologist* 58 (1995): 26–35; Thomas E. Levy et al., "Egyptian-Canaanite Interaction at Nahal Tillah, Israel (ca. 4500–3000 B.C.E.): An Interim Report on the 1994–1995 Excavations," *Bulletin of the American Schools of Oriental Research* 307 (1997): 1–51.

408 Since Egyptian kingship traditionally began with Dynasty 1, it was necessary for archaeologists to create a Dynasty 0 (3150–3050 B.C.) to accommodate earlier rulers such as Scorpion and Narmer. Also see Ian Shaw, ed., *The Oxford History of Ancient Egypt* (Oxford University Press, 2000); Peter A. Clayton, *Chronicle of the Pharaohs: The Reign-by-Reign Record of the Rulers and Dynasties of Ancient Egypt* (Thames and Hudson, London, 1994).

409 Manetho's 30 dynasties (the framework still in use for Egyptian chronology) are given in William G. Waddell, *Manetho* (Harvard University Press, 1940),

and in Sir Alan Gardiner, *Egypt of the Pharaohs: An Introduction* (Oxford University Press, 1978).

410 Discussions of the Egyptian deities and their pantheon can be found in Henri Frankfort, *Ancient Egyptian Religion: An Interpretation* (Harper & Row, New York, 1961); Byron E. Shafer, ed., *Religion in Ancient Egypt: Gods, Myths, and Personal Practice* (Cornell University Press, 1991); Siegfried Morenz, *Egyptian Religion* (Cornell University Press, 1992); Stephen Quirke, *Ancient Egyptian Religion* (British Museum Press, London, 1992).

410 For good discussions of the *ka*, the *ba*, and the *akh*, see Henri Frankfort, *Kingship and the Gods: A Study of Ancient Near Eastern Religion as the Integration of Society and Nature* (University of Chicago Press, 1948); James P. Allen et al., eds., *Religion and Philosophy in Ancient Egypt* (Department of Near Eastern Languages, Yale University, 1989); A. Jeffrey Spencer, *Death in Ancient Egypt* (Penguin Books, New York, 1982); Stephen Quirke, *Ancient Egyptian Religion* (see previous reference).

411 Several sources discuss Imhotep, the architect who designed Zoser's Step Pyramid, a feat so important that he was later deified; see Mark Lehner, *The Complete Pyramids: Solving the Ancient Mysteries* (Thames and Hudson, London, 1997); I. E. S. Edwards, *The Pyramids of Egypt* (Viking Press, New York, 1986); Kathryn A. Bard, *An Introduction to the Archaeology of Ancient Egypt* (Blackwell, Malden, Mass., 2008).

411 Mark Lehner, *The Complete Pyramids;* I. E. S. Edwards, *The Pyramids of Egypt* (see previous reference).

411 The work gangs that built the pyramids are discussed in Ann M. Roth, "Egyptian Phyles in the Old Kingdom: The Evolution of a System of Social Organization," *Studies in Ancient Oriental Civilization* 48 (Oriental Institute, University of Chicago, 1991). See also Christopher J. Eyre, "Work and the Organization of Work in the Old Kingdom," in Marvin A. Powell, ed., *Labor in the Ancient Near East* (American Oriental Society, New Haven, Conn., 1987).

412 Overviews of bureaucratic offices in the Egyptian state can be found in Sir Alan Gardiner, *Egypt of the Pharaohs* (see previous reference); Klaus Baer, *Rank and Title in the Old Kingdom* (University of Chicago Press, 1960); Barbara S. Lesko, "Rank, Roles, and Rights," in Leonard H. Lesko, ed., *Pharaoh's Workers: The Villagers of Deir El Medina* (Cornell University Press, 1994), 15–39; Jaroslav Černý, *A Community of Workmen at Thebes in the Ramesside Period* (Bibliotheque d'Etude Institut Français, Archeologie Orientale, Cairo, 1973).

412 How Uni worked his way up the bureaucratic ladder of success is discussed in James Henry Breasted, *Ancient Records of Egypt,* 5 vols. (University of Chicago Press, 1906–1907); also see Henri Frankfort, *Ancient Egyptian Religion* (see previous reference).

413 Tokens for bread are discussed and illustrated in Barry J. Kemp, *Ancient Egypt* (see previous reference). For a study of the barracks where pyramid workers resided, see Mark Lehner, "Of Gangs and Graffiti: How Ancient Egyptians Organized Their Labor Force," *Aeragram* 7 (2004): 11–13 (Newsletter of the Ancient Egypt Research Associates, Cambridge, Mass.).

415 Leslie A. White, "Ikhnaton: The Great Man vs. The Culture Process," *Journal of the American Oriental Society* 68 (1948): 91–114.

416 There are countless books about King Tut's tomb. Among them are Carl Nicholas Reeves, *The Complete Tutankhamun: The King, The Tomb, The Royal Treasure* (Thames and Hudson, London, 1990); Christiane Desroches-Noblecourt, *Tutankhamen: Life and Death of a Pharaoh,* 4th printing (New York Graphic Society, New York, 1978); and the report of the discoverer of the tomb, Howard Carter, *The Tomb of Tut-ankh-Amen,* 3 vols. (Cassell, New York, 1923–1933).

419 The life and times of Hatshepsut are discussed by Donald B. Redford, *History and Chronology of the Eighteenth Dynasty of Egypt: Seven Studies* (University of Toronto Press, 1967); Gay Robins, *Women in Ancient Egypt* (British Museum Press, London, 1993); Lana Troy, "Patterns of Queenship in Ancient Egyptian Myth and History" (*Acta Universitatis Upsaliensis* 14, Uppsala, Sweden, 1986); Eric Uphill, "A Joint Sed Festival of Thutmose III and Queen Hatshepsut," *Journal of Near Eastern Studies* 20 (1961): 248–251; and Joyce Marcus, "Breaking the Glass Ceiling: The Strategies of Royal Women in Ancient States," in Cecelia F. Klein, ed., *Gender in Pre-Hispanic America* (Dumbarton Oaks, Washington, D.C., 2001), 305–340.

CHAPTER 20: BLACK OX HIDES AND GOLDEN STOOLS

422 David Phillipson, in *The Later Prehistory of Eastern and Southern Africa* (Heinemann, London, 1977), discusses the Bantu migration.

423 The rise of the Dlamini clan is described in Hilda Kuper, *An African Aristocracy: Rank among the Swazi* (Oxford University Press, 1947). See also Andrew S. Goudie and D. Price Williams, "The Atlas of Swaziland," *Occasional Papers* 4 (The Swaziland National Trust Commission, Mbabane, 1983).

426 Hilda Kuper, *An African Aristocracy* (see previous reference).

433 Arthur A. Saxe, "Social Dimensions of Mortuary Practices" (PhD diss., University of Michigan, 1970).

435 Thurstan Shaw, *Igbo-Ukwu* (Faber, London, 1970).

435 Thomas C. McCaskie, "Denkyira in the Making of Asante," *Journal of African History* 48 (2007): 1–25; Thomas C. McCaskie, *State and Society in Pre-Colonial Asante* (Cambridge University Press, 1995); Ivor Wilks, *As-*

ante in the Nineteenth Century: The Structure and Evolution of a Political Order (Cambridge University Press, 1975).

436 Naomi Chazan, "The Early State in Africa: The Asante Case," in Shmuel N. Eisenstadt, Michel Abitbol, and Naomi Chazan, eds., *The Early State in African Perspective: Culture, Power and Division of Labor* (E. J. Brill, Leiden, Netherlands, 1988), 60–97.

441 Robert S. Rattray, *Ashanti* (Clarendon Press, Oxford, 1923). Also see Robert S. Rattray, *Religion and Art in Ashanti* (Clarendon Press, Oxford, 1927), and *Ashanti Law and Constitution* (Clarendon Press, Oxford, 1929).

447 Baden-Powell is quoted by Robert S. Rattray in *Ashanti* (see previous reference).

CHAPTER 21: THE NURSERY OF CIVILIZATION

450 Robert McC. Adams, "Agriculture and Urban Life in Early Southwestern Iran," *Science* 136 (1962): 109–122; Frank Hole, ed., *The Archaeology of Western Iran* (Smithsonian Institution Press, Washington, D.C., 1987).

450 Pinhas P. Delougaz and Helene J. Kantor, "Chogha Mish, vol. 1: The First Five Seasons of Excavations, 1961–71," *Oriental Institute Publications*, vol. 101 (University of Chicago Press, 1996).

451 Frank Hole, "Archaeology of the Village Period," in Frank Hole, ed., *The Archaeology of Western Iran*, 29–78; Frank Hole, "Settlement and Society in the Village Period," in Frank Hole, ed., *The Archaeology of Western Iran*, 79–105.

452 Gregory A. Johnson, "The Changing Organization of Uruk Administration on the Susiana Plain," in Frank Hole, ed., *The Archaeology of Western Iran*, 107–139.

452 Anatol Rapaport, in "Rank-Size Relations," in David Sills, ed., *International Encyclopedia of the Social Sciences*, vol. 13 (Macmillan, New York, 1968), 319–329, provides a short introduction to rank-size models. Gregory A. Johnson, in "Rank-size Convexity and System Integration: A View from Archaeology," *Economic Geography* 56 (1980): 234–247, shows the relevance of the models to archaeology.

453 Gregory A. Johnson, "The Changing Organization of Uruk Administration on the Susiana Plain" (see previous reference).

453 Henry T. Wright and Gregory A. Johnson, "Population, Exchange, and Early State Formation in Southwestern Iran," *American Anthropologist* 77 (1975): 267–289.

454 Gregory A. Johnson, "Local Exchange and Early State Development in Southwestern Iran," *Anthropological Papers* 51 (Museum of Anthropology, University of Michigan, 1973).

455 Hans J. Nissen, "Grabung in den Quadraten K/L XII in Uruk-Warka," *Baghdader Mitteilungen* 5 (1970): 102–191 (Deutsches Archäologisches Institut, Baghdad, Iraq); Gregory A. Johnson, "Local Exchange and Early State Development in Southwestern Iran" (see previous reference).

456 Gregory A. Johnson, "The Changing Organization of Uruk Administration on the Susiana Plain"; Pinhas P. Delougaz and Helene J. Kantor, "Chogha Mish, vol. 1: The First Five Seasons of Excavations, 1961–71" (see previous reference).

457 T. Cuyler Young Jr., "Excavations at Godin Tepe," *Occasional Papers, Art and Archaeology,* no. 17 (Royal Ontario Museum, Toronto, 1969); T. Cuyler Young Jr. and Louis D. Levine, "Excavations of the Godin Tepe Project: Second Progress Report," *Occasional Papers, Art and Archaeology,* no. 26 (Royal Ontario Museum, Toronto, 1974).

457 One of the best overviews of the archaeological sequence leading to the early Mesopotamian state is Marcella Frangipane, *La Nascita dello Stato nel Vicino Oriente* (Editori Laterza, Rome, 1996). It is unfortunate that this book has not been translated into English.

457 Robert McC. Adams, *Land Behind Baghdad: A History of Settlement on the Diyala Plains* (University of Chicago Press, 1965), and *Heartland of Cities: Surveys of Ancient Settlement and Land Use on the Central Floodplain of the Euphrates* (University of Chicago Press, 1981); Robert McC. Adams and Hans J. Nissen, *The Uruk Countryside: The Natural Setting of Urban Societies* (University of Chicago Press, 1972).

458 Gregory A. Johnson, "Locational Analysis and the Investigation of Uruk Local Exchange Systems," in Jeremy A. Sabloff and C. C. Lamberg-Karlovsky, eds., *Ancient Civilization and Trade* (University of New Mexico Press, 1975), 285–339; "Spatial Organization of Early Uruk Settlement Systems," *Colloques Internationaux du Centre National de la Recherche Scientifique* 580 (1980): 233–263 (Editions C.N.R.S., Paris).

459 Robert McC. Adams and Hans J. Nissen, *The Uruk Countryside* (see previous reference); Hans J. Nissen, "The City Wall of Uruk," in Peter J. Ucko, Ruth Tringham, and Geoffrey W. Dimbleby, eds., *Man, Settlement and Urbanism* (Gerald Duckworth & Co., London, 1972), 793–798.

460 Hans J. Nissen, in *An Early History of the Ancient Near East* (University of Chicago Press, 1988), provides an overview of past work at Uruk. An earlier, still useful overview was that of Ann Louise Perkins, "The Comparative Archaeology of Early Mesopotamia," *Studies in Ancient Oriental Civilization* 25 (Oriental Institute, University of Chicago, 1949). Most of the original descriptions of Uruk's public buildings were published in German in Berlin, often by the Prussian Academy of Sciences.

463 A list of recognizable nouns in Late Uruk/Jemdet Nasr writing was assembled by Robert McC. Adams in "Level and Trend in Early Sumerian Civilization" (PhD diss., University of Chicago, 1956), and in *The Evolution of Urban Society* (Aldine Press, Chicago, 1966).

464 Benno Landsberger, "Three Essays on the Sumerians," *Sources and Monographs: Monographs on the Ancient Near East,* vol. 1, fascicle 2 (Undena, Los Angeles, 1974).

465 Marcella Frangipane, "Centralization Processes in Greater Mesopotamia: Uruk 'Expansion' as the Climax of Systemic Interactions among Areas of the Greater Mesopotamian Region," in Mitchell S. Rothman, ed., *Uruk Mesopotamia & Its Neighbors* (School of American Research Press, Santa Fe, N. Mex., 2001), 307–347.

466 Arthur Tobler, in *Excavations at Tepe Gawra,* vol. 2 (University of Pennsylvania Press, 1950), and Ann Louise Perkins, in "The Comparative Archaeology of Early Mesopotamia," discuss the original excavations at Tepe Gawra. Mitchell S. Rothman, in "Tepe Gawra: The Evolution of a Small, Prehistoric Center in Northern Iraq," *University Museum Monographs 112* (Museum of Archaeology and Anthropology, University of Pennsylvania, 2002), has completely reanalyzed Levels XII–VIII of Gawra. Rothman's book includes a reanalysis of Gawra's burials by Brian Peasnall.

467 Marcella Frangipane, "Centralization Processes in Greater Mesopotamia" (see previous reference). Also see Marcella Frangipane, "Arslantepe-Malatya: External Factors and Local Components in the Development of an Early State Society," in Linda Manzanilla, ed., *Emergence and Change in Early Urban Societies* (Plenum Press, New York, 1997), 43–58.

468 Eva Strommenger, *Habuba Kabira: Eine Stadt vor 5000 Jahren* (Phillip von Zabern, Mainz am Rhein, Germany, 1980); G. van Driel and Carol van Driel-Murray, "Jebel Aruda, 1977–78," *Akkadica* 12 (1979): 2–8; "Jebel Aruda, the 1982 Season of Excavations," *Akkadica* 33 (1983): 1–26. Both Habuba Kabira and Jebel Aruda are put in perspective by Joan Oates in "Trade and Power in the Fifth and Fourth Millennia B.C.: New Evidence from Northern Mesopotamia," *World Archaeology* 24 (1993): 403–422, and by Guillermo Algaze, *The Uruk World System: The Dynamics of Expansion of Early Mesopotamian Civilization* (University of Chicago Press, 1993). Also see Guillermo Algaze, *Ancient Mesopotamia at the Dawn of Civilization* (University of Chicago Press, 2008). Gregory A. Johnson, in "Late Uruk in Greater Mesopotamia: Expansion or Collapse?" *Origini* 14 (1988–1989): 595–613, raises the possibility that some Uruk factions left Southern Mesopotamia as the result of political conflict.

469 Gil J. Stein, "Indigenous Social Complexity at Hacinebi (Turkey) and the Organization of Uruk Colonial Contact," in Mitchell S. Rothman, *Uruk*

Mesopotamia & Its Neighbors (see previous reference), 265–305. Also see Gil J. Stein, *Rethinking World Systems: Diasporas, Colonies, and Interaction in Uruk Mesopotamia* (University of Arizona Press, 1999).

470 Max E. L. Mallowan, "Excavations at Brak and Chagar Bazar: 3rd campaign," *Iraq* 9 (1947): 1–259; Joan Oates, "Tell Brak: The 4th Millennium Sequence and Its Implications," in J. Nicholas Postgate, ed., *Artefacts of Complexity: Tracking the Uruk in the Near East* (British School of Archaeology in Iraq, London, 2002), 111–122; Geoffrey Emberling and Helen McDonald, "Excavations at Tell Brak 2001–2002: Preliminary Report," *Iraq* 65 (2003): 1–75.

471 McGuire Gibson and Muhammad Maktash, "Tell Hamoukar: Early City in Northeastern Syria," *Antiquity* 74 (2000): 477–478.

472 Guillermo Algaze, *Ancient Mesopotamia at the Dawn of Civilization* (see previous reference).

472 Robert McC. Adams, "Agriculture and Urban Life in Early Southwestern Iran" (see previous reference).

CHAPTER 22: GRAFT AND IMPERIALISM

475 Igor M. Diakonoff, "Structure of Society and State in Early Dynastic Sumer," *Sources and Monographs: Monographs of the Ancient Near East,* vol. 1, fascicle 3 (Undena Press, Los Angeles, 1974).

476 Sandra L. Olsen, "Early Horse Domestication on the Eurasian Steppe," in Melinda A. Zeder, Daniel G. Bradley, Eve Emschwiller, and Bruce D. Smith, eds., *Documenting Domestication: New Genetic and Archaeological Paradigms* (University of California Press, 2006), 245–269. Caroline Grigson, in "The Earliest Domestic Horses in the Levant? New Finds from the Fourth Millennium of the Negev," *Journal of Archaeological Science* 20 (1993): 645–655, sees little evidence for the domestic horse in Mesopotamia before 2500 B.C., centuries after donkeys had arrived.

477 A classic description of the Mesopotamian cosmos is found in Thorkild Jacobsen, "Mesopotamia," in Henri Frankfort et al., eds., *The Intellectual Adventure of Ancient Man* (Phoenix Books, University of Chicago Press, 1977), 125–219.

478 Samuel Noah Kramer, in *The Sumerians: Their History, Culture, and Character* (University of Chicago Press, 1963), gives one version of Gudea's dream and the rebuilding of Eninnu. For an account of temple personnel, also see Adam Falkenstein, "The Sumerian Temple City," *Monographs in History: Ancient Near East,* vol. 1, fascicle 1 (Undena Press, Los Angeles, 1974).

479 Robert McC. Adams, in *The Evolution of Urban Society* (University of Chicago Press, 1966), and Samuel Noah Kramer, in *The Sumerians*, discuss Sumerian kinship and marriage.

480 For a discussion of Sumerian land and social groups, see both Igor M. Diakonoff, "Structure of Society and State in Early Dynastic Sumer" (see previous reference) and A. I. Tyumenev, "The Working Personnel of the Estate of the Temple of Ba-U in Lagaš during the Period of Lugalanda and Urukagina," in *Ancient Mesopotamia: Socio-economic History* (Nauka, Moscow, 1969), 88–126.

481 Benno Landsberger, "Three Essays on the Sumerians," *Sources and Monographs: Monographs of the Ancient Near East,* vol. 1, fascicle 2 (Undena Press, Los Angeles, 1974).

482 Pinhas P. Delougaz, "The Temple Oval at Khafajah," *Publication* 53 (Oriental Institute, University of Chicago, 1940).

484 Igor M. Diakonoff, "Structure of Society and State in Early Dynastic Sumer" (see previous reference).

484 Our population estimates for the province of Lagash are based on Robert McC. Adams, *Heartland of Cities* (University of Chicago Press, 1981); A. I. Tyumenev, "The Working Personnel of the Estate of the Temple of Ba-U in Lagaš during the Period of Lugalanda and Urukagina"; Igor M. Diakonoff, "Structure of Society and State in Early Dynastic Sumer"; and Jerrold S. Cooper, "Reconstructing History from Ancient Inscriptions: The Lagash-Umma Border Conflict," *Sources from the Ancient Near East,* vol. 2, fascicle 1 (Undena Publications, Malibu, 1983).

486 For the mythological prologue to the Sumerian kings, see Samuel Noah Kramer, *The Sumerians.* For later kings and dynasties, see Joan Oates, *Babylon: Revised Edition* (Thames and Hudson, 1986), and J. Nicholas Postgate, *Early Mesopotamia: Society and Economy at the Dawn of History* (Routledge, London, 1992). For the royal inscriptions in which Early Dynastic kings claimed hegemony over other provinces, see Jerrold S. Cooper, *Sumerian and Akkadian Royal Inscriptions, vol. I: Presargonid Inscriptions* (The American Oriental Society, New Haven, Conn., 1986).

487 C. Leonard Woolley, *Ur Excavations 2: The Royal Cemetery* (British Museum, London, and University of Pennsylvania Museum, Philadelphia, 1934).

490 Igor M. Diakonoff, "Structure of Society and State in Early Dynastic Sumer"; A. I. Tyumenev, "The Working Personnel of the Estate of the Temple of Ba-U in Lagaš during the Period of Lugalanda and Urukagina" (see previous reference).

491 Samuel Noah Kramer, in *The Sumerians,* documents many cases of Early Dynastic corruption; he, Diakonoff, and Tyumenev also discuss Urukagina's reforms.

492 Jerrold S. Cooper, "Reconstructing History from Ancient Inscriptions"; Robert McC. Adams, *Heartland of Cities.*

494 Brian Lewis, "The Sargon Legend: A Study of the Akkadian Text and the Tale of the Hero Who Was Exposed at Birth," *Dissertation Series,* no. 4 (American Schools of Oriental Research, Cambridge, Mass., 1980); Joan Oates, *Babylon: Revised Edition;* J. Nicholas Postgate, *Early Mesopotamia;* Samuel Noah Kramer, *The Sumerians.*

494 Jerrold S. Cooper and Wolfgang Heimpel, "The Sumerian Sargon Legend," *Journal of the American Oriental Society* 103 (1983): 67–82.

495 Jerrold S. Cooper, in *Sumerian and Akkadian Royal Inscriptions, vol. I: Presargonid Inscriptions,* provides many examples of Early Dynastic rulers who claimed to control several cities or provinces.

496 J. Nicholas Postgate, *Early Mesopotamia.*

496 Sargon's boast is given by George A. Barton in *Archaeology and the Bible: 3rd Edition* (American Sunday School Union, Philadelphia, 1920).

497 Joan Oates, *Babylon: Revised Edition;* J. Nicholas Postgate, *Early Mesopotamia.*

498 Marc van de Mieroop, "Society and Enterprise in Old Babylonian Ur," *Berliner Beiträge zum Vorderen Orient,* vol. 12 (Dietrich Reimer Verlag, Berlin, 1992).

499 Piotr Michalowski, *The Lamentation over the Destruction of Sumer and Ur* (Eisenbrauns, Winona Lake, Ind., 1989).

501 Michael Hudson, "Privatization: A Survey of the Unresolved Controversies," in Michael Hudson and Baruch A. Levine, eds., "Privatization in the Ancient Near East and Classical World," *Peabody Museum Bulletin* no. 5 (Harvard University, 1996), 1–32; Michael Hudson, "The Dynamics of Privatization, from the Bronze Age to the Present," in Michael Hudson and Baruch A. Levine, "Privatization in the Ancient Near East and Classical World," 33–57.

CHAPTER 23: HOW NEW EMPIRES LEARN FROM OLD

504 William T. Sanders, Jeffrey R. Parsons, and Robert S. Santley, *The Basin of Mexico: Ecological Processes in the Evolution of a Civilization* (Academic Press, New York, 1979); René F. Millon, "Teotihuacan Studies from 1950 to 1990 and Beyond," in Janet C. Berlo, ed., *Art, Ideology, and the City of Teotihuacan* (Dumbarton Oaks, Washington, D.C., 1992), 339–429; Linda Manzanilla, ed., *Anatomía de un Conjunto Residencial Teotihuacano en Oztoyahualco,* 2 vols. (Universidad Nacional Autónoma de México, Instituto de Investigaciones Antropológicas, Mexico City, 1993); Saburo Sugiyama, *Human Sacrifice, Militarism, and Rulership: Materialization of*

State Ideology at the Feathered Serpent Pyramid, Teotihuacan (Cambridge University Press, 2005).

505 Kenneth G. Hirth, *Archaeological Research at Xochicalco,* 2 vols. (University of Utah Press, 2000); Janet C. Berlo, "Early Writing in Central Mexico," in Richard A. Diehl and Janet C. Berlo, eds., *Mesoamerica After the Decline of Teotihuacan, AD 700–900* (Dumbarton Oaks, Washington, D.C., 1989), 19–47; Norberto González C. and Silvia Garza T., "Xochicalco," *Arqueología Mexicana* 2 (1994): 70–74.

506 Angel García Cook and Beatriz L. Merino C., "Cantona: Urbe Prehispánica en el Altiplano Central de México," *Latin American Antiquity* 9 (1998): 191–216.

506 Diana López de Molina and Daniel Molina, *Cacaxtla* (Instituto Nacional de Antropología e Historia, Mexico City, 1980); Claudia Brittenham, "The Cacaxtla Painting Tradition: Art and Identity in Epiclassic Mexico" (PhD diss., Yale University, 2008); Ellen T. Baird, "Stars and War at Cacaxtla," in Richard A. Diehl and Janet C. Berlo, *Mesoamerica After the Decline of Teotihuacan, AD 700–900,* 105–122 (see previous reference).

507 Clara Díaz, *Chingú: Un Sitio Clásico del Área de Tula, Hgo.* (Instituto Nacional de Antropología e Historia, Mexico City, 1980); Alba Guadalupe Mastache and Robert H. Cobean, "The Coyotlatelco Culture and the Origins of the Toltec State," in Richard A. Diehl and Janet C. Berlo, eds., *Mesoamerica After the Decline of Teotihuacan, AD 700–900,* 49–67.

507 Francisco del Paso y Troncoso, *Papeles de Nueva Espana: Segunda Serie, Geografía y Estadística,* 7 vols. (Tipográfico "Sucesores de Rivadeneyra," Madrid, 1905–1906); René Acuña, ed., *Relaciones Geográficas del Siglo XVI,* 9 vols. (Universidad Autónoma de México, Mexico City, 1984–1987).

507 For Jiménez Moreno's key suggestion that the site of Tula (in the state of Hidalgo) was the Toltec capital, see Wigberto Jiménez Moreno, "Tula y Los Toltecas Según Las Fuentes Históricas," *Revista Mexicana de Estudios Antropológicos* 5 (1941): 79–83, and "La Migración Mexica," *Atti del XL Congresso Internazionale Degli Americanisti* 1 (1973): 163–173. Also see Lawrence H. Feldman, "Tollan in Hidalgo: Native Accounts of the Central Mexican Tolteca," in Richard A. Diehl, ed., "Studies of Ancient Tollan: A Report of the University of Missouri Tula Archaeological Project," *Monograph* 1 (University of Missouri, 1974), 130–149.

508 The story of Mixcoatl and Ce Acatl Topiltzin Quetzalcoatl is discussed in Henry B. Nicholson, "Topiltzin Quetzalcoatl of Tollan: A Problem in Mesoamerican Ethnohistory" (PhD diss., Harvard University, 1957); Lawrence H. Feldman, "Tollan in Hidalgo: Native Accounts of the Central Mexican Tolteca," in Richard A. Diehl, ed., *Studies of Ancient Tollan,* 130–149; Dan M. Healan, ed., *Tula of the Toltecs: Excavations and Survey* (University of

Iowa Press, 1989); Alba Guadalupe Mastache, Robert H. Cobean, and Dan M. Healan, *Ancient Tollan: Tula and the Toltec Heartland* (University Press of Colorado, 2002); Nigel Davies, *The Toltecs until the Fall of Tula* (University of Oklahoma Press, 1977).

509 Patricia L. Crown and W. Jeffrey Hurst, "Evidence of Cacao Use in the Prehispanic American Southwest," *Proceedings of the National Academy of Sciences* 106 (2009): 2110–2113.

510 More about the legend of Tezcatlipoca driving Quetzalcoatl from Tula is given in Henry B. Nicholson, "Topiltzin Quetzalcoatl of Tollan" (PhD diss., Harvard University, 1957); also see Nigel Davies, *The Toltecs until the Fall of Tula* (see previous reference).

510 For the legend of the Aztec leaving Aztlan and changing their name to Mexica, see the *Tira de la Peregrinación (Codex Boturini)* (Librería Anticuaria, Mexico City, 1944); Nigel Davies, *The Aztecs: A History* (University of Oklahoma Press, 1973); Diego Durán, *The Aztecs: The History of the Indies of New Spain* (Orion Press, New York, 1964).

511 For information on the Aztec living as vassals of Azcapotzalco and sacrificing the Culhuacan princess, see *Crónica Mexicayotl* (Imprenta Universitaria, Mexico City, 1949); Richard F. Townsend, *The Aztecs* (Thames and Hudson, London, 2000); Eduardo Matos Moctezuma, *Tenochtitlan* (Fondo de Cultura Económica, Mexico City, 2006); Rudolf van Zantwijk, *The Aztec Arrangement: The Social History of Pre-Spanish Mexico* (University of Oklahoma Press, 1985); Diego Durán, *The Aztecs* (see previous reference).

512 The legendary events of Coatepec Hill are described by Eduardo Matos Moctezuma in "The Temple Mayor of Tenochtitlan: History and Interpretation," in Johanna Broda, Davíd Carrasco, and Eduardo Matos Moctezuma, eds., *The Great Temple of Tenochtitlan: Center and Periphery in the Aztec World* (University of California Press, 1987), 15–60, and in Alfredo López Austin and Leonardo López Luján, *Monte Sagrado-Templo Mayor: El Cerro y La Pirámide en la Tradición Religiosa Mesoamericana* (Instituto Nacional de Antropología e Historia and Universidad Nacional Autónoma de México, Mexico City, 2009).

512 The story of Acamapichtli is given in Diego Durán, *The Aztecs,* and in Nigel Davies, *The Aztecs* (see previous reference).

512 Maxtla's driving of Nezahualcoyotl into exile is described in Fernando de Alva Ixtlilxochitl, *Obras Históricas,* 2 vols. (Editora Nacional, Mexico City, 1952); the lament composed by Nezahualcoyotl is given in Miguel León Portilla, *Fifteen Poets of the Aztec World* (University of Oklahoma Press, 1992), 90–91.

513 Diego Durán, in *The Aztecs,* Rudolf van Zantwijk, in *The Aztec Arrangement,* Richard F. Townsend, in *The Aztecs,* and Nigel Davies, in *The Aztecs,* discuss the winning of Aztec political independence.

513 William T. Sanders, Jeffrey R. Parsons, and Robert S. Santley, *The Basin of Mexico* (see previous reference).

514 For discussions of the sixteenth-century sources on pilli, tecuhtli, macehualli, and mayeque, see Mercedes Olivera, *Pillis y Macehuales* (La Casa Chata, Mexico City, 1978); Charles Gibson, *The Aztecs under Spanish Rule: A History of the Indians of the Valley of Mexico, 1519–1810* (Stanford University Press, 1964); and James Lockhart, *The Nahuas After the Conquest: A Social and Cultural History of the Indians of Mexico* (Stanford University Press, 1992).

515 For a discussion of the calpulli, see James Lockhart, *The Nahuas After the Conquest;* Pedro Carrasco, "La Casa y Hacienda de un Señor Tlahuica," *Estudios de Cultura Náhuatl* 10 (1972): 235–244; Frederic Hicks, "Tetzcoco in the Early 16th Century: The State, the City and the Calpolli," *American Ethnologist* 9 (1982): 230–249; Charles Gibson, *The Aztecs under Spanish Rule;* and Rudolf van Zantwijk, *The Aztec Arrangement* (see previous reference).

516 Key references on the pochteca are found in Frances F. Berdan and Patricia R. Anawalt, eds., *The Codex Mendoza,* 4 vols. (University of California Press, 1992); Frances F. Berdan et al., *Aztec Imperial Strategies* (Dumbarton Oaks, Washington, D.C., 1996); and Ross Hassig, *Trade, Tribute, and Transportation: The Sixteenth Century Political Economy of the Valley of Mexico* (University of Oklahoma Press, 1985). For additional information on Xicalango and ports of trade, see Anne Chapman, "Port of Trade Enclaves in Aztec and Maya Civilizations," in Karl Polanyi, Conrad M. Arensberg, and Harry W. Pearson, eds., *Trade and Market in the Early Empires* (Free Press, Glencoe, Ill., 1957), 114–153; also see Lorenzo Ochoa S. and Ernesto Vargas P., "Xicalango, Puerto Chontal de Intercambio: Mito y Realidad," *Anales de Antropología* 25 (1986): 95–114.

517 The chinampa system is discussed in Pedro Armillas, "Gardens on Swamps," *Science* 174 (1971): 653–661; Teresa Rojas, "Evolución Histórica del Repertorio de Plantas Cultivadas en las Chinampas de la Cuenca de México," in Teresa Rojas, ed., *La Agricultura Chinampera: Compilación Histórica* (Universidad Autónoma de Chapingo, Mexico City, 1982), 181–214; and Edward E. Calnek, "Settlement Pattern and Chinampa Agriculture at Tenochtitlan," *American Antiquity* 37 (1972): 104–115.

517 For an eyewitness description of downtown Tenochtitlan, see Bernal Díaz del Castillo, *The Conquest of New Spain* (Penguin Books, New York, 1963). Also see Edward E. Calnek, "The Internal Structure of Tenochtitlan," in Eric R. Wolf, ed., *The Valley of Mexico: Studies of Pre-Hispanic Ecology and Society* (University of New Mexico Press, 1976), 287–302.

519 Nigel Davies, *The Aztecs;* Rudolf van Zantwijk, *The Aztec Arrangement;* Susan D. Gillespie, *The Aztec Kings: The Constitution of Rulership in*

Mexica History (University of Arizona Press, 1989); Joyce Marcus, "Aztec Military Campaigns against the Zapotecs: The Documentary Evidence," in Kent V. Flannery and Joyce Marcus, eds., *The Cloud People: Divergent Evolution of the Zapotec and Mixtec Civilizations* (Academic Press, New York, 1983), 314–318; Kent V. Flannery, "Zapotec Warfare: Archaeological Evidence for the Battles of Huitzo and Guiengola," in Kent V. Flannery and Joyce Marcus, eds., *The Cloud People,* 318–322.

521 Aztec perspectives on the Spanish conquest are given by Arthur J. O. Anderson and Charles E. Dibble in *The War of Conquest: How It Was Waged Here in Mexico* (University of Utah Press, 1978), and in Miguel León-Portilla, *The Broken Spears: The Aztec Account of the Conquest of Mexico* (Beacon Press, Boston, 1962). For the Spaniards' perspective, see Hernando Cortés, *His Five Letters of Relation to the Emperor Charles V* (A. H. Clark, Cleveland, 1908), or Hernando Cortés, *Five Letters of Cortés to the Emperor* (W.W. Norton, New York, 1962).

521 For the Moche and their empire, see Jeffrey Quilter and Luis Jaime Castillo, eds., *New Perspectives on Moche Political Organization* (Dumbarton Oaks, Washington, D.C., 2010), and Luis Jaime Castillo et al., eds., *Arqueología Mochica: Nuevos Enfoques* (Fondo Editorial de la Pontificia Universidad Católica del Perú and Instituto Francés de Estudios Andinos, Lima, 2008). For the Nasca state, see Helaine Silverman, *Ancient Nasca Settlement and Society* (University of Iowa Press, 2002), and Helaine Silverman and Donald A. Proulx, *The Nasca* (Blackwell, Malden, Mass., 2002).

521 Richard S. MacNeish et al., *Prehistory of the Ayacucho Basin, Peru,* vol. 2: *Excavations and Chronology* (University of Michigan Press, 1981).

522 William H. Isbell and Gordon F. McEwan, eds., *Huari Administrative Structure: Prehistoric Monumental Architecture and State Government* (Dumbarton Oaks, Washington, D.C., 1991); Katharina J. Schreiber, "Wari Imperialism in Middle Horizon Peru," *Anthropological Paper* 87 (University of Michigan Museum of Anthropology, 1992);Justin Jennings, "Understanding Middle Horizon Peru: Hermeneutic Spirals, Interpretive Traditions, and Wari Administrative Centers," *Latin American Antiquity* 17 (2006): 265–286.

522 Marcia Ascher and Robert Ascher, *Code of the Quipu* (University of Michigan Press, 1981).

523 Pikillaqta is described by William T. Sanders in "The Significance of Pikillacta in Andean Culture History," *Occasional Papers in Anthropology* 8: 380–428 (Pennsylvania State University, 1973), and in William H. Isbell and Gordon F. McEwan, eds., *Huari Administrative Structure* (see previous reference).

524 Charles Stanish, *Ancient Titicaca: The Evolution of Social Power in the Titicaca Basin of Peru and Bolivia* (University of California Press, 2003). Another important publication on Tiwanaku is Alan Kolata, ed., *Tiwanaku and Its Hinterland II: Urban and Rural Archaeology* (Smithsonian Institution Press, Washington, D.C., 2003).

524 Pukara is described by José María Franco in "Arqueología Sudperuana: Informe Sobre los Trabajos Arqueológicos de la Misión Kidder en Pukara, Peru (enero a julio de 1939)," *Revista del Museo Nacional* 9 (1940): 128–142, and in Elizabeth Klarich, "From the Mundane to the Monumental: Defining Early Leadership Strategies at Late Formative Pukara, Peru" (PhD diss., University of California at Santa Barbara, 2005).

524 Carlos Ponce Sanginés, *El Templete Semisubterraneo de Tiwanaku* (Editorial Juventud, La Paz, Bolivia, 1990); Alan Kolata, *The Tiwanaku* (Blackwell, Cambridge, UK, 1993); Juan Albarracín-Jordan, *Tiwanaku: Arqueología Regional y Dinámica Segmentaria* (Editores Plural, La Paz, Bolivia, 1996); John Wayne Janusek, *Ancient Tiwanaku* (Cambridge University Press, 2008); Justin Jennings and Nathan Craig, "Polity Wide Analysis and Imperial Political Economy: The Relationship between Valley Political Complexity and Administrative Centers in the Wari Empire of the Central Andes," *Journal of Anthropological Archaeology* 20 (2001): 479–502.

525 Donna J. Nash and Patrick Ryan Williams, "Wari Political Organization: The Southern Periphery," in Joyce Marcus and Patrick Ryan Williams, eds., *Andean Civilization: A Tribute to Michael E. Moseley* (UCLA Cotsen Institute of Archaeology Press, Los Angeles, 2009), 257–276. Also see Ryan Williams, "Cerro Baúl: A Wari Center on the Tiwanaku Frontier," *Latin American Antiquity* 12 (2001): 67–83, and Michael E. Moseley et al., "Burning Down the Brewery: Establishing and Evacuating an Ancient Imperial Colony at Cerro Baúl, Peru," *Proceedings of the National Academy of Sciences* 102 (2005): 17264–17271.

525 Sergio Chávez, "The Arapa and Thunderbolt Stelae: A Case of Stylistic Identity with Implications for Pucara Influences in the Area of Tiahuanaco," *Ñawpa Pacha* 13 (1975): 3–26, and "La Piedra del Rayo y La Estela de Arapa: Un Caso de Identidad Estilística, Pucara-Tiahuanaco," *Arte y Arqueología* 8–9 (1984): 1–27.

525 For information on the breakdown of the Tiwanaku region to pucaras after A.D. 1200, see chapter 9 in Charles Stanish, *Ancient Titicaca* (see previous reference).

527 Garth Bawden, "Galindo: A Study in Cultural Transition during the Middle Horizon," in Michael E. Moseley and Kent C. Day, eds., *Chan Chan: Andean Desert City* (University of New Mexico Press, 1982), 285–320.

527 Key data on Chan Chan are contained in Michael E. Moseley and Kent C. Day, *Chan Chan,* including chapters on the royal compounds by Kent C. Day, on the burial platforms by Geoffrey W. Conrad, on the compounds of the lesser elite by Alexandra M. Ulana Klymyshyn, and on commoner residences by John R. Topic Jr., as well as other important themes.

527 Chimú split inheritance is discussed by Geoffrey W. Conrad in "Cultural Materialism, Split Inheritance, and the Expansion of Ancient Peruvian Empires," *American Antiquity* 46 (1981): 3–42, and in "The Burial Platforms of Chan Chan: Some Social and Political Implications," in Michael E. Moseley and Kent C. Day, *Chan Chan,* 87–117.

529 Carol Mackey, "Chimú Statecraft in the Provinces," in Joyce Marcus and Patrick Ryan Williams, eds., *Andean Civilization,* 325–349; also see Michael E. Moseley and Alana Cordy-Collins, eds., *The Northern Dynasties: Kingship and Statecraft in Chimor* (Dumbarton Oaks, Washington, D.C., 1990).

531 R. Alan Covey, *How the Inca Built Their Heartland: State Formation and the Innovation of Imperial Strategies in the Sacred Valley, Peru* (University of Michigan Press, 2006); Brian S. Bauer, *The Development of the Inca State* (University of Texas Press, 1992). Also see Brian S. Bauer and R. Alan Covey, "Processes of State Formation in the Inca Heartland (Cuzco, Peru)," *American Anthropologist* 10 (2002): 846–864, and Brian S. Bauer, *Ancient Cuzco: Heartland of the Inca* (University of Texas Press, 2004).

531 The Inka rulers are described in sixteenth- and seventeenth-century documents, including Juan de Betanzos, *Suma y Narración de los Incas* (Ediciones Atlas, Madrid, 1987); Juan Polo de Ondegardo, "Del Linaje de los Ingas y Como Conquistaron," in *Colección de Libros y Documentos Referentes a la Historia del Perú* 4: 45–94 (Sanmartí Press, Lima, Peru, 1917); Pedro Sarmiento de Gamboa, "Historia de Los Incas," *Biblioteca de Autores Españoles,* vol. 135 (Ediciones Atlas, Madrid, 1965); Miguel Cabello Balboa, *Miscelánea Antártica: Una Historia del Perú Antiguo* (Universidad Nacional Mayor de San Marcos, Lima, 1951); Antonio Vázquez de Espinosa, "Compendio y Descripción de las Indias Occidentales," *Biblioteca de Autores Españoles,* vol. 231 (Ediciones Atlas, Madrid, 1969). Also see María Rostworowski, *Pachacutec Inca Yupanqui* (Torres Aguirre Press, Lima, Peru, 1953), and *History of the Inca Realm* (Cambridge University Press, 1999). For information on the Chanka, see Brian S. Bauer, Lucas C. Kellett, and Miriam Aráoz, *The Chanka: Archaeological Research in Andahuaylas (Apurimac), Peru* (UCLA Cotsen Institute of Archaeology Press, Los Angeles, 2010).

532 See Richard L. Burger and Lucy C. Salazar, eds., *The 1912 Yale Peruvian Scientific Expedition Collections from Machu Picchu* (Yale University

Press, New Haven, Conn., 2003), and Richard L. Burger and Lucy C. Salazar, eds., *Machu Picchu: Unveiling the Mystery of the Incas* (Yale University Press, 2004). The 2003 volume includes a chapter by biological anthropologist John Verano, who is able to show that the individuals buried at Machu Picchu came from different parts of the Inka Empire. See also Johan Reinhard, *Machu Picchu: Exploring an Ancient Sacred Center,* 4th ed. (UCLA Cotsen Institute of Archaeology Press, Los Angeles, 2007).

532 The ayllu is discussed by R. Tom Zuidema, *Inca Civilization in Cuzco* (University of Texas Press, 1990); María Rostworowski de Diez Canseco, *Historia del Tahuantinsuyu* (Instituto de Estudios Peruanos, Lima, 1988); John V. Murra, "The Economic Organization of the Inca State" (PhD diss., University of Chicago, 1956); and Karen Spalding, *Huarochirí: An Andean Society under Inca and Spanish Rule* (Stanford University Press, 1984).

533 Darrell E. La Lone, "The Inca as a Nonmarket Economy: Supply on Command versus Supply and Demand," in Jonathon E. Ericson and Timothy K. Earle, eds., *Contexts for Prehistoric Exchange* (Academic Press, New York, 1982), 291–316; Charles Stanish, "Nonmarket Imperialism in the Prehispanic Americas: The Inka Occupation of the Titicaca Basin," *Latin American Antiquity* 8 (1997): 195–216.

533 The terms *Inka, Sapa Inka, Curaca,* and *Qoya* are discussed in John H. Rowe, "The Inca Culture at the Time of the Spanish Conquest," in Julian H. Steward, ed., *Handbook of South American Indians,* vol. 2 (Bureau of American Ethnology, Smithsonian, Washington, D.C., 1946), 183–330; María Rostworowski, *History of the Inca Realm,* and John V. Murra, "The Economic Organization of the Inca State" (see previous reference).

533 The litter-bearers and road-sweepers from the Province of Rucanas are described by John H. Rowe in "The Inca Culture at the Time of the Spanish Conquest." Also see John V. Murra, "The Economic Organization of the Inca State."

534 Information on royal mummies is given in Brian S. Bauer, *Ancient Cuzco: Heartland of the Inca* (University of Texas Press, 2004), and in John V. Murra, "The Economic Organization of the Inca State."

535 Downtown Cusco is described by Inca Garcilaso de la Vega in *Royal Commentaries of the Incas and General History of Peru,* parts 1 and 2 (University of Texas Press, 1966), and by Pedro Pizarro in *Relation of the Discovery and Conquest of the Kingdoms of Peru* (The Cortés Society, New York, 1921). Also see Brian S. Bauer, *Ancient Cuzco,* chapter 10. Chapter 11 of Bauer's book covers the Coricancha and the ceque system. Information on the ceques is also given by R. Tom Zuidema in *The Ceque System of Cuzco: The Social Organization of the Capital of the Inca* (E. J. Brill, Leiden, Netherlands, 1964).

538 Tampus and collcas are discussed in John Hyslop, *The Inka Road System* (Academic Press, Orlando, Fla., 1984); Terry Y. Levine, ed., *Inka Storage Systems* (University of Oklahoma Press, 1992); Craig Morris, "Storage in Tawantinsuyu" (PhD diss., University of Chicago, 1967); and Craig Morris, "Storage, Supply, and Redistribution in the Economy of the Inka State," in John Murra, Nathan Wachtel, and Jacques Revel, eds., *Anthropological History of Andean Polities* (Cambridge University Press, 1986), 59–68. The Inca decimal system is described by Catherine J. Julien in "Inca Decimal Administration in the Lake Titicaca Region," in George A. Collier, Renato I. Rosaldo, and John D. Wirth, eds., *The Inca and Aztec States 1400–1800* (Academic Press, New York, 1982), 119–151, and by John H. Rowe in "The Inca Culture at the Time of the Spanish Conquest."

538 Miguel Cabello Balboa, *Miscelánea Antártica* (see previous reference).

539 The Chincha Valley sites are discussed by Craig Morris in "Links in the Chain of Inka Cities: Communication, Alliance, and the Cultural Production of Status, Value, and Power," in Joyce Marcus and Jeremy A. Sabloff, eds., *The Ancient City: New Perspectives on Urbanism in the Old and New World* (School for Advanced Research Press, Santa Fe, N. Mex., 2008), 299–319, and by Craig Morris and Julián I. Santillana in "The Inka Transformation of the Chincha Capital," in Richard L. Burger, Craig Morris, and Ramiro Matos M., eds., *Variations in the Expression of Inka Power* (Dumbarton Oaks, Washington, D.C., 2007), 135–163.

539 Joyce Marcus, *Excavations at Cerro Azul: The Architecture and Pottery* (UCLA Cotsen Institute of Archaeology Press, Los Angeles, 2008).

540 Craig Morris and Donald E. Thompson, *Huánuco Pampa: An Inca City and Its Hinterland* (Thames and Hudson, London, 1985). For a description of Huánuco Pampa in A.D. 1553, see Pedro Cieza de León, *The Incas* (University of Oklahoma Press, 1959).

542 Pedro Sarmiento de Gamboa, *Historia de Los Incas;* María Rostworowski, *History of the Inca Realm;* John H. Rowe, "The Inca Culture at the Time of the Spanish Conquest" (see previous reference).

Sources of Illustrations

With the exception of Figure 54, all illustrations in this book are drawings done by John Klausmeyer (JK) or Kay Clahassey (KC) for Kent Flannery and Joyce Marcus. Many of the drawings were inspired by old photographs, taken before globalization had irreversibly modified the society depicted. Other illustrations have been redrawn, with modification, from earlier works.

Figure 1 (JK): The seal diagram was inspired by a sketch in Frans Van de Velde, "Les Règles du Partage des Phoques pris par la Chasse aux Aglus," *Anthropologica* 3 (1956): 5–14.

Figure 2 (JK): This diagram is based on data given on page 70 of John E. Yellen, *Archaeological Approaches to the Present* (Academic Press, New York, 1977).

Figure 3 (JK): The plan of the Andaman encampment was redrawn, with modification, from a diagram in A. R. Radcliffe-Brown, *The Andaman Islanders* (Cambridge University Press, 1922). The drawing of the girl was inspired by a photo taken between 1906 and 1908 and published in the same book.

Figure 4 (JK): This drawing is a montage, based on three different 100-year-old photos taken by Baldwin Spencer and F. J. Gillen, *The Northern Tribes of Central Australia* (Macmillan & Co., London, 1904).

Figure 5 (JK): This drawing was inspired by a 100-year-old photo taken by Baldwin Spencer and F. J. Gillen, *The Northern Tribes of Central Australia* (Macmillan & Co., London, 1904).

Figure 6 (JK): This drawing was inspired by a 100-year-old photo taken by George Hunt at Jewitt's Lake, British Columbia. See page 259 of Peter Nabokov and Robert Easton, *Native American Architecture* (Oxford University Press, 1989).

Figure 7 (JK): This diagram is based on data given by Philip Drucker, "The Northern and Central Nootkan Tribes," *Bulletin* 114 (Bureau of American Ethnology, Smithsonian Institution, Washington, D.C., 1951).

Figure 8 (JK): This drawing is a montage. The Tlingit chief was inspired by Photo SITK-3926 in the archives of Sitka National Historical Park, Alaska (taken by Elbridge Warren Merrill between 1919 and 1922). The cedar screen and carved post were inspired by photos of a Tlingit house torn down in the late nineteenth century. Color paintings of this house were published by George T. Emmons, "The Whale House of the Chilkat," *Anthropological Papers of the American Museum of Natural History* 19 (1916): 1–33.

Figure 9 (JK): This drawing is a montage. It was inspired by two different photos, taken during the 1960s, by Andrew Strathern, *The Rope of Moka* (Cambridge University Press, 1971).

Figure 10 (JK): This drawing was inspired by a 100-year-old photograph taken by T. C. Hodson, "Head-hunting among the Hill Tribes of Assam," *Folklore* 20 (1909): 132–143.

Figure 11 (KC): These plans of men's houses were redrawn, with modification, from diagrams in James P. Mills, *The Rengma Nagas* (Macmillan & Co., London, 1937).

Figure 12 (KC): This map was redrawn, with modification, from an illustration in Maureen Anne Mackenzie, *Androgynous Objects: String Bags and Gender in New Guinea* (Harwood Academic Publishers, Melbourne, Australia, 1991).

Figure 13 (JK): This drawing was inspired by a photo taken between 1938 and 1939 by Douglas L. Oliver, *A Solomon Island Society* (Harvard University Press, 1955).

Figure 14, top (JK): Redrawn, with modification, from Stefan Carol Kozlowski, "M'lefaat: Early Neolithic Site in Northern Iraq," *Cahiers de l'Euphrate* 8 (1998): 179–273.

Figure 14, bottom (JK): Redrawn, with modification, from Nigel Goring-Morris and Anna Belfer-Cohen, "Structures and Dwellings in the Upper and Epi-Paleolithic (ca. 42–10 k BP) Levant: Profane and Symbolic Uses," in S. A. Vasil'ev, Olga Soffer, and J. Kozlowski, eds., "Perceived Landscapes and Built Environments," *BAR International Series* 1122 (Archaeopress, Oxford, UK, 2003), 65–81.

Figure 15, top (KC): This drawing was inspired by photographs in Klaus Schmidt, *Sie Bauten die Ersten Tempel: Das Rätselhafte Heiligtum der Steinzeitjäger* (Verlag C. H. Beck, Munich, 2006).

Figure 15, bottom (KC): This drawing was inspired by photographs in Harald Hauptmann, "Ein Kultgebäude in Nevali Çori," in Marcella Frangipane et al., eds., *Between the Rivers and Over the Mountains* (Università di Roma "La Sapienza," Rome, 1993), 37–69.

Figure 16, top (JK): Redrawn, with modifications, from Andrew M. T. Moore, Gordon C. Hillman, and Anthony J. Legge, *Village on the Euphrates* (Oxford University Press, 2000).

Figure 16, bottom (JK): Redrawn, with modifications, from Mehmet Özdoğan and A. Özdoğan, "Çayönü: A Conspectus of Recent Work," *Paléorient* 15 (1989): 65–74.

Figure 17, left (JK): This drawing was inspired by a photo in Kathleen Kenyon, *Archaeology in the Holy Land, Third Edition* (Praeger, New York, 1970).

Figure 17, right (JK): This drawing was inspired by a photo in Gary O. Rollefson, Alan H. Simmons, and Zeidan Kafafi, "Neolithic Cultures at 'Ain Ghazal, Jordan," *Journal of Field Archaeology* 19 (1992): 443–470.

Figure 18 (JK): Redrawn, with modifications, from Mehmet Özdoğan and A. Özdoğan (see reference to Figure 16, bottom).

Figure 19, top (KC): Redrawn, with modification, from Joyce Marcus and Kent V. Flannery, *Zapotec Civilization* (Thames and Hudson, London, 1996).

Figure 19, bottom (KC): Redrawn, with modification, from Kent V. Flannery and Joyce Marcus, "Early Formative Pottery of the Valley of Oaxaca, Mexico," *Memoir* 27 (Museum of Anthropology, University of Michigan, Ann Arbor, 1994).

Figure 20, top (KC): Redrawn, with modification, from Terence Grieder et al., *La Galgada, Peru* (University of Texas Press, 1988).

Figure 20, bottom (KC): Redrawn, with modification, from Seiichi Izumi, "The Development of the Formative Culture in the Ceja de Montaña: A Viewpoint Based on the Materials from the Kotosh Site," in Elizabeth P. Benson, ed., *Dumbarton Oaks Conference on Chavín* (Dumbarton Oaks, Washington, D.C., 1971), 49–72.

Figure 21 (KC): Redrawn, with modification, from Alfonso Ortiz, *The Tewa World* (University of Chicago Press, 1969).

Figure 22 (JK): This drawing was inspired by a 100-year-old photograph in the Archives of the Smithsonian Institution. See page 409 of Peter Nabokov and Robert Easton, *Native American Architecture* (Oxford University Press, 1989).

Figure 23 (JK): This drawing was inspired by Photograph #0239–075 of the State Historical Society of North Dakota. See Patrick Springer, "Medicine Bundles Help Keep Stories from 'Dream Time,'" *The Forum* (Forum Communications, Fargo, N.Dak., 2003), 1–3.

Figure 24 (JK): Redrawn, with modification, from Simon J. Harrison, *Stealing People's Names* (Cambridge University Press, 1990).

Figure 25 (JK): Redrawn, with modification, from Edmund R. Leach, *Political Systems of Highland Burma* (G. Bell & Sons, London, 1954).

Figure 26 (JK): This drawing was inspired by two different photos taken in the 1930s by Christoph von Fürer-Haimendorf, *The Konyak Nagas* (Holt, Rinehart and Winston, New York, 1969).

Figure 27 (JK): Redrawn, with modification, from Joyce Marcus and Kent V. Flannery, *Zapotec Civilization* (Thames and Hudson, London, 1996).
Figure 28 (JK): Original drawing by John Klausmeyer.
Figure 29 (KC): Redrawn, with modification, from Elsa M. Redmond and Charles S. Spencer, "Rituals of Sanctification and the Development of Standardized Temples in Oaxaca, Mexico," *Cambridge Archaeological Journal* 18 (2008): 230–266.
Figure 30 (KC): Redrawn, with modification, from Ruth Shady Solís, *La Ciudad Sagrada de Caral-Supe en los Albores de la Civilización en el Perú* (Universidad Nacional Mayor de San Marcos, Lima, 1997).
Figure 31 (KC): This drawing was inspired by a photograph by George Steinmetz, in John F. Ross, "First City in the New World?" *Smithsonian* 33 (2002): 57–64.
Figure 32 (KC): This drawing is based on photographs taken by Joyce Marcus at Cerro Sechín, Peru, during the 1980s.
Figure 33 (KC): This drawing is a montage. The temple plan was redrawn, with modification, from Richard L. Burger, *Chavín and the Origins of Andean Civilization* (Thames and Hudson, London, 1992). The carved stone monument was redrawn, with modification, from Julio C. Tello, *Chavín: Cultura Matriz de la Civilización Andina* (Universidad Nacional Mayor de San Marcos, Lima, 1960).
Figure 34 (JK): This drawing was inspired by two different photos taken during the 1940s by Christoph von Fürer-Haimendorf, *The Apa Tanis and Neighbours* (Routledge & Kegan Paul, London, 1962).
Figure 35 (KC): Redrawn, with modification, from Seton Lloyd and Fuad Safar, "Tell Hassuna," *Journal of Near Eastern Studies* 4 (1945): 255–289.
Figure 36 (KC): This drawing synthesizes the information from several seasons at Tell es-Sawwan. It was redrawn, with modification, from Vadim M. Masson, *Pervye Tsivilizatsii* (Nauka, Leningrad, 1989).
Figure 37 (JK): This drawing was inspired by a photo in Joan Oates, "Religion and Ritual in Sixth-Millennium B.C. Mesopotamia," *World Archaeology* 10 (1978): 117–124.
Figure 38 (JK): This drawing was inspired by two different photos in Nikolai Y. Merpert and Rauf M. Munchaev, "Burial Practices of the Halaf Culture," in Norman Yoffee and Jeffrey J. Clark, eds., *Early Stages in the Mesopotamian Civilization: Soviet Excavations in Northern Iraq* (University of Arizona Press, 1993), 207–223.
Figure 39, top (KC): This drawing was inspired by a photograph in Nikolai Y. Merpert and Rauf M. Munchaev, "Yarim Tepe III: The Halaf Levels," in Norman Yoffee and Jeffrey J. Clark, eds., *Early Stages in the Mesopotamian Civilization: Soviet Excavations in Northern Iraq* (University of Arizona Press, 1993), 163–205.

Figure 39, bottom (KC): Redrawn, with modification, from Max E. L. Mallowan and J. Cruikshank Rose, "Excavations at Tall Arpachiyah, 1933," *Iraq* 2 (1935): 1–178.

Figure 40 (KC): Redrawn, modified, and reassembled from three different illustrations in Max E. L. Mallowan and J. Cruikshank Rose, "Excavations at Tall Arpachiyah, 1933" *Iraq* 2 (1935), 1–178.

Figure 41 (KC): Redrawn, with modification, from Arthur Tobler, *Excavations at Tepe Gawra,* vol. 2 (University of Pennsylvania Museum, Philadelphia, 1950).

Figure 42 (KC): Redrawn, with modification, from Arthur Tobler, *Excavations at Tepe Gawra,* vol. 2 (University of Pennsylvania Museum, Philadelphia, 1950).

Figure 43, top (KC): This drawing was inspired by a photograph in Fuad Safar, Mohammad Ali Mustafa, and Seton Lloyd, *Eridu* (Iraqi Ministry of Culture and Information, Baghdad, 1981).

Figure 43, bottom (KC): Redrawn, with modification, from Fuad Safar, Mohammad Ali Mustafa, and Seton Lloyd, *Eridu* (Iraqi Ministry of Culture and Information, Baghdad, 1981).

Figure 44 (KC): Redrawn, with modification, from Seton Lloyd and Fuad Safar, "Tell Uqair: Excavations by the Iraq Government Directorate of Antiquities in 1940–1941," *Journal of Near Eastern Studies* 2 (1943): 131–155.

Figure 45 (KC): Redrawn, with modification, from Sabah Abboud Jasim, "Excavations at Tell Abada: A Preliminary Report," *Iraq* 45 (1983): 165–186.

Figure 46, left (JK): Redrawn, with modification, from Vernon James Knight Jr., "Moundville as a Diagrammatic Ceremonial Center," in Vernon James Knight Jr. and Vincas P. Steponaitis, eds., *Archaeology of the Moundville Chiefdom* (Smithsonian Institution Press, Washington, D.C., 1998), 44–62.

Figure 46, right (JK): Redrawn, with modification, from Vernon James Knight Jr. and Vincas P. Steponaitis, "A New History of Moundville," in Vernon James Knight Jr. and Vincas P. Steponaitis, eds., *Archaeology of the Moundville Chiefdom* (Smithsonian Institution Press, Washington, D.C., 1998), 1–25.

Figure 47 (JK): This drawing was inspired by a map and a photograph in Adam King, *Etowah: The Political History of a Chiefdom Capital* (University of Alabama Press, 2003).

Figure 48 (JK): Redrawn, with modification, from Will Carleton McKern, "Archaeology of Tonga," *Bulletin* 60 (Bernice P. Bishop Museum, Honolulu, 1929).

Figure 49 (JK): Redrawn, with modification, from two sources: Philip Drucker, Robert F. Heizer, and Robert J. Squier, "Excavations at La Venta, Tabasco, 1955," *Bulletin* 170 (Bureau of American Ethnology, Smithsonian Institution, Washington, D.C., 1959), and Rebecca González Lauck, "La Venta: An Olmec Capital," in Elizabeth P. Benson and Beatriz de la Fuente, eds., *Olmec Art of Ancient Mexico* (National Gallery of Art, Washington, D.C., 1996), 73–81.

Figure 50 (JK): Redrawn, with modification, from Charles S. Spencer and Elsa M. Redmond, "Militarism, Resistance and Early State Development in Oaxaca, Mexico," *Social Evolution and History* 2 (2003): 25–70.

Figure 51 (KC): This drawing is based on photos taken by Joyce Marcus at Monte Albán.

Figure 52 (JK): This painting was inspired by a photo in Alfonso Caso and Ignacio Bernal, "Urnas de Oaxaca," *Memoria* 2 (Instituto Nacional de Antropología e Historia, Mexico City, 1952).

Figure 53 (JK): This drawing was inspired by a color painting in Walter Alva and Christopher B. Donnan, *Royal Tombs of Sipán* (Fowler Museum of Cultural History, UCLA, Los Angeles, 1993).

Figure 54: Detail from a drawing in Christopher B. Donnan and Donna McClelland, *Moche Fineline Painting: Its Evolution and Its Artists* (Fowler Museum of Cultural History, UCLA, Los Angeles, 1999). Reproduced by permission of Christopher B. Donnan and the estate of Donna McClelland.

Figure 55 (KC): Redrawn, with modification, from William J. Folan, Joyce Marcus, and W. Frank Miller, "Verification of a Maya Settlement Model through Remote Sensing," *Cambridge Archaeological Journal* 5 (1995): 277–282.

Figure 56 (KC): Redrawn, with modification, from William J. Folan et al., "Calakmul: New Data from an Ancient Maya City in Campeche, Mexico," *Latin American Antiquity* 6 (1995): 310–334.

Figure 57 (KC): Redrawn, with modification, from Linda Schele and Mary E. Miller, *The Blood of Kings* (Kimbell Art Museum, Forth Worth, Tex., 1986). See also Joyce Marcus, "Identifying Elites and Their Strategies," in Christina M. Elson and R. Alan Covey, eds., *Intermediate Elites in Pre-Columbian States and Empires* (University of Arizona Press, 2006), 212–246.

Figure 58 (KC): This map combines information from a variety of sources. The coverage of Upper Egypt is partly inspired by Barry J. Kemp, *Ancient Egypt* (Routledge, London, 1989).

Figure 59 (KC): This version of the Narmer palette is redrawn from Joyce Marcus, *Mesoamerican Writing Systems* (Princeton University Press, 1992).

Figure 60 (JK): This drawing was inspired by photographs in Christiane Desroches-Noblecourt, *Tutankhamen: Life and Death of a Pharaoh* (New York Graphic Society, New York, 1978).

Figure 61 (KC): Redrawn, with modification, from Hilda Kuper, *An African Aristocracy: Rank among the Swazi* (Oxford University Press, 1947).

Figure 62 (KC): Redrawn, with modification, from Hilda Kuper, *An African Aristocracy: Rank among the Swazi* (Oxford University Press, 1947).

Figure 63 (JK): This drawing was inspired by two different photos in Malcolm D. McLeod, *The Asante* (British Museum, London, 1981).

Figure 64 (KC): This map combines information from a variety of sources and is partly inspired by Gregory A. Johnson, "Late Uruk in Greater Mesopotamia: Expansion or Collapse?" *Origini* 14 (1988–1989): 595–613.

Figure 65 (KC): This illustration is a montage. It combines building plans redrawn, with modification, from Hans J. Nissen, *An Early History of the Ancient Near East* (University of Chicago Press, 1988), and Ann Louise Perkins, "The Comparative Archaeology of Early Mesopotamia," *Studies in Ancient Oriental Civilization* 25 (Oriental Institute, University of Chicago, 1949).

Figure 66 (JK): The drawing of the temple oval is loosely based on an original work by David West Reynolds, which is the property of Flannery and Marcus. The diagram of the high priest's residence is redrawn, with modification, from Kent V. Flannery, "The Ground Plans of Archaic States," in Gary M. Feinman and Joyce Marcus, eds., *Archaic States* (School of American Research Press, Santa Fe, N. Mex., 1998), 15–57.

Figure 67 (JK): Redrawn, with modification, from C. Leonard Woolley, *Ur Excavations 2: The Royal Cemetery* (British Museum, London, and University of Pennsylvania Museum, Philadelphia, 1934).

Figure 68 (JK): This drawing is based on a photograph taken by Joyce Marcus.

Figure 69 (JK): This drawing was inspired by a photograph in Richard A. Diehl, *Tula: The Toltec Capital of Ancient Mexico* (Thames and Hudson, London, 1983).

Figure 70 (KC): Redrawn, with modification, from the sixteenth-century Codex Mendoza; see Frances F. Berdan and Patricia R. Anawalt, *The Essential Codex Mendoza* (University of California Press, 1997).

Figure 71 (KC): Redrawn, with modification, from William H. Isbell, "Mortuary Preferences: A Wari Culture Case Study from Middle Horizon Peru," *Latin American Antiquity* 15 (2004): 3–32.

Figure 72 (JK): This illustration is a montage. It combines ground plans redrawn, with modification, from several different chapters in Michael E. Moseley and Kent C. Day, eds., *Chan Chan: Andean Desert City* (University of New Mexico Press, 1982).

Figure 73 (KC): Redrawn, with modification, from the sixteenth-century author Felipe Guamán Poma de Ayala, *El Primer Nueva Corónica y Buen Gobierno,* 3 vols. (Siglo Veintiuno, Mexico City, 1980).

Index